CO_2 Mineralization for Construction Materials and Products

CO_2 Mineralization for Construction Materials and Products is a comprehensive exploration of this emerging field, offering a detailed overview of the science, technology, and applications of CO_2 mineralization techniques in the construction industry. The book covers the basics of mineralization: principles, microstructure, and techniques, as well as mineralization of a range of cementitious products. It also provides unique practical applications and case studies from both academic research and industrial practice to demonstrate how CO_2 mineralization techniques are being used in practice around the world.

This book is essential for graduate students and academic researchers pursuing studies in sustainable construction and engineering, as well as innovative concrete professionals aiming to apply CO_2 mineralization techniques to construction projects.

Caijun Shi is Distinguished Professor at Hunan University, China, and Director of the Key Laboratory of Building Safety and Energy Efficiency of the Chinese Ministry of Education.

Xiang Hu is Associate Professor at Hunan University, China.

Yuguang Mao received his PhD from Hunan University, China, and is a research associate at Imperial College London.

CO_2 Mineralization for Construction Materials and Products

Caijun Shi, Xiang Hu and Yuguang Mao

CRC Press
Taylor & Francis Group
London Boca Raton New York

CRC Press is an imprint of the
Taylor & Francis Group, an **informa** business

Front cover image: carterdayne/Getty Images

First edition published 2026
by CRC Press
4 Park Square, Milton Park, Abingdon, Oxon, OX14 4RN

and by CRC Press
2385 NW Executive Center Drive, Suite 320, Boca Raton FL 33431

CRC Press is an imprint of Taylor & Francis Group, LLC

British Library Cataloguing-in-Publication Data
A catalogue record for this book is available from the British Library

ISBN: 978-1-041-03783-5 (hbk)
ISBN: 978-1-041-04250-1 (pbk)
ISBN: 978-1-003-62753-1 (ebk)

DOI: 10.1201/9781003627531

Typeset in Times
by Deanta Global Publishing Services, Chennai, India

Contents

Preface

The global construction industry currently faces a pivotal challenge in balancing the escalating demand for infrastructure development with the imperative to significantly reduce its carbon footprint. As one of the largest contributors to global CO_2 emissions, the construction sector must embrace innovative, sustainable technologies to mitigate its environmental impact. Among the most promising solutions is the development of CO_2 mineralization techniques, which not only capture and store CO_2 but also transform it into valuable construction materials. The book, *CO_2 Mineralization for Construction Materials and Products*, is a comprehensive exploration of this emerging field, offering a detailed overview of the science, technology, and applications of CO_2 mineralization in the construction industry.

The concept of CO_2 mineralization is not new; its roots can be traced back to the late 19th century, when early pioneers first experimented with CO_2 mineralization techniques for artificial stone and concrete. However, it is only in recent decades, driven by the escalating climate crisis and advancements in materials science, that CO_2 mineralization has gained significant attention as a viable pathway toward sustainable construction. This book aims to provide a thorough understanding of the principles, processes, and potential of CO_2 mineralization, drawing on the latest research and practical applications in the field.

The journey of writing this book has been both challenging and rewarding. As researchers and practitioners in the field of construction materials, we have witnessed firsthand the rapid evolution of CO_2 mineralization techniques. From early laboratory experiments to the commercialization of CO_2-mineralized concrete blocks and other products, the progress has been remarkable. Yet, there is still much to be explored and optimized. This book is our attempt to consolidate existing knowledge, highlight the current state of the art, and identify future directions for research and development.

The book is structured to provide a systematic and in-depth understanding of CO_2 mineralization techniques and their applications. It begins with an introduction to the historical development and significance of CO_2 mineralization, followed by detailed discussions on the fundamental chemistry of CO_2 mineralization, CO_2 mineralization products, and the microstructure of CO_2-mineralized materials. Subsequent chapters explore the CO_2 mineralization of calcium silicate minerals, thermodynamic modeling, experimental verifications of the process, and the application of CO_2 mineralization in various construction products, including precast concrete, recycled aggregates, and steel slag. The book also addresses critical issues such as the durability of CO_2-mineralized materials, methods for determining mineralized CO_2, and the standardization of these materials for widespread use.

One of the unique features of this book is its focus on practical applications and real-world case studies. We have included examples from both academic research and industrial practice, demonstrating how CO_2 mineralization can be integrated into the production of construction materials. These examples not only illustrate the technical feasibility of CO_2 mineralization but also highlight its potential to reduce the carbon footprint of the construction industry.

This book is intended for a broad audience, including researchers, engineers, policymakers, and students in the fields of civil engineering, materials science, and environmental sustainability. We hope that it will serve as a valuable resource for those seeking to deepen their understanding of CO_2 mineralization and its role in the transition to a low-carbon economy. For academics, it provides a comprehensive review of the latest research and a foundation for future studies. For industry professionals, it offers practical insights into the application of CO_2 mineralization techniques in construction projects. And for students, it presents a clear and accessible introduction to a cutting-edge field with immense potential for innovation and impact.

We would like to express our gratitude to the colleagues and collaborators who have contributed to the finalization of this book. Their insights, feedback, and support have been invaluable. We also extend our thanks to the publishers for their guidance and encouragement throughout the process. Finally, we acknowledge the pioneering work of researchers and industry leaders who have paved the way for the advancements discussed in this book. Their contributions have laid the foundation for a more sustainable future in construction.

As we look to the future, it is clear that CO_2 mineralization will play an increasingly important role in the global effort to combat climate change. By transforming CO_2 from a waste product into a valuable resource, this technology offers a pathway to carbon-neutral and even carbon-negative construction materials. We hope that this book will inspire further research, innovation, and collaboration in this vital field, contributing to a more sustainable and resilient built environment.

Caijun Shi

Xiang Hu

Yuguang Mao

1 Introduction

1.1 BACKGROUND AND SIGNIFICANCE OF CO_2 MINERALIZATION TECHNIQUES

The increasing global concern over climate change and the urgent need to reduce carbon emissions have driven significant interest in innovative technologies for carbon capture, utilization, and storage (CCUS). Among these, CO_2 mineralization, also known as carbonation, has emerged as a promising approach to mitigate CO_2 emissions while simultaneously producing valuable materials. CO_2 mineralization involves the chemical reaction of CO_2 with alkaline or alkaline-earth metals, such as calcium (Ca) and magnesium (Mg), to form stable carbonate minerals. This process not only sequesters CO_2 in a solid form but also has the potential to produce construction materials, such as carbonate binders, which can be used in various industrial applications.

Carbonate binders, which are produced through the reaction of CO_2 with calcium or magnesium compounds, represent a significant advancement in the field of sustainable construction materials. Unlike traditional cementitious binders that rely on hydration reactions with water, carbonate binders are formed through carbonation reactions, offering a unique opportunity to utilize CO_2 as a raw material. This approach aligns with the global push toward carbon neutrality and the circular economy, as it transforms CO_2 from a waste product into a valuable resource.

The concept of CO_2 mineralization is not new; it has been studied and applied in various forms for over a century. However, recent advancements in materials science, coupled with the growing urgency to address climate change, have reignited interest in this technology. Researchers and industries are now exploring ways to optimize CO_2 mineralization processes, improve the performance of carbonate binders, and scale up their production for commercial use. This chapter provides an overview of the historical development of CO_2 mineralization techniques, with a focus on their application in the production of carbonate materials, and discusses the current state of research and future directions in this field.

1.2 HISTORICAL DEVELOPMENT OF CO_2 MINERALIZATION TECHNIQUES

The use of CO_2 in the production of construction materials dates back to the late 19th century. Table 1.1 provides an overview of the key milestones achieved throughout the development of carbonate binders. The earliest recorded application of CO_2 mineralization was patented by James L. Rowland in 1870 (Rowland, 1870), who described a method for accelerating the hardening of artificial stone using CO_2. This pioneering work laid the foundation for subsequent research and development in

DOI: 10.1201/9781003627531-1

TABLE 1.1
Milestone development in the CO_2 treatment of cementitious materials

Year	Authors	Contributions	References
1870	James L. Rowland	CO_2 mineralization technique was first used for making concrete blocks.	(Rowland, 1870)
1972	Berger R. L. et al.	The mechanisms of CO_2 mineralization of calcium silicate, including C_3S, β-C_2S, γ-C_2S, and CS.	(Klemm, W. and Berger, R., 1972)
1992	Souza M. R. et al.	Application of carbonate binder for the rapid production of cement particleboard.	(M.R. Souza, 1992)
1998	Hermawan D. et al.	Supercritical CO_2 was used to mineralize cement particle board.	(Hermawan and Hata, 1998)
2007	Shi C. & Wu Y.	A pre-conditioning technology was developed to increase the CO_2 uptake rate and quantity.	(Shi and Wu, 2007)
2011	Solidia	The CO_2 mineralization of Solidia cement (CS and C_3S_2).	(Ondrey,2013)
2018	Shi C. et al.	CO_2 mineralization was used to treat hardened cement paste powder.	(Lu et al.,2018)

the field of CO_2 mineralization. Throughout the late 1800s and early 1900s, several patents were issued for the production of cementitious materials using CO_2 mineralization techniques (Heinzerling, 1897; Owen, 1903; Rowland, 1872, 1874; Sprogle, 1873), including blocks, walls, floors, and other structural elements.

Despite these early innovations, the widespread adoption of CO_2 mineralization was hindered by technical and economic challenges. The natural carbonation process, which relies on the reaction of CO_2 with alkaline earth hydroxides such as slaked lime, was known to result in slow strength development. However, researchers discovered that the carbonation process could be significantly accelerated by exposing materials to high concentrations of CO_2 or by applying elevated pressures. This breakthrough led to renewed interest in CO_2 mineralization during the 1970s, particularly in the context of cement and concrete production (Berger and Klemm, 1972; Goodbrake et al., 1979; J.F. Young et al., 1974; Klemm, W. and Berger, R., 1972).

In the 1970s, Berger et al. (Klemm and Berger 1972) conducted pioneering studies on the accelerated carbonation of cementitious materials, focusing on the mechanisms of CO_2 mineralization for various calcium silicate phases, including tricalcium silicate (C_3S) and dicalcium silicate (C_2S). Their work demonstrated that increasing the cement-to-sand ratio and compaction pressure could enhance the compressive strength of carbonated materials. Additionally, they found that post-hydration of carbonated samples could further improve strength by 30–45% (He et al., 2016; Shi et al., 2012). These findings marked a significant milestone in the development of

CO_2 mineralization techniques and paved the way for further research in this area (Klemm and Berger, 1972).

The 1990s saw the application of accelerated CO_2 mineralization in the production of wood-cement boards, where CO_2 was injected into the furnish during the pressing process (Cabral et al., 2018; Souza et al., 1992). This method not only reduced the pressing time but also improved the compatibility between wood and cement, resulting in composite materials with properties comparable to those produced using conventional cold-pressing methods (Simatupang et al., 1995). Further advancements were made with the use of supercritical CO_2, which, due to its high pressure and temperature, significantly increased the CO_2 mineralization rate and produced more silica gel and calcium carbonate (García-González et al., 2006). These materials interlocked with wood fibers, resulting in high-strength composites. However, the high energy consumption associated with supercritical CO_2 mineralization limited its widespread adoption (Knopf et al., 1999).

In the early 2000s, Shi et al. (Shi and Wu, 2008) proposed a novel method to improve the mineralization degree and compressive strength of CO_2-mineralized cementitious materials, known as pre-conditioning. The specimens were placed in a dry environment for different periods of time before CO_2 mineralization. During pre-conditioning, the specimens lost some water and reached a lower water-to-cement (w/c) ratio, which was beneficial for the penetration of CO_2. As a result, the compressive strength of concrete blocks subjected to 2 hours of pre-conditioning in a dry environment increased by around 83% compared to those in a wet environment before 2 hours of CO_2 mineralization. The compressive strength and CO_2 consumption increased with the moisture loss. This method has been successfully applied to a wide range of materials, including lightweight concrete, recycled aggregate concrete, and alkali-activated cement, leading to significant improvements in compressive strength and CO_2 uptake. In 2018, Shi et al. (Lu et al.,2018) also succeeded in using CO_2 mineralization to recycle waste concrete powder. Generally, the powder is molded and mineralized or treated in a dry gaseous CO_2 environment, leading to the formation of carbonate binders that can be incorporated into concrete. Besides, CO_2 mineralization can also take place in an aqueous environment, where carbonatable compounds, such as recycled concrete powder (RCP) and steel slag powders, react with dissolved CO_2 in the liquid medium (Mao et al.,2023). The aqueous CO_2 mineralization process, also known as wet carbonation, has been extensively studied recently to treat and transform cement waste and industrial by-products into supplementary cementitious materials (SCMs).

1.3 CURRENT STATUS AND FUTURE DIRECTIONS

Today, CO_2 mineralization techniques are being actively researched and developed by both academia and industry. Several companies, such as CarbonCure and Solidia, have commercialized these techniques for the production of low-carbon cement-based materials. These methods not only reduce the carbon footprint of construction materials but also enhance their mechanical properties and durability. For example, CO_2-mineralized concrete blocks have been shown to exhibit improved pore structure and reduced permeability, making them more resistant to environmental degradation.

Recent research has focused on increasing the efficiency of CO_2 sequestration, optimizing the CO_2 mineralization process, and developing new applications for carbonate binders. One promising area of research is the use of industrial by-products, such as steel slag and fly ash, as raw materials for CO_2 mineralization. These materials, which are rich in calcium and magnesium, can react with CO_2 to form stable carbonates, thereby reducing the environmental impact of industrial waste. Additionally, researchers are exploring the use of wet carbonation to produce SCMs from cement waste powders, further contributing to the sustainability of the construction industry.

Another important area of research is the life cycle assessment (LCA) of CO_2-mineralized materials and products. By analyzing the carbon emissions associated with the production and use of carbonate binders, researchers can identify opportunities for further reducing the environmental impact of these materials. For example, studies have shown that CO_2-mineralized concrete blocks can achieve negative carbon emissions when the CO_2 used in the production process is sourced from industrial waste gases.

In conclusion, CO_2 mineralization techniques represent a promising pathway for reducing carbon emissions and producing sustainable construction materials. While significant progress has been made in the development of these technologies, further research is needed to optimize their performance, reduce costs, and scale up their production for widespread commercial use. As the world continues to grapple with the challenges of climate change, CO_2 mineralization is likely to play an increasingly important role in the transition to a low-carbon economy.

1.4 STRUCTURE OF THE BOOK

This book is organized into several chapters that provide a comprehensive overview of CO_2 mineralization techniques and their applications in construction materials and products. Chapter 2 delves into the fundamental chemistry and mechanisms of CO_2 mineralization techniques. Chapter 3 discusses the thermodynamic modeling of CO_2 mineralization and experimental verifications. Chapter 4 explores the carbonation of calcium silicate minerals. Chapter 5 examines the products and microstructure of CO_2 mineralized materials and products. Chapter 6 discusses the CO_2 surface treatment of precast concrete products. Chapter 7 introduces the mineralization by recycled concrete aggregate. Chapter 8 introduces the CO_2 mineralization by recycled concrete fines. Chapter 9 introduces the CO_2 mineralization of steel slag. Chapter 10 introduces the CO_2 mineralization in magnesium-based materials. Chapter 11 explores the durability of CO_2 mineralized materials and products. Chapter 12 examines the methods for the determination of mineralized CO_2. Finally, Chapter 13 discusses the applications and standardization of CO_2 mineralized materials and products.

By providing a detailed and up-to-date overview of CO_2 mineralization techniques for construction materials and products, this book aims to serve as a valuable resource for researchers, engineers, and policymakers working toward a more sustainable and carbon-neutral future.

REFERENCES

Berger, R.L., Klemm, W., 1972. Accelerated curing of cementitious systems by carbon dioxide: Part II. Hydraulic calcium silicates and aluminates. *Cement and Concrete Research* 2(6), 647–652.

Cabral, M.R., Nakanishi, E.Y., dos Santos, V., Gauss, C., dos Santos, S.F., Fiorelli, J., 2018. Evaluation of accelerated carbonation curing in cement-bonded balsa particleboard. *Materials and Structures* 51(2), 1–14.

García-González, C.A., Hidalgo, A., Andrade, C., Alonso, M.C., Fraile, J., López-Periago, A.M., Domingo, C., 2006. Modification of composition and microstructure of Portland cement pastes as a result of natural and supercritical carbonation procedures. *Industrial & Engineering Chemistry Research* 45(14), 4985–4992.

Goodbrake, C.J., Young, J.F., Berger, R.L., 1979. Reaction of beta-dicalcium silicate and tricalcium silicate with carbon dioxide and water vapor. *Journal of the American Ceramic Society* 62(3–4), 168–171.

He, P., Shi, C., Tu, Z., Poon, C.S., Zhang, J., 2016. Effect of further water curing on compressive strength and microstructure of CO_2-cured concrete. *Cement and Concrete Composites* 72, 80–88.

Heinzerling, C., 1897. *Artificial stone*. U. S. Patent 591168. United States Patent and Trademark Office.

Hermawan, D., Hata, T., 1998. *Development technology of rapid production of high-strength cement-bonded particleboard by using gaseous or supercritical carbon dioxide curing*. Proceedings of Inorganic-Bonded Wood Fiber Composite Materials, pp. 70–87.

Klemm, W., Berger, R., 1972. Accelerated curing of cementitious systems by carbon dioxide: Part I. Portland cement. *Cement and Concrete Research* 2(5), 567–576.

Knopf, F.C., Roy, A., Samrow, H.A., Dooley, K.M., 1999. High-pressure molding and carbonation of cementitious materials. *Industrial Engineering Chemistry Research* 38(7), 2641–2649.

Lu, B., Shi, C., Zhang, J., Wang, J., 2018. Effects of carbonated hardened cement paste powder on hydration and microstructure of Portland cement. *Construction and Building Materials* 186, 699–708.

Mao, Y.G., He, P.P., Drissi, S., Zhang, J., Hu, X., Shi, C., 2023. Effect of conditions on wet carbonation products of recycled cement paste powder. *Cement and Concrete Composites* 144, 105307.

Owen, W., 1903. *Process of manufacturing artificial stone*. U. S. Patent 731608. United States Patent and Trademark Office.

Ondrey G., 2013. Cement increases strength and lowers emissions footprint. *Chemical Engineering* 120, 12–20.

Rowland, J.L., 1870. *Improvement in the manufacture of artificial stone*. U. S. Patent 109669. United States Patent and Trademark Office.

Rowland, J.L., 1872. *Improvement in hardening artificial-stone walls, concrete*. U. S. Patent 128980. United States Patent and Trademark Office.

Rowland, J.L., 1874. *Improvement in the manufacture of artificial stone*. U. S. Patent 149682. United States Patent and Trademark Office.

Shi, C., Wang, D., He, F., Liu, M., 2012. Weathering properties of CO_2-cured concrete blocks. *Resources, Conservation Recycling* 65, 11–17.

Shi, C., Wu, Y., 2007. Effect of pre-conditioning on CO_2 curing of concrete products. *Journal of Civil Engineering Materials* 26(1), 257–267.
Shi, C., Wu, Y., 2008. Studies on some factors affecting CO2 curing of lightweight concrete products. *Resources, Conservation and Recycling* 52(8), 1087–1092.
Simatupang, M.H., Habighorst, C., Lange, H., Neubauer, A., 1995. Investigations on the influence of the addition of carbon dioxide on the production and properties of rapidly set wood-cement composites. *Cement and Concrete Composites* 17(3), 187–197.
Souza, M.R., Moslemi, A.A., Simatupang, M.H., 1992. *Carbon dioxide application for rapid production of cement particleboard.* Inorganic Bonded Wood and Fiber Composite Materials Conference, Spokane, WA.
Sprogle, D.M., 1873. *Improvement in processes of making artificial stone.* U. S. Patent 139274. United States Patent and Trademark Office.
Young, J.F., Berger, R.L., Breese, J., 1974. Accelerated curing of compacted calcium silicate mortars on exposure to CO_2. *Journal of the American Ceramic Society* 57(9), 394–397.

2 CO_2 mineralization techniques

2.1 INTRODUCTION

The CO_2 mineralization of cement-based materials has advanced significantly with the development of CO_2 mineralization technology. The earliest recorded technology dates back to a 1870 patent by Rowland for a CO_2 curing chamber designed to accelerate the hardening of concrete products (Rowland, 1870). However, due to the low diffusion rate of CO_2 in the pore solution, early CO_2 mineralization of cement-based materials often required high temperatures and pressures. As a result, the technology was not widely adopted at the time due to its high cost. It was not until the mid-2000s that Shi and Wu (C. Shi, 2018) introduced an innovative pretreatment technique known as pre-conditioning, as illustrated in Figure 2.1, which significantly accelerated CO_2 diffusion in cement-based materials. This approach involved pre-conditioning the specimens to remove excess water from capillary pores before CO_2 curing, thereby enhancing and accelerating CO_2 diffusion and achieving a high degree of mineralization in a very short period. Over the following two decades, driven by the implementation of low-carbon policies in various countries, CO_2 mineralization technologies for cement-based materials have developed rapidly. This chapter first introduces the fundamental principles of CO_2 mineralization techniques for cement-based materials. Subsequently, the two most common CO_2 mineralization technologies currently used for cement-based materials are CO_2 curing and accelerated mineralization. It then describes the processes, mechanisms, and equipment associated with these technologies and reviews the factors currently affecting their performance.

2.2 MECHANISM OF CO_2 MINERALIZATION TECHNIQUES FOR CEMENT-BASED MATERIALS

CO_2 mineralization in cement-based materials mimics and accelerates the natural mineralization and weathering processes of rocks in nature, using CO_2 to react chemically with alkaline earth metal ions (Ca^{2+}/Mg^{2+}) in natural minerals or solid waste to form stable carbonates ($CaCO_3/MgCO_3$) for permanent sequestration. In general, the mineralization process occurs spontaneously in the gas/liquid/solid three-phase region, with the main steps of CO_2 uptake and CO_3^{2-} formation, leaching of alkaline-earth metals into the fluid, chemical reactions, and precipitation of insoluble $CaCO_3$ within the pore structure. From a thermodynamic point of view, the

DOI: 10.1201/9781003627531-2

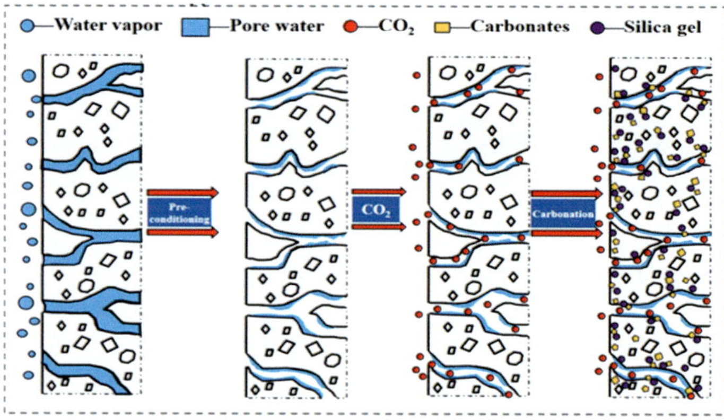

FIGURE 2.1 Schematic diagram of pre-conditioning and CO_2 curing of concrete products.

reaction is a process from a high-energy state to a low-energy state, and the standard Gibbs free energy of the reaction product, carbonate, is lower than that of CO_2, which is theoretically capable of proceeding spontaneously under natural conditions and produces a more stable carbonate form. Therefore, CO_2 mineralization is considered to be one of the most important technologies for the stable sequestration of CO_2 in the construction industry. The mechanism of CO_2 mineralization of common mineral components in cementitious materials is described in the following sections.

2.2.1 Reaction of Cement Clinker with CO_2

C_3S and β-C_2S are the primary mineral components in ordinary silicate cement, accounting for approximately 80% of its composition. Liu et al. (2010) found that the ability of the mineral compositions of cement clinker to CO_2 mineralization was in the descending order: $C_3S > C_2S > C_3A > C_4AF$. CO_2 mineralization of cement-based materials typically occurs during the early stages of cement hydration, with C_3S and β-C_2S particles that have not yet participated in hydration reacting directly with CO_2 to form C-S-H and $CaCO_3$, as shown in Equations (2.1) and (2.2).

$$3CaO \cdot SiO_2 + (3-x)CO_2 + nH_2O \rightarrow xCaO \cdot SiO_2 \cdot nH_2O + (3-x)CaCO_3 \quad (2.1)$$

$$2CaO \cdot SiO_2 + (2-x)CO_2 + nH_2O \rightarrow xCaO \cdot SiO_2 \cdot nH_2O + (2-x)CaCO_3 \quad (2.2)$$

The degree of mineralization of C_3S and β-C_2S is related to their respective activation energies. It has been reported that the activation energy of the reaction between C_3S and CO_2 is lower (184 kJ/mol) than that of β-C_2S (347 kJ/mol) (Goodbrake et al., 1979a, 1979b), resulting in a higher degree of mineralization for C_3S during the early stages, leading to higher initial strength. However, the rapid chemical reaction in the early stage of C_3S mineralization generates micro-cracks within the cementitious material, which can negatively affect strength in the later stages. In contrast, the early mineralization of β-C_2S is relatively small, resulting in low early strength. However, as the curing time extends, the late strength of β-C_2S increases steadily and may even surpass that of C_3S. Additionally, there is an ongoing debate regarding the end products of C_3S and β-C_2S mineralization. Some studies (Young et al., 1974) have found that the end products of the mineralization reactions of C_3S and β-C_2S are low-calcium C-S-H and calcite-type $CaCO_3$. It is assumed that the low-calcium C-S-H is structurally similar to the calcium silicate hydration product and calcium silicate hydrate, and thus it is hypothesized that mineralization may accelerate the hydration of calcium silicate. However, other researchers (Ashraf and Olek, 2016) have suggested that the end products of these reactions are calcium-modified silica and $CaCO_3$.

In addition to C_3S and β-C_2S, C_3A and C_4AF can also react with CO_2 in cement clinker particles that have not participated in the hydration reaction. However, compared to C_3S and β-C_2S, the reactivity of C_3A and C_4AF with CO_2 is much lower. It is primarily their hydration products, AFt and AFm, that contribute to the mineralization reaction in cementitious materials. The reactivity of C_3A is enhanced in the presence of gypsum ($CaSO_4{\cdot}2H_2O$), but it remains weaker than that of C_3S and β-C_2S. The reaction is shown in Equation (2.3):

$$\begin{aligned}&3\left(3CaO\cdot Al_2O_3\right)+6\left(CaSO_4\cdot 2H_2O\right)+\left(3-x\right)CO_2+\left(n+32\right)H_2O\rightarrow\\&3CaO\cdot Al_2O_3\cdot 3CaSO_4\cdot 32H_2O+3CaO\cdot Al_2O_3\cdot 3CaSO_4\cdot 12H_2O\\&\quad+\left(3-x\right)CaCO_3+xCaO\cdot Al_2O_3+nH_2O\end{aligned}\tag{2.3}$$

2.2.2 Reaction of calcium hydroxide with CO_2

During the mineralization of cement-based materials, calcium hydroxide ($Ca(OH)_2$) is the most reactive of all calcium-containing compounds with CO_2, owing to its high water solubility and elevated pH (Arias et al., 2022; Hassani et al., 2020; Materic and Smedley, 2011; Goodbrake et al., 1979b). The reaction is shown in Equation (2.4).

$$Ca(OH)_2+CO_2\rightarrow CaCO_3+H_2O\tag{2.4}$$

The mineralization reaction in cement paste is known to occur within capillary pores that are filled with a pore solution at high pH, and $Ca(OH)_2$ mostly exists in a solid state (Wang and Chang, 2019). The pore solution consists of many cations, such as sodium and potassium, with hydroxide ions existing in equilibrium; the presence of calcium ions is relatively low in quantity. Hence, the high pH of the pore solution in cement paste is highly dependent on the concentration of Na^+ and K^+ that are in equilibrium with OH^-. CO_2 from external sources enters and dissolves in the pore solution to form a carbonate (H_2CO_3) that behaves as an acid, as shown in Equation (2.5).

$$CO_2+H_2O\rightarrow H_2CO_3\rightarrow 2H^++CO_3^{2-}\tag{2.5}$$

Carbonic acid reacts with the available alkali ions, such as sodium and potassium ions; however, the resulting products—sodium carbonate and potassium carbonate with high solubility in water (0.0181 g/g and 0.525 g/g at 20°C, respectively)—initiate the dissociation of the ions. As a result, the concentration of alkali metal ions such as sodium and potassium in the pore solution remains unchanged because of the reversible reaction. Nevertheless, calcium hydroxide dissolved in the pore solution can react with carbonic acid to produce a calcium carbonate precipitate with low solubility (see Eq. (2.3)), and such an irreversible reaction contributes to lowering the concentration of Ca^{2+} in the pore solution.

$$Ca(OH)_2 + 2H_2CO_3 \rightarrow CaCO_3 + 2H_2O \tag{2.6}$$

Furthermore, this reaction promotes the decomposition of calcium hydroxide in a solid state to maintain the equilibrium concentration between the alkali metal ions and OH^- in the pore solution. As the reaction continues, the calcium hydroxide content in the cement paste decreases, and the production of calcium carbonate increases, subsequently reducing the pH of the hardened paste.

2.2.3 Reaction of calcium silicate hydrates with CO_2

Calcium silicate hydrates (C-S-H), as another major hydration product of cement-based materials, determines the strength development, shrinkage, creep, and service life of cement-based materials, and is called the "gene" of cement-based materials. It is generally believed that C-S-H reacts with CO_2 only after a large amount of $Ca(OH)_2$ is consumed (Sevelsted and Skibsted, 2015), and the reaction is shown in Equation (2.7).

$$x\mathrm{CaO}\cdot \mathrm{SiO_2}\cdot n\mathrm{H_2O} + x\mathrm{CO_2} \rightarrow x\mathrm{CaCO_3} + \mathrm{SiO_2}\cdot n\mathrm{H_2O} \tag{2.7}$$

Because this reaction involves the decalcification of C-S-H, which has the greatest influence on the structural integrity of the hardened cement paste, mineralization often leads to a reduction in mechanical strength (Li et al., 2020). Furthermore, the molar volume of the reaction product silicate gel is smaller than that of the C-S-H before mineralization, which can potentially induce shrinkage. However, the production of $CaCO_3$ precipitates compensates for this loss in volume, and as a result, the overall volume of the cement paste remains constant.

2.2.4 Reaction of calcium sulfoaluminate hydrates with CO_2

C_3A and C_4AF produce ettringite and AFm (Al_2O_3-Fe_2O_3-mono) phases as a reaction product of hydration. The content of the ettringite and AFm phases varies with the duration of hydration, while the concentration of SO_4^{2-} is the most influential factor. The AFm phases in matured Portland cement paste generally consist of monosulfate (Kuzel, 1996). The stoichiometry of the AFm phases is expressed as Ca_2(Al, Fe)$(OH)_6X\cdot nH_2O$, and here, X denotes a single-charged or half of a double-charged anion. The AFm phases are formed by a layered structure consisting of a positively charged primary layer $[Ca_2(Al, Fe)(OH)_6]^+$ and a negatively charged interlayer $[X \cdot nH_2O]^-$, and here, X can be replaced by the possible anions SO_4^{2-}, OH^-, Cl^-, and CO_3^{2-} (Matschei et al., 2007). Hence, the fixation of CO_2 in cement paste can be in the form of monocarbonate $3CaO \cdot (Al_nFe_{1-n})_2O_3 \cdot (CaCO_3)mH_2O$ and hemicarbonate $3CaO \cdot (Al_nFe_{1-n})2O_3 \cdot (CaCO_3)0.5 \cdot (Ca(OH)_2)0.5 \cdot mH_2O$, where $n = 0$ to 1, replacing the CO_3^{2-} in the interlayer of the AFm (Dilnesa et al., 2011). The mineralization of ettringite can potentially occur, which decomposes ettringite and produces gypsum, calcite, alumina gel, and amorphous gel, including CO_2 as new reaction products.

2.2.5 Reaction of Magnesium-Based Materials with CO_2

An alternative preparation method for activated magnesium oxide is the low-temperature sintering of magnesium silicate. Because the raw material does not contain carbon and a low sintering temperature, activated magnesium oxide-based cement is considered a low-carbon alternative to ordinary Portland cement (Gartner and Sui, 2018; Walling and Provis, 2016). The mineralization mechanism of activated magnesium oxide is similar to that of C_3A and C_4AF. MgO must be hydrated first to obtain brucite ($Mg(OH)_2$) (Equation (2.8)). And then the hydration product brucite is carbonated by CO_2 (Equation (2.9)). The possible mineralization products of brucite include nesquehonite ($Mg(HCO_3\bullet)(OH)\bullet 2(H_2O)$), hydromagnesite ($Mg_5(CO_3)_4(OH)_2 \bullet 4H_2O$), and dypingite ($Mg_5(CO_3)_4(OH)_2 \bullet 5H_2O$). Brucite is considered a highly carbon-reactive phase such as $Ca(OH)_2$ and can theoretically be completely carbonated (Grünhäuser Soares and Castro-Gomes, 2021). Ruan et al. (2019) gave a schematic representation of the mineralization process of reactive magnesia oxide. The mineralization process of brucite, such as mineral mineralization of Portland cement, can be divided into a phase-boundary-controlled stage and a diffusion-controlled stage.

$$MgO + H_2O \rightarrow Mg(OH)_2 \tag{2.8}$$

$$Mg(OH)_2 + CO_2 + (n-1)H_2O \rightarrow MgCO_3 \cdot nH_2O \tag{2.9}$$

2.3 CO_2 MINERALIZATION TECHNIQUES FOR CEMENT-BASED MATERIALS

2.3.1 CO_2 Curing

CO_2 curing is a technique used to mineralize and sequester CO_2 in cement-based materials. This technology is primarily applied to cement-based products after they have undergone early hydration and molding. During CO_2 curing, the reactive ingredients and hydration products within the cement-based material interact with CO_2, resulting in the formation of stable carbonate compounds and amorphous silica gel. This process leads to a denser pore structure within the material. Compared to conventional standard curing, CO_2 curing can achieve higher strength in a short period of time (Lu et al., 2018, 2022). In addition, the cement-based products cured with CO_2 exhibit improved durability (Song et al., 2022a, 2022b).

Figure 2.2(a) illustrates a typical CO_2 curing setup for cement-based products, specifically a closed CO_2 curing system. This equipment maintains controlled pressure, temperature, and humidity using a vacuum pump, as well as pressure, temperature, and humidity controls. Besides the closed CO_2 curing system, another type is the gas flow CO_2 curing equipment, shown in Figure 2.2(b). This setup features a gas flow pipe attached to the curing vessel, allowing cement-based products to be cured under atmospheric pressure, room temperature, and humidity. The gas flow system is simpler and more cost-effective compared to the closed CO_2 curing equipment.

The general CO_2 curing process is outlined in Figure 2.3 and typically involves four stages: preparation of cement-based products, demolding and pre-conditioning, the CO_2 curing process, and subsequent further curing (Chen et al., 2022). After casting, the cement-based products undergo pre-conditioning before and after

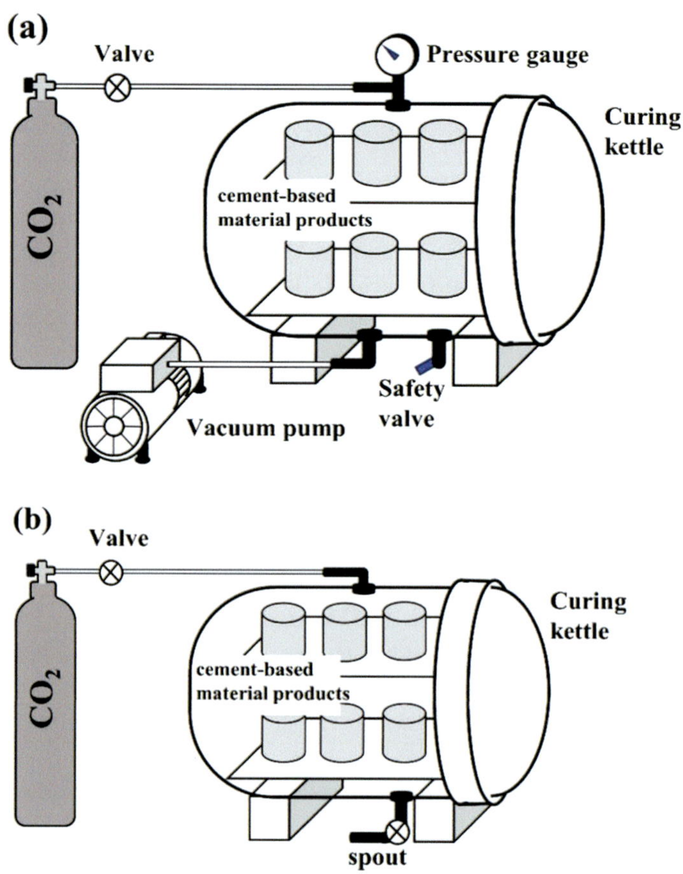

FIGURE 2.2 Schematic diagram of a typical CO_2 curing equipment: (a) a closed CO_2 curing equipment and (b) CO_2 curing equipment of a flow-through type.

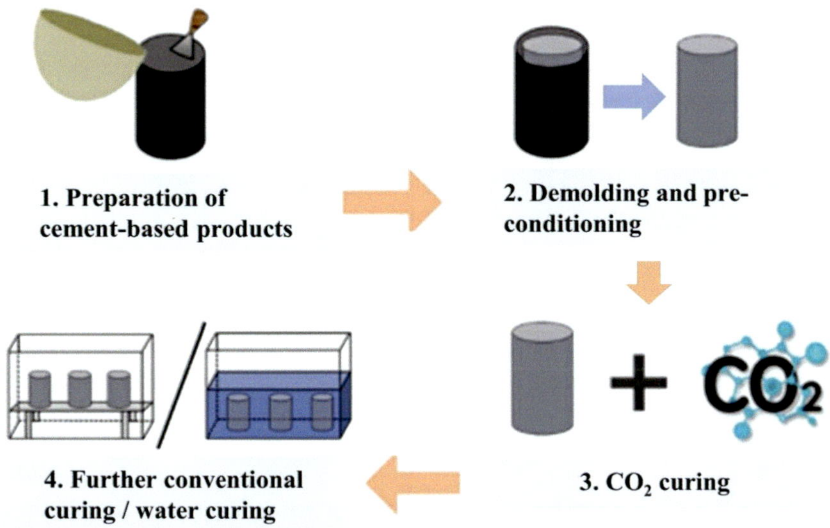

FIGURE 2.3 General Steps of CO_2 curing of cement-based products.

demolding. This step is crucial because, under high pressures, temperatures, and humidity, water may block the voids between the particles of the carbonated cement-based material, and a thicker layer of reaction-generated $CaCO_3$ can form on the surface of the cement particles, hindering further reactions (Shi et al., 2012b). The permeation rate of CO_2 gas in saturated micropores is significantly lower than in unsaturated micropores, making it almost negligible. As a result, the reaction degree in CO_2 curing using low water-cement ratio molding is quite low. As mentioned above, to address this issue, an earlier study by the authors of this book (Shi et al., 2012a) proposed pre-conditioning prior to CO_2 curing (Figure 2.1). We identified that the moisture content of the specimens after pre-conditioning is a key factor in the CO_2 curing process. Essentially, pre-conditioning is a process of water loss. While the duration of pre-conditioning has less impact on CO_2 curing, the degree of CO_2 curing is primarily enhanced by adjusting the water content (Shi et al., 2012b). There exists an optimal residual water ratio during pre-conditioning that maximizes the degree of curing and compressive strength of the concrete.

Regarding the CO_2 curing process for cement-based products, the efficiency of CO_2 curing is significantly influenced by the curing environment, including CO_2 concentration, pressure, temperature, and humidity (Chen et al., 2022; Shi et al., 2012b). These factors are discussed in detail in the next section. Additionally, because cement minerals that do not react with CO_2 continue to hydrate, subsequent curing can further enhance the strength of CO_2-conditioned cement-based products. This strength improvement is more pronounced with additional water curing compared to conventional curing. In such cases, subsequent water curing is superior to further conventional curing (Lu et al., 2021).

2.3.2 Accelerated mineralization

Due to the slow kinetics of the CO_2 reaction, cement-based materials exhibit very limited natural CO_2 sequestration during their service life. To enhance the capacity and efficiency of CO_2 sequestration, accelerated mineralization technology has gained significant attention in recent years. This technology is particularly valuable for processing cement-based solid wastes and industrial by-products, such as fly ash, slag, and recycled cement paste powders. By accelerating the mineralization process, these materials can more effectively capture and mineralize CO_2, contributing to improved sustainability and reduced environmental impact.

2.3.2.1 Dry mineralization

Dry mineralization is the most commonly employed method for accelerated mineralization. As illustrated in Figure 2.4, typical dry mineralization equipment includes a mineralization chamber, a CO_2 gas cylinder, a vacuum pump, and associated piping. Dry mineralization can be further categorized into two types based on relative humidity (RH): dry mineralization and semi-dry mineralization (RH greater than 80%). In the context of this book, semi-dry mineralization is considered a variant of dry mineralization due to its relatively high RH.

During the dry mineralization of cement-based materials, CO_2 mineralization reactions occur at multiple local mineralization fronts. As these fronts advance into the interior of the particles, CO_2 is rapidly consumed by a limited number of reaction sites, which hinders further diffusion of CO_2 gas. At this stage, the ionic saturation and concentration levels in the pore solution of the cement-based material limit the rate of CO_2 mineralization and influence the crystallinity and polymerization of the mineralization products.

The dry mineralization process of cement-based materials consists of two phases. In the first phase, which occurs within a few minutes after the onset of CO_2

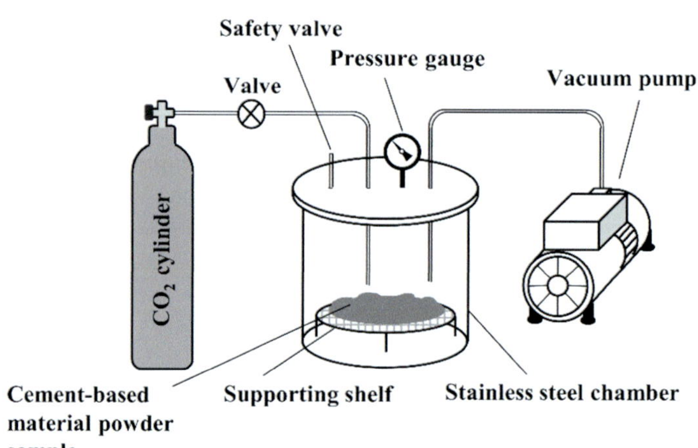

FIGURE 2.4 Typical dry mineralization equipment.

mineralization, the CO_2 gas reacts with all calcium-containing phases present. The mineralization rate of calcium hydroxide (CH) is faster than that of other phases, such as C_3S, beta-C_2S, and C-S-H. During this phase, $Ca(OH)_2$ on the surface of the particles is rapidly dissolved and transported to the CO_2 mineralization front, where it reacts with CO_2 dissolved in the interstitial solution. The exothermic nature of this reaction causes continuous evaporation of water from the cement-based materials, which hinders further CO_2 entry into the pores. Consequently, CO_2 diffusion becomes the limiting factor for mineralization. In the late stage of this phase, a dense $CaCO_3$ layer forms over the $Ca(OH)_2$, limiting the dissolution of remaining CH within the particles. As a result, the mineralization rate sharply decreases after the rapid depletion of $Ca(OH)_2$, leading to a significant reduction in $CaCO_3$ precipitation.

In the second stage of dry mineralization of cement-based materials, the CO_2 mineralization front progresses into the interior of the particles. At this point, Ca^{2+} is primarily removed and diffuses through micropores, forming zigzag $CaCO_3$ layers around other calcium-containing phases, excluding $Ca(OH)_2$. Although the water released from mineralization during the early stages enhances the dissolution dynamics of the cement-based material particles, the exothermic nature of the reaction accelerates water loss. The research results (Mao et al. 2023) show that even if the CO_2 mineralization is extended to 28 days, the complete CO_2 mineralization cannot be achieved. Therefore, the water content in the CO_2 mineralization environment seriously restricts the CO_2 mineralization power of cement-based materials, and the degree of mineralization is low, and the silica gel content in the CO_2 mineralization products is less.

2.3.2.2 Wet mineralization

Wet mineralization is a CO_2 mineralization technique in which cement-based material is immersed in a liquid environment to facilitate the CO_2 mineralization reaction through carbonate ion diffusion. Figure 2.5 shows a schematic of

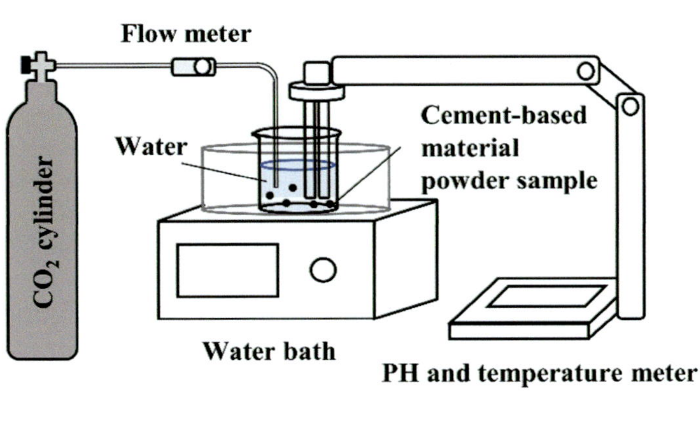

FIGURE 2.5 Typical wet mineralization equipment.

typical laboratory-scale equipment for wet mineralization. The equipment consists of a vessel containing a powdered sample of cement-based material and a liquid, a thermostatic water bath (to control the temperature and stirring rate during the wet mineralization process), a tank containing CO_2, piping with a flow control device, and a measuring meter to detect the temperature and pH during the wet mineralization.

Wet mineralization occurs in a solution environment where CO_2 is not limited by gaseous diffusion but is instead dissolved in large quantities. In this setting, the calcium-containing phases in cement-based materials are undersaturated, which promotes continuous Ca^{2+} leaching and accelerates the precipitation of mineralization products. Consequently, numerous studies (Mao et al. 2023, Zajac et al., 2020b, Jiang et al., 2023) have demonstrated that wet mineralization significantly enhances both the degree and efficiency of CO_2 mineralization compared to dry mineralization, with cement-based materials achieving a high level of mineralization within just a few hours.

The physical phase reaction sequence of wet mineralization in cement-based materials differs from that of dry mineralization. Due to its high solubility in solution, decalcification of other calcium-containing phases begins only after $Ca(OH)_2$ is nearly exhausted at the onset of the CO_2 mineralization reaction. The wet mineralization process can be divided into four stages. In the first stage, CO_2 dissolves in water and ionizes to produce CO_3^{2-} and H^+, with Ca^{2+} primarily released by $Ca(OH)_2$. The rate of CO_2 mineralization in this phase is higher, accompanied by a slight increase in pH, indicating that the dissolution rate of cement-based material particles exceeds the dissolution rate of CO_2. The dissolution of CO_2 significantly influences the progress of the CO_2 mineralization reaction (Peng et al., 2023). It is important to note that, in an alkaline environment, the increase in Al and Si concentration in the first stage (Zajac et al., 2020a) may impact $CaCO_3$ precipitation. Subsequently, either the dissolution of CO_2 or the precipitation of $CaCO_3$ can constrain the CO_2 mineralization process. In the second stage of wet mineralization, Ca^{2+} originates from the decalcification of $Ca(OH)_2$ dissolution with other calcium-containing phases. The gradual enrichment of $CaCO_3$ on the particle surface of cement-based materials limits Ca^{2+} transport from the interior of the particles to the surface, where the pH decreases more rapidly. This lower dissolution rate limits the CO_2 mineralization reaction. In the third stage, the CO_2 mineralization process gradually slows down. The growing thickness of the $CaCO_3$ layer on the surface of cement-based material particles restricts decalcification of the internal calcium-containing phases. The reaction rate is influenced by the dissolution of hydrates. In the fourth stage, amorphous silica gel and alumina gel form rapidly. The $CaCO_3$ layer becomes homogeneous, and the core shrinks and becomes porous until it disintegrates. Although the $CaCO_3$ layer no longer impedes the diffusion of Ca^{2+} within the cement-based material particles, the decalcification of C-S-H with a low Ca/Si ratio is challenging, Ca^{2+} diffusion is slow [46], and the CO_2 mineralization reaction tends to plateau.

2.4 FACTORS INFLUENCING CO_2 MINERALIZATION FOR CEMENT-BASED MATERIALS

2.4.1 Factors influencing CO_2 curing

2.4.1.1 Parameters during pre-conditioning

Moisture on the surface of cement-based products is the primary factor that hinders the transmission of CO_2 into the interior of these products during the CO_2 curing process. Pre-conditioning is essentially a process designed to remove the surface moisture, thereby facilitating more effective CO_2 penetration (Shi et al., 2012a).

2.4.1.1.1 Water content after pre-conditioning

A significant correlation was observed between the dissipation of surface moisture from cement-based products and the CO_2 sequestration effectiveness of CO_2 curing, as well as the compressive strength of the products, as shown in Figure 2.6. The loss of surface moisture exhibited a strong positive correlation with the CO_2 sequestration ratio of ordinary cement-based products (Shi et al., 2012a). Specifically, as the surface moisture loss increased from 0.7% to 4.8%, the CO_2 sequestration ratio of the cement-based products rose from 10% to 25%, and their compressive strength also improved. However, when the surface moisture loss further increased from 4.8% to 5.2%, both the CO_2 sequestration ratio and compressive strength of the cement-based products decreased slightly. Figure 2.7 illustrates the relationship between the CO_2 sequestration ratio and compressive strength of CO_2-cured concrete blocks and their water content after pre-conditioning. It is evident that an optimal water content, approximately 16%, maximizes the CO_2 sequestration ratio and compressive strength of CO_2-cured concrete blocks (Shi et al., 2012a). This finding aligns with the general consensus in current research, which identifies the optimal water content of cement-based materials post-pre-conditioning to be in the range of 10–20%.

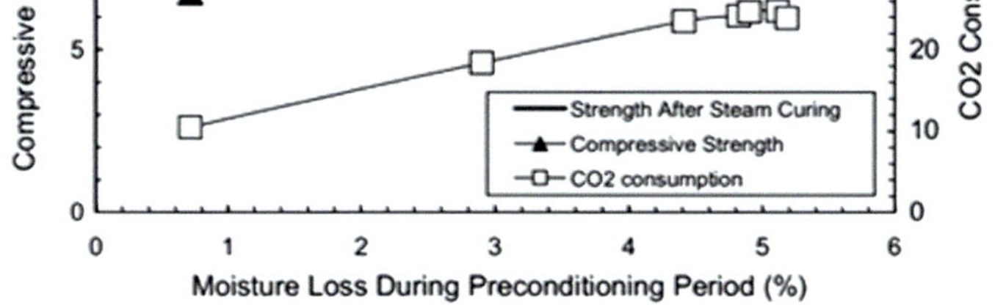

FIGURE 2.6 Effect of moisture loss on compressive strength and CO_2 sequestration ratio after pre-conditioning (Shi et al., 2012a).

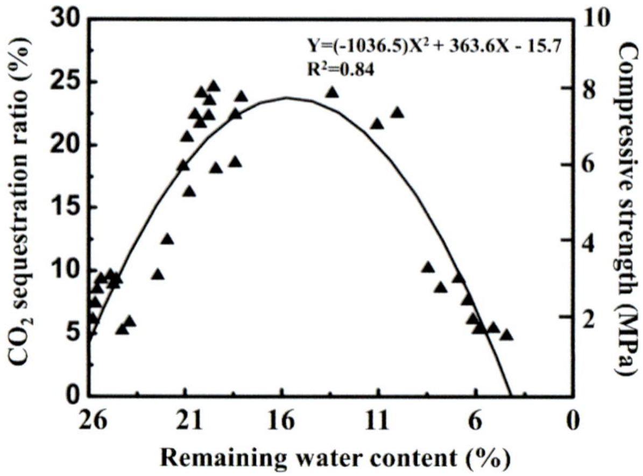

FIGURE 2.7 Relationship between residual water content of concrete blocks after pre-conditioning and CO_2 sequestration ratio and compressive strength of concrete blocks after CO_2 curing (Shi et al., 2012a).

2.4.1.1.2 *Methods of Pre-conditioning*

The impact of the pre-conditioning method on the CO_2 curing effectiveness of cement-based products is fundamentally linked to its influence on the moisture loss rate in these materials. As shown in a previous study (Shi et al., 2012a), the relationship between moisture loss and environmental factors such as temperature, wind speed, and relative humidity is depicted in Figure 2.8. The findings indicate that higher temperatures, lower ambient relative humidity, and increased wind speed all contribute to a greater rate of moisture loss.

2.4.1.1.3 *Effect of preconditioning period on residual water content*

Figure 2.9 presents the compressive strength of lightweight concrete blocks subjected to various pre-conditioning periods followed by CO_2 curing (Shi et al., 2012a). The results indicate that specimens pre-conditioned for durations ranging from 2 to 18 hours exhibit similar strength after 2 hours of CO_2 curing, with the maximum compressive strength observed in specimens pre-conditioned for 8 hours. This strength was only slightly lower than that of the reference specimens that underwent steam curing for 20–24 hours. Furthermore, Figure 2.10 shows that extending the pre-conditioning period from 2 to 8 hours consistently increases the CO_2 sequestration ratio of the CO_2 cured lightweight concrete blocks. However, for pre-conditioning times exceeding 8 hours, the rate of increase in CO_2 sequestration ratio tends to plateau, with only minimal gains observed. This suggests that an extended pre-conditioning period is not necessary for effective CO_2 sequestration. The compressive strength and CO_2 sequestration ratio of CO_2 cured cement-based products with a 2–4 hour pre-conditioning period are essentially comparable to those with longer pre-conditioning times, offering potential economic benefits by reducing the duration of the pre-conditioning phase.

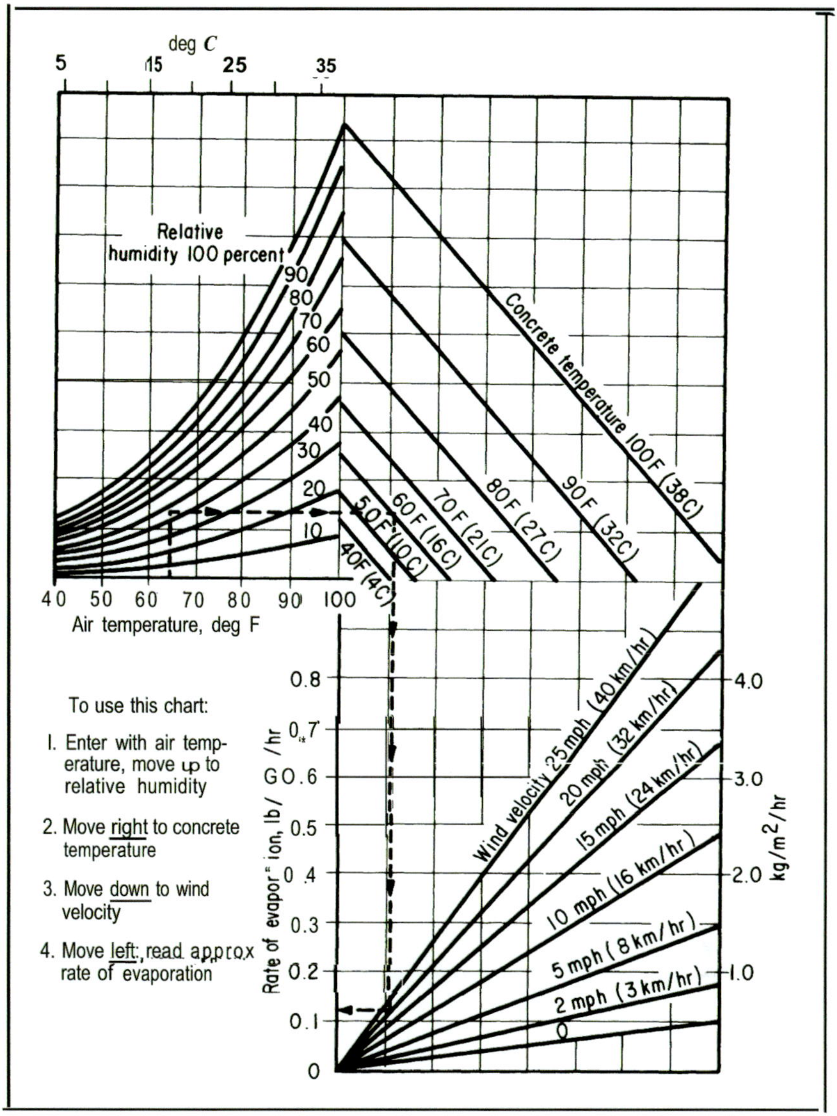

FIGURE 2.8 The variation of moisture loss with ambient temperature, wind speed, and relative humidity (Shi et al., 2012a).

2.4.1.2 Parameters during CO_2 curing

2.4.1.2.1 CO_2 concentration

CO_2 concentration has an important effect on the degree of CO_2 curing and mechanical properties of cementitious cement-based products after CO_2 curing. The mechanism is shown in Figure 2.11. At low CO_2 concentrations, water within the

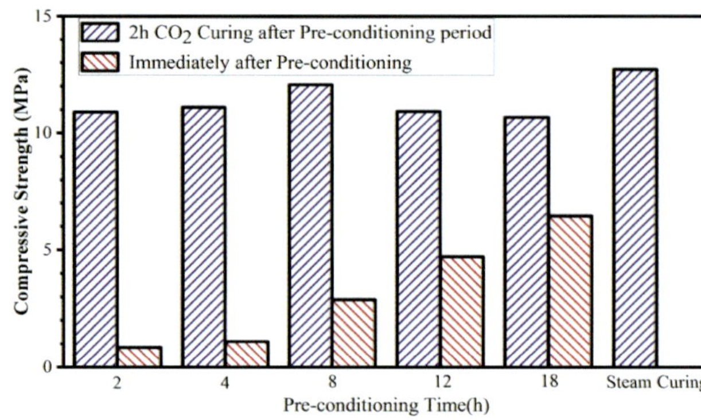

FIGURE 2.9 Relationship between pre-conditioning period and compressive strength of concrete blocks after CO_2 curing (Shi et al., 2012a).

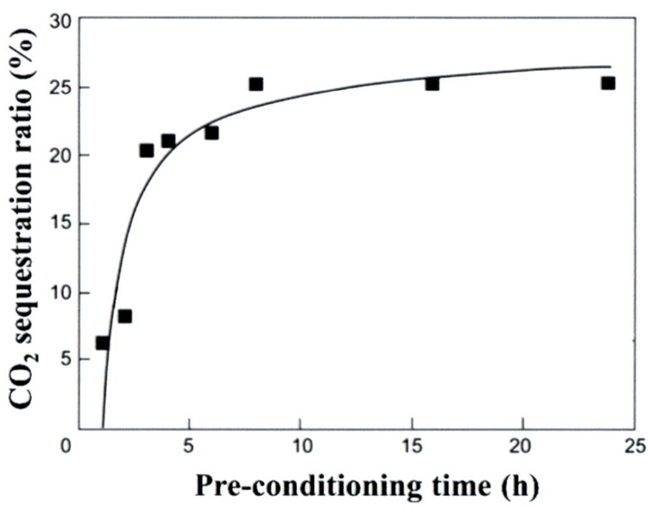

FIGURE 2.10 Relationship between pre-conditioning period and CO_2 sequestration ratio of concrete blocks after CO_2 curing (Shi et al., 2012a).

cement-based product and atmospheric water can reach equilibrium; in fact, under natural conditions with CO_2 concentrations below 3–4%, this process does not change the composition of the cement-based products formed during mineralization and retains a similar C-S-H phase with a relatively low Ca/Si. On the contrary, exposure to higher CO_2 concentrations increased the internal humidity due to water formation. The effect of CO_2 concentration on the compressive strength and CO_2 uptake of the products is almost the same for a given conditioning parameter (Dung et al., 2021; Lu et al., 2021). Figure 2.12 shows the compressive strength of cement-based products after CO_2 curing with different CO_2 concentrations. The compressive

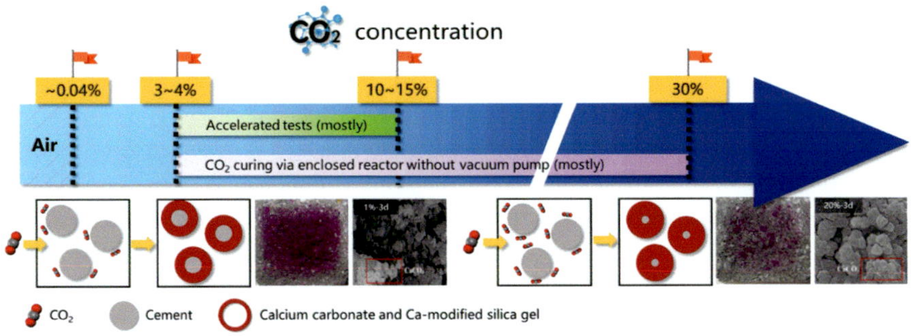

FIGURE 2.11 The mineralization behaviors with various CO_2 concentrations.

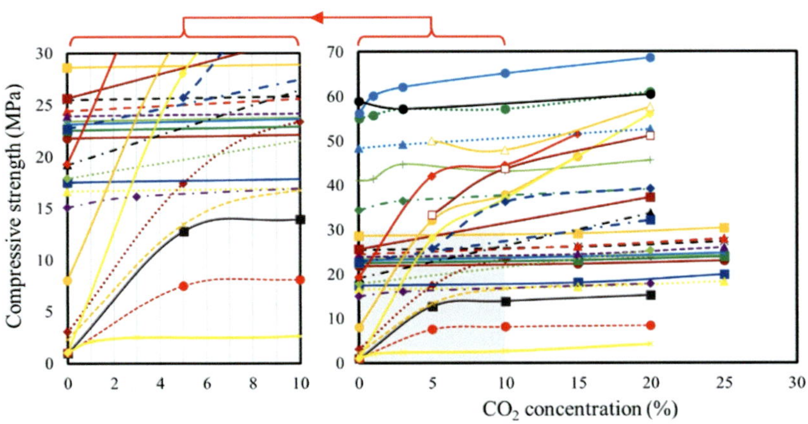

FIGURE 2.12 Variation of compressive strength of CO_2-cured cement-based products with CO_2 concentration (Chen et al., 2022).

strength of cement-based products increases significantly with the increase of CO_2 concentration until about 5% concentration. However, as the CO_2 concentration continues to increase to about 10%, the compressive strength of cement-based products basically ceases to increase with the increase in CO_2 concentration, which indicates that the rate of carbon sequestration tends to be 0, and the amount of CO_2 sequestration reaches the maximum value. However, there are also a large number of previous findings indicating that the optimum CO_2 concentration in the CO_2 curing process is about 20%, i.e., with the increase of CO_2 concentration up to 20%, an obvious increase in CO_2 mineralization can be observed, but this increase gradually slows down with the further increase in CO_2 concentration.

The observed trends can be explained as follows: (1) Acceleration of CO_2 Diffusion: Higher CO_2 concentrations accelerate the inward movement of CO_2 through the pores of cement-based products, leading to a higher degree of mineralization. (2) Formation of Dense Mineral Layers: CO_2 mineralization produces dense minerals, which block the transport paths of incoming CO_2 gas, thereby slowing down the mineralization process for subsequent layers. As these mineralization products accumulate within the pore spaces, the pore structure is altered, resulting in a denser outer layer that hinders the diffusion of CO_2 to deeper, unmineralized regions. This creates clear microstructural differences between CO_2-mineralized and unmineralized regions: unmineralized samples form large, curved pores, while mineralized samples exhibit a denser structure. Both pore size and porosity decrease with increasing CO_2 concentration, as sufficient CO_2 is available to produce $CaCO_3$, filling these pores. (3) Limitation by Ca^{2+} Availability: When the CO_2 concentration reaches high levels (e.g., above 20%), the reaction between CO_3^{2-} and the hydration products becomes limited by the availability of Ca^{2+} in the hydration products rather than by CO_2 concentration itself.

2.4.1.2.2 *Temperature*

The curing temperature plays a crucial role in the CO_2 curing process for cementitious materials. Increasing the temperature enhances CO_2 diffusion by accelerating ion migration and the evaporation of water from the pore solution. Research indicates that as temperature rises from 20°C to 80°C, permeability increases by approximately 13–62% (Jooss and Reinhardt, 2002). However, higher temperatures also reduce CO_2 solubility, leading to an increased pH in the liquid phase and a decreased dissolution rate of Ca^{2+} (see Figure 2.13) (Chen and Gao, 2019). Consequently, CO_2 curing temperatures are typically set between 20°C and 30°C, as CO_2 mineralization is an exothermic reaction (Liu and Meng, 2021). Temperature significantly impacts both the compressive strength and CO_2 sequestration ratio of cement-based materials. For instance, Lu et al. (2022) observed that the compressive strength of cement paste specimens increased from 16.9 MPa to 36.2 MPa (~114.2%) when the temperature was raised from 5°C to 50°C. In contrast, Zhan et al. (2016) reported that

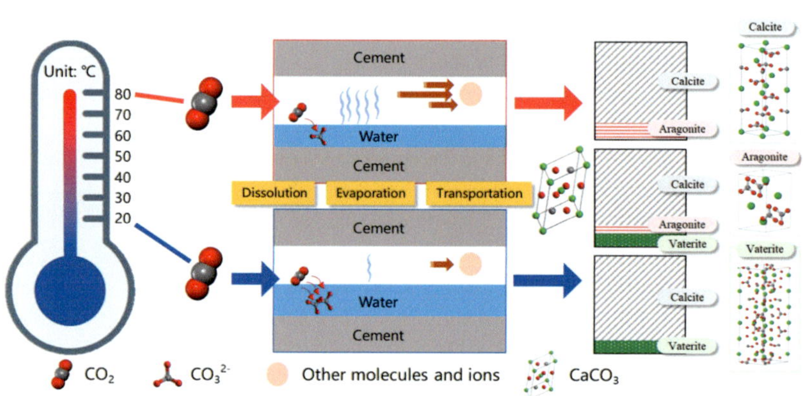

FIGURE 2.13 The mineralization behaviors with various curing temperature (Chen et al., 2022).

increasing the curing temperature from 20°C to 80°C led to a slight decrease in the compressive strength of concrete blocks. This discrepancy highlights the need for further studies to elucidate the underlying mechanisms. Additionally, the trend of CO_2 uptake generally parallels the trend in compressive strength up to 100°C. Beyond this temperature, CO_2 uptake values either plateau or decrease gradually, likely due to the temperature exceeding the boiling point of water [13].

2.4.1.2.3 Curing period

The CO_2 curing period is a crucial parameter in the CO_2 curing process. Figure 2.14 provides an example of a CO_2-cured aerated concrete specimen block. The compressive strength of these aerated concrete specimens increased with the duration of CO_2 mineralization up to 24 hours. Specifically, after 5 hours of CO_2 curing, the compressive strength reached approximately 73.9 MPa, which is close to the maximum achievable value. This trend indicates that while the compressive strength of CO_2-cured cement-based materials improves with an extended curing period, the rate of this increase gradually diminishes over time.

A similar trend is observed in the variation of the CO_2 sequestration ratio with changes in the CO_2 curing period. Ahmad et al. (2017) found that the mass gain of concrete samples exposed to CO_2 continued to increase at a decreasing rate throughout a 10-hour curing period, with a higher curing pressure (60 psi) further reducing this rate. Conversely, another study (Shi et al., 2012b) indicated that 4–5 hours of CO_2 curing was sufficient to achieve the peak CO_2 sequestration ratio in pressed and formed cement paste products, with additional curing time offering minimal further enhancement. However, for samples with higher porosity, such as those with a water-to-solids ratio of 0.14 and 0.10, the CO_2 mineralization effect was more pronounced. In these cases, 7–8 hours of CO_2 curing were required to reach peak CO_2 sequestration levels, with the highest observed CO_2 sequestration ratio of 9.56%.

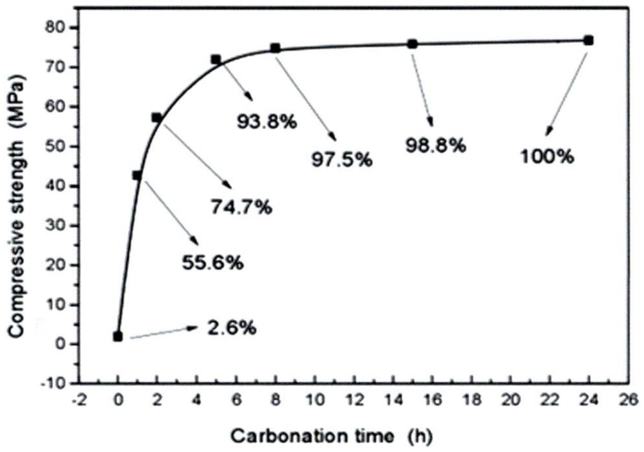

FIGURE 2.14 Compressive strength versus CO_2 mineralization period (Ahmad et al., 2017).

2.4.1.2.4 CO_2 pressure

Gas pressure directly affects the entry, diffusion, and dissolution of CO_2 in the pore solution of the cementitious material during CO_2 curing, releasing carbonate ions. Relatively high pressures have been shown to favor CO_2 gas transporting and leaching of alkali metal ions, which are closely related to CO_2 curing efficiency. Figures 2.15 and 2.16 show this effect using a typical example. Cement paste products, pre-conditioned in a dry environment for 4 hours, were subjected to CO_2 curing at pressures ranging from 0.1 MPa to 0.5 MPa. The trends in the CO_2 sequestration ratio and compressive strength with varying CO_2 pressures were measured. From Figure 2.15, it can be seen that the CO_2 sequestration ratio and the compressive strength of the specimens increased gradually with the increase in CO_2 gas pressure. Since there is a pressure difference, the gas is able to penetrate into the specimen. As the pressure of the CO_2 gas increases, this pressure difference increases, making it easier for the gas to diffuse into the specimen. Therefore, an increase in CO_2 gas pressure leads to an increase in the overall reaction rate. Under the same curing period, the increase in CO_2 gas pressure leads to an increase in carbon sequestration ratio of the specimen. At the same time, it can be seen that when the pressure reaches 0.4 MPa, the increase in gas pressure is not obvious to the increase in degree of CO_2 curing, which may be because the excessive pressure makes the mineralization product layer formed quickly, blocking the pore space, which leads to insufficient diffusion of the gas. Figure 2.16 shows that the compressive strength of the specimens also increases with CO_2 gas pressure. This trend correlates with the increased CO_2 sequestration rate, which, in turn, improves the compressive strength of the specimens.

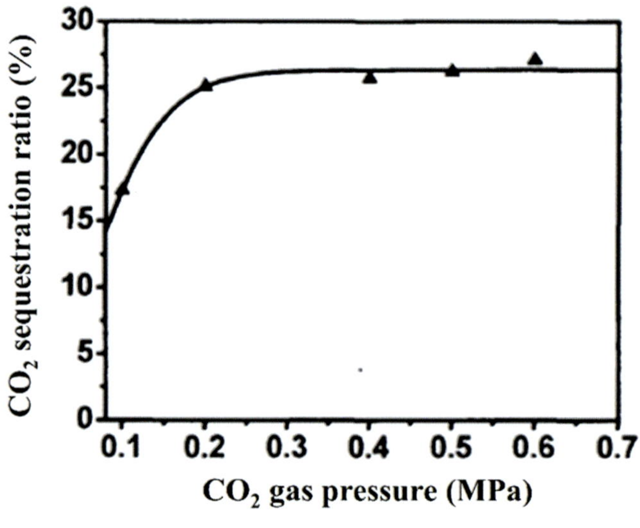

FIGURE 2.15 CO_2 pressure versus specimen CO_2 sequestration ratio.

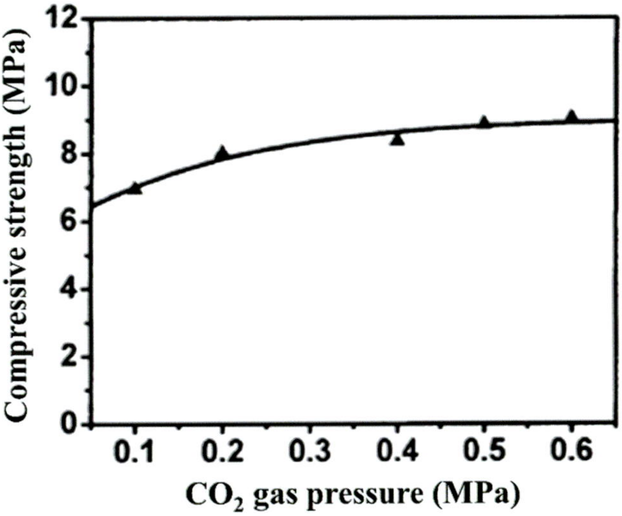

FIGURE 2.16 Relationship between CO_2 pressure and compressive strength of specimens.

2.4.1.2.5 CO_2 flow rate

Some previous studies have reported the effect of CO_2 flow rate on the degree of curing during CO_2 curing. According to Zhan et al. (2016), a higher CO_2 flow rate promotes the penetration of CO_2 into the matrix of cement-based materials, thereby increasing their degree of curing and strength. On the other hand, Xuan et al. (2018) suggested that prepared samples without a pre-drying treatment have relatively high initial evaporated water. The increased gas flow rate favored the removal of internal water, leading to an immediate decrease in free water content. This links the effect of CO_2 flow rate to the aforementioned residual moisture in the matrix during the pre-conditioning stage. The idea that CO_2 flow rate affects the CO_2 curing efficiency of cementitious materials by influencing water evaporation is supported by several earlier studies (Kashef-Haghighi and Ghoshal, 2010; Kashef-Haghighi et al., 2015; Liang et al., 2020).

2.4.1.2.6 Relative humidity

Another critical factor influencing the CO_2 curing process is the relative humidity (RH) of the CO_2 curing chamber. Previous studies (Chen et al., 2022) have shown that the CO_2 sequestration ratio and compressive strength of CO_2-cured cement-based materials first increase and then decrease as the RH increases from 0% to 100%. These values peak within an RH range of 50% to 65% (Xuan et al., 2018). This trend can be attributed to the balance between gas diffusion and the kinetics of the mineralization reaction (Liang et al., 2020). At lower RH levels, rapid water evaporation increases pore space, which facilitates the inward diffusion of CO_2 and enhances the rate of early CO_2 mineralization (Bouchaala et al., 2011). However, when RH decreases below a certain threshold (around 7%), the lack of sufficient water limits the extent of CO_2 mineralization as hydration products cannot react further. Conversely, excessively high RH levels can inhibit gas diffusion, reducing the efficiency of the CO_2 curing process (Morandeau et al., 2014).

2.4.1.3 Subsequent curing process parameters

Removal of water during the molding process can inhibit the hydration of some unhydrated components within the cement-based material. Therefore, subsequent curing (i.e., further curing after CO_2 curing) is essential to compensate for this loss of moisture and to ensure that hydration reactions continue. This additional curing helps to improve the pore structure and mechanical properties of the material (Liu and Meng, 2021; Zhang et al., 2017). Figure 2.17 illustrates the impact of various curing techniques on compressive strength, overall reaction degree, and microstructural evolution of cement pastes. This indicates that more significant increases in compressive strength during post-curing are observed for specimens that underwent shorter initial CO_2 curing periods. However, regardless of the initial mineralization duration, the ultimate values achieved after 28 days of curing were consistent (Zhan et al., 2016). Additionally, He al. (2016) observed that during post-curing, the decomposition of carbonate coatings occurs, leading to the formation of calcium monocarbonaluminate hydrate from C_3A.

2.4.2 Factors influencing accelerated mineralization

2.4.2.1 Factors influencing dry mineralization

Relative humidity (RH) plays a crucial role in the dry mineralization process of cement-based materials. Optimal RH levels facilitate CO_2 diffusion into the material, enhancing the mineralization reaction. Conversely, very high RH levels can hinder CO_2 diffusion and slow down the mineralization process. Galan et al. (2015)

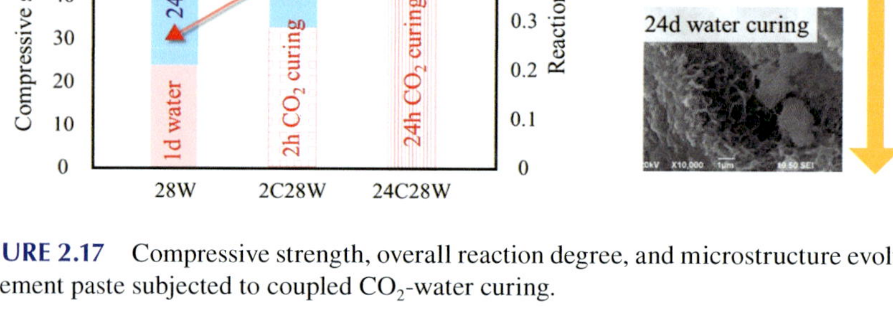

FIGURE 2.17 Compressive strength, overall reaction degree, and microstructure evolutions of cement paste subjected to coupled CO_2-water curing.

observed that the diffusion coefficient of CO_2 decreases significantly as water saturation increases, dropping by seven orders of magnitude from a dry to a near-saturated state. This highlights the importance of controlling RH in dry mineralization processes. Typically, pretreatment ensures that RH remains within an optimal range of 50–70% to balance CO_2 diffusion and mineralization efficiency.However, Zajac et al. (2022) reported that the degree of CO_2 mineralization increased significantly with RH in the range of 80–100%, surpassing the mineralization achieved at lower RH levels (50–70%). This increase is attributed to higher water content, which enhances calcium dissolution in the cement-based materials. In fact, semi-dry mineralization with RH = 100% can lead to higher initial mineralization reaction rates compared to wet mineralization, demonstrating that higher RH can sometimes be more effective in promoting the CO_2 mineralization reaction (Thiery et al., 2007) .

Increased temperature can promote CO_2 diffusion (Drouet et al., 2015), which leads to an overall accelerated trend in the rate of CO_2 mineralization, thus increasing the degree of CO_2 mineralization. However, powdered cement-based materials have a large specific surface area, and CO_2 mineralization is susceptible to water evaporation at high temperatures. Once the temperature exceeds the threshold, the rate of CO_2 mineralization decreases instead (Drouet et al., 2019). Wu et al (2022) concluded that the temperature threshold of cement-based materials may be near 100°C. When the temperature is between the interval of 20°C and 100°C, the CO_2 mineralization rate of cement-based materials increases with temperature. However, as the temperature increases further and water evaporation intensifies, the CO_2 mineralization rate shows a decreasing trend. Therefore, initial water supplementation is necessary for the dry mineralization of cement-based materials at high temperatures.

In addition, higher CO_2 concentration and pressure can also significantly increase the degree and efficiency of dry mineralization of cement-based materials. This is attributed to the fact that higher CO_2 concentrations and mineralization pressures allow CO_2 to diffuse deeper into the cement-based material under a larger concentration or pressure difference (Wang et al. 2017), which facilitates the second stage of dry CO_2 mineralization. However, when the duration of the CO_2 mineralization reaction is sufficient, the difference between the degree of CO_2 mineralization affected by CO_2 concentration will gradually decrease. This is because with the continuation of mineralization time, the mineralization products layer keeps accumulating, which hinders the diffusion of CO_2 gas, and the CO_2 mineralization rate gradually slows down or even terminates the reaction. Recent studies (He et al., 2020) found that the flue gas is rich in CO_2 (concentration of about 15–30%), and with residual temperature, the utilization of flue gas can eliminate the process of separating CO_2 from the captured gas, and at the same time can obtain the heat energy required for mineralization. Therefore, the utilization of flue gas has great prospects for industrial applications. However, it should be noted that the flue gas contains a small amount of NO_2 and SO_2, and the mechanism of the influence of these impurity gases on mineralization has not been sufficiently studied, which needs to be explored in depth in order to lay a foundation for its industrial promotion.

2.4.2.2 Factors influencing wet mineralization

The wet mineralization process of cement-based materials is governed by dissolution-precipitation mechanisms. Consequently, in wet mineralization, the dissolution rates of CO_2 and Ca^{2+}, as well as the precipitation of CO_2 mineralization products, can be optimized by adjusting temperature, CO_2 concentration, and mineralization time to alter the degree and efficiency of CO_2 mineralization. The effects of temperature and time on the rate and degree of wet mineralization are similar to those observed in dry mineralization, though excessively high temperatures primarily hinder the rate of CO_2 mineralization by reducing the solubility of CO_2 and ions (Zhan et al., 2016). In wet mineralization, CO_2 concentration can be controlled by adjusting the CO_2 flow rate. Mao et al. (2023) found that increasing the CO_2 flow rate from 3 L/min to 5 L/min significantly enhanced the $CaCO_3$ content during the first 10 minutes of wet mineralization of recycled cement paste powder, with the impact of flow rate on CO_2 mineralization being more pronounced than that of temperature increase. Additionally, the agitation rate during wet mineralization may influence the mineralization rate by affecting the dissolution of cement-based material particles and particle size, though systematic studies on this effect are currently lacking.

2.5 CURRENT LIMITATIONS AND FUTURE DIRECTIONS OF CO_2 MINERALIZATION TECHNIQUES FOR CEMENT-BASED MATERIALS

Currently, numerous studies have focused on the mechanism and theoretical relationship between the influence of individual environmental factors on the CO_2 curing of cement-based materials, with research on single factors being relatively well-established. However, in practical applications, CO_2 curing of cement-based materials is often influenced by the combined effects of multiple factors. For example, during transportation, CO_2 concentration and relative humidity can form gradients within cementitious materials, which may help achieve the target mineralization degree, thereby improving both the performance of the materials and their carbon sequestration potential. However, the mineralization of cement-based materials with low water-to-cement ratios at higher temperatures may lead to lower CO_2 solubility due to reduced water content, which could negatively impact mineralization and carbon sequestration. Therefore, future research should further explore the mineralization behavior and influencing factors of different cementitious materials under coupled conditions, to identify the optimal combination of mineralization and curing conditions. Additionally, current research on the mechanisms of mineralization and curing of cementitious materials remains limited. The control of $CaCO_3$ crystal morphology, the structure of $SiO_2 \cdot nH_2O$, and the polymerization mechanisms of mineralization products under varying conditions are still unclear. More in-depth studies are needed to reveal the spatial and temporal evolution of the reactants and products of cementitious materials at different stages in the presence of CO_2, and to develop a mineralization model that can effectively predict both mineralization depth and carbon sequestration ratios. Currently, high-concentration CO_2 is commonly used for the mineralization of cement-based materials. However, capturing and enriching high-concentration CO_2 incurs relatively high operational costs, limiting its practical application. Future research could explore the direct use of emission tail gases from industries such as steel and cement for mineralization and curing. Additionally, utilizing waste heat and steam from these tail gases could accelerate the mineralization reaction and enhance both the performance and carbon sequestration rate of cementitious materials. At present, most "carbon sources" used for CO_2 mineralization and curing of cement-based materials are limited to CO_2 gas, and research primarily focuses on prefabricated components, which restricts its application to on-site casting. This limitation significantly reduces the broader applicability of CO_2 mineralization technology. In the future, the range of carbon sources could be expanded, such as by using carbonated water. In addition, there are limited studies on the effect of CO_2 mineralization on reinforced concrete, and it remains unclear whether the presence of CO_2 directly weakens the durability of reinforced concrete. Therefore, future research should model the phenomena involved in the mineralization of reinforced concrete to better understand the corrosion behavior of steel and to optimize the use of CO_2 mineralization technology. Currently, the application of CO_2 mineralization in cementitious materials is limited to ordinary silicate cement, with fewer studies focusing on the mineralization of sulfur-aluminate cement, fly-ash silicate cement, and other types. Future research could explore the integration of recycled aggregates, recycled powders, steel slag, fly ash, magnesium cement, various tailing slags, and aerated concrete with CO_2 mineralization. This would help the construction industry progress toward achieving higher levels of carbon neutrality.

The effect of temperature on the accelerated CO_2 mineralization of cement-based materials has been investigated in previous studies, but the underlying mechanisms remain insufficiently understood. Temperature effects often vary for industrial wastes with different chemical compositions, particle sizes, and microstructures. Numerous studies have shown that increasing the temperature from ambient levels to a threshold range of 60–80°C enhances both the CO_2 sequestration capacity and the mineralization rate of the material. However, beyond this threshold, further temperature increases appear to have little effect or may even hinder mineral CO_2 sequestration. Conversely, at higher temperatures, CO_2 uptake capacity and efficiency can increase dramatically. Therefore, further studies are needed to systematically explain these complex temperature effects. Additionally, there are still very few studies on the use of carbonated materials as replacements for aggregates or cement-based materials in concrete. Previous researches on accelerated mineralization treatments, such as for RCA, fly ash, and slag, has focused on identifying the optimal conditions for maximizing CO_2 uptake. However, for specific materials, there may be a CO_2 absorption threshold, beyond which increased CO_2 uptake can be detrimental to the strength enhancement of cementitious composites containing accelerated CO_2 mineralization materials. Future research should investigate the optimal CO_2 uptake that leads to the most significant strength enhancement.

2.6 SUMMARY

Current CO_2 mineralization techniques for cement-based materials can be classified into two categories: CO_2 curing and accelerated mineralization. CO_2 curing is suitable for larger-volume cement-based materials, such as concrete bricks and cement products, while accelerated mineralization is more appropriate for powder cement-based materials. The CO_2 curing process involves three steps: pre-conditioning, CO_2 curing, and subsequent further curing. Key factors influencing the degree of CO_2 curing and the properties of cement-based materials include the residual moisture content after pre-conditioning, temperature, humidity, duration, CO_2 pressure, and CO_2 concentration during the CO_2 curing phase. In contrast, accelerated mineralization can be further divided into wet mineralization and dry mineralization. The factors affecting the dry mineralization process are similar to those in CO_2 curing, whereas the wet mineralization process is influenced not only by these parameters but also by CO_2 flow rate and additives.

REFERENCES

Ahmad, S., Assaggaf, R.A., Maslehuddin, M., Al-Amoudi, O.S.B., Adekunle, S.K., Ali, S.I., 2017. Effects of carbonation pressure and duration on strength evolution of concrete subjected to accelerated carbonation curing. *Construction and Building Materials* 136, 565–573.

Arias, B., A. Criado, Y., Pañeda, B., Abanades, J.C., 2022. Carbonation kinetics of Ca(OH)2 under conditions of entrained reactors to capture CO2. *Industrial & Engineering Chemistry Research* 61(9), 3272–3277.

Ashraf, W., Olek, J., 2016. Carbonation behavior of hydraulic and non-hydraulic calcium silicates: potential of utilizing low-lime calcium silicates in cement-based materials. *Journal of Materials Science* 51(13), 6173–6191.

Bouchaala, F., Payan, C., Garnier, V., Balayssac, J.P., 2011. Carbonation assessment in concrete by nonlinear ultrasound. *Cement and Concrete Research* 41(5), 557–559.

Chen, K.-Y., Xia, J., Wu, R.-J., Shen, X.-Y., Chen, J.-J., Zhao, Y.-X., Jin, W.-I., 2022. An overview on the influence of various parameters on the fabrication and engineering properties of CO2-cured cement-based composites. *Journal of Cleaner Production* 366, 132968.

Chen, T., Gao, X., 2019. Effect of carbonation curing regime on strength and microstructure of Portland cement paste. *Journal of CO2 Utilization* 34, 74–86.

Dilnesa, B.Z., Lothenbach, B., Le Saout, G., Renaudin, G., Mesbah, A., Filinchuk, Y., Wichser, A., Wieland, E., 2011. Iron in carbonate containing AFm phases. *Cement and Concrete Research* 41(3), 311–323.

Drouet, E., Poyet, S., Torrenti, J.M., 2015. Temperature influence on water transport in hardened cement pastes. *Cement and Concrete Research* 76, 37–50.

Drouet, E., Poyet, S., Bescop, P., Torrenti, J.M., Bourbon, X., 2019. Carbonation of hardened cement pastes: influence of temperature. *Cement and Concrete Research* 115, 445–459.

Dung, N.T., Hay, R., Lesimple, A., Celik, K., Unluer, C., 2021. Influence of CO2 concentration on the performance of MgO cement mixes. *Cement and Concrete Composites* 115, 103826.

Galan, I., Glasser, F.P., Baza, D., Andrade C., 2015. Assessment of the protective effect of carbonation on portlandite crystals. *Cement and Concrete Research* 74: 68–77.

Gartner, E., Sui, T., 2018. Alternative cement clinkers. *Cement and Concrete Research* 114, 27–39.

Goodbrake, C.J., Young, J.F., Berger, R.L., 1979a. Reaction of beta-dicalcium silicate and tricalcium silicate with carbon dioxide and water vapor. *Journal of the American Ceramic Society* 62(3–4), 168–171.

Goodbrake, C.J., Young, J.F., Berger, R.L., 1979b. Reaction of hydraulic calcium silicates with carbon dioxide and water. *Journal of the American Ceramic Society* 62(9–10), 488–491.

Grünhäuser Soares, E., Castro-Gomes, J., 2021. Carbonation curing influencing factors of Carbonated Reactive Magnesia Cements (CRMC) – a review. *Journal of Cleaner Production* 305, 127210.

Hassani, E., Feyzbar-Khalkhali-Nejad, F., Rashti, A., Oh, T.-S., 2020. Carbonation, regeneration, and cycle stability of the mechanically activated Ca(OH)2 sorbents for CO2 capture: an in Situ X-ray diffraction study. *Industrial & Engineering Chemistry Research* 59(25), 11402–11411.

He, P., Shi, C., Tu, Z., Poon, C.S., Zhang, J., 2016. Effect of further water curing on compressive strength and microstructure of CO2-cured concrete. *Cement and Concrete Composites* 72, 80–88.

He, Z., Wang, S., Mahoutian, M., Shao, Y, 2020. Flue gas carbonation of cement-based building products. *Journal of CO2 Utilization* 37, 309–319.

Jooss, M., Reinhardt, H.W., 2002. Permeability and diffusivity of concrete as function of temperature. *Cement and Concrete Research* 32(9), 1497–1504.

Kashef-Haghighi, S., Ghoshal, S., 2010. CO2 sequestration in concrete through accelerated carbonation curing in a flow-through reactor. *Industrial & Engineering Chemistry Research* 49(3), 1143–1149.

Kashef-Haghighi, S., Shao, Y., Ghoshal, S., 2015. Mathematical modeling of CO2 uptake by concrete during accelerated carbonation curing. *Cement and Concrete Research* 67, 1–10.

Kuzel, H.J., 1996. Initial hydration reactions and mechanisms of delayed ettringite formation in Portland cements. *Cement and Concrete Composites* 18(3), 195–203.

Jiang, Y., Peng, L., Ma, Z., Lu, J.-x., Shen, P., Poon, C.S., 2023. Enhancing the treatment efficiency of recycled concrete fines with aqueous carbonation. *Cement and Concrete Research* 174, 107338.

Li, Y., Liu, W., Xing, F., Wang, S., Tang, L., Lin, S., Dong, Z., 2020. Carbonation of the synthetic calcium silicate hydrate (C-S-H) under different concentrations of CO2: chemical phases analysis and kinetics. *Journal of CO2 Utilization* 35, 303–313.

Liang, C., Pan, B., Ma, Z., He, Z., Duan, Z., 2020. Utilization of CO2 curing to enhance the properties of recycled aggregate and prepared concrete: a review. *Cement and Concrete Composites* 105, 103446.

Liu, T., Lin, B., Qin, J., 2010. Carcinoma-associated fibroblasts promoted tumor spheroid invasion on a microfluidic 3D co-culture device. *Lab on a Chip* 10(13), 1671–1677.

Liu, Z., Meng, W., 2021. Fundamental understanding of carbonation curing and durability of carbonation-cured cement-based composites: a review. *Journal of CO2 Utilization* 44, 101428.

Lu, B., Drissi, S., Liu, J., Hu, X., Song, B., Shi, C., 2022. Effect of temperature on CO_2 curing, compressive strength and microstructure of cement paste. *Cement and Concrete Research* 157, 106827.

Lu, B., He, P., Liu, J., Peng, Z., Song, B., Hu, X., 2021. Microstructure of Portland cement paste subjected to different CO2 concentrations and further water curing. *Journal of CO2 Utilization* 53, 101714.

Lu, B., Shi, C., Hou, G., 2018. Strength and microstructure of CO2 cured low-calcium clinker. *Construction and Building Materials* 188, 417–423.

Mao, Y., He, P., Drissi, S., Zhang, J., Hu, X., Shi, C., 2023. Effect of conditions on wet carbonation products of recycled cement paste powder. *Cement and Concrete Composites* 105307.

Materic, V., Smedley, S.I., 2011. High temperature carbonation of Ca(OH)2. *Industrial & Engineering Chemistry Research* 50(10), 5927–5932.

Matschei, T., Lothenbach, B., Glasser, F.P., 2007. The AFm phase in Portland cement. *Cement and Concrete Research* 37(2), 118–130.

Morandeau, A., Thiéry, M., Dangla, P., 2014. Investigation of the carbonation mechanism of CH and C-S-H in terms of kinetics, microstructure changes and moisture properties. *Cement and Concrete Research* 56, 153–170.

Peng, L., Jiang, Y., Ban, J., Shen, Y., Ma, Z., Zhao, Y., Shen, P., Poon, C.-S., 2023. Mechanism underlying early hydration kinetics of carbonated recycled concrete fines-ordinary portland cement (CRCF-OPC) paste. *Cement and Concrete Composites* 144, 105275.

Rowland, J.L., 1870. Improvement in the manufacture of artificial stone, in: Patent, U.S. (ed.).

Ruan, S., Qiu, J., Weng, Y., Yang, Y., Yang, E.-H., Chu, J., Unluer, C., 2019. The use of microbial induced carbonate precipitation in healing cracks within reactive magnesia cement-based blends. *Cement and Concrete Research* 115, 176–188.

Sevelsted, T.F., Skibsted, J., 2015. Carbonation of C–S–H and C–A–S–H samples studied by 13C, 27Al and 29Si MAS NMR spectroscopy. *Cement and Concrete Research* 71, 56–65.

Shi, C., He, F., Wu, Y., 2012a. Effect of pre-conditioning on CO2 curing of lightweight concrete blocks mixtures. *Construction and Building Materials* 26(1), 257–267.

Shi, C., He, F., & Wu, Y., 2018. Effect of pre-conditioning on CO2 curing of concrete products. *Journal of Building Material* 11, 116–120.

Shi, C., Liu, M., He, P., Ou, Z., 2012b. Factors affecting kinetics of CO_2 curing of concrete. *Journal of Sustainable Cement-Based Materials* 1(1–2), 24–33.

Song, B., Hu, X., Liu, S., Shi, C., 2022a. Chloride binding of early CO2-cured Portland cement-fly ash-GGBS ternary pastes. *Cement and Concrete Composites* 134, 104793.

Song, B., Liu, S., Hu, X., Ouyang, K., Li, G., Shi, C., 2022b. Compressive strength, water and chloride transport properties of early CO2-cured Portland cement-fly ash-slag ternary mortars. *Cement and Concrete Composites* 134, 104786.

Thiery, M., Villain, G., Dangl, A.P., Platret, G., 2007. Investigation of the carbonation front shape on cementitious materials: effects of the chemical kinetics. *Cement and Concrete Research*, 37, 1047–1058.

Walling, S.A., Provis, J.L., 2016. Magnesia-based cements: a journey of 150 years, and cements for the future? *Chemical Reviews* 116(7), 4170–4204.

Wang, D., Chang, J., 2019. Comparison on accelerated carbonation of β-C2S, Ca(OH)2, and C4AF: reaction degree, multi-properties, and products. *Construction and Building Materials* 224, 336–347.

Wang, T., Huang, H., Hu, X., Fang, M., Luo, Z, Guo, R, 2017. Accelerated mineral carbonation curing of cement paste for CO2 sequestration and enhanced properties of blended calcium silicate. *Chemical Engineering Journal* 323, 320–329.

Wu, Y., Mehdizadeh, H., Mo, K., Ling, T.C., 2022. High-temperature CO_2 for accelerating the carbonation of recycled concrete fines. *Journal of Building Engineering* 52, 104526.

Xuan, D., Zhan, B., Poon, C.S., 2018. A maturity approach to estimate compressive strength development of CO2-cured concrete blocks. *Cement and Concrete Composites* 85, 153–160.

Young, J.F., Berger, R.L., Breese, J., 1974. Accelerated curing of compacted calcium silicate mortars on exposure to CO2. *Journal of the American Ceramic Society* 57(9), 394–397.

Zajac, M., Skibsted, J., Bullerjahn, F., Skocek, J., 2022. Semi-dry carbonation of recycled concrete paste. *Journal of CO2 Utilization* 63, 102111.

Zajac, M., Skibsted, J., Durdzinski, P., Bullerjahn, F., Skocek, J., Ben Haha, M., 2020a. Kinetics of enforced carbonation of cement paste. *Cement and Concrete Research* 131, 106013.

Zajac, M., Skibsted, J., Skocek, J., Durdzinski, P., Bullerjahn, F., Ben Haha, M., 2020b. Phase assemblage and microstructure of cement paste subjected to enforced, wet carbonation. *Cement and Concrete Research* 130, 105990.

Zhan, B.J., Xuan, D.X., Poon, C.S., Shi, C.J., 2016. Effect of curing parameters on CO2 curing of concrete blocks containing recycled aggregates. *Cement and Concrete Composites* 71, 122–130.

Zhang, D., Ghouleh, Z., Shao, Y., 2017. Review on carbonation curing of cement-based materials. *Journal of CO2 Utilization* 21, 119–131.

3 Products and microstructure of CO_2 mineralized materials and products

3.1 INTRODUCTION

The CO_2 mineralization of cement-based materials has been introduced in previous chapters, showing that all cement-based materials produce two main types of mineralization products during CO_2 mineralization: carbonates (including calcium carbonate, ($CaCO_3$) and magnesium carbonate ($MgCO_3$) and amorphous silica gel. The performance of CO_2 mineralized cement-based materials and products, particularly their microstructure, depends on the properties and effects of these mineralization products. Therefore, understanding these products is essential. In this section, we first introduce the properties and effects of CO_2 mineralized cement-based materials and products in detail. Subsequently, we review the current research progress on the distribution of CO_2 mineralization products and their influence on the microstructure of CO_2 mineralized materials and products.

3.2 $CACO_3$ IN CO_2 MINERALIZED MATERIALS AND PRODUCTS

3.2.1 PROPERTIES OF $CACO_3$

$CaCO_3$ is the primary mineralization product in CO_2 mineralized materials and products, comprising an estimated 65–75% in fully CO_2 mineralized materials and products (Mao et al., 2023; Shen et al., 2022b). It commonly exists in one or more polymorphic forms, including amorphous $CaCO_3$, calcite (rhombohedral, Figure 3.1(a)), aragonite (acicular or rod-like, Figure 3.1(b)), and vaterite (globular, Figure 3.1(c)). Figure 3.2 shows the morphology of amorphous $CaCO_3$ observed under transmission electron microscopy (TEM) and the scanning electron microscopy (SEM) morphology of the three types of crystalline $CaCO_3$.

Amorphous $CaCO_3$ consists of calcium-rich nanoscale frameworks, which are thermodynamically unstable and spontaneously transform into crystals with lower energy requirements through direct physical connections between the growing and contracting phases. The formation of $CaCO_3$ can occur through two main types of nucleation processes: heterogeneous and homogeneous nucleation. Homogeneous nucleation occurs solely in the solution phase with supersaturated conditions, where

DOI: 10.1201/9781003627531-3

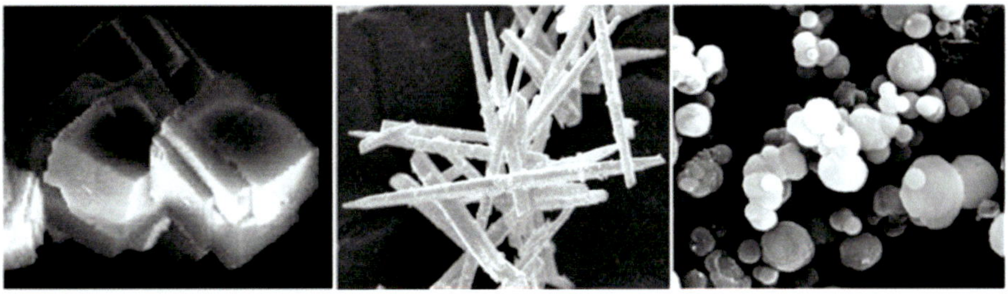

FIGURE 3.1 Polymorphism of crystalline $CaCO_3$.

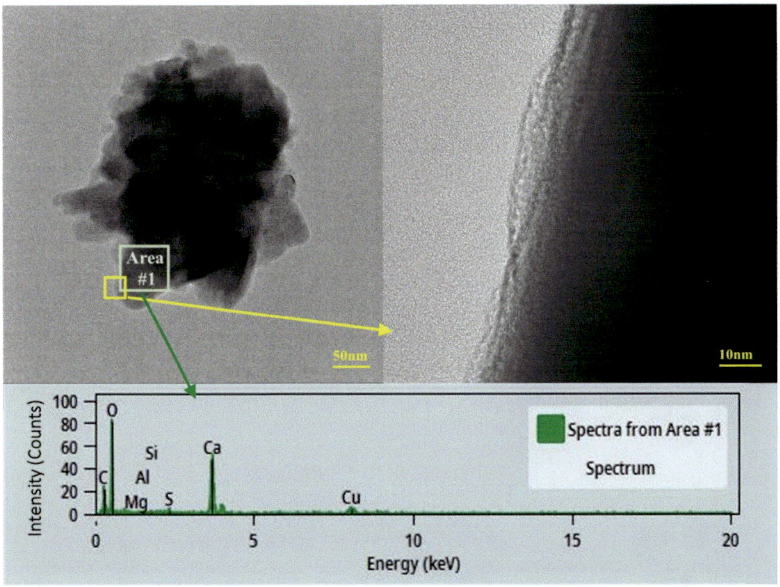

FIGURE 3.2 TEM images and element analysis of amorphous $CaCO_3$.

highly elevated calcium and carbonate ions come together to form clusters. These clusters then grow and aggregate to form larger crystals. However, in heterogeneous nucleation, the formation of $CaCO_3$ begins on a pre-existing surface or a foreign substance called a nucleation site. These nucleation sites can be impurities, rough surfaces, or even organic additives, which provide a template for $CaCO_3$ crystals to form more easily, by lowering interfacial energy and thus the energy barrier for nucleation. This can also result in faster crystal growth. Therefore, interfacial free energy often controls heterogeneous nucleation of $CaCO_3$ on mineral surfaces. Due to the presence of solid phase that presents surfaces or templates for $CaCO_3$ nucleation, heterogeneous nucleation is typically favored during the mineralization of cement-based materials.

The mechanisms of the nucleation and growth of $CaCO_3$ polymorphs, with and without additives, remain debatable. Upon reaction, calcium cations bond with carbonate ions forming $CaCO_3$ embryos in solution or on a substrate. The embryos subsequently develop into a crucial nucleus, which, under favorable local conditions, proliferates to form a primary nanoparticle. These nanoparticles can convert into a single crystal with the addition of calcium and carbonate ions or can assemble into iso-oriented (or equiaxed) crystals. When additives are present in the reaction, these nanoparticles can transform into mesoscale crystals. In addition, the assembly of embryos can lead to the formation of an amorphous phase using collision and coalescence, which can form a crystal phase. Due to the multiple nucleation and growth pathways of $CaCO_3$, various distinct precrystalline entities, liquid droplets, amorphous precursors, and mesocrystals would be present during the crystallization process. Despite ongoing debate on the transformation from amorphous $CaCO_3$ into crystalline $CaCO_3$, a full abiotic transformation pathway from amorphous $CaCO_3$ via vaterite to calcite was elucidated by Pieter et al. (Bots et al., 2012), as illustrated in Figure 3.3. In the first stage, amorphous $CaCO_3$ nucleates from a highly supersaturated solution. In the second stage, the initially formed amorphous $CaCO_3$ phase dehydrates, and the local order within the amorphous $CaCO_3$ increases. High-level supersaturation allows continuous vaterite nucleation and spherulitic growth, followed by an intermediate stage where vaterite continues to form from the dissolving amorphous $CaCO_3$. Once all of the amorphous $CaCO_3$ is consumed, the vaterite crystallite size continues to increase via surface-controlled Ostwald ripening (stages

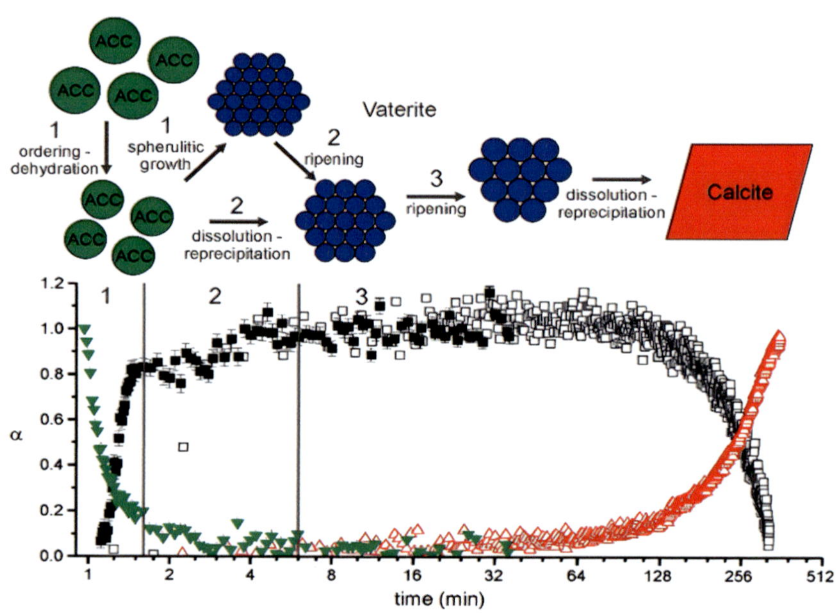

FIGURE 3.3 Schematic representation of multistage ACC → vaterite → calcite crystallization pathway (top) with the underlying combined reaction progress (Bots et al., 2012).

TABLE 3.1
Properties of the anhydrous crystalline forms of $CaCO_3$

Polymorphs	Calcite	Aragonite	Vaterite
Crystal structure	Hexagonal	Orthorhombic	Hexagonal
Illustration of crystal structure	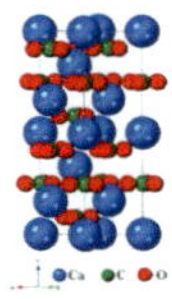	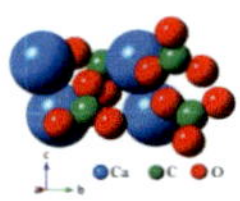	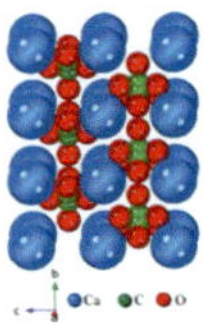
Database code (AMCSD)	0001327	0018860	0009279
Solubility product (Ksp) at 25°C	3.319×10^{-9}	4.623×10^{-9}	12.242×10^{-9}
Density (kg/m^3)	2710	2930	2650
Molar volume (cm^3/mol)	36.9	34.13	39.37

AMCSD: American Mineralogist Crystal Structure Database.

two and three). In the third and final stage, the unstable vaterite dissolves and recrystallizes into the most stable crystalline form, calcite.

Table 3.1 presents the properties of calcite, vaterite, and aragonite. Rhombohedral calcite is the predominant crystalline form of $CaCO_3$ and is highly thermodynamically stable at temperatures ranging from 0°C to 90°C. Aragonite, which is sub-stable and acicular or rod-like, is slightly less thermodynamically stable under certain controlled temperatures. Vaterite is the least thermodynamically stable form of $CaCO_3$ and tends to transform into calcite in low-temperature and humid environments. It belongs to the hexagonal crystal system and typically appears as spherical particles. The polymorph and morphology of $CaCO_3$ are controlled by the thermodynamic surface free energy of the crystal surface, as well as the kinetics of crystal nucleation, growth, and phase transition. Therefore, thermodynamically unstable vaterite and metastable aragonite can also exist under ambient conditions. In general, the fastest growing crystals are oriented perpendicular to the crystal face with the highest surface energy, while the low surface energy face remains until crystals form. Additives, such as surfactants, can be adsorbed selectively on the crystals, resulting in a decrease or increase in the surface energy of the crystal faces, thus changing the final morphology of the crystals. The surface free energy also varies with temperature. Besides, the activation energy of crystal nucleation, growth, and phase transition kinetically controls the $CaCO_3$ polymorph, which is linked to the changes in mineral solubility and supersaturation. As the driving force of nucleation, supersaturation plays a crucial role in $CaCO_3$ formation and is influenced by many factors, such as temperature and pH value of the solution. Vaterite forms from solutions with high levels of supersaturation and low pH owing to the favorable transformation of ACC into vaterite during the early stage of the crystallization, while calcite forms preferentially at lower supersaturation levels and high pH values. When one of the reactants is extraordinarily excessive, the value of Ca^{2+}/CO_3^{2-} may also determine the $CaCO_3$ polymorphs. Moreover, the formation conditions, such as addition rate, the degree of agitation, and ultrasound irradiation, can also impact the nucleation and growth process, thus kinetically controlling the formation of $CaCO_3$. Also, changes in crystal solubility result in changes in the degree of supersaturation, which in turn affects the driving force and growth rate of crystallization. It should be noted that the nature of minerals may also impact $CaCO_3$ formation. During the mineralization of hydrated cementitious materials, mineralization of portlandite mainly results in calcite and aragonite formations, whereas mineralization of C-S-H primarily yields vaterite rather than calcite.

The polymorphism of $CaCO_3$ can significantly influence the properties of CO_2 mineralized cement-based materials due to its unique morphology. For instance, spherical vaterite particles may positively affect the early rheological properties of cement paste. Aragonite, with its relatively large length-to-diameter ratio, is prone to forming whisker-like structures, which can act similarly to fiber reinforcement, enhancing the toughness and flexural strength of the material—an important factor for cement-based materials having low tensile strength and toughness. Precise regulation of $CaCO_3$ polymorphism during CO_2 mineralization allows for selective control of material properties, thereby expanding the application of this sustainable product and promoting a shift from traditional energy-intensive processes (e.g., high-temperature steam curing) to greener and more sustainable CO_2 curing processes. Consequently, research in this area has gained increasing attention in recent years. Studies have shown that producing fibrous aragonite can be achieved using inorganic or organic crystal stabilizers to control the mineralization products of the cement system, with wet mineralization systems being more suitable than dry mineralization for better dispersion of these stabilizers. Some studies have successfully induced the formation of aragonite by adding $MgCl_2$ to cement paste, achieving satisfactory regulatory results. However, the involvement of inorganic ions in chemical reactions can reduce the availability of effective ionic components for polymorphism control and may cause unpredictable chemical changes within the system. This issue is particularly pronounced in cement-based materials, where ions like Mg^{2+} can convert C-S-H to M-S-H, reducing its cementation capacity, or convert $CaCO_3$ to $Mg(OH)_2$ or dolomite, thereby reducing the total aragonite content. The authors suggest that using more chemically stable organic additives may provide an effective alternative for adjusting the crystalline shape of $CaCO_3$ during CO_2 mineralization of cement-based materials.

Figure 3.4 summarizes the effects of conditions, chemical additives, and physical treatments on the formation of $CaCO_3$ polymorphs. Among the mineralization conditions, temperature holds a pivotal position in influencing the formation and stability of metastable $CaCO_3$. Elevated temperatures, especially above 60°C, favor the formation of aragonite. However, a significant drawback of dry mineralization at elevated temperatures is the evaporation of water, which subsequently constrains the mineralization reaction. Conversely, wet mineralization has superior efficiency, easier temperature modulation, and better control over aragonite formation. Moisture is indispensable for both the mineralization reaction and polymorph transformation. A low-humidity environment during the CO_2 mineralization of hardened cementitious

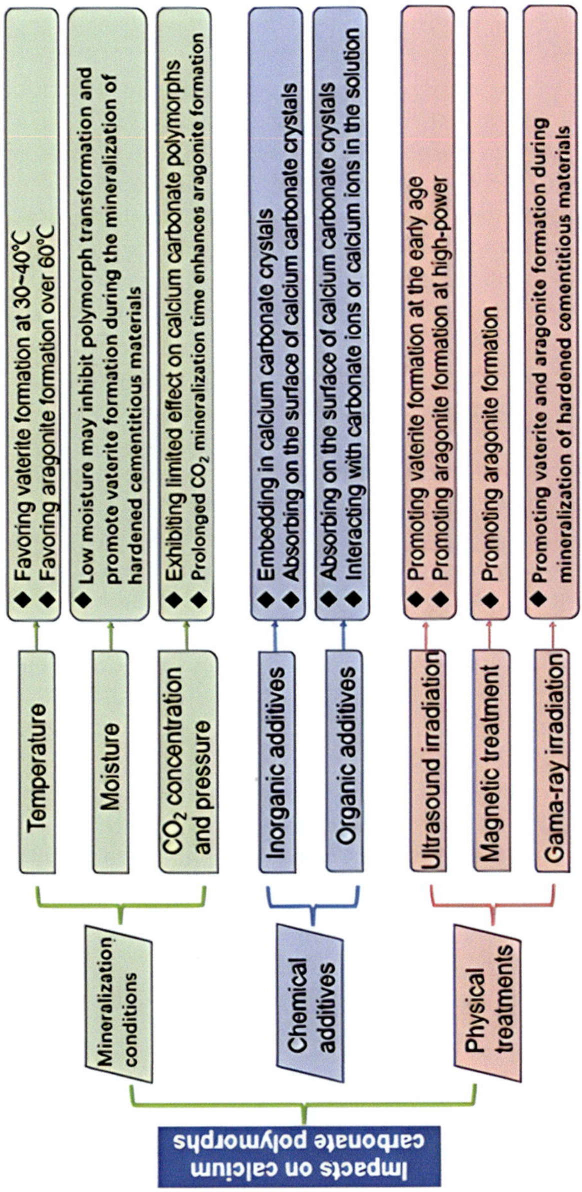

FIGURE 3.4 Impacts of carbonation conditions, chemical additives, and physical treatments on $CaCO_3$ polymorphs.

materials may potentially impede the transformation of vaterite into calcite. Notably, C-S-H and CH, the primary components of recycled cement paste, can react with CO_2 to produce water, potentially increasing internal moisture levels. This phenomenon necessitates further exploration. In contrast, CO_2 concentration and pressure exert minor influences on polymorphic transformation, although extended CO_2 mineralization durations may trigger aragonite formation.

Inorganic additives primarily influence the polymorphs of $CaCO_3$ by modifying its solubility or crystallization behavior (Chen and Xiang, 2009). Metal divalent ions, such as Mg^{2+} and Cu^{2+}, can displace Ca^{2+} ions in calcite, forming an unstable calcite solid solution $(M_x, C_{a1-x})CO_3$. As the impurity content gradually increases, the thermodynamic stability of the solid solution gradually diminishes, whereas aragonite, due to its higher stability, begins to precipitate and becomes the dominant phase. Organic additives, on the other hand, do not embed in $CaCO_3$ crystals but absorb on their surfaces or interact with calcium and carbonate ions (Štajner et al., 2018). The CO_2 mineralization process of cementitious materials is more complex than the synthesis of $CaCO_3$, involving the diffusion and dissolution of CO_2, the leaching of alkaline ions, and the precipitation of carbonates. Cementitious materials possess complex chemical and mineralogical compositions, and ion mobility in the pore solution of CO_2 mineralized cementitious products may be restricted, especially as carbonate formation blocks pores. Consequently, chemical additives (e.g., l-arginine) may be ineffective or inefficient in regulating $CaCO_3$ polymorphs during the CO_2 mineralization of cementitious materials due to complex reactions or restricted ion movement, which needs further investigation.

Physical treatments induce a series of physical and chemical effects that collectively influence the crystallization and transformation of $CaCO_3$. Considerable progress is still needed to promote and facilitate the use of physical treatment techniques in the CO_2 mineralization process. New technologies and equipment must be developed in conjunction with CO_2 mineralization units, ensuring safety and reliability. Besides, the effectiveness of physical treatments in controlling the polymorph of $CaCO_3$ during the CO_2 curing process remains to be verified, as current results primarily stem from synthetic processes or natural carbonation. Further research is required to investigate the impact of various physical treatment methods on the mineralization efficiency and mineralization products of cementitious materials.

3.2.2 Role of $CaCO_3$ played in CO_2 mineralized materials and products

3.2.2.1 Filler effect

The particle size of $CaCO_3$ in CO_2 mineralized cement-based materials typically falls within the range of approximately 0.1 μm to 2 μm, aligning with the size range of capillary pores in hardened cement paste. This suggests that the smaller particle size of $CaCO_3$ in CO_2 mineralized cement-based materials can disperse into these capillary pores (Peng et al., 2023), exerting a filler effect. The filler effect in densifying cement paste pores causes the pores within cement paste containing CO_2 mineralized cement-based materials to be smaller compared to those in ordinary portland cement (OPC) paste. Previous studies have indicated that the filler effect of $CaCO_3$ intensified as the particle size decreased (Wang et al., 2019; Zajac et al., 2020b). Smaller particles can fill smaller pores, resulting in a denser microstructure. However, it has been observed that variations in mineralization parameters during the mineralization process do not notably change the particle size of $CaCO_3$ or mineralized cement-based materials (Mao et al., 2024a). Moreover, the filler effect becomes more pronounced with higher $CaCO_3$ content in the mineralized cement-based materials (Lu et al., 2018; Mao et al., 2024a). The influence of the mineralization process on the $CaCO_3$ content (degree of mineralization) has been extensively reviewed (Aquino Rocha and Toledo Filho, 2023; Poon et al., 2023; Zajac et al., 2023). On the other hand, unfortunately, the influence of $CaCO_3$ polymorphism on the filler effect has not been thoroughly investigated. Many studies have primarily focused on the filler effect of calcite (the most common and thermally stable crystalline form of $CaCO_3$). Given the fibrous structure of aragonite and its higher aspect ratio (Liendo et al., 2022), it is conceivable that aragonite may better fill the interstitial spaces of cement paste particles while enhancing the toughness and flexural properties of the cement paste matrix (Shen et al., 2023; Shen et al., 2022a). Therefore, future research in this area should explore the influence of different $CaCO_3$ polymorphisms on the filler effect to provide a comprehensive understanding of its role in cement paste properties.

3.2.2.2 Nucleation effect

The $CaCO_3$ particles present in CO_2 mineralized materials and products serve as nucleation sites for hydrated C-S-H, thereby promoting cement hydration (Mao et al., 2024a; Ouyang et al., 2020; Zajac et al., 2020b). While previous works mainly focused on the nucleation effect of calcite, recent research (Berodier and Scrivener, 2014; Stark et al., 2007) suggests additional mechanisms beyond merely increasing the specific surface area of particles. Specifically, calcite promotes the dissolution of the calcium phase, and it provides a site "template" for C-S-H nucleation by arranging calcium atoms on the calcite surface. These mechanisms enable the growth of C-S-H along an alignment perpendicular to the particle surface of calcite (Figure 3.5(a)) (Mao et al., 2024a; Ouyang et al., 2020; Zajac et al., 2020b). This growth pattern contrasts significantly with the disorganized growth direction of C-S-H on the C_3S surface (Figure 3.5(b)). Moreover, the growth rate of C-S-H on calcite is linearly correlated with the cube of supersaturation (Berodier and Scrivener, 2014; Scherer and Bellmann, 2018). A recent study (Zhao et al., 2023a) found that the nucleation effect of calcite significantly surpasses that of aragonite and vaterite for cement paste hydration (Figure 3.6(a)) despite the latter two having larger specific surface areas (Figure 3.6(b)). More hydration products are generated on the surface of calcite (Figure 3.6(c)) compared to aragonite (Figure 3.6(d)) and vaterite (Figure 3.6(e)), indicating that calcite is better suited to serve as a nucleation site (Bentz et al., 2013; Li et al., 2019). This may result from the intrinsic differences in the structure of calcite and the other two $CaCO_3$ polycrystals, whose atomic arrangements lead to different interfacial properties between the $CaCO_3$ polycrystals related to the hydrated phases (mainly C-S-H) and, thus, different energy barriers for nucleation to occur (Oey et al., 2013).

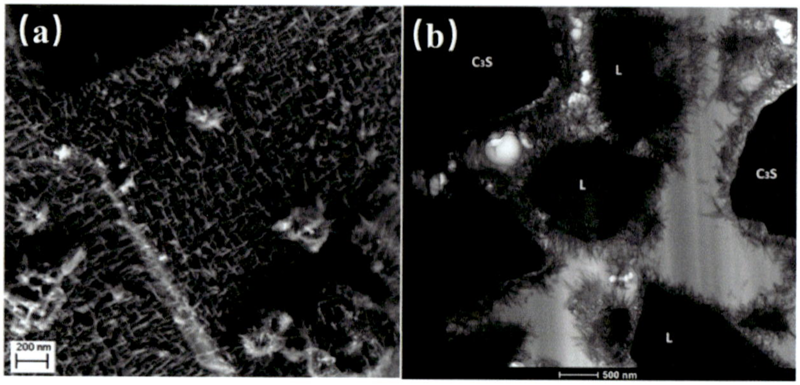

FIGURE 3.5 (a) Microstructure on calcite in limestone–C_3S paste at 3 h. (b) Comparison of the morphology of C-S-H on the surface of C_3S particles and limestone (L, the main component is calcite) particles (Berodier and Scrivener, 2014).

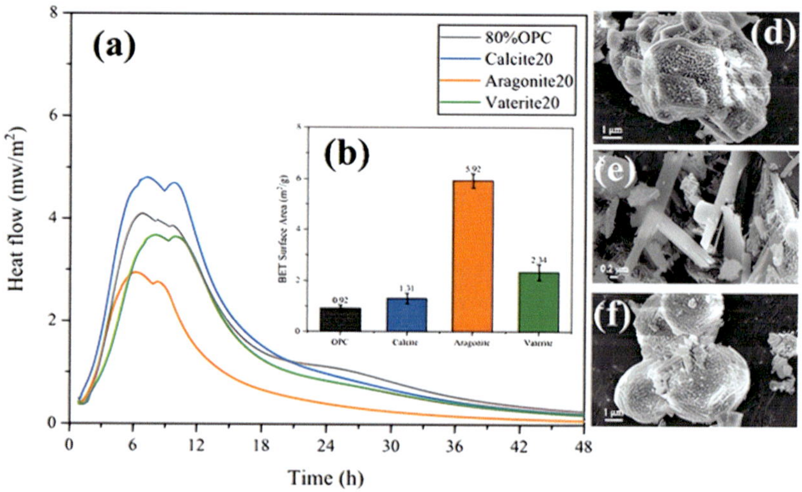

FIGURE 3.6 (a) The heat flow of cement paste containing 20% $CaCO_3$ polymorphs, (b) the surface area of OPC and $CaCO_3$ polymorphs, nucleation and growth of hydrated phases on the surface of (c) calcite, (d) aragonite, and (e) vaterite at 4 h (Zhao et al., 2023a). (OPC represents ordinary cement paste without $CaCO_3$, Calcite20, Aragonite20, and Vaterite20 represent cement paste containing 20% Calcite, Aragonite, and Vaterite, respectively).

3.2.2.3 Chemical effects of the reaction between $CaCO_3$ and C_3A

$CaCO_3$ in mineralized cement-based materials can stabilize the AFt that would otherwise be consumed by C_3A through its reaction with C_3A. These stabilized AFt phases contribute to a denser microstructure of the cement paste (Lu et al., 2018; Mao et al., 2024a). As we know, the $CaCO_3$ could react with C_3A, resulting in the formation of calcium aluminate hemi-carbonate (CAHC, Eq. (3.1)) and calcium aluminate mono-carbonate (CAMC, Eq. (3.2)) through the following two reactions:

$$C_3A + 0.5C\underline{C} + 0.5CH + 11.5H \rightarrow C_4A\underline{C_{0.5}}H_{12} \tag{3.1}$$

$$C_3A + C\underline{C} + 11H \rightarrow C_4A\bar{C}H_{11} \tag{3.2}$$

Several previous studies (Lu et al., 2018; Mao et al., 2024a; Zajac et al., 2020b) indicated that during the early stages of hydration (before seven days) of cement pastes containing $CaCO_3$, the reaction to generate CAHC is dominant, which is attributed to the lower dissolution rate of CAHC at higher pH and its faster formation rate (Lu et al., 2018; Zajac et al., 2014). As the hydration period extends, the proportion of CAMC in carbon-doped $CaCO_3$ gradually exceeds that of CAHC because the former is more stable than the latter (Lu et al., 2018; Zajac et al., 2020b).

Reducing the particle size of $CaCO_3$ and increasing its incorporation can enhance the reaction of $CaCO_3$ with C_3A and ensure the stable presence of AFt (Wang et al., 2019). On the other hand, the reactions between $CaCO_3$ and C_3A exhibited by the polymorphism of $CaCO_3$ do not make a significant difference in the stabilization of AFt (Zhao et al., 2023a). Nevertheless, these effects appear to result in differences in the contents of CAHC and CAMC in cement pastes containing $CaCO_3$ polymorphs at 28 days. Increasing the incorporation of calcite and aragonite leads to a higher percentage of CAHC, while the incorporation of vaterite results in a higher presence of CAMC.

3.2.2.4 Surface potential effect

When CO_2 mineralized cement-based materials are used as supplementary cementitious materials, attention also needs to be paid to their surface potential effect, as this can influence the rheology of the cement paste after incorporation of CO_2 mineralized cement-based materials. Previous studies have shown that $CaCO_3$ in mineralized cement-based materials influences the surface potential of the paste. When calcite is immersed in an aqueous solution, it causes charge separation at the mineral–solution interface (Al Mahrouqi et al., 2017; Ouyang et al., 2020). The excess charge on the calcite surface is balanced by areas of equal but opposite charge in the neighboring solution, resulting in the formation of the electrical double layer, consisting of the stern layer and diffuse layer (Figure 3.7(a)). The zeta potential is the potential in the shear plane (Figure 3.7(b)). In some bilayer models, it is assumed that the shear plane and the outer Helmholtz planes (OHP) are identical. In such cases, the ζ potential also represents the potential at OHP (Heberling et al., 2011).

The zeta potential of calcite is determined by the balance of positive and negative ions on its shear surface. In aqueous environments, if the adsorption of carbonate ions exceeds that of calcium ions in the calcite shear layer, a negative surface potential ensues (Ouyang et al., 2020). This negative zeta potential is observed in most calcite samples (Yuan et al., 2018; Zhao et al., 2023b) (Figure 3.7(c)), and the absolute value of the zeta potential increases with higher pH value (Eriksson et al., 2007; Gunnelius

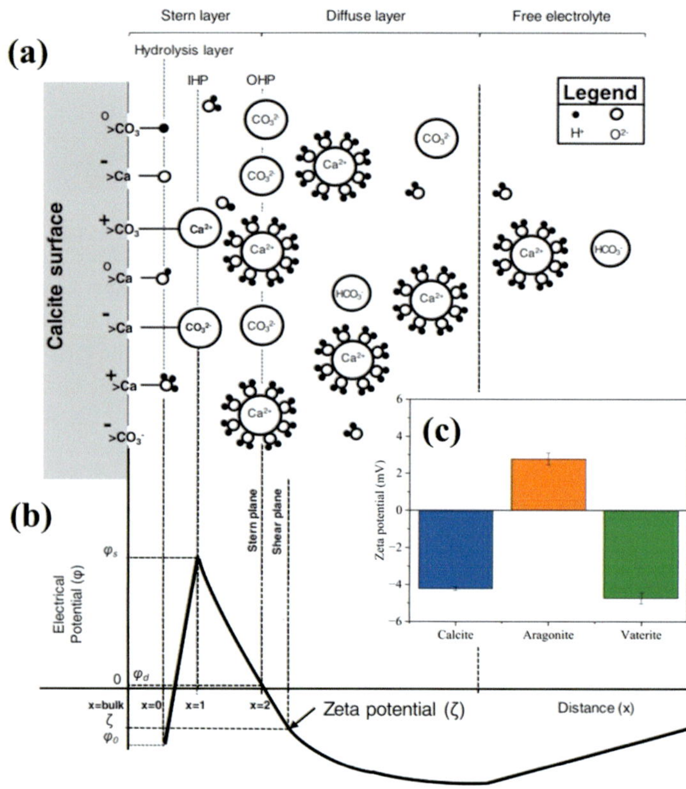

FIGURE 3.7 (a) Calcite–water interfacial bilayer, (b) with a schematic representation of the variation of zeta potential with distance from the calcite surface (Al Mahrouqi et al., 2017); (c) Zeta potential of $CaCO_3$ polymorphs in deionized water (Zhao et al., 2023b).

et al., 2014; Mahani et al., 2013). For cement slurries with a water–cement ratio of less than 1.0, the zeta potential typically exhibits a negative value with a relatively small absolute value (Al Mahrouqi et al., 2017; Ouyang et al., 2020). Consequently, the incorporation of calcite into cement paste generally enhances inter-particle charge repulsion. On the other hand, it is essential to recognize that beyond the electrical double layer, the potential diminishes with increasing distance from the mineral surface until it reaches zero (Al Mahrouqi et al., 2017; Glover et al., 2012). The difference in surface potential effects due to the $CaCO_3$ polymorphs must be taken into consideration. Future work should first focus on whether the $CaCO_3$polymorphs precipitated on the surface of mineralized cement-based materials will spill off into the paste. If this does not occur, the effect of $CaCO_3$ polymorphism will be minimal. However, if spalling occurs, the influence of the content, shape, and surface potential of calcite, vaterite, and aragonite on the rheology of the paste will need to be thoroughly investigated. Like calcite (−4.1 mV), vaterite exhibits a negative surface potential (−4.7 mV) in deionized water at pH 7 (Figure 3.7(c)) (Zhao et al., 2023b). Aragonite displays a positive potential (+2.6 mV) under these conditions, suggesting a potentially different effect on paste potential. However, no reports are available on the potentials of aragonite and vaterite in an alkaline medium such as cement paste to elucidate their impact on charge repulsion between paste particles.

3.3 SILICA GEL IN CO_2 MINERALIZED MATERIALS AND PRODUCTS

3.3.1 Properties of silica gel

Another mineralization product, amorphous silica gel, is formed by continuous decalcification of the C-S-H gel, and its content increases with the degree of CO_2 mineralization. Many studies (Jiang et al., 2023; Mao et al., 2023; Zajac et al., 2024) have shown that aluminum-containing phases (e.g., C-A-S-H gels and AFm) in CO_2 mineralized cement-based materials produce amorphous gels containing aluminum phases, such as alumina-silica gels. Given the extremely low content of alumina in silica gels and its physical mixing with silica gels (Mao, 2024; Shen et al., 2022b), aluminum-containing silica gels are considered normal silica gels in the present book. Silica gel is considered a source of pozzolanic activity in CO_2 mineralized cement-based materials because silica gel can react with $Ca(OH)_2$ in the cement paste to form new C-S-H, i.e., secondary hydration reaction. The pozzolanic activity of silica gel is closely related to its degree of polymerization. Characterization of the chemical environment of Si within a typical CO_2 mineralized cement-based material (CO_2 mineralized recycled cement paste powder) by ^{29}Si-NMR spectroscopy (Figure 3.8 (Ouyang et al., 2020)) shows that, unlike the Q^1 and Q^2 structures in the C-S-H gels of the cement hydration products, the Si in the silica gels of the mineralization products is predominantly in the Q^3 and Q^4 structures (representing the surface and internal silica-oxygen tetrahedra, respectively), with only a small content of Q^1 and Q^2 structures, suggesting that the silica gel is a three-dimensional network structure of interconnected silica-oxygen tetrahedra, and Figure 3.9 demonstrates two amorphous silica gels with different shapes observed using transmission electron microscopy (TEM). In addition, the large number of hydrophilic hydroxyl groups on the surface of the silica gel allows it to absorb large amounts of water. The pozzolanic and water-absorbing effects of silica gel are described in detail in the next section.

3.3.2 Role of silica gel played in CO_2 mineralized materials and products

3.3.2.1 Pozzolanic effect

The pozzolanic reactivity of silica gel is critically important. Two previous works evaluated the pozzolanic reactivity of silica gel in mineralized cement-based materials (CO_2 mineralized recycled cement paste powder): one through thermodynamic simulations and the other by reacting silica gel extracted from mineralized cement-based materials with $Ca(OH)_2$. According to the thermodynamic simulations by

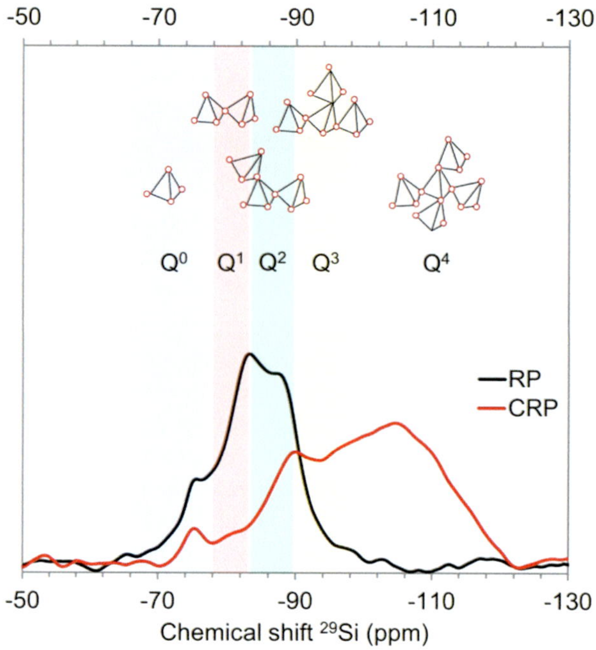

FIGURE 3.8 ^{29}Si-NMR spectra of unmineralized recycled cement paste powder (RP) and CO_2 mineralized recycled cement paste powder (CRP) (Ouyang et al., 2020).

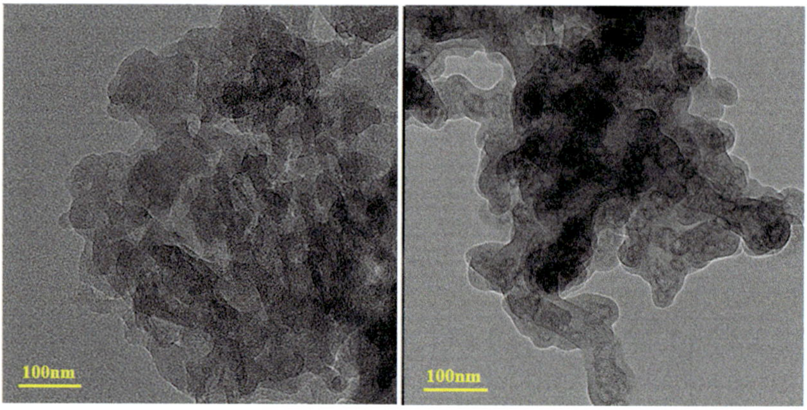

FIGURE 3.9 Amorphous silica gel in CO_2 mineralized cement-based materials observed using transmission electron microscopy.

Zajac et al. (2024), silica gel in the mineralized cement-based materials was consumed within 7–14 days after hydration, with the rapid pozzolanic reaction leading to the formation of C-S-H with a low Ca–Si ratio (Figure 3.10). Our previous study (Mao et al., 2024a) showed an even swifter pozzolanic reaction of silica gel

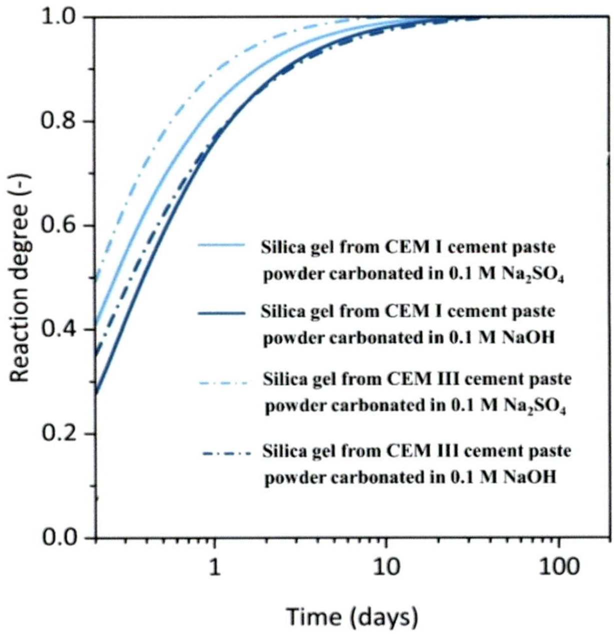

FIGURE 3.10 Degrees of reaction for silica gel in the mineralized cement-based materials (Zajac et al., 2024).

in wet mineralized cement-based materials, with complete transformation to C-S-H within 24 hours of hydration. These results indicate that the reactivity of silica gel in mineralized cement-based materials is significantly higher than that of SiO_2 in conventional SCMs. Interestingly, according to the hydration heat results of the pozzolanic reaction, the pozzolanic reaction of silica gel observed in these studies commenced approximately 6 hours after contact with $Ca(OH)_2$, likely attributed to the formation of larger clusters of spherical particles during wet mineralization. This morphology complicates dispersion in solution, rendering calcium ions diffusion into these clusters (Lin et al., 2011). The pozzolanic effect of silica gel in mineralized cement-based materials depends on the quantity of silica gel generated and its pozzolanic activity. Increasing the mineralization degree of cement-based materials can lead to a higher amount of silica gel formation (Fang et al., 2021a, 2021b; Mao et al., 2023b). However, it's important to note that only a maximum of about 0.1 g of silica gel can be extracted from 1 g of mineralized cement-based materials, limiting the contribution of the pozzolanic effect. Moreover, the pozzolanic reactivity of the extracted silica gel from mineralized cement-based materials increases with higher temperatures, longer mineralization times, and greater CO_2 flow rates during wet mineralization. This is attributed to the enhancement of silica gel polymerization with these mineralization parameters (Maia Neto et al., 2024; Mao et al., 2023; Zajac et al., 2020a).

3.3.2.2 Water absorption

Silica gel exhibits strong hydrophilic properties owing to the abundance of hydrophilic hydroxyl groups on its surface (Benesi and Jones, 1959; Vansant et al., 1993). This may have an influence on the rheology of the cement paste when CO_2 mineralized cement-based materials are used as SCM. Notably, the structural hydroxyl groups within the silica gel do not contribute to its water-absorbing capacity (Vansant et al., 1993). Therefore, when examining the relationship between hydroxyl group quantity and water absorption, acid–base titration, which specifically determines surface hydroxyl groups, is preferred over the thermogravimetric method, which measures total hydroxyl group content (Vansant et al., 1995).

In a previous study (Mao et al., 2024b), silica gel vacuum dried to a moisture content of approximately 1.5% was tested for its ability to absorb pore solution (Figure 3.11), which resembles conditions in cement paste more closely than pure water. It was observed that some liquid absorbed early by the silica gel was later desorbed during the test. This phenomenon can be attributed to calcium ions in the pore solution binding to hydroxyl groups within the silica gel. The chemical bonding formed by calcium leads to (1) a reduction in the effective charge density of the anionic group (Liu et al., 2021; Mechtcherine et al., 2018) and (2) shrinkage and cracking of the polymer network (Kang et al., 2017). Consequently, the absorbed pore solution escapes from the polymer network. The amount of water absorbed and released by silica gel correlates with the number of surface hydroxyl groups, which in turn is closely linked to the specific surface area of the silica gel (Benesi and Jones, 1959; Vansant et al., 1993). A study (Mao et al., 2024b) suggests that silica gel obtained from wet CO_2 mineralized recycled cement paste powder subjected to higher CO_2 flow rates demonstrates increased maximum water absorption and

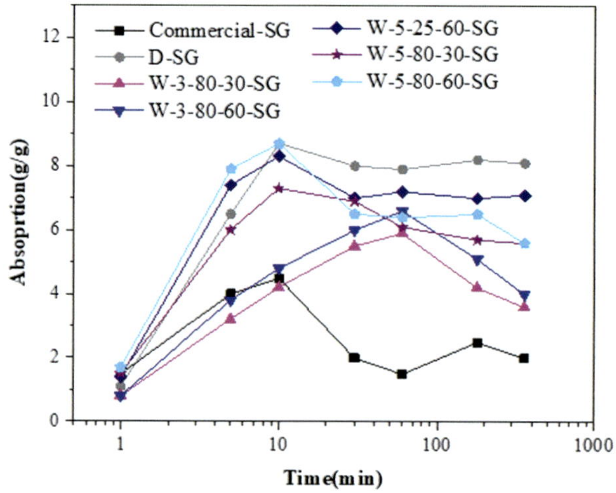

FIGURE 3.11 Pore solution absorption behavior of silica gel samples extracted from CO_2 mineralized materials and products (Mao et al., 2024b).

absorption rates. On the other hand, the increase in CO_2 flow (from 3 L/min to 5 L/min) during mineralization also reduced the amount of subsequently absorbed pore solution that was released. This is attributed to the fact that the increased CO_2 flow rate leads to the formation of more capillary pores in the silica gel (Mao et al., 2023). Consequently, it enhances the specific surface area (increased 49–100% BET surface area as CO_2 flow increased from 3 L/min to 5 L/min) and the total number of surface hydroxyl groups (increased 54–103% total number of surface hydroxyl groups as CO_2 flow increased from 3 L/min to 5 L/min), resulting in greater water absorption efficiency and retention capacity.

3.3.2.3 Surface potential effect

It is well known that amorphous silica gel often exhibits a negative surface potential in solution due to the presence of surface hydroxyl groups (Sonnefeld et al., 2001). However, there are no studies reporting the exact surface potential values of amorphous silica gel in CO_2 mineralized materials and products. A previous study (Mao, 2024) found that the presence of silica gel in CO_2 mineralized cement-based materials significantly reduces the absolute value of the surface potential of calcite. This suggests that silica gel in CO_2 mineralized cement-based materials should have a much smaller absolute value of surface potential than calcite. Such a small absolute value of potential is not expected to contribute significantly to charge repulsion between particles.

3.3 OTHER PRODUCTS IN CO_2 MINERALIZED MATERIALS AND PRODUCTS

Magnesium carbonate hydrate is the most important mineralization product in the CO_2 mineralized cementitious material made from magnesium material because the first step in the mineralization process of activated magnesium oxide is the hydration of activated magnesium oxide, so its mineralization product often contains more than one water molecule. Common hydrated magnesium carbonates are magnesium carbonate trihydrate ($MgCO_3{\cdot}3H_2O$), magnesite hydrate ($Mg_5(CO_3)_4(OH)_2{\cdot}4H_2O$), and magnesite ($Mg_5(CO_3)_4(OH)_2{\cdot}5H_2O$). Figure 3.12 shows the morphology of these

(a) Magnesium carbonate (b) Magnesite hydrate (c) Magnesite

FIGURE 3.12 Three common types of hydrated magnesium carbonate.

three hydrated magnesium carbonates as observed by SEM. It can be seen that all three types of hydrated magnesium carbonates are in the form of fibers or short rods, which allows them to fill well into the interstices of the particles of CO_2 mineralized cement-based materials made of magnesium, resulting in a dense matrix microstructure. According to the morphology of the three common hydrated magnesium carbonates, magnesite hydrate has the best contribution to the matrix microstructure, followed by magnesium carbonate trihydrate and magnesite. As to which type of hydrated magnesium carbonate is produced after hydration and mineralization of activated magnesium oxide, this is determined by a variety of factors, including CO_2 concentration, temperature, CO_2 mineralization period, relative humidity, the nature of the activated magnesium oxide, the raw material mixing ratio, and pH value. Magnesium carbonate hydrate can produce effects similar to $CaCO_3$ and will not be repeated here.

3.4 DISTRIBUTION OF MINERALIZATION PRODUCTS AND MICROSTRUCTURE OF CO_2 MINERALIZED MATERIALS AND PRODUCTS

The microstructure of CO_2 mineralized cement-based materials produced via conventional CO_2 mineralization techniques, such as CO_2 curing and dry accelerated mineralization, demonstrates consistent characteristics. As illustrated in Figure 3.13, a representative example reveals unreacted calcium silicate particles enveloped by a thin silica gel layer approximately 1–3 μm thick, as determined by grayscale analysis in backscattered electron images of polished samples. The microstructure comprises

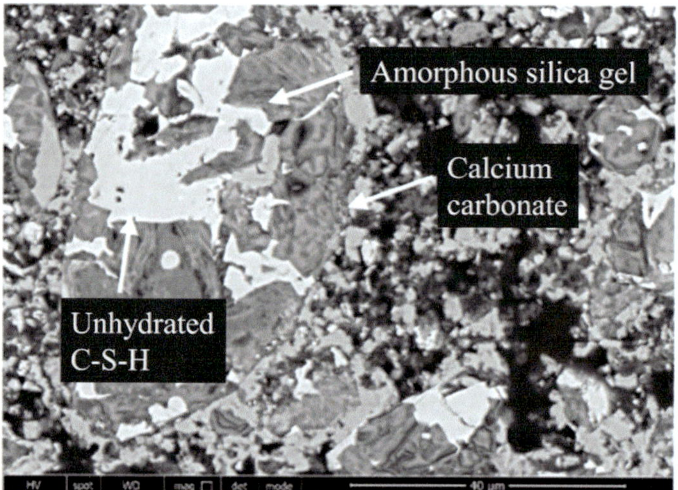

FIGURE 3.13 Typical BSE-EDX micrographs of CO_2 mineralized calcium silicate carbonate (Zajac et al., 2023).

unreacted calcium silicate particles acting as discrete "islands," embedded within a continuous crystalline $CaCO_3$ phase, with highly polymerized silica gel serving as the binding agent between these islands. Crystalline $CaCO_3$ particles occupy the interstitial spaces, forming the matrix skeleton. At the interface between unreacted calcium silicate particles and $CaCO_3$, the silica gel functions as a passivation layer, impeding the transport of substances. Any remaining void space ultimately evolves into pores of varying sizes. This layered architecture bears a resemblance to naturally occurring silicate minerals, such as wollastonite and olivine, which have undergone extensive CO_2 mineralization over geological timescales. Within the calcium silicate mineralization matrices, pore sizes are predominantly distributed around 10 nm and 1 μm. These two distinct pore size regimes can be attributed to the formation of silica gel and the residual voids in the CO_2-mineralized briquettes. Prolonged CO_2 exposure enriches the nanoporosity while reducing the volume of micropores. This microstructural arrangement and the distribution pattern of CO_2 mineralization products can be explained as follows: the spatial distribution of $CaCO_3$ is influenced by the source of calcium-silicate-hydrate (C-S-H), whether derived from the inner or outer regions, with the inner C-S-H exhibiting a denser morphology. It is commonly observed that $CaCO_3$ precipitation in dry-mineralized recycled cement paste (RCP) occurs predominantly within the RCP particles, whereas $CaCO_3$ and decalcified C-S-H are intermixed in the outer C-S-H region. This phenomenon arises because the dense structure of the inner C-S-H layer restricts $CaCO_3$ crystal growth, prompting the migration of Ca^{2+} ions to the outer product region, where crystallization occurs more readily.

A denser microstructure is typically achieved in cement-based composites following CO_2 mineralization due to the precipitation of $CaCO_3$. CO_2 mineralization reduces both the capillary and total porosity of the cement-based matrix, resulting in decreased permeability and water absorption, thereby enhancing the transport properties of the CO_2-mineralized cement-based composites (Jang and Lee, 2016; Zhan et al., 2019). Notably, CO_2 mineralization is more effective in reducing larger pores than smaller pores within the cement-based matrix, whereas hydration exerts a greater influence on the reduction of smaller pores. This densification effect varies depending on the mineral composition of the clinker. For instance, Jang and Lee (2016) reported that CO_2 mineralization of belite resulted in lower pore volume and connectivity compared to that of alite, although both mineralized phases exhibited smaller pore sizes relative to the non-mineralized hydration reference. Figure 3.14 illustrates the pore structure of a CO_2 mineralized cement paste, as measured by mercury intrusion porosimetry (MIP) (Zhang and Shao, 2018). Compared to non-mineralized cement paste, the CO_2 mineralized cement paste exhibited an increase in pores smaller than 50 nm and a significant reduction in pores larger than 50 nm. This indicates that CO_2 mineralization contributes to the densification of the cementitious matrix. During post-conditioning, further hydration continues to densify the microstructure by filling pores with hydration products (He et al., 2016). Another study also observed that, regardless of whether CO_2 mineralization or moisture curing was applied, the overall apparent porosities of cement paste were similar after 28 days, although CO_2 mineralized samples exhibited a denser surface.

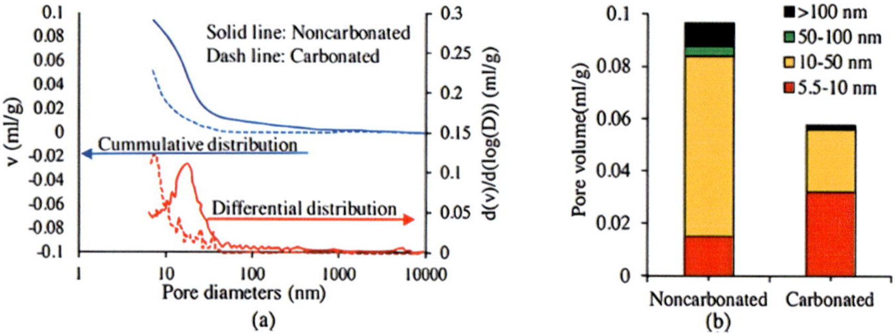

FIGURE 3.14 Characteristics of pore structures of CO_2 mineralized cementitious paste at one year identified with MIP: (a) differential and cumulative distribution of pore sizes, (b) pore volume regarding various pore diameter ranges.

Recent studies on the wet accelerated mineralization technique have revealed that the microstructure of CO_2-mineralized cementitious materials produced through wet mineralization, particularly the relative distribution of $CaCO_3$ and silica gel, differs significantly from that of materials obtained via dry accelerated mineralization, as illustrated in Figure 3.13. In wet mineralization, silica gel and $CaCO_3$ are uniformly mixed and distributed without distinct layering (Zajac et al., 2023), in contrast to materials produced by dry mineralization, where $CaCO_3$ is typically covered by a silica gel layer, resulting in a pronounced layered structure (Zajac et al., 2023). This distinction is attributed to the higher ion concentration in the wet mineralization solution, which facilitates the diffusion of carbonate ions into the interior of the matrix.

Current research on CO_2 mineralized cement-based materials and products predominantly focuses on the microstructural evolution of the matrix during the mineralization process. The microstructure of cement paste undergoes significant changes during CO_2 mineralization due to the chemical alterations resulting from $CaCO_3$ precipitation. Obtained by using scanning electron microscopy (SEM) and energy-dispersive X-ray spectroscopy (EDX), Figures 3.16 and 3.17 illustrate the distinct microstructures and elemental compositions of cement pastes subjected to 24 hours of CO_2 mineralization and 24 hours of water curing, respectively. As shown in Figure 3.16(a), a substantial amount of grainy calcite crystals is observed in the cement paste cured under CO_2 for 24 hours, with the corresponding elemental composition shown in Figure 3.16(b). The EDX spectrum in Figure 3.16(b) reveals prominent carbon and calcium peaks, while the silicon peak is significantly lower compared to that in Figure 3.17(b), indicating the formation of $CaCO_3$ crystals following CO_2 mineralization. In contrast, Figure 3.17(a) displays typical hydration products, including platy $Ca(OH)_2$, fibrillar C-S-H, and needle-like ettringite, which are commonly observed in water-cured samples but rarely detected in CO_2 mineralized samples (Figure 3.16(a)). A comparison between Figures 3.8(a) and 3.17(a) suggests that the 24 hour CO_2 mineralized samples exhibit higher compactness and

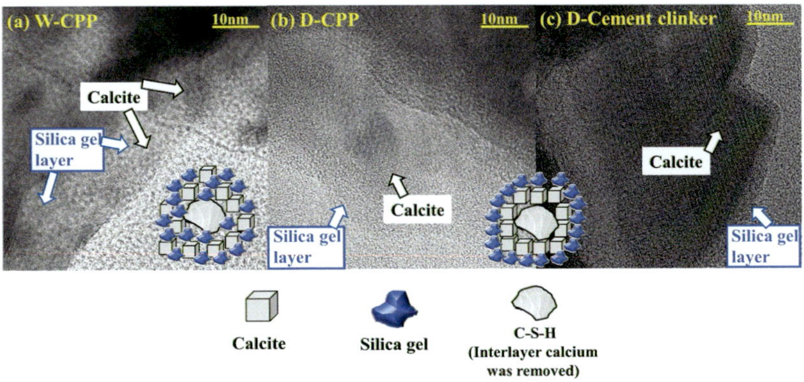

FIGURE 3.15 TEM images of CO_2 mineralized cement-based material.

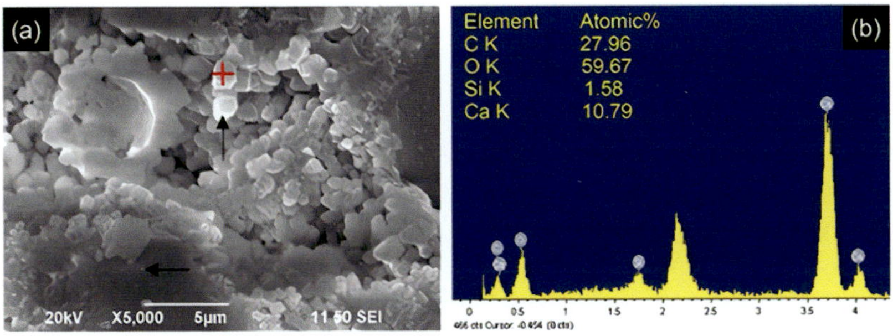

FIGURE 3.16 (a) Typical SEM photomicrographs and (b) EDX spectrum of the cement paste after 24 hours of CO_2 mineralization (Zhan et al., 2019).

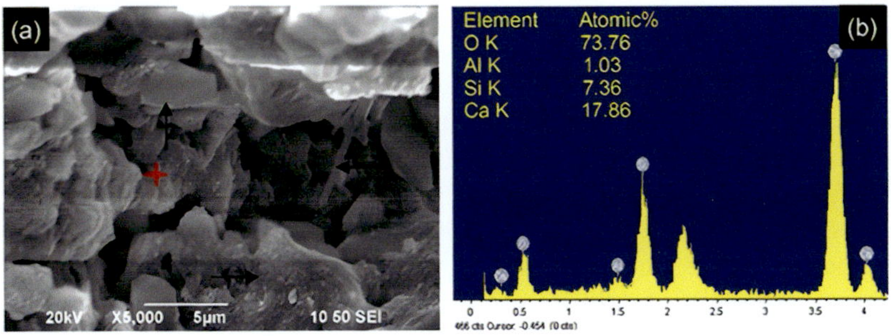

FIGURE 3.17 (a) Typical SEM photomicrographs and (b) EDX spectrum of the cement paste after 24 hours of water curing (Zhan et al., 2019).

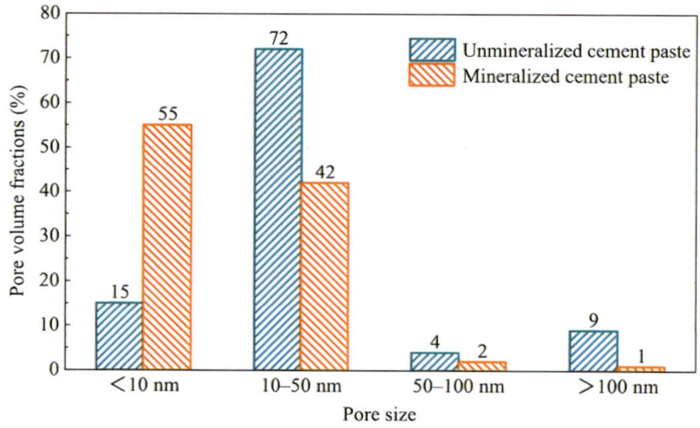

FIGURE 3.18 Pore volume fractions according to pore size (Johannesson and Utgenannt, 2001).

lower porosity. In general, the mineralization of C_3S, C_2S, C-S-H, and $Ca(OH)_2$ increases the solid phase volume by 92.5%, 108.7%, 23.1%, and 11.5%, respectively (Shi et al., 2016). Shi et al. (2012) investigated the microstructure of mineralized concrete and reported that the filling effect of these mineralization products significantly reduced the total porosity, particularly for pores with sizes ranging from 50 nm to 1000 nm. Similarly, Johannesson et al. (Johannesson and Utgenannt, 2001) observed a decrease in total porosity but an increase in the volume of pores with sizes between 2 nm and 7 nm. Figure 3.18 further illustrates the volume fractions of pores of varying sizes in mineralized and unmineralized cement pastes. In unmineralized cement paste, pores with sizes of 10–50 nm dominate, accounting for 72% of the total porosity. In contrast, the dominant pore size in mineralized cement paste shifts to less than 10 nm. These microstructural changes significantly influence the mechanical properties, permeability, and durability of the samples.

3.5 SUMMARY

The primary CO_2 mineralization products of cement-based materials are $CaCO_3$ and silica gel. The particle size and polymorphism of $CaCO_3$, i.e., calcite, aragonite, and vaterite, significantly influence its various effects, including the filler effect, nucleation effect, chemical interactions with C_3A, and surface potential effect. For silica gel, its pozzolanic activity is the most critical property during the CO_2 mineralization of cement-based materials, as it significantly impacts the microstructure and mechanical properties of these materials. Magnesium-containing phases in cement-based materials also react during CO_2 mineralization to form various hydrated magnesium carbonates, including magnesium carbonate trihydrate, magnesite hydrate, and magnesite, which exhibit effects similar to those of $CaCO_3$. Additionally, CO_2 mineralization can lead to a denser microstructure in cement-based materials. CO_2

curing and dry accelerated mineralization typically produce a distinctive microstructure characterized by layered $CaCO_3$ and silica gel, whereas wet mineralization results in a mixed morphology of $CaCO_3$ and silica gel.

REFERENCES

Al Mahrouqi, D., Vinogradov, J., Jackson, M.D., 2017. Zeta potential of artificial and natural calcite in aqueous solution. *Advances in Colloid and Interface Science* 240, 60–76.

Aquino Rocha, J.H., Toledo Filho, R.D., 2023. The utilization of recycled concrete powder as supplementary cementitious material in cement-based materials: a systematic literature review. *Journal of Building Engineering* 76, 107319.

Benesi, H.A., Jones, A.C., 1959. An infrared study of the water-silica gel system. *The Journal of Physical Chemistry* 63(2), 179–182.

Bentz, D.P., Ardani, A., Barrett, T., Jones, S.Z., Lootens, D., Peltz, M.A., Sato, T., Stutzman, P.E., Tanesi, J., Weiss, W.J., 2013. Multi-scale investigation of the performance of limestone in concrete. *Construction and Building Materials* 75, 1–10.

Berodier, E., Scrivener, K., 2014. Understanding the filler effect on the nucleation and growth of C-S-H. *Journal of the American Ceramic Society* 97(12), 3764–3773.

Bots, P., Benning, L.G., Rodriguez-Blanco, J.-D., Roncal-Herrero, T., Shaw, S., 2012. Mechanistic Insights into the Crystallization of Amorphous Calcium Carbonate (ACC). *Crystal Growth & Design* 12(7), 3806–3814.

Chen, J., Xiang, L., 2009. Controllable synthesis of calcium carbonate polymorphs at different temperatures. *Powder Technology* 189(1), 64–69.

Eriksson, R., Merta, J., Rosenholm, J.B., 2007. The calcite/water interface: I. Surface charge in indifferent electrolyte media and the influence of low-molecular-weight polyelectrolyte. *Journal of Colloid and Interface Science* 313(1), 184–193.

Fang, X., Xuan, D., Zhan, B., Li, W., Poon, C.S., 2021a. Characterization and optimization of a two-step carbonation process for valorization of recycled cement paste fine powder. *Construction and Building Materials* 278, 122343.

Fang, X., Xuan, D., Zhan, B., Li, W., Poon, C.S., 2021b. A novel upcycling technique of recycled cement paste powder by a two-step carbonation process. *Journal of Cleaner Production* 290, 125192.

Glover, P.W.J., Walker, E., Jackson, M.D., 2012. Streaming-potential coefficient of reservoir rock: a theoretical model. *Geophysics* 77(2), D17-D43.

Gunnelius, K.R., Lundin, T.C., Rosenholm, J.B., Peltonen, J., 2014. Rheological characterization of cement pastes with functional filler particles. *Cement and Concrete Research* 65, 1–7.

He, P., Shi, C., Tu, Z., Poon, C.S., Zhang, J., 2016. Effect of further water curing on compressive strength and microstructure of CO2-cured concrete. *Cement and Concrete Composites* 72, 80–88.

Heberling, F., Trainor, T.P., Lützenkirchen, J., Eng, P., Denecke, M.A., Bosbach, D., 2011. Structure and reactivity of the calcite–water interface. *Journal of Colloid and Interface Science* 354(2), 843–857.

Jang, J.G., Lee, H.K., 2016. Microstructural densification and CO2 uptake promoted by the carbonation curing of belite-rich Portland cement. *Cement and Concrete Research* 82, 50–57.

Jiang, Y., Peng, L., Ma, Z., Lu, J., Shen, P., Poon, C.S., 2023. Enhancing the treatment efficiency of recycled concrete fines with aqueous carbonation. *Cement and Concrete Research* 174, 107338.

Johannesson, B., Utgenannt, P., 2001. Microstructural changes caused by carbonation of cement mortar. *Cement and Concrete Research* 31(6), 925–931.

Kang, S.-H., Hong, S.-G., Moon, J., 2017. Absorption kinetics of superabsorbent polymers (SAP) in various cement-based solutions. *Cement and Concrete Research* 97, 73–83.
Li, L., Cao, M., Yin, H., 2019. Comparative roles between aragonite and calcite calcium carbonate whiskers in the hydration and strength of cement paste. *Cement and Concrete Composites* 104, 103350.
Liendo, F., Arduino, M., Deorsola, F.A., Bensaid, S., 2022. Factors controlling and influencing polymorphism, morphology and size of calcium carbonate synthesized through the carbonation route: a review. *Powder Technology* 398, 117050.
Lin, Q., Xu, Z., Lan, X., Ni, Y., Lu, C., 2011. The reactivity of nano silica with calcium hydroxide. *Journal of Biomedical Materials Research Part B: Applied Biomaterials* 99B(2), 239–246.
Liu, J., Ma, X., Shi, C., Drissi, S., 2021. Internal curing of blended cement pastes with ultra-low water-to-cement ratio: Absorption/desorption kinetics of superabsorbent polymer. *Journal of the American Ceramic Society* 104(7), 3603–3618.
Lu, B., Shi, C., Zhang, J., Wang, J., 2018. Effects of carbonated hardened cement paste powder on hydration and microstructure of Portland cement. *Construction and Building Materials* 186, 699–708.
Mahani, H., Keya, A.L., Berg, S., Bartels, W.-B., Nasralla, R., Rossen, W.R., 2013. Insights into the mechanism of wettability alteration by low-salinity flooding (LSF) in Carbonates. *Energy & Fuels* 29(3), 1352–1367.
Maia Neto, F., Snellings, R., Skibsted, J., 2024. Aqueous carbonation of aged blended Portland cement pastes: impact of the Al/Si ratio on the structure of the alumina-silica gel. *Cement and Concrete Research* 177, 107428.
Mao, Y., Drissi, S., He, P., Hu, X., Zhang, J., Shi, C., 2024a. Quantifying the effects of wet carbonated recycled cement paste powder on the properties of cement paste. *Cement and Concrete Research* 175, 107381.
Mao, Y., Drissi, S., Hu, X., Zhang, J., Shi, C., 2024b. Effect of wet carbonated recycled cement paste powder on the rheology of cement paste. *Cement and Concrete Research* 181, 107553.
Mao, Y., He, P., Drissi, S., Zhang, J., Hu, X., Shi, C., 2023. Effect of conditions on wet carbonation products of recycled cement paste powder. *Cement and Concrete Composites* 144, 105307.
Mechtcherine, V., Snoeck, D., Schröfl, C., De Belie, N., Klemm, A.J., Ichimiya, K., Moon, J., Wyrzykowski, M., Lura, P., Toropovs, N., Assmann, A., Igarashi, S.-i., De La Varga, I., Almeida, F.C.R., Erk, K., Ribeiro, A.B., Custódio, J., Reinhardt, H.W., Falikman, V., 2018. Testing superabsorbent polymer (SAP) sorption properties prior to implementation in concrete: results of a RILEM Round-Robin Test. *Materials and Structures* 51(1), 28.
Oey, T., Kumar, A., Bullard, J.W., Neithalath, N., Sant, G., 2013. The filler effect: the influence of filler content and Surface Area on cementitious reaction rates. *Journal of the American Ceramic Society* 96(6), 1978–1990.
Ouyang, X., Wang, L., Xu, S., Ma, Y., Ye, G., 2020. Surface characterization of carbonated recycled concrete fines and its effect on the rheology, hydration and strength development of cement paste. *Cement and Concrete Composites* 114, 103809.
Peng, L., Jiang, Y., Ban, J., Shen, Y., Ma, Z., Zhao, Y., Shen, P., Poon, C.-S., 2023. Mechanism underlying early hydration kinetics of carbonated recycled concrete fines-ordinary portland cement (CRCF-OPC) paste. *Cement and Concrete Composites* 144, 105273.
Poon, C.S., Shen, P., Jiang, Y., Ma, Z., Xuan, D., 2023. Total recycling of concrete waste using accelerated carbonation: a review. *Cement and Concrete Research* 173, 107284.
Scherer, G.W., Bellmann, F., 2018. Kinetic analysis of C-S-H growth on calcite. *Cement and Concrete Research* 103, 226–233.

Shen, P., Jiang, Y., Zhang, Y., Liu, S., Xuan, D., Lu, J., Zhang, S., Poon, C.S., 2023. Production of aragonite whiskers by carbonation of fine recycled concrete wastes: an alternative pathway for efficient CO2 sequestration. *Renewable and Sustainable Energy Reviews* 173, 113079.

Shen, P., Lu, J., Zhang, Y., Jiang, Y., Zhang, S., Poon, C.S., 2022a. Preparation aragonite whisker-rich materials by wet carbonation of cement: towards yielding micro-fiber reinforced cement and sequestrating CO2. *Cement and Concrete Research* 159, 106891.

Shen, P., Zhang, Y., Jiang, Y., Zhan, B., Lu, J., Zhang, S., Xuan, D., Poon, C.S., 2022b. Phase assemblance evolution during wet carbonation of recycled concrete fines. *Cement and Concrete Research* 154, 106733.

Shi, C., He, F., Wu, Y., 2012. Effect of pre-conditioning on CO2 curing of lightweight concrete blocks mixtures. *Construction and Building Materials* 26(1), 257–267.

Shi, C., Li, Y., Zhang, J., Li, W., Chong, L., Xie, Z., 2016. Performance enhancement of recycled concrete aggregate – A review. *Journal of Cleaner Production* 112, 466–472.

Sonnefeld, J., Löbbus, M., Vogelsberger, W., 2001. Determination of electric double layer parameters for spherical silica particles under application of the triple layer model using surface charge density data and results of electrokinetic sonic amplitude measurements. *Colloids and Surfaces A: Physicochemical and Engineering Aspects* 195(1), 215–223.

Štajner, L., Kontrec, J., Njegić Džakula, B., Maltar-Strmečki, N., Plodinec, M., Lyons, D.M., Kralj, D., 2018. The effect of different amino acids on spontaneous precipitation of calcium carbonate polymorphs. *Journal of Crystal Growth* 486, 71–81.

Stark, J., Möser, B., Bellmann, F., 2007. Nucleation and growth of C-S-H phases on mineral admixtures, in: Grosse, C.U. (ed.), *Advances in construction materials 2007*. Springer, pp. 531–538.

Vansant, E.F., Van Der Voort, P., Vrancken, K.C., 1993. Chapter 5 The distribution of the silanol types and their desorption energies, in: Vansant, E.F., Van Der Voort, P., Vrancken, K.C. (eds.), *Studies in surface science and catalysis*. Elsevier, pp. 93–126.

Vansant, E.F., Van Der Voort, P., Vrancken, K.C., 1993. Chapter 2 Physical characterization of the silica surface, in: Vansant, E.F., Van Der Voort, P., Vrancken, K.C. (eds.), *Studies in surface science and catalysis*. Elsevier, pp. 31–58.

Van Der Voort, P., Vansant, E.F., Vrancken, K.C., 1995. Chapter 2 Physical characterization of the silica surface, in: Vansant, E.F., Van Der Voort, P., Vrancken, K.C. (eds.), *Studies in Surface Science and Catalysis*, Elsevier, 1995, pp. 31–58.

Wang, D., Shi, C., Farzadnia, N., Jia, H., Zeng, R., Wu, Y., Lao, L., 2019. A quantitative study on physical and chemical effects of limestone powder on properties of cement pastes. *Construction and Building Materials* 204, 58–69.

Yuan, Q., Zhou, D., Li, B., Huang, H., Shi, C., 2018. Effect of mineral admixtures on the structural build-up of cement paste. *Construction and Building Materials* 160, 117–126.

Zajac, M., Maruyama, I., Iizuka, A., Skibsted, J., 2023. Enforced carbonation of cementitious materials. *Cement and Concrete Research* 174, 107283.

Zajac, M., Rossberg, A., Le Saout, G., Lothenbach, B., 2014. Influence of limestone and anhydrite on the hydration of Portland cements. *Cement and Concrete Composites* 46, 99–108.

Zajac, M., Skibsted, J., Skocek, J., Durdzinski, P., Bullerjahn, F., Ben Haha, M., 2020a. Phase assemblage and microstructure of cement paste subjected to enforced, wet carbonation. *Cement and Concrete Research* 130, 105990.

Zajac, M., Skocek, J., Durdzinski, P., Bullerjahn, F., Skibsted, J., Ben Haha, M., 2020b. Kinetics of enforced carbonation of cement paste. *Cement and Concrete Research* 134, 106090.

Zajac, M., Song, J., Ullrich, P., Skocek, J., Ben Haha, M., Skibsted, J., 2024. High early pozzolanic reactivity of alumina-silica gel: A study of the hydration of composite cements with carbonated recycled concrete paste. *Cement and Concrete Research* 175, 107343.

Zhan, B.J., Xuan, D.X., Poon, C.S., Shi, C.J., 2019. Mechanism for rapid hardening of cement pastes under coupled CO2-water curing regime. *Cement and Concrete Composites* 97, 78–88.

Zhang, D., Shao, Y., 2018. Surface scaling of CO2-cured concrete exposed to freeze-thaw cycles. *Journal of CO2 Utilization* 27, 137–144.

Zhao, D., Williams, J.M., Li, Z., Park, A.-H.A., Radlińska, A., Hou, P., Kawashima, S., 2023a. Hydration of cement pastes with calcium carbonate polymorphs. *Cement and Concrete Research* 173, 107270.

Zhao, D., Williams, J.M., Park, A.-H.A., Kawashima, S., 2023b. Rheology of cement pastes with calcium carbonate polymorphs. *Cement and Concrete Research* 172, 107214.

4 CO_2 mineralization by calcium silicate and Portland cement materials

4.1 INTRODUCTION

Anhydrous calcium silicate encompasses a series of mineral phases, namely tricalcium silicate (C_3S), dicalcium silicate (β-C_2S and γ-C_2S), tricalcium disilicate (C_3S_2), and monocalcium silicate (CS). C_3S and β-C_2S, characterized by high hydraulic reactivity, constitute the primary components of Portland cement, finding ubiquitous application in the construction and building materials industries. With the growing demand for low-carbon building materials, research on low-lime calcium silicate minerals, including γ-C_2S, C_3S_2, and CS, has captured increasing attention. Whether hydraulic or non-hydraulic, these calcium silicate minerals exhibit remarkable carbonation reactivity, creating a new path for the development of low-carbon cementitious materials through CO_2 activation. The carbon mineralization of calcium silicate minerals is an intricate process, being subject to the influence of a plethora of factors. These factors include, but are not limited to, temperature, moisture content, CO_2 concentration, CO_2 pressure, and additives. Additionally, the properties of the carbonation products play a decisive role in the final hardening performance of the carbonated materials. This chapter explores the carbonation reactivity and the fundamental carbonation mechanism of calcium silicate minerals, analyzes various factors influencing the carbonation degree, and reveals the properties and control strategies of carbonation products.

4.2 CHARACTERISTICS OF CALCIUM SILICATE MINERALS

Calcium silicate minerals are mainly classified into four types: C_3S, C_2S, C_3S_2, and CS, depending on the ratio of CaO to SiO_2. The crystal structure and coordination polyhedron of Ca for different calcium silicate minerals are shown in Table 4.1 (Wang, Q.et al., 2017). Among these calcium silicates, C_3S and β-C_2S with a large calcium-to-silicate (Ca–Si) ratio are hydraulic and can hydrate with water to produce calcium silicate hydrate (C-S-H) gel, contributing to the setting and hardening of cement. Conventional Portland cements contain approximately 80% of C_3S and β-C_2S. These minerals have relatively high calcium content, requiring a larger amount of limestone as raw material for their production. Moreover, the calcination

DOI: 10.1201/9781003627531-4

TABLE 4.1

Crystal structure and morphology of calcium silicate minerals

Minerals	Crystal structure	Coordination polyhedron	Morphology	Coordination number of Ca	Connection
C_3S		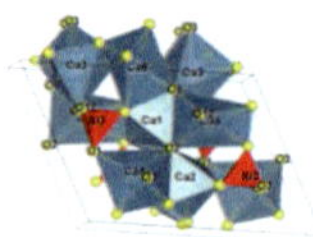	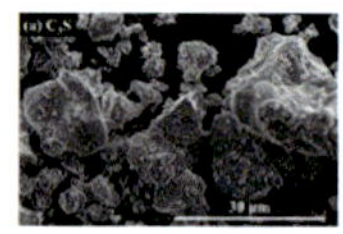	4, 6	Co-apex/ edge/ planar
β-C_2S	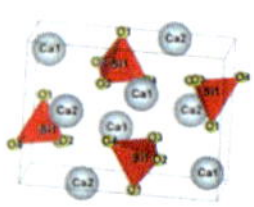	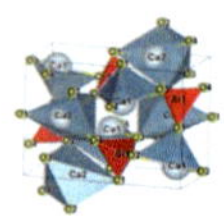		6, 7	Co-apex/ edge/ planar
γ-C_2S	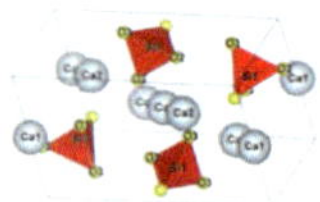	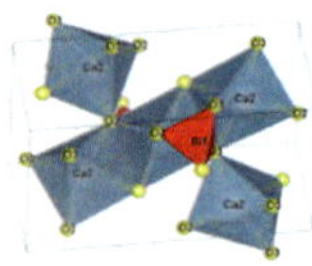	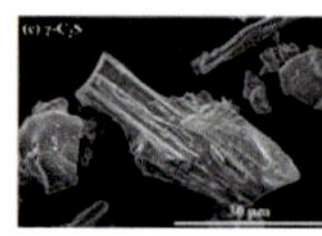	6	Co-edge
C_3S_2	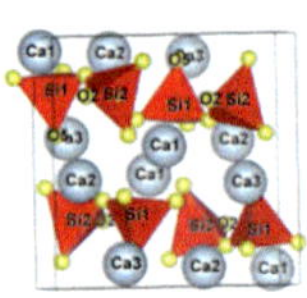	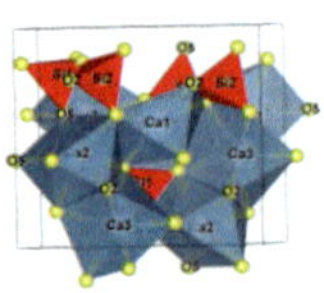	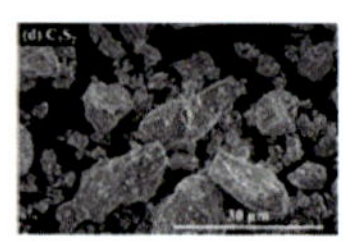	4, 6	Co-edge/ planar
CS	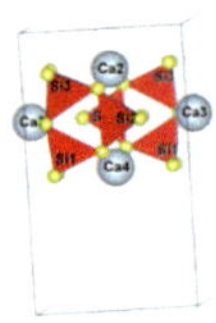	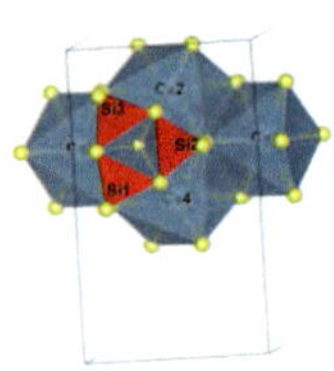	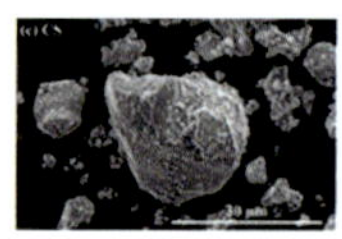	4, 8	Co-edge

Source: Mu et al. (2019); Wang, Q.Q. et al. (2017).

temperature for producing hydraulic calcium silicate compound is relatively higher. To reduce the CO_2 footprint of cementitious materials, the production and utilization of low-lime calcium silicates, such as CS, C_3S_2, and γ-C_2S, have become a major research focus.

4.2.1 Tricalcium silicate

C_3S (tricalcium silicate) belongs to the island-type silicate structure, where isolated [SiO4] tetrahedra are connected by Ca–O polyhedra. As the temperature increases, C_3S exhibits seven distinct polymorphs in the triclinic (T, including T1, T2, T3), monoclinic (M, including M1, M2, M3), and rhombohedral (R) crystal systems.

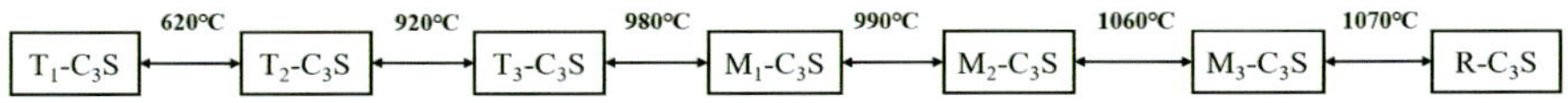

FIGURE 4.1 Transformation diagram of C_3S polymorphs (Maki and Kato, 1982).

These polymorphs share similar structures but differ in terms of disorder and symmetry due to the varying orientations of the [SiO_4] tetrahedra.

Figure 4.1 illustrates the transformation relationships between the seven polymorphs of C_3S (Maki and Kato, 1982). At room temperature, pure C_3S is stable in the T1 polymorph, which exhibits relatively low hydraulic activity. The polymorphs M1 and M3, obtained through rapid cooling, have higher hydraulic activity and are commonly found in cement clinker. As the primary component of Portland cement, C_3S, also known as alite, typically constitutes around 50% of the total cement clinker. Its hydration process plays a crucial role in the initial setting and strength development of cementitious materials. The hydration of C_3S releases a significant amount of heat, which largely determines the early strength of Portland cement.

4.2.2 Dicalcium silicate

C_2S exists in five polymorphs: γ, β, α'L, α'H, and α. Figure 4.2 illustrates the transformation relationships between these polymorphs (Wen et al., 2015), which differ significantly in structure and performance. The high-temperature phases α-C_2S, α'L-C_2S, α'H-C_2S, and β-C_2S can transform into γ-C_2S during cooling, which has minimal hydration activity (Zhao, 2012). In γ-C_2S, Ca^{2+} ions exhibit regular coordination, while in the high-temperature and β phases, Ca^{2+} coordination is irregular, leading to vacancy defects and lattice distortion, which enhance hydration reactivity.

The most common polymorphs of C_2S are the β-phase and the high-temperature α-phase. β-C_2S, also called belite, is a key component of Portland cement, typically comprising 15–30% of the clinker. Compared to C_3S, β-C_2S hydrates more slowly, resulting in a delayed contribution to the overall strength of cement paste. However, it plays a crucial role in the long-term strength and durability of cement-based materials. Rapid cooling during clinker production is used to prevent the β-to-γ phase transformation, which can also be inhibited by doping. The β-to-γ transformation can cause up to 12% volumetric expansion due to density differences, and the associated stresses can induce self-pulverization, producing powdered γ-C_2S (Jiang et al., 2023).

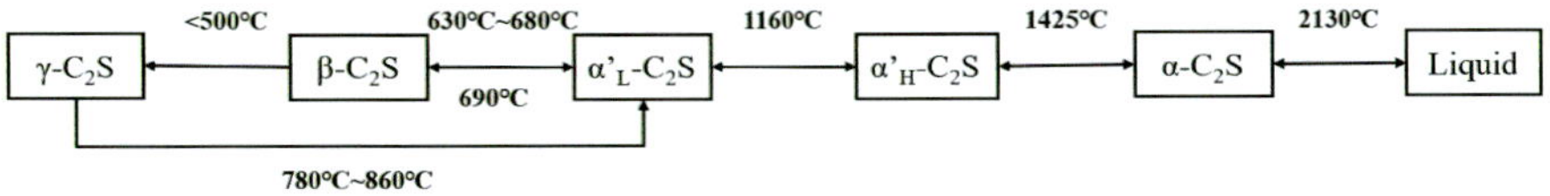

FIGURE 4.2 Transformation diagram of C_2S polymorphs (Wen et al., 2015).

4.2.3 Tricalcium disilicate

C_3S_2 currently exists in two known polymorphs: kilchoanite, which belongs to the orthorhombic crystal system, and rankinite, which belongs to the monoclinic crystal system. C_3S_2 is relatively rare in nature, and its traditional preparation methods require high calcination temperatures (1380–1460°C), which are higher than those required for the formation of other common calcium silicate phases, such as C_3S (tricalcium silicate) (Fang et al., 2016; Chen, 2018). C_3S_2 is a non-hydraulic, low-lime calcium silicate phase, but due to its high-temperature and mechanical stability, it is commonly used in advanced materials, including refractories and durable ceramics.

4.2.4 Monocalcium silicate

CS exists in two polymorphs: α-CS and β-CS, both of which are non-hydraulic. α-CS forms above 1125°C and is primarily found in high-temperature phases such as metallurgical slag, cement, and ceramics. When cooled slowly to 1125°C, α-CS transforms into β-CS. β-CS has two variants: wollastonite and pseudo-wollastonite, which coexist abundantly in nature (Seryotkin et al., 2012). In wollastonite's crystallographic structure, the SiO_4 chain and CaO_6 octahedral columns form the basic building block of its crystal structure. This unique crystal structure determines its properties; when crushed into fine particles, it retains a needle-like morphology.

4.3 CARBONATION REACTIVITY AND MECHANISM OF CALCIUM SILICATE MINERALS

4.3.1 Carbonation reactivity of calcium silicate minerals

The carbon mineralization of calcium silicate minerals can be generally summarized by Equation (4.1). Both hydraulic and non-hydraulic calcium silicate compounds exhibit remarkable carbonation reactivity and can actively react with CO_2 under natural conditions. Ashraf et al. (Ashraf and Olek, 2016) compared the carbonation behaviors of non-hydraulic phases (such as γ-C_2S, C_3S_2, and CS) and typical hydraulic phases (such as C_3S and β-C_2S), observing that carbonation reactions, based on total CaO content, are more effective than hydration reactions. Figure 4.3 shows the carbonation degree of calcium silicate phases exposed to different conditions (Ashraf and Olek, 2016). It can be clearly seen that the reactivity of C_3S_2 and CS samples was lower than that of C_3S and C_2S samples under normal conditions. However, when the temperature increased to 55°C and the CO_2 concentration rose to 100%, their carbonation rates increased rapidly, ultimately reaching complete carbonation (100% carbonation). The carbonation reactivity of β-C_2S was significantly higher than that of C_3S. Since carbonation is an exothermic process, tracking temperature shifts during carbonation indicates that the order of carbonation reactivity of several minerals from high to low is: γ-C_2S > C_3S_2 > CS (Mu et al., 2019).

From the perspective of carbonation activation energy, the carbonation reaction can be divided into two distinct controlling stages: interfacial control (Stage 1) and

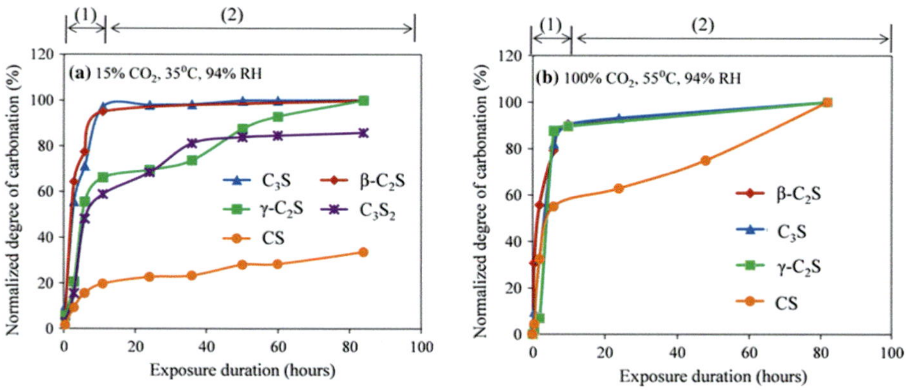

(a) 15% CO_2, 35°C, 94% RH (b) 100% CO_2, 55°C, 94% RH

FIGURE 4.3 Carbonation degree of calcium silicate phases exposed to different conditions (Ashraf and Olek, 2016).

product layer diffusion control (Stage 2) (Ashraf and Olek, 2016). For non-hydraulic calcium silicates (γ-C_2S, C_3S_2, and CS), the carbonation activation energy in Stage 1 ranges from 25 kJ/mol to 35 kJ/mol. In contrast, for C_3S and β-C_2S, the activation energies are determined to be 5.5 kJ/mol and 7.5 kJ/mol, respectively. The combination of hydration and carbonation reactions in C_3S and β-C_2S results in a reduction of their carbonation activation energy. Regarding the stage of carbonation controlled by product layer diffusion, the activation energy was found to range between 30 kJ/mol and 40 kJ/mol, suggesting that this stage of carbonation is primarily influenced by the reaction products formed. As the product layer thickens, the diffusion of CO_2 and ions is hindered by the reaction products, causing the reaction rate to slow down. Furthermore, Zhao et al. (Zhao et al., 2020a) conducted an in-depth investigation into the carbonation reaction of γ-C_2S and put forward a multistage carbonation control mechanism. As the matrix gradually densifies, CO_2 diffusion is increasingly restricted. Concurrently, the scarcity of Ca^{2+} ions imposes an additional constraint on the carbonation reaction. As a result, the carbonation reaction is jointly governed by the diffusion of CO_2 and the dissolution of Ca^{2+} ions.

$$Ca_xSi_yO_{x+2y} + xCO_2 \rightarrow xCaCO_3 + ySiO_2 \quad (4.1)$$

4.3.2 Carbon mineralization of hydraulic calcium silicate

The carbonation rates of C_3S and β-C_2S are higher than the hydration rates due to the higher heat flow intensity (Hu et al., 2022a; Ibáñez et al., 2007). As the major mineral phase present in Portland cement and other industrial by-products such as steel slag, hydraulic calcium has been extensively studied for its the carbon mineralization behavior (Berger and Klemm, 1972; Na et al., 2021; Wang et al., 2020). Generally, the carbon mineralization of calcium silicate mineral is a dissolution-ion

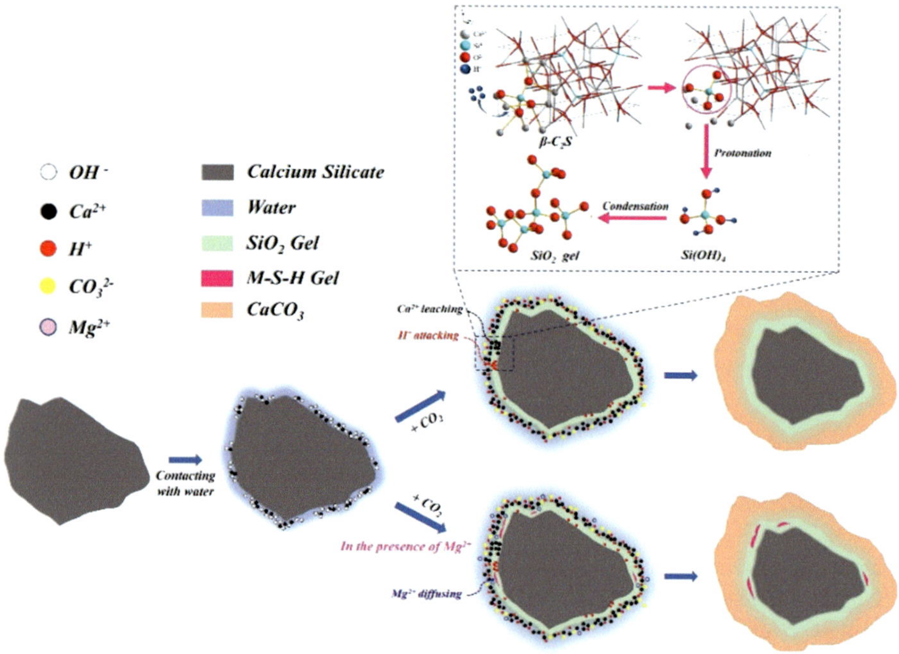

FIGURE 4.4 Schematic representation of the carbonation processes of typical calcium silicate minerals (Liu et al., 2022).

diffusion-precipitation process. As shown in Figure 4.4 (Liu et al., 2022), in the presence of water, CO_2 dissolves and ionizes to form CO_3^{2-} and H^+. Then Ca^{2+} is released from the calcium silicate mineral phase in acidic environment and diffuses out of the surface of calcium silicate. With the continuous dissolution of Ca^{2+} and CO_3^{2-}, a supersaturated solution is gradually formed and finally results in the precipitation of $CaCO_3$. The dissolved CO_2 gradually penetrates through the Si-rich area and facilitates the leaching of Ca^{2+}. After the leaching of calcium, $[SiO_4]$ is formed and thereafter leads to the formation of orthosilicic acid ($Si(OH)_4$) due to the protonation of $[SiO_4]$. Generally, the stability of ($Si(OH)_4$) is low and may rapidly condense to form silica gel (SiO_2). During the carbon mineralization of calcium silicate, the mineral composition and carbonation condition may affect the crystal growth of calcium carbonate and the polymerization of silica gel (Wang et al., 2020). Therefore, different calcium carbonate polymers (calcite, aragonite, and vaterite) and silica gel with different degrees of polymerization can be generated.

Saito et al. (2010) suggested that during the carbon mineralization of calcium silicate, Ca^{2+} was dissolved from these particles, and $Ca(OH)_2$ formed rapidly. Then $Ca(OH)_2$ reacted with CO_3^{2} to form $CaCO_3$. The main carbonation product was calcite. Besides, a small amount of vaterite and aragonite was also detected in some specific carbonation conditions (Chen et al., 2022a). Apart from crystalline calcium carbonate, amorphous calcium carbonate was also detected during the carbon mineralization of calcium silicate, which showed lower decomposition temperature (450–625°C) compared to that of well-crystallized calcium carbonate, i.e., calcite (700–850°C), aragonite, and vaterite (650–700°C) (Ashraf, Warda and Olek, Jan 2018; Ashraf and Olek, 2018a; Chang et al., 2016; El-Hassan et al., 2013; Wang et al., 2019). This poorly crystallized calcium carbonate transformed to more stable calcite during subsequent carbonation and hydration processes (El-Hassan and Shao, 2015; El-Hassan et al., 2013).

For studies on the production of carbonate binders by cement-based materials, early gaseous CO_2 curing for fresh-state cement paste/mortar/concrete and carbonation treatment of hardened ones are widely applied (Liu, Z. et al., 2023; Zhang et al., 2023). During the process of CO_2 curing and/or carbonation treatment, CO_2 gas experiences the process of diffusion in air and dissolution to liquid and finally reacts with the hydraulic calcium silicate minerals or hydration products within cement-based materials to produce carbonate binders, during which the migration, dissolution, and reaction processes all control the carbonation rate and efficiency (Pu et al., 2023). However, the complex structure and moisture distribution within cement-based materials significantly limit the enhancement of carbonation rate and degree, and afterward the application of these techniques in the production of carbonate binders.

4.3.3 Carbon mineralization of non-hydraulic calcium silicate

The carbonation mechanisms of non-hydraulic calcium silicates are different from those of hydraulic calcium silicates. During the carbon mineralization of γ-C_2S and CS, water is absorbed on the surface of these particles and forms a water film (Figure 4.5). Then, CO_2 is dissolved in water and forms ions including H^+, HCO_3^-, and CO_3^{2-}. Meantime, Ca^{2+} is formed due to the dissociation of calcium silicate and leads to the formation and precipitation of $CaCO_3$ in solution. A Si-rich layer is formed on the particle surface and transforms into Ca-modified silica gel in the presence of water. Ca-modified silica gel is normally formed around the unreacted calcium carbonate particles, while calcium carbonate is precipitated in the external space. Finally, Ca-modified silica gel becomes thick to inhibit the contact between Ca^{2+} and CO_3^{2-}, and the reaction slows down (Mu et al., 2018a). C_3S, β-C_2S, and

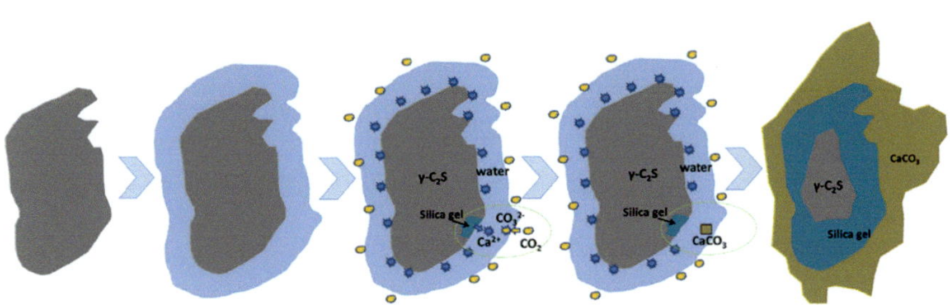

FIGURE 4.5 Carbonization process of γ-C_2S particles (De Silva et al., 2006).

γ-C_2S powder showed similar carbonation rate when subjected to CO_2 curing, while CS showed the lowest carbonation rate (Ashraf and Olek, 2018b).

C_3S_2 prepared by calcining a C-S-H gel precursor at 1300°C reached a compressive strength of 62.5 MPa after 24 hours of carbonation (Wang, K. et al., 2018). The carbonation products are calcite, vaterite, and highly polymerized amorphous silica gel, and no aragonite is detected. A rim can usually be observed surrounding calcium silicate particles after carbonation, which consists of Ca-modified silica gel or a combination of silica gel and $CaCO_3$ products. Smigelskyte (Smigelskyte and Siauciunas, 2019; Smigelskyte et al., 2020) systematically studied the influence of different factors on the carbonation activity of C_3S_2. Higher carbonation temperature also increased the carbonation degree of C_3S_2. The carbonation products of pastes and mortars with C_3S_2 were different. The carbonation products of pastes consisted of three polymorphs of calcium carbonate, including calcite, vaterite, and aragonite, while the only polymorph of calcium carbonate observed in the mortar samples was calcite. It was supposed that the presence of the quartz aggregates was conducive to the formation of calcite rather than other polymorphs.

CS with different crystal forms has different carbonation activity, and α-CS has the highest carbonation activity (Liu et al., 2021). High carbonation temperature and CO_2 pressure can increase the carbonation degree of CS (Kashim et al., 2020) and promote the formation of needle-like aragonite with high bending strength (Tan et al., 2022). Svensson et al. (2018, 2019) investigated the carbonation production of natural wollastonite, which is a sort of inorganic natural mineral composed of calcium silicate with a Ca–Si ratio of 1:1. It was found during the carbonation that a larger amount of calcite and lower amounts of X-ray amorphous were observed. According to their results, the use of wollastonite to produce CS cement as a carbonate binder was possible. Similar to C_2S, the incorporation of impurity elements can also impact the carbon mineralization of CS. Zhang et al. (2022) proposed a method to stabilize high-temperature α-CS and studied the effects of sodium on the carbonation behavior of α-CS. As sodium doping increased, the early carbonation hardening properties were slightly lower than those of pure α-CS but resulted in a significant increase in compressive strength and carbonation degree at later stages (Zhang et al., 2022).

4.4 INFLUENCING FACTORS OF CARBONATION DEGREE

4.4.1 CO_2 ENVIRONMENT

The diffusivity of CO_2 is directly related to its concentration and pressure. Figure 4.6 shows the carbonation degree of cement paste subjected to different CO_2 concentrations. At elevated CO_2 concentrations, the gas diffuses at an accelerated pace toward the pore air-pore–solution interface. More CO_2 can dissolve in the pore solution, which accelerates the carbonation reaction (Li and Ling, 2020; Visser, 2014). Therefore, the carbonation degree increases as the CO_2 concentration rises (Castellote et al., 2009). However, when the CO_2 concentration surpasses the 20% threshold, the impact of further concentration augmentation on the carbonation

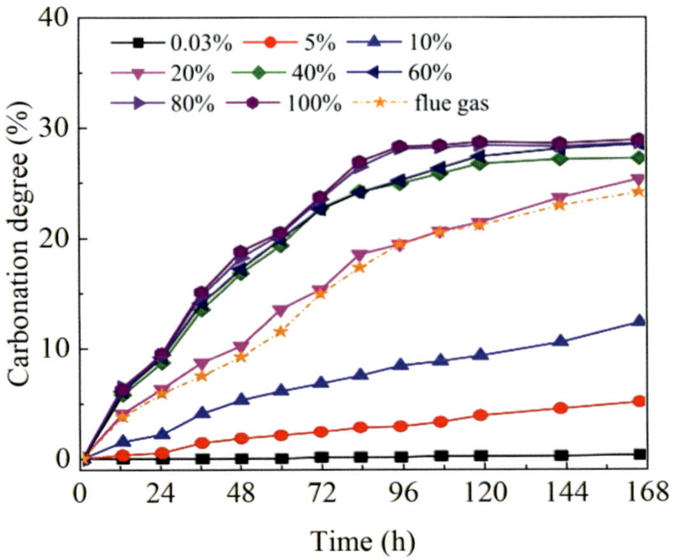

FIGURE 4.6 Effects of CO_2 concentrations on carbonation degree (Xuan et al., 2018).

degree becomes markedly diminished. This phenomenon can be ascribed to the fact that, as the reaction progresses, the newly generated $CaCO_3$ begins to fill the pores, effectively erecting a barrier that impedes the infiltration of CO_2 required for further reactions. Pu et al. (2021) found that the carbonation degree at a CO_2 concentration of 5% was approximately 5.1%, while it sharply increased to 25.3% by promoting the CO_2 concentration to 20%. Yet, a rather muted response was observed when CO_2 concentration increased from 20% to 100%, with the carbonation degree inching up only from 24.3% to 28.9%. Similarly, Kashef-Haghighi et al. (2015) discerned that an increase in CO_2 gas concentration from 20% to 40% doubled the rate of CO_2 uptake by cement. Intriguingly, however, the carbonation degree did not register a significant alteration due to the formation of solid carbonation products on reactive surfaces and in the pore space. In contrast, Xuan et al. (2018) reported that an exponential increase in carbonation degree corresponds to the growth of CO_2 concentration from 10% to 100% in the concrete blocks. This divergence from the norm might potentially be ascribed to the weak filling effect of carbonation products in concrete blocks, which, by virtue of possessing a more porous structure, allowed for continued ingress of CO_2 even in the face of product accumulation.

Figure 4.7 illustrates the correlation between CO_2 pressure and carbonation degree of cement paste. An elevation in pressure augments the CO_2 diffusivity and increases the quantity of CO_2 molecules dissolved in the pore water, thereby accelerating the rate of the carbonation reaction (Carroll et al., 1991; Mo et al., 2015). Shi et al. (Shi and Wu, 2008) found that the consumed CO_2 per mass of cement under 10, 20, and 60 psi CO_2 pressures were 10.5%, 13.1%, and 17.5%, respectively. Similarly, Harrison et al. (2013) reported that the rise in pressure from 0.5 MPa

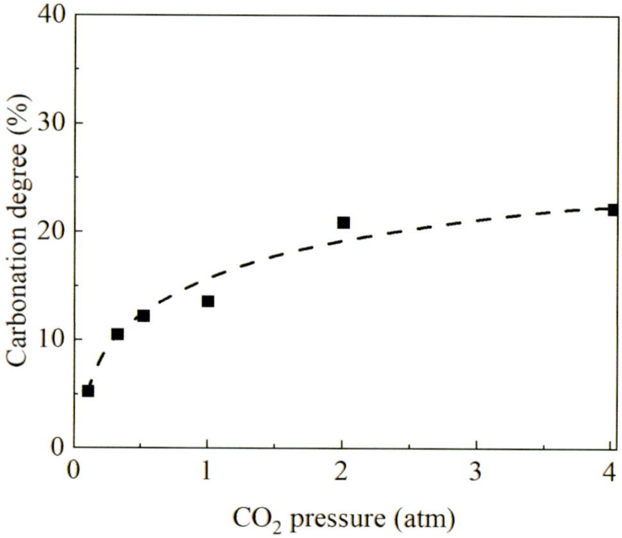

FIGURE 4.7 Relationship between carbonation degree and CO_2 pressure (Zhan, Xuan et al., 2016).

to 2.5 MPa is accompanied by the CO_2 uptake from 13.27% to 18.34%. However, the rate of increase in the carbonation degree decelerates as the pressure escalates. Once the pressure exceeds 0.5 MPa, it exhibits no significant impact on either the carbonation degree or the compressive strength (Zhan et al., 2016b). Moreover, an excessively high CO_2 gas pressure may be detrimental to the carbon mineralized materials/products since the substantial heat generation heightens the risk of microcracking (Bukowski, J. and Berger, R.L., 1979). A moderate CO_2 gas pressure suffices to foster the strength development of concrete blocks (Zhan et al., 2016b). Even under lower CO_2 pressure, an extended exposure duration can still contribute to a remarkable enhancement in strength.

4.4.2 Water-to-binder ratio

The diffusion of CO_2 within the matrix is intimately associated with the moisture content and pore structure, both of which depend on the water-to-binder ratio (w/b) (Lo et al., 2008). CO_2 diffuses more easily in cement-based materials with higher w/b, which possess higher porosity and more open pores (Ashraf et al., 2019). Additionally, cement-based materials prepared with higher w/b showed greater cement hydration and produced a higher amount of hydration products, hence facilitating a better carbonation reaction. Lo et al. (2008) reported that the carbonation reaction decreases (as indicated by its carbonation depth) with a reduction in the w/b due to fewer open pores. Siddique et al. (2020) reported that a high w/b enhanced the hydration of belite-rich cement; however, the presence of excessive free water

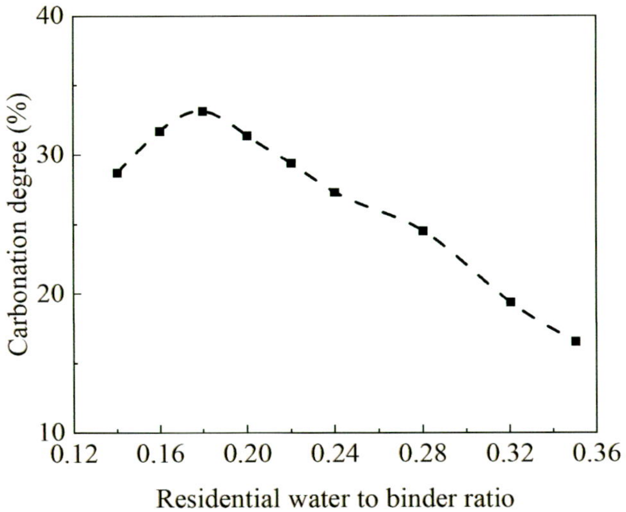

FIGURE 4.8 Effect of residential w/b to the carbonation degree of concrete (Shi C. et al., 2014).

could create larger pores in the system and in turn negatively impact the mechanical strength of cement paste (Ashraf et al., 2019; Siddique et al., 2020). Shi et al. (2012a) investigated the effects of pre-conditioning on the CO_2 curing of lightweight concrete block mixtures. The findings revealed that there existed an optimal level of moisture loss at which the specimens attained the highest strength and CO_2 consumption. Figure 4.8 shows the effect of residential w/b to the carbonation degree of concrete. As seen, concrete with a residual w/c ratio of 0.18 exhibits the highest carbonation degree after CO_2 curing (He et al., 2016; Shi et al., 2014). However, it should be noted that the moisture evaporation rate must be regulated to avoid plastic shrinkage cracking.

4.4.3 Relative humidity

Relative humidity (RH) also exerts an influence on the water content within the matrix. The moisture present in the carbonation environment not only impacts the diffusion of CO_2 but also supplies the necessary water for the carbonation reaction (Papadakis et al., 1989). Figure 4.9 depicts the relation between RH and carbonation degree (Elsalamawy et al., 2019). At low RH levels, an insufficient amount of water may be available to sustain the carbonation reaction. Ashraf et al. (2019) found that a lower RH results in a reduced amount of the stored CO_2 during CO_2 curing of calcium silicate minerals, as the samples tend to dry out more rapidly. When an adequate water content is ensured, the maximum carbonation degree can be achieved even at relatively low temperatures and low CO_2 concentration (Ashraf and Olek, 2016). Nevertheless, as the RH rises, the pores of the matrix gradually become filled

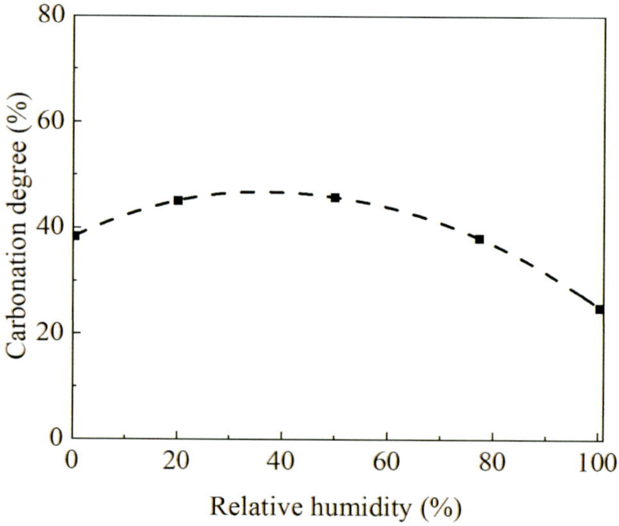

FIGURE 4.9 Relation between RH from 40% to 80% and relative carbonation depth (Elsalamawy et al., 2019).

with water, resulting in the difficulty of CO_2 diffusion. Consequently, an optimal RH of 50–70% is typically selected during CO_2 mineralization, which provides adequate water for the reaction of carbonation while still permitting a considerable degree of CO_2 diffusivity (Galan et al., 2013; Papadakis et al., 1989).

4.4.4 Temperature

Temperature is a pivotal parameter that significantly influences carbonation efficiency, and its impact on the carbonation reaction is complex and multifaceted. Higher temperatures can accelerate the leaching of Ca^{2+} ions in the pore solution and enhance the diffusivity of CO_2, thereby facilitating the chemical reaction. Nevertheless, elevated temperatures can simultaneously reduce the solubility of CO_2 in the solution and hasten the rate of water evaporation, both of which have an adverse effect on carbonation efficiency (Abdullahi et al., 2016). Figure 4.10 shows the relationship between temperature and CO_2 uptake (Wang, T. et al., 2017). Lu et al. (2022) reported that an increase in temperature from 5°C to 50°C enhanced the CO_2 uptake of cement pastes. The carbonation depth of the cement pastes after 28 days of CO_2 curing at 5°C, 20°C, and 50°C was 12.2 mm, 18.5 mm, and 21.8 mm, respectively. Curing temperature above 60°C may hinder the carbonation reaction due to the fast evaporation of water and diminished solubility of CO_2 in water (Drouet et al., 2019; Humbert et al., 2019). Liu et al. (2001) reported that CO_2 uptake increased from 2% to 14% as the curing temperature increased from 20°C to 60°C, while decreasing from 14% to 6% as the temperature increased from 60°C to 100°C.

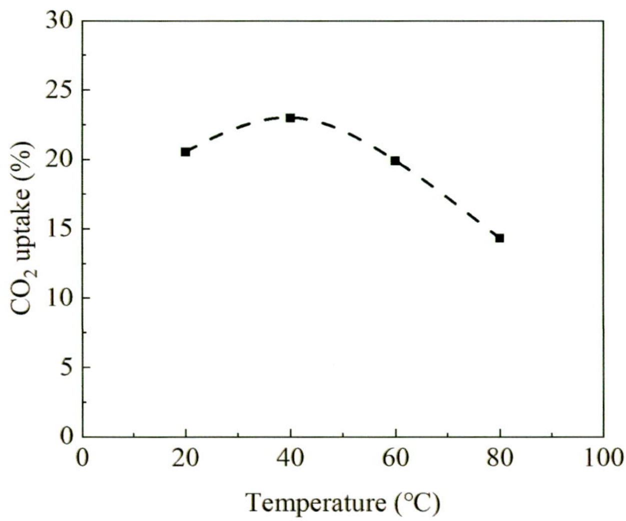

FIGURE 4.10 Relationship between temperature and CO_2 uptake (Wang, T. et al., 2017).

4.4.5 Supplementary cementitious materials

Supplementary cementitious materials (SCMs), such as fly ash, silica fume, slag, and metakaolin, are widely used as pozzolans or fillers to partially replace cement, which have proven to be effective in emission reduction and performance improvement (Wang, D. et al., 2018). The incorporation of SCM might interfere with cement carbonation behavior despite their little contribution to the carbonate reaction (Joshaghani et al., 2018). Vanoutrive et al. (2022) focused on carbonation, specifically the effect of the replacement level of PC by ground granulated blast-furnace slag (GGBFS) on the reaction products and CO_2 buffer capacity and found that by increasing the curing period to 7 or 28 days, 40 wt% replacement by GGBFS showed considerably slower decalcification of C-S-H compared to 70 wt% replacement by GGBFS. Wang et al. (2017) found that the blended CS significantly enhanced the reaction rate and CO_2 uptake capacity, accompanied by the intensified carbonation conversion of cement components.

The incorporation of supplementary cementitious materials, including limestone (Tu et al., 2016; Tu et al., 2018; Zhang et al., 2020), metakaolin (Hameed et al., 2020), biochar (Chen et al., 2022b; Yang and Wang, 2021), pulverized coal combustion and circulating fluidized bed combustion (Kim et al., 2022), fly ash (Gupta et al., 2021; Zhang and Shao, 2016), quartz (Lin and Wang, 2021), coral powder calcite seeds (Zhao et al., 2020b), and nano-TiO_2 (Moro et al., 2021) can provide more space for the growth of carbonation products due to their nucleation effect (Lioliou et al., 2007), or introduced more channels for the penetration of CO_2, such as porous coral powder (Zhang et al., 2022) and rice husk ash (Hu et al., 2022b), or increased the CO_2 uptake like ground volcanic ash (Seo et al., 2018). With the same water-to-binder ratio, the effective water-to-cement ratio increases as SCMs are added because of the reduction of OPC content. A higher water-to-cement ratio leads to a more porous structure, providing higher possibilities for the reaction with CO_2 (Zhang et al., 2016). Moreover, some researchers used coral powder to increase the compressive strength and carbonation degree of γ-C_2S as its porous structure could provide more channels for the penetration of CO_2 (Mu et al., 2022).

4.4.6 Chemical additives

Various amines or amino acid salts have been employed to expedite the absorption of CO_2 and facilitate the multiphase reactions of CO_2 mineralization by an aqueous phase to produce carbonate or bicarbonate ions (Vinoba et al., 2014), such as monoethanolamine, MEA (Zhang, W. et al., 2022), diethanolamine, DEA (Arti et al., 2017), methyldiethanolamine, MDEA (Arti et al., 2017), and piperazine (Rochelle et al., 2011). Amine solvents have the capacity to raise the pH of the solution and engage in a neutralization reaction with CO_2, promoting the dissolution of CO_2 in the solution (Luis, 2016; Ramezani et al., 2017). Several novel carbonation methods, such as incorporating CO_2-dissolved amine solvents during the mixing process (Liu, J. et al., 2023) and submerging cement-based samples in CO_2-dissolved amine solution (Han et al., 2023), have been proposed to increase carbonation degree and improve the properties of cementitious materials.

Alkaline reagents (e.g., sodium hydroxide (NaOH), ammonium hydroxide (NH_4OH)) hold the potential for enhancing carbonation efficiency since it can elevate the pH value and make it suitable for CO_2 absorption (Naraharisetti et al., 2019). By promoting the dissolution of Ca^{2+} within calcium silicate minerals, acidic compounds can supply more reactive ions for the carbonation reaction, thereby expediting the carbonation process. Nevertheless, in the practical application of CO_2 curing, the research and utilization of acidic compounds are confronted with substantial limitations. During the entire curing period, acidic compounds are subject to permanent and continuous depletion. This not only renders it arduous to sustain a consistent reaction-enhancing effect but also leads to increased costs. Ethylenediaminetetraacetic acid (EDTA, $C_{10}H_{16}N_2O_8$), functioning as an effective catalyst, exhibits distinctive merits in accelerating the carbonation reaction. The complex interaction between EDTA and Ca^{2+} leads to an increase in soluble Ca^{2+}, accelerating the carbonation process, which is favorable to nuclei formation and leads to the formation of superfine particles (Xiang et al., 2002). The catalytic role of EDTA endows it with the potential for application in the CO_2 curing process.

4.4.7 Other factors

The carbonation activity of calcium silicate minerals is intrinsically determined by their crystal structure. The introduction of impurity elements can alter the mineral crystal structure, subsequently exerting a profound impact on its carbonation reactivity. Mu et al. (2018b) found that the Ba doping with appropriate contents could retard the initial carbonation reaction but promote the carbonation degree of γ-C_2S.

The addition of Mg, Ti, Ni, and Cu ions was also found to improve compressive strength and carbonation degree of γ-C_2S. Zhang et al. (2022) investigated the effect of sodium (Na) doping on the properties of α-CS minerals and found that as sodium doping increased, the early carbonation hardening properties were slightly lower than those of pure α-CS, but there was a significant increase in compressive strength and carbonation degree in the later stages. This is because sodium doping increases the alkalinity of the paste and delays the dissolution of calcium ions from the α-CS. Furthermore, the incorporation of Al_2O_3 into γ-C_2S clinker is capable of diminishing the heat release that occurs during the carbonation process, thereby facilitating a continuous progression of the carbonation reaction.

Mineral microbial treatment is also capable of remarkably expediting the carbonation reaction, which is achieved by promoting the adsorption of CO_2 and the generation of carbonate ions. The carbonation reactivity of γ-C_2S was reported to be improved by biomineralization, as the pre-existing calcite formed by biomineralization acts as nucleation seeds (Zhao et al., 2020b). Jin et al. (2019) found that the addition of bacteria could effectively accelerate the carbon mineralization of β-C_2S. It did not change the crystallized form of calcium carbonate but had great effects on morphology and size of the crystals.

4.5 PROPERTIES AND CONTROL OF CARBONATION PRODUCTS

4.5.1 Calcium carbonate

The reaction between CO_2 and calcium silicate minerals generally results in the formation of calcium carbonate and silica/aluminum gel. The typical forms of calcium carbonate include amorphous calcium carbonate (ACC) and three crystalline polymorphs, namely calcite, aragonite, and vaterite. ACC consists of a calcium-rich nanoscale framework with pores containing water and carbonate ions (Goodwin et al., 2010). The amount of structural water is a crucial factor influencing the ACC structure, up to approximately 1.6 mol of water per mole of calcium carbonate. Less hydrated and even anhydrous forms of ACC have also been reported (Rodriguez-Navarro et al., 2015; Tlili et al., 2002). ACC is thermodynamically unstable and will spontaneously transform into more energetically favorable crystals (Nielsen et al., 2014; Rodriguez-Blanco et al., 2011). Calcite, generally rhombohedral particles, is the highest thermodynamically stable calcium carbonate polymorph. It exhibits the least solubility over a temperature range of 0–90°C among other polymorphs (Ogino et al., 1987). The aragonite phase crystallizes in the orthorhombic crystal system, typically in a needle-like or rod-like shape (Chang et al., 2017). With thermodynamic stability intermediate between calcite and vaterite, aragonite is more stable at high temperatures and pressures. Vaterite is the least thermodynamically stable form of crystalline calcium carbonate, with the highest solubility and the lowest density. The particles of vaterite usually cluster into spherical particles, but flake-like, flower-like, and plate-like morphologies were also observed (Wang et al., 2015; Xu et al., 2006).

The morphology, polymorphs, and size of calcium carbonate changed with time during the carbon mineralization of C_3S and β-C_2S (Zheng et al., 2023). Wang et al.

(2020) investigated the growth of calcite during the carbonation process of C_3S and proposed the formation mechanism. According to the results of heat flow, the carbonation process can be separated into three stages. In the primary heat-releasing stage (i.e., stages I and II), rhombic calcite formed from the transformation of amorphous calcite carbonate (ACC), with a slight increase in calcite particle sizes. In the leveled stage (i.e., stage III), the particle size of calcite increased significantly. During the carbonation, the crystal size of calcite increased from 55 nm to 121 nm. Moreover, the crystallinity of calcite also increased with the carbonation time as its decomposition temperature increased from 698°C to 810°C. They also investigated the development of crystal size of calcite during the carbon mineralization of β-C_2S, and the carbonation process was also divided into three stages (Wang et al., 2019). Different from the carbon mineralization of C_3S, the crystal size of calcite sharply increased within the first several minutes for the carbon mineralization of β-C_2S. Then poorly crystalline calcite was formed during the first 12 hours and grew from 50 nm to about 120 nm after 12 hours. Besides the calcium silicate minerals, the hydration products of Portland cement can also be carbonated. According to the ^{29}Si-MAS-NMR spectroscopy results, two steps were included during the decomposition process of C-S-H gel caused by carbonation (Sevelsted and Skibsted, 2015). The interlayer Ca ions of the C-S-H first participated in the carbonation reaction, then the calcium atoms within the interlayer and silicate chains of C-S-H would be removed successively (Chen et al., 2022a).

Considering the various mineral structures of calcium silicate and other factors on carbonation, the formation of carbonation products still has no consensus. By using two models of calcium silicate polymorphs—wollastonite (chain-structured) and pseudowollastonite (ring-structured)—Plattenberger et al. (2018) reported the mechanism of calcium carbonate formation with different crystal structures of calcium silicate. As shown in Figure 4.11, the release of Ca^{2+} and Si^{4+} in the system was different and also played an important role in the morphology of the carbonation products. Generally, the reaction between wollastonite and CO_2(aq) in liquid formed amorphous silica and calcium carbonate while the reaction of CO_2 with pseudowollastonite produced platelike crystalline calcium silicate phases, along with $CaCO_3$ and amorphous silica.

Mu et al. (2019) comparatively studied the carbon mineralization of calcium silicates concentrating on the carbonation products and microstructure of carbonated minerals. As shown in Figure 4.11, unreacted calcium silicate, crystalline calcium carbonate, and amorphous phase were three kinds of main phases observed in the carbonated calcium silicate minerals. Calcite and aragonite are the preferred polymorphs of calcium carbonate in the carbonated C_3S, γ-C_2S, β-C_2S, and C_3S_2, while vaterite only exists in the carbonated CS. The content of these carbonation products in different systems is also different. Due to the least stable of vaterite among the three polymorphs of calcium carbonate, the contents of vaterite in the carbonation products are generally negligible, and it is a precursor to the other two polymorphs.

The morphology and polymorphs of calcium carbonate played an important role in the intrinsic properties of carbonated products (Ashraf, Warda and Olek, Jan

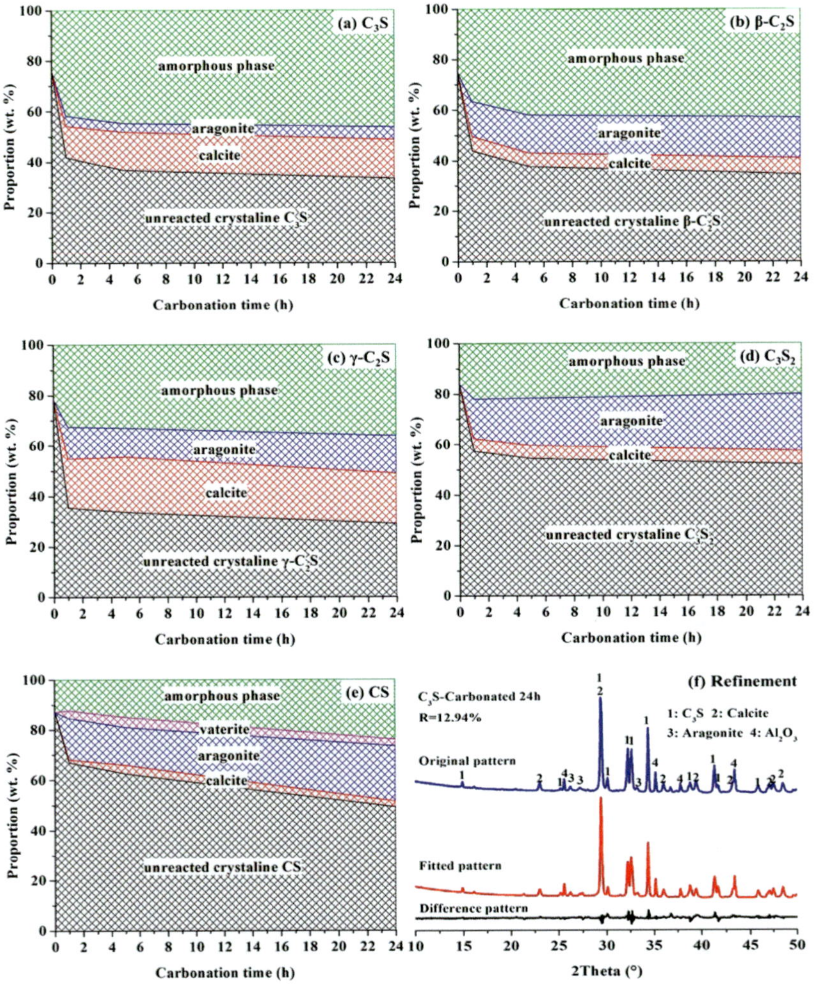

FIGURE 4.11 Phase assemblage of different calcium silicates carbonated for various durations (a–e), and example of XRD curve fitting by Rietveld refinement (f) (Mu et al., 2019).

2018). For example, the stiffnesses of vaterite, aragonite, and calcite are 39.13 GPa, 67 GPa, and 72.83 GPa, respectively (Dhami et al., 2016). It was reported that a 20% to 40% variation in the mechanical performance of carbonated composites generated in a similar carbonation condition may happen due to the difference in calcium carbonate morphology (Ashraf, Warda and Olek, Jan 2018). Many factors affect the formation of calcium carbonate, and different types of crystallized calcium carbonate may form even for the carbon mineralization of the same type of calcium silicate and Portland cement (You et al., 2023). The factors affecting the crystal type of calcium carbonate include: (1) Ca/Si molar ratio in calcium silicates, (2) initial w/c ratio, and (3) CO_2 curing conditions such as CO_2 concentration, temperature, and curing time (Lu et al., 2022; Tan et al., 2022; Zajac et al., 2023). Besides, the pH value of the pore solution also affects the formation of crystals (Rodriguez-Blanco et al., 2012; Shtepenko et al., 2006). Moreover, the addition of some admixtures may also alter the polymorphs of calcium carbonate (You et al., 2023).

Given the significant impact of the properties of calcium carbonate on the performance of carbonated products, it is crucial to explore effective control methods. The control of carbonation products, covering aspects such as the content of carbonation products, the crystallization form of calcium carbonate, and the polymerization degree of silica gel, has spurred the proposal and investigation of diverse methods. These methods include carbonation condition optimization, the addition of chemical agents, and biologically induced carbonation (Liu and Gadikota, 2019; Rivera and Van Gerven, 2020). During the carbon mineralization of calcareous materials or Portland cement-based materials, the relative humidity (Wang et al., 2022; Yahui et al., 2021), CO_2 concentration (Li et al., 2020), carbonation pressure (Wang, T. et al., 2017), and temperature (Lu et al., 2022) will all affect the carbonation rate and degree. Therefore, the studies on the optimization of carbonation conditions currently focused on the enhancement of carbonation product formation to improve the performance of carbonate binders and carbonated concrete products. In recent years, bacteria-induced carbonate precipitation has been considered an environmentally friendly method to accelerate carbonation reactions. Some microorganisms can accelerate the precipitation of calcium carbonate in the Earth's circulation system through a series of physiological activities (respiration, metabolism, reproduction, catalysis, etc.), which is a widely existing complex phenomenon (Qian et al., 2021; Zhu and Dittrich, 2016).

For the chemical methods, the addition of inorganic and organic agents can, on one hand, accelerate the dissolution of calcium silicate minerals and CO_2 by adjusting the pH value and increasing the nucleation sites, thus enhancing the carbonation rate and formation of carbonation products (Park, 2018). On the other hand, the addition of chemical agents can also modify the $CaCO_3$ polymorphs by altering the thermodynamic property and monitoring the transformation of calcium carbonate with different polymorphs. It was reported that the addition of a chelating agent with a low Ca-Ligand stability constant could accelerate the separation of Ca from the complex and the transformation of vaterite to calcite (Kim and Jeon, 2020). Besides, formic acid, acetic acid, and lactic acid have also been applied to accelerate the Ca leaching from minerals (Ghoorah et al., 2014). Alkaline reagents (e.g., sodium hydroxide (NaOH), ammonium hydroxide (NH_4OH)) are then introduced to the leachate to elevate the pH value and make it suitable for CO_2 absorption and $CaCO_3$/$MgCO_3$ precipitation (Naraharisetti et al., 2019). To increase the fracture toughness of carbonate binders, studies have been conducted to produce more fibrous aragonite instead of calcite in carbonate binders (Lu et al., 2021). The divalent cations (such as Mg^{2+}, Sr^{2+}, Fe^{2+}, Zn^{2+}, and Cu^{2+}) in some inorganic agents can be absorbed by calcite and increase the Gibbs free energy, thus lowering the stability of calcite (Wada et al., 1995). The addition of $MgCl_2$ agents to facilitate the formation of aragonite has been extensively investigated (Lee et al., 2015; Meiron et al., 2011). Organic agents, mainly amines such as polyacrylic acid (PAA), polyacrylamide (PAM), polyvinyl alcohol (PVA), and monoethanolamine (MEA), could also adsorb on the surface of $CaCO_3$ particles to change the interfacial energy and critical nuclear energy, and thus tune the morphology of the newly formed $CaCO_3$ (Ouhenia et al., 2008).

4.5.2 Amorphous silica gel

In most cases, only 20–50% calcium silicates in specimens can be carbonated. Therefore, amorphous silica or silica-rich calcium silicate hydrates (C-S-H) can usually be detected (Ashraf, Warda and Olek, Jan, 2018). The silica-rich C-S-H was formed due to the incomplete decalcification of original C-S-H from the normal hydration of the cement. It is similar to the original C-S-H gel but will eventually transform to silica gel after progressive carbonation according to SEM-EDX examination (J.F. Young et al., 1974). Calcium silicates with a high Ca–Si ratio generate a silica-rich C-S-H gel with higher Ca content after carbonation. However, the Q3(OH)/Q4 ratio of the carbonated products was nearly the same for all the carbonated calcium silicate. Moreover, the calculated Q3/Q4 content ratio indicated that C_3S showed the highest value, which indicated the lowest polymerization of silica gel. Similar results were also found in other references (Ashraf and Olek, 2016; Lu et al., 2018). γ-C_2S can be more easily carbonated compared to β-C_2S, and the silica-rich phase formed for the former had higher polymerization compared to that of the latter (Fang et al., 2020). Lu et al. (2022) found that high temperature increased the polymerization of silica gel. During further water curing, $Ca(OH)_2$ was not detected, which might be due to the reaction between $Ca(OH)_2$ and silica gel (He et al., 2016; Lu et al., 2023; Wang and Gillott, 1991). C-S-H gel also formed during further hydration after the carbonation and showed lower polymerization compared to silica gel (Khoshnazar and Shao, 2018).

After the carbonation curing, the anhydrous minerals within Portland cement can hydrate, and the carbonation products participate in the hydration reactions during further water curing. The presence of calcium carbonate can accelerate the hydration of calcium silicates (Kakali et al., 2000; Péra et al., 1999). Some researchers found the presence of calcium carbonate accelerated the transformation of ettringite to monosulfate (Ramachandra and Chun-Mei, 1986; Ramachandran, 1988), while others observed an opposite phenomenon (Kakali et al., 2000). Monocarboaluminate ($3CaO{\cdot}Al_2O_3{\cdot}CaCO_3{\cdot}11H_2O$) (Mc) and hemicarboaluminate (Hc) were detected after further water curing according to XRD examination (He et al., 2016; Lu et al., 2023; Wang et al., 2021) due to the hydration of C_3A with the presence of calcite (Matschei et al., 2007). Hc was unstable and converted to Mc within 24 hours. The morphology of Mc changed from early thick flake to long rod shape and fine-needle shape after 28 days of hydration (Xiao et al., 2010). John Plank et al. (2016) studied the stability of single phase C_3A hydrates under pressurized CO_2 conditions (7 bar, 20–50°C). During the carbonation, the ettringite was first carbonated to form gypsum, CaCO3, and $Al(OH)_3$, followed by C-A-H gels. After 15 days of carbonation curing, monosulfoaluminate (AFm) started to be carbonated and transformed into $[Ca_2Al(OH_6)]_2(SO_4)_{1-x}(CO_3)_x$-LDH. The stability of Katoite was the best and remained unchanged even after 30 days of carbonation curing.

4.6 PERFORMANCE OF CARBON MINERALIZED CALCIUM SILICATE MATERIALS

Extensive research (El-Hassan and Shao, 2014; Li and Wu, 2022; Qian et al., 2018) has been conducted on the carbon mineralization of different calcareous materials, especially with the Portland cement system. Due to the carbonatable phases contained in Portland cement, the production of cement concrete products by carbonation has been widely investigated and already applied. Shi et al. (2012a, 2012b; Shi and Wu, 2008) systematically studied the effect of carbonation curing on the performance of lightweight cement concrete blocks. The results showed that CO_2-cured blocks demonstrated lower drying shrinkage and water absorption than steam-cured ones. Pervious concrete is regarded as a better recipient for CO_2 curing due to its highly interconnected void content (Chen, Tiefeng and Gao, Xiaojian, 2020; Lei et al., 2022). It also shows deeper and more homogeneous carbonation depth than conventional concrete. Moreover, CO_2 curing could significantly improve its freeze-thaw and leaching resistance.

The microscopic morphology of the hydrated and carbonated calcium silicate minerals is shown in Figure 4.12. In carbonated systems, polymerized silica gel and $CaCO_3$ were identified as the main reaction products. The carbonated Portland cement showed lower porosity due to the continuous formation of calcium carbonate and silica gel, which have a much larger volume compared to raw calcium silicates (He et al., 2016; Jiang, L. et al., 2023; Monkman et al., 2018; Su et al., 2022). Both the carbonation and hydration decreased the porosity of cement (Xian and Shao, 2021; Zhu et al., 2022). It was found that the carbonation reaction decreased the percentage of large capillary pores, especially at 0.1–4 μm (Kashef-Haghighi and Ghoshal, 2013; Zhan et al., 2018), while hydration decreased the amount of capillary pores (Chen and Gao, 2019). The connectivity of pores in cement specimens also decreased, and the paste with a higher w/c ratio of 0.5 showed more pore closure (Jang and Lee, 2016; Siddique et al., 2020). Carbonation curing could also decrease the porosity of cement paste within the interfacial transitional zone (ITZ) (Chen, Tiefeng and Gao, Xiaojian 2020). The increase in porosity of concrete subjected to longer CO_2 curing time was also reported in Junior et al. (2015). The reason behind this phenomenon is still not clear. Subsequent water curing could further decrease the porosity of the product (He et al., 2016) and the most probable diameters of pores (Lu et al., 2021).

Carbonated calcium silicates can usually demonstrate very high compressive strength within several hours, especially with the pre-conditioning process before carbonation. Carbonated low calcium clinker after a couple of days of carbonation can even demonstrate compressive strength higher than ordinary Portland cement after 28 days of standard curing (Liu et al., 2021). In a recent study (Wang et al., 2019), the compressive strength of the carbonated β-C_2S specimen reached as high as 120 MPa when the carbonation time was prolonged to 168 hours. A compressive strength of 68 MPa and carbonation degree of 60% were also obtained when the CO_2 pressure was increased to 2 MPa for CO_2-cured γ-C_2S (Bukowski, J.M. and Berger, R.l., 1979). Guan et al. (2016; LIU Songhui, 2016) found that γ-C_2S mortar could be

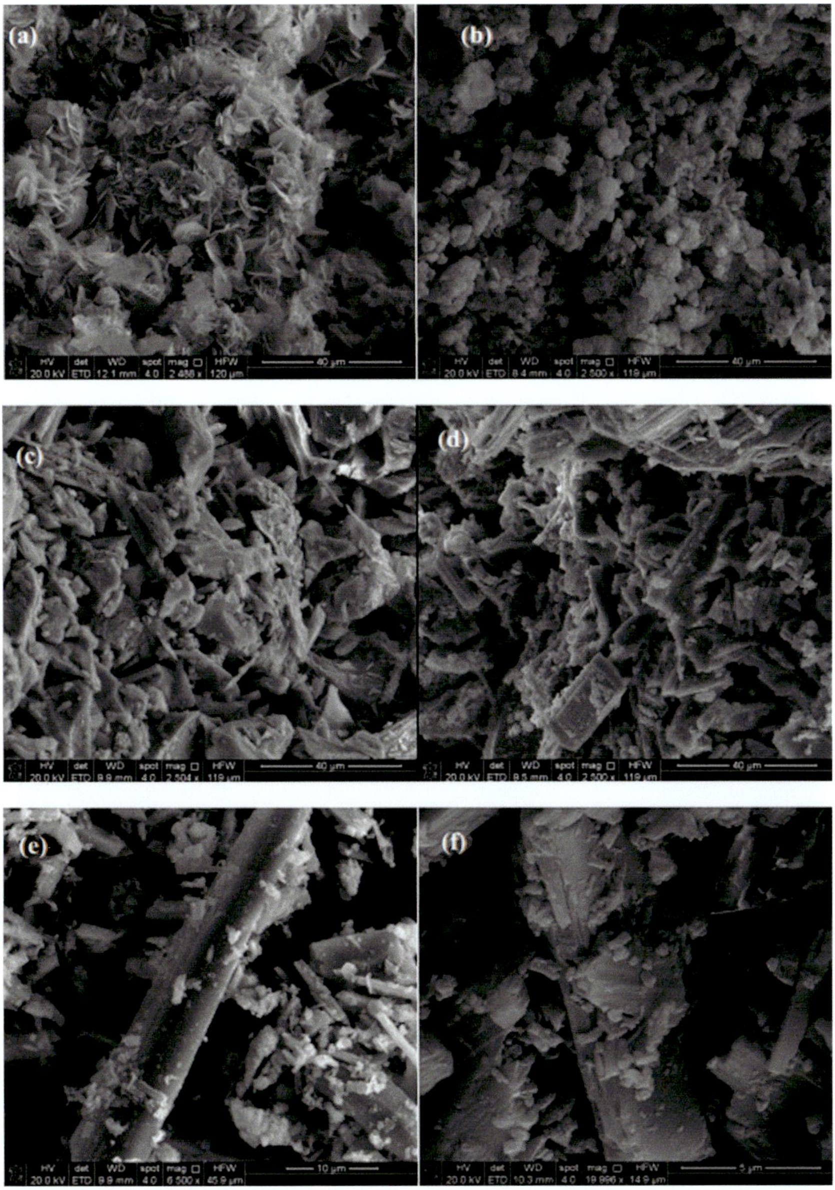

FIGURE 4.12 Secondary images of: (a) hydrated C_3S, (b) carbonated C_3S, (c) hydrated γ-C_2S, (d) carbonated γ-C_2S, (e) hydrated wollastonite, and (f) carbonated wollastonite after 24 hours of corresponding reactions (Ashraf et al., 2015).

completely hardened in 24 hours with a compressive strength as high as 61.82 MPa. Chang et al. (2016) found that the compressive strength of β-C_2S and γ-C_2S after carbonation for 2 hours at 0.2 MPa was 77.3 MPa and 30.4 MPa, respectively.

Compacted C_3S_2 specimens could reach 10.2 MPa and 21.3 MPa compressive strengths after carbonation at 0.5 MPa CO_2 for 3 hours and 12 hours (Zhang et al., 2018). Lu et al. (Hou et al., 2020; LU Bao, 2016) found that with the increase in the water-to-binder ratio, the carbonation weight gain rate and compressive strength of C_3S_2 specimens exhibited a pattern of change first increasing and then decreasing. The optimal water-to-binder ratio for carbon mineralization of compact C_3S_2 specimens under the condition of 5 MPa was 0.14, which led to a weight gain of 13.5% and a compressive strength of 33.6 MPa after 3 days carbonation. The carbonation activity of CS was lower than that of other calcium silicate minerals, leading to lower compressive strength after carbonation. Mu et al. (2019) found that the compressive strength of β-CS carbonized for 24 hours at a CO_2 pressure of 0.5 MPa was less than 20 MPa.

In general, a higher carbonation degree results in a denser and stronger calcium silicate carbonate mineral matrix. The compressive strength of CO_2-cured Portland cement-based products tends to be proportional to the carbonation degree, as shown in Figure 4.13. It can be seen that the carbonation degree of Portland cement is less than 40%. Meanwhile, if the blocks are kept in a moist environment after carbonation curing, these uncarbonated clinker minerals will hydrate and contribute to the continued strength increase with time.

Figure 4.14 presents the frequency distribution of the elastic modulus values obtained from nanoindentation experiments carried out on the carbonated calcium silicate pastes (Ashraf, Warda, and Olek, Jan 2018). The microstructural phases in the carbonated calcium silicate paste include calcium-modified silica gel, calcium carbonate, and unreacted particles. Among all the carbonated calcium silicate

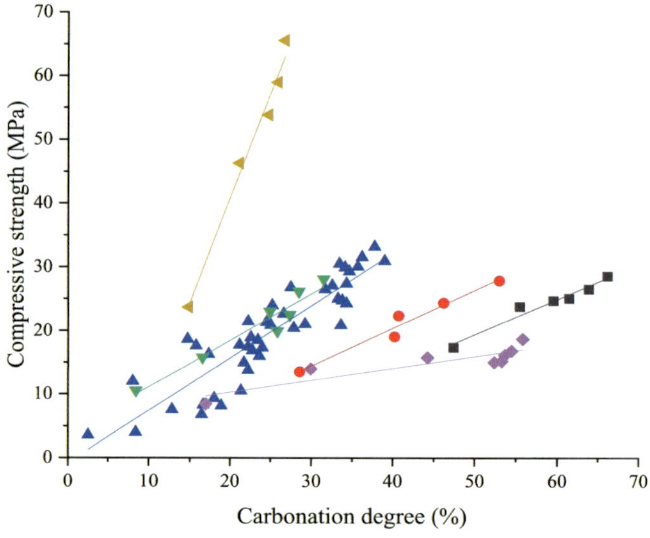

FIGURE 4.13 Relationship between compressive strength and carbonation degree of CO_2-cured Portland cement-based products. Data from reference (Lu et al., 2018; Xuan et al., 2018; Zhan et al., 2016a; Zhan et al., 2016).

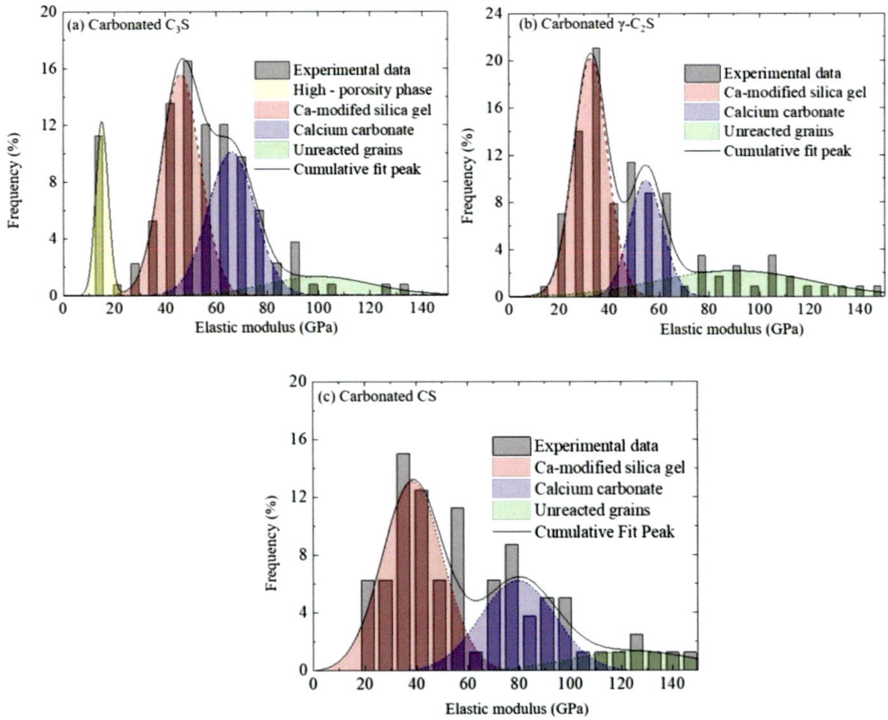

FIGURE 4.14 Frequency distributions of elastic modulus values of the microscopic phases present in carbonated calcium silicates: (a) C_3S, (b) γ-C_2S, and (c) CS (Ashraf, Warda and Olek, Jan 2018).

pastes, the average elastic modulus of calcium-modified silica gel ranges from 32 GPa to 42 GPa. By contrast, the average elastic modulus of calcium carbonate displays a broader span (54–79 GPa), which hinges on the particular type of calcium silicate phase present in the initial phase of carbonation. Calcium carbonate formed during the carbon mineralization of C_3S and γ-C_2S phases contains approximately 20% of metastable $CaCO_3$ polymorphs, including aragonite, vaterite, and amorphous calcium carbonate (ACC). The assortment of calcium carbonate types is responsible for the more substantial alterations in the elastic modulus of the calcium carbonate phase. Additionally, as the carbonation degree increases, it can augment the elastic modulus of the resultant silica gel, owing to an elevated polymerization degree (Ashraf et al., 2016). In contrast, the elastic modulus of calcium carbonate exhibits only marginal alterations.

4.7 SUMMARY

(1) Irrespective of hydraulic reactivity, calcium silicate minerals exhibit a high level of carbonation reactivity. Among them, the reactivity of C_3S_2 and CS is lower compared to that of C_3S and C_2S. The simultaneous occurrence

of hydration and carbonation reactions potentially reduces the carbonation activation energy for C_3S and β-C_2S. As the matrix gradually densifies, the carbonation reaction gradually shifts from being controlled by the phase boundary to being jointly controlled by the diffusion of CO_2 and the dissolution of calcium ions.

(2) The carbon mineralization of calcium silicate mineral is a dissolution-ion diffusion-precipitation process. For hydraulic minerals, the early-stage carbonation is also influenced by the hydration reaction that initially forms C-S-H gel and $Ca(OH)_2$, which then react with CO_2 to produce calcium carbonate. In contrast, non-hydraulic minerals directly react with CO_2 to form calcium carbonate.

(3) The carbonation degree of calcium silicate minerals is affected by multiple factors, including carbonation conditions, admixtures/chemical additives, and mineral structure. Augmenting the CO_2 concentration and pressure enhances the carbonation rate; however, such an enhancement becomes unremarkable once they exceed a certain threshold. There exists an optimal RH, residual w/b, and temperature under which calcium silicate minerals can achieve the highest carbonation degree. Enhancing the dissolution of CO_2 and facilitating the leaching of Ca^{2+} are pivotal determinants in augmenting the carbonation degree of cement-based materials.

(4) Calcium carbonate and silica gel are the primary products in carbonated calcium silicate minerals. The morphology and distribution of calcium carbonate vary depending on the conditions of the carbonation reaction, and these variations significantly impact the properties of carbon mineralized products. Precisely controlling the polymorphs of calcium carbonate and optimizing the polymerization degree of silica gel are undoubtedly crucial breakthrough directions for the development of high-performance carbon mineralized binders.

(5) Calcium silicate minerals and Portland cement-based materials exhibit remarkable mechanical properties merely within a few hours of CO_2 curing. Calcium carbonate and silica gel reduce porosity and boost the structural integrity, thereby improving the performance of carbon mineralized materials. The compressive strength of CO_2-cured Portland cement-based products demonstrates a strong propensity to be directly proportional to the carbonation degree.

REFERENCES

Abdullahi, M., Odigure, J., Kovo, A., Abdulkareem, A., 2016. Characterization and predictive reaction model for cement-sand-kaolin composite for CO_2 sequestration. *Journal of CO2 Utilization* 16, 169–181.

Arti, M., Youn, M.H., Park, K.T., Kim, H.J., Kim, Y.E., Jeong, S.K., 2017. Single process for CO_2 capture and mineralization in various alkanolamines using calcium chloride. *Energy & Fuels* 31(1), 763–769.

Ashraf, W., Olek, J., 2016. Carbonation behavior of hydraulic and non-hydraulic calcium silicates: potential of utilizing low-lime calcium silicates in cement-based materials. *Journal of Materials Science* 51(13), 6173–6191.

Ashraf, W., Olek, J., 2018a. Carbonation activated binders from pure calcium silicates: reaction kinetics and performance controlling factors. *Cement and Concrete Composites* 93, 85–98.

Ashraf, W., Olek, J., 2018b. Elucidating the accelerated carbonation products of calcium silicates using multi-technique approach. *Journal of CO2 Utilization* 23, 61–74.

Ashraf, W., Olek, J., Atakan, V., 2015. *A comparative study of the reactivity of calcium silicates during hydration and carbonation reactions.* 14th International Congress on Cement Chemistry, Beijing, China. pp. 13–16.

Ashraf, W., Olek, J., Sahu, S., 2019. Phase evolution and strength development during carbon mineralization of low-lime calcium silicate cement (CSC). *Construction and Building Materials* 210, 473–482.

Ashraf, W., Olek, J., Tian, N., 2016. Multiscale characterization of carbonated wollastonite paste and application of homogenization schemes to predict its effective elastic modulus. *Cement and Concrete Composites* 72, 284–298.

Berger, R.L., Klemm, W., 1972. Accelerated curing of cementitious systems by carbon dioxide: Part II. Hydraulic calcium silicates and aluminates. *Cement and Concrete Research* 2(6), 647–652.

Bukowski, J., Berger, R.L., 1979. Reactivity and strength development of CO2 activated non-hydraulic calcium silicates. *Cement and Concrete Research* 9(1), 57–68.

Bukowski, J.M., Berger, R.l., 1979. Reactivity and strength development of CO_2 activated non-hydraulic calcium silicates. *Cement and Concrete Research* 9, 57–68.

Carroll, J.J., Slupsky, J.D., Mather, A.E., 1991. The solubility of carbon dioxide in water at low pressure. *Journal of Physical and Chemical Reference Data* 20(6), 1201–1209.

Castellote, M., Fernandez, L., Andrade, C., Alonso, C., 2009. Chemical changes and phase analysis of OPC pastes carbonated at different CO_2 concentrations. *Materials and Structures* 42(4), 515–525.

Chang, J., Fang, Y., Shang, X., 2016. The role of β-C2S and γ-C2S in carbon capture and strength development. *Materials and Structures* 49(10), 4417–4424.

Chang, R., Choi, D., Kim, M.H., Park, Y., 2017. Tuning crystal polymorphisms and structural investigation of precipitated calcium carbonates for CO_2 mineralization. *ACS Sustainable Chemistry & Engineering* 5(2), 1659–1667.

Chen, T., Gao, X., 2019. Effect of carbonation curing regime on strength and microstructure of Portland cement paste. *Journal of CO2 Utilization* 34, 74–86.

Chen, J., 2018. Study on the composition and properties of C_3S_2 low calcium clinker [D]. *Jiangsu University.*

Chen, T., Gao, X., 2020. Use of carbonation curing to improve mechanical strength and durability of pervious concrete. *ACS Sustainable Chemistry & Engineering* 8(9), 3872–3884.

Chen, T., Xu, P., Gao, X., Wang, T., Qin, L., 2022a. New sights in early carbon mineralization of calcium silicates: performance, mechanism and nanostructure. *Construction and Building Materials* 314, 125622.

Chen, T., Zhao, L., Gao, X., Li, L., Qin, L., 2022b. Modification of carbonation-cured cement mortar using biochar and its environmental evaluation. *Cement and Concrete Composites* 134.

De Silva, P., Bucea, L., Moorehead, D.R., Sirivivatnanon, V., 2006. Carbonate binders: reaction kinetics, strength and microstructure. *Cement and Concrete Research* 28(7), 613–620.

Dhami, N.K., Mukherjee, A., Reddy, M.S., 2016. Micrographical, minerological and nanomechanical characterisation of microbial carbonates from urease and carbonic anhydrase producing bacteria. *Ecological Engineering* 94, 443–454.

Drouet, E., Poyet, S., Le Bescop, P., Torrenti, J.-M., Bourbon, X., 2019. Carbon mineralization of hardened cement pastes: influence of temperature. *Cement and Concrete Research* 115, 445–459.

El-Hassan, H., Shao, Y., 2014. Carbon storage through concrete block carbonation curing. *Journal of Clean Energy Technologies* 2(3), 287–291.

El-Hassan, H., Shao, Y., 2015. Early carbonation curing of concrete masonry units with Portland limestone cement. *Cement and Concrete Composites* 62, 168–177.

El-Hassan, H., Shao, Y., Ghouleh, Z., 2013. Reaction products in carbonation-cured lightweight concrete. *Journal of Materials in Civil Engineering* 25(6), 799–809.

Elsalamawy, M., Mohamed, A.R., Kamal, E.M., 2019. The role of relative humidity and cement type on carbonation resistance of concrete. *Alexandria Engineering Journal* 58(4), 1257–1264.

Fang, Y., Chang, J., Cao, M., 2016. Influence of compaction pressure on the accelerated carbon mineralization of calcium hydroxide. *Journal of Wuhan University of Technology – Materials Science Edition* 31(6), 1187–1192.

Fang, Y., Liu, Z., Wang, Q., Zhang, Y., Zhang, M., Wang, Q., 2020. Strength development and products evolution of β-C_2S and γ-C_3S induced by accelerated carbonation curing. *Journal of Wuhan University of Technology – Materials Science Edition* 35(6), 1053–1060.

Galan, I., Andrade, C., Castellote, M., 2013. Natural and accelerated CO_2 binding kinetics in cement paste at different relative humidities. *Cement and Concrete Research* 49, 21–28.

Ghoorah, M., Dlugogorski, B.Z., Balucan, R.D., Kennedy, E.M., 2014. Selection of acid for weak acid processing of wollastonite for mineralisation of CO2. *Fuel* 122, 277–286.

Goodwin, A.L., Michel, F.M., Phillips, B.L., Keen, D.A., Dove, M.T., Reeder, R.J., 2010. Nanoporous structure and medium-range order in synthetic amorphous calcium carbonate. *Chemistry of Materials* 22(10), 3197–3205.

Guan, X., Liu, S., Feng, C., Qiu, M., 2016. The hardening behavior of γ-C2S binder using accelerated carbonation. *Construction and Building Materials* 114, 204–207.

Gupta, S., Kashani, A., Mahmood, A.H., Han, T., 2021. Carbon sequestration in cementitious composites using biochar and fly ash–Effect on mechanical and durability properties. *Construction and Building Materials* 291, 123363.

Hameed, R., Seo, J., Park, S., Amr, I.T., Lee, H., 2020. CO_2 uptake and physicochemical properties of carbonation-cured ternary blend portland cement–metakaolin–limestone pastes. *Materials* 13(20), 4656.

Han, S.H., Jun, Y., Kim, J.H., 2023. The use of monoethanolamine CO2 solvent for the CO2 curing of cement-based materials. *Sustainable Materials and Technologies* 35, e00559.

Harrison, A.L., Power, I.M., Dipple, G.M., 2013. Accelerated carbon mineralization of brucite in mine tailings for carbon sequestration. *Environmental Science & Technology* 47(1), 126–134.

He, P., Shi, C., Tu, Z., Poon, C.S., Zhang, J., 2016. Effect of further water curing on compressive strength and microstructure of CO_2-cured concrete. *Cement and Concrete Composites* 72, 80–88.

Hou, G., Chen, J., Lu, B., Chen, S., Cui, E., Naguib, H.M., Guo, M.-Z., Zhang, Q., 2020. Composition design and pilot study of an advanced energy-saving and low-carbon rankinite clinker. *Cement and Concrete Research* 127, 105926.

Hu, L., Jia, Y., Chen, Z., Yao, Y., Sun, J., Xie, Q., Yang, H., 2022a. An insight of carbonation-hydration kinetics and microstructure characterization of cement paste under accelerated carbonation at early age. *Cement and Concrete Composites* 134, 104763.

Hu, L., Yang, H., He, Z., Chen, Z., Hu, J., Zhang, S., 2022b. Effect of rice husk ash on carbon sequestration, mechanical property and microstructure evolution of cement-based materials with early-age carbonation treatment. *Cement and Concrete Composites* 133, 104672.

Humbert, P., Castro-Gomes, J., Savastano Jr, H., 2019. Clinker-free CO_2 cured steel slag based binder: optimal conditions and potential applications. *Construction and Building Materials* 210, 413–421.

Ibáñez, J., Artús, L., Cuscó, R., López, Á., Menéndez, E., Andrade, M.C., 2007. Hydration and carbon mineralization of monoclinic C2S and C3S studied by Raman spectroscopy. *Journal of Raman Spectroscopy: An International Journal for Original Work in all Aspects of Raman Spectroscopy, Including Higher Order Processes, and also Brillouin and Rayleigh Scattering* 38(1), 61–67.

Jang, J.G., Lee, H.-K., 2016. Microstructural densification and CO_2 uptake promoted by the carbonation curing of belite-rich Portland cement. *Cement and Concrete Research* 82, 50–57.

Jiang, L., Wu, Q., Huo, Z., Zhu, Z., Wu, F., Lu, B., 2023. An approach to improve compressive strength of cement paste at low temperature by carbonation curing. *Construction and Building Materials* 365, 130128.

Jiang, T., Cui, K., Chang, J., 2023. Development of low-carbon cement: carbon mineralization of compounded C2S by β-C2S and γ-C2S. *Cement and Concrete Composites* 139, 105071.

Jin, P., Wang, R., Su, Y., Dong, H., Wang, Q., 2019. Study on carbonation process of β-C2S under microbial enzymatic action. *Construction and Building Materials* 228, 117110.

Joshaghani, A., Moeini, M.A., Balapour, M., Moazenian, A., 2018. Effects of supplementary cementitious materials on mechanical and durability properties of high-performance non-shrinking grout (HPNSG). *Journal of Sustainable Cement-Based Materials* 7(1), 38–56.

Junior, A.N., Toledo Filho, R.D., Fairbairn, E.d.M.R., Dweck, J., 2015. The effects of the early carbonation curing on the mechanical and porosity properties of high initial strength Portland cement pastes. *Construction and Building Materials* 77, 448–454.

Kakali, G., Tsivilis, S., Aggeli, E., Bati, M., 2000. Hydration products of C_3A, C_3S and Portland cement in the presence of $CaCO_3$. *Cement and Concrete Research* 30(7), 1073–1077.

Kashef-Haghighi, S., Ghoshal, S., 2013. Physico–chemical processes limiting CO_2 uptake in concrete during accelerated carbonation curing. *Industrial & Engineering Chemistry Research* 52(16), 5529–5537.

Kashef-Haghighi, S., Shao, Y., Ghoshal, S., 2015. Mathematical modeling of CO2 uptake by concrete during accelerated carbonation curing. *Cement and Concrete Research* 67, 1–10.

Kashim, M.Z., Tsegab, H., Rahmani, O., Abu Bakar, Z.A., Aminpour, S.M., 2020. Reaction mechanism of wollastonite in situ mineral carbonation for CO_2 sequestration: effects of saline conditions, temperature, and pressure. *ACS Omega* 5(45), 28942–28954.

Khoshnazar, R., Shao, Y., 2018. Characterization of carbonation-cured cement paste using X-ray photoelectron spectroscopy. *Construction and Building Materials* 168, 598–605.

Kim, H., Sharma, R., Pei, J., Jang, J.G., 2022. Effect of carbonation curing on physicochemical properties of mineral admixture blended belite-rich cement. *Journal of Building Engineering*, 104771.

Kim, M.-J., Jeon, J., 2020. Effects of Ca-ligand stability constant and chelating agent concentration on the CO2 storage using paper sludge ash and chelating agent. *Journal of CO2 Utilization* 40, 101202.

Lee, J., Hwang, J., Lee, H., Son, B., Oh, J., 2015. The effect of chloride additives and pH on direct aqueous carbon mineralization of cement paste. *Journal of the Mineralogical Society of Korea* 28(1), 39–49.

Lei, M., Deng, S., Liu, Z., Wang, F., Hu, S., 2022. Development of a sustainable CO2 solidified aerated concrete. *ACS Sustainable Chemistry & Engineering* 10(12), 3990–4001.

Li, L., Wu, M., 2022. An overview of utilizing CO_2 for accelerated carbonation treatment in the concrete industry. *Journal of CO2 Utilization* 60, 102000.

Li, X., Ling, T.-C., 2020. Instant CO_2 curing for dry-mix pressed cement pastes: consideration of CO_2 concentrations coupled with further water curing. *Journal of CO2 Utilization* 38, 348–354.

Li, Y., Liu, W., Xing, F., Wang, S., Tang, L., Lin, S., Dong, Z., 2020. Carbon mineralization of the synthetic calcium silicate hydrate (C-S-H) under different concentrations of CO_2: chemical phases analysis and kinetics. *Journal of CO2 Utilization* 35, 303–313.

Lin, R.-S., Wang, X.-Y., 2021. Effects of cement types and addition of quartz and limestone on the normal and carbonation curing of cement paste. *Construction and Building Materials* 305, 124799.

Lioliou, M.G., Paraskeva, C.A., Koutsoukos, P.G., Payatakes, A.C., 2007. Heterogeneous nucleation and growth of calcium carbonate on calcite and quartz. *Journal of Colloid and Interface Science* 308(2), 421–428.

Liu, J., Ruan, H., Lu, B., Chen, Z., Shi, C., 2023. A new enhanced carbonation curing method using monoethanolamine (MEA) solution: their effects on hydration and microstructure of cement-based materials. *Construction and Building Materials* 396, 132172.

Liu, L., Ha, J., Hashida, T., Teramura, S., 2001. Development of a CO_2 solidification method for recycling autoclaved lightweight concrete waste. *Journal of Materials Science Letters* 20(19), 1791–1794.

Liu, M., Gadikota, G., 2019. Integrated CO2 capture, conversion, and storage to produce calcium carbonate using an amine looping strategy. *Energy & Fuels* 33(3), 1722–1733.

Liu, P., Zhang, M., Mo, L., Zhong, J., Xu, M., Deng, M., 2022. Probe into carbonation mechanism of steel slag via FIB-TEM: the roles of various mineral phases. *Cement and Concrete Research* 162, 106991.

Liu, S., Zhang, L., Xuan, D., Shen, P., Zhu, J., Guan, X., Shi, C., 2021. Enhanced carbonation reactivity of wollastonite by rapid cooling process: towards an ultra-low calcium CO2 sequestration binder. *Construction and Building Materials* 299, 124336.

LIU Songhui, G.X., QIU Man, FENG Chunhua., 2016. Activation of γ-C_2S Mineral by Accelerated Carbonation. *Journal of the Chinese Ceramic Society* 44(05), 658–662.

Liu, Z., Van den Heede, P., Zhang, C., Shi, X., Wang, L., Li, J., Yao, Y., Lothenbach, B., De Belie, N., 2023. Carbon mineralization of blast furnace slag concrete at different CO_2 concentrations: carbonation rate, phase assemblage, microstructure and thermodynamic modelling. *Cement and Concrete Research* 169, 107161.

Lo, T., Tang, W., Nadeem, A., 2008. Comparison of carbon mineralization of lightweight concrete with normal weight concrete at similar strength levels. *Construction and Building Materials* 22(8), 1648–1655.

Lu, B., Drissi, S., Liu, J., Hu, X., Song, B., Shi, C., 2022. Effect of temperature on CO_2 curing, compressive strength and microstructure of cement paste. *Cement and Concrete Research* 157, 106827.

Lu, B., He, P., Liu, J., Peng, Z., Song, B., Hu, X., 2021. Microstructure of Portland cement paste subjected to different CO2 concentrations and further water curing. *Journal of CO2 Utilization* 53, 101714.

Lu, B., Huo, Z., Xu, Q., Hou, G., Wang, X., Liu, J., Hu, X., 2023. Characteristics of CSH under carbonation and its effects on the hydration and microstructure of cement paste. *Construction and Building Materials* 364, 129952.

Lu, B., Shi, C., Hou, G., 2018. Strength and microstructure of CO2 cured low-calcium clinker. *Construction and Building Materials* 188, 417–423.
Lu Bao, G.X., Jiang Weili, 2016. The influence of water solid ratio to the carbonization of rankinite. *Materials Reports* 30(02), 114–117+122.
Luis, P., 2016. Use of monoethanolamine (MEA) for CO2 capture in a global scenario: consequences and alternatives. *Desalination* 380, 93–99.
Maki, I., Kato, K., 1982. Phase identification of alite in Portland cement clinker. *Cement and Concrete Research* 12(1), 93–100.
Matschei, T., Lothenbach, B., Glasser, F.P., 2007. The role of calcium carbonate in cement hydration. *Cement and Concrete Research* 37(4), 551–558.
Meiron, O.E., Bar-David, E., Aflalo, E.D., Shechter, A., Stepensky, D., Berman, A., Sagi, A., 2011. Solubility and bioavailability of stabilized amorphous calcium carbonate. *Journal of Bone and Mineral Research* 26(2), 364–372.
Mo, L., Zhang, F., Deng, M., 2015. Effects of carbonation treatment on the properties of hydrated fly ash-MgO-Portland cement blends. *Construction and Building Materials* 96, 147–154.
Monkman, S., Kenward, P.A., Dipple, G., MacDonald, M., Raudsepp, M., 2018. Activation of cement hydration with carbon dioxide. *Journal of Sustainable Cement-Based Materials* 7(3), 160–181.
Moro, C., Francioso, V., Velay-Lizancos, M., 2021. Impact of nano-TiO_2 addition on the reduction of net CO_2 emissions of cement pastes after CO_2 curing. *Cement and Concrete Composites* 123, 104160.
Mu, Y., Liu, Z., Wang, F., 2019. Comparative study on the carbonation-activated calcium silicates as sustainable binders: reactivity, mechanical performance, and microstructure.*ACS Sustainable Chemistry & Engineering* 7(7), 7058–7070.
Mu, Y., Liu, Z., Wang, F., 2022. Study on heat resistance of carbonated γ-C_2S binder: strength, phase and microstructure evolution. *Construction and Building Materials* 329, 127049.
Mu, Y., Liu, Z., Wang, F., Huang, X., 2018a. Carbonation characteristics of γ-dicalcium silicate for low-carbon building material. *Construction and Building Materials* 177, 322–331.
Mu, Y., Liu, Z., Wang, F., Huang, X., 2018b. Effect of barium doping on carbonation behavior of γ-C2S. *Journal of CO2 Utilization* 27, 405–413.
Na, H., Wang, Y., Zhang, X., Li, J., Zeng, Y., Liu, P., 2021. Hydration activity and carbonation characteristics of dicalcium silicate in steel slag: a review. *Metals* 11(10), 1580.
Naraharisetti, P.K., Yeo, T.Y., Bu, J., 2019. New classification of CO2 mineralization processes and economic evaluation. *Renewable and Sustainable Energy Reviews* 99, 220–233.
Nielsen, M.H., Aloni, S., De Yoreo, J.J., 2014. In situ TEM imaging of $CaCO_3$ nucleation reveals coexistence of direct and indirect pathways. *Science* 345(6201), 1158–1162.
Ogino, T., Suzuki, T., Sawada, K., 1987. The formation and transformation mechanism of calcium carbonate in water. *Geochimica et Cosmochimica Acta* 51(10), 2757–2767.
Ouhenia, S., Chateigner, D., Belkhir, M.A., Guilmeau, E., Krauss, C., 2008. Synthesis of calcium carbonate polymorphs in the presence of polyacrylic acid. *Journal of Crystal Growth* 310(11), 2832–2841.
Papadakis, V.G., Vayenas, C.G., Fardis, M., 1989. A reaction engineering approach to the problem of concrete carbonation. *AIChE Journal* 35(10), 1639–1650.
Park, S., 2018. CO2 reduction-conversion to precipitates and morphological control through the application of the mineral carbonation mechanism. *Energy* 153, 413–421.
Péra, J., Husson, S., Guilhot, B., 1999. Influence of finely ground limestone on cement hydration. *Cement and Concrete Composites* 21(2), 99–105.
Plank, J., Zhang-Preße, M., Ivleva, N., Niessner, R., 2016. Stability of single phase C3A hydrates against pressurized CO2. *Construction and Building Materials* 122, 426–434.

Plattenberger, D.A., Ling, F.T., Tao, Z., Peters, C.A., Clarens, A.F., 2018. Calcium silicate crystal structure impacts reactivity with CO_2 and precipitate chemistry. *Environmental Science & Technology Letters* 5(9), 558–563.
Pu, Y., Li, L., Shi, X., Wang, Q., Abomohra, A., 2023. Recent advances in accelerated carbonation for improving cement-based materials and CO2 mitigation from a life cycle perspective. *Construction and Building Materials* 388, 131695.
Pu, Y., Li, L., Wang, Q., Shi, X., Fu, L., Zhang, G., Luan, C., Abomohra, A.E.-F., 2021. Accelerated carbonation treatment of recycled concrete aggregates using flue gas: a comparative study towards performance improvement. *Journal of CO2 Utilization* 43.
Qian, C., Ren, X., Rui, Y., Wang, K., 2021. Characteristics of bio-CaCO3 from microbial bio-mineralization with different bacteria species. *Biochemical Engineering Journal* 176, 108180.
Qian, X., Wang, J., Fang, Y., Wang, L., 2018. Carbon dioxide as an admixture for better performance of OPC-based concrete. *Journal of CO2 Utilization* 25, 31–38.
Ramachandra, V., Chun-Mei, Z., 1986. Hydration kinetics and microstructural development in the 3 CaO. Al_2O_3-$CaSO_4$. $2H_2O$-$CaCO_3$-H_2O system. *Materials and Structures* 19(6), 437–444.
Ramachandran, V.S., 1988. Thermal analyses of cement components hydrated in the presence of calcium carbonate. *Thermochimica Acta* 127, 385–394.
Ramezani, R., Mazinani, S., Di Felice, R., Darvishmanesh, S., Van der Bruggen, B., 2017. Selection of blended absorbents for CO2 capture from flue gas: CO2 solubility, corrosion and absorption rate. *International Journal of Greenhouse Gas Control* 62, 61–68.
Rivera, R.M., Van Gerven, T., 2020. Production of calcium carbonate with different morphology by simultaneous CO2 capture and mineralisation. *Journal of CO2 Utilization* 41, 101241.
Rochelle, G., Chen, E., Freeman, S., Van Wagener, D., Xu, Q., Voice, A., 2011. Aqueous piperazine as the new standard for CO_2 capture technology. *Chemical Engineering Journal* 171(3), 725–733.
Rodriguez-Blanco, J., Shaw, S., Bots, P., Roncal-Herrero, T., Benning, L.G., 2012. The role of pH and Mg on the stability and crystallization of amorphous calcium carbonate. *Journal of Alloys Compounds* 536, S477–S479.
Rodriguez-Blanco, J.D., Shaw, S., Benning, L.G., 2011. The kinetics and mechanisms of amorphous calcium carbonate (ACC) crystallization to calcite, via vaterite. *Nanoscale* 3(1), 265–271.
Rodriguez-Navarro, C., Kudłacz, K., Cizer, Ö., Ruiz-Agudo, E., 2015. Formation of amorphous calcium carbonate and its transformation into mesostructured calcite. *CrystEngComm* 17(1), 58–72.
Saito, T., Sakai, E., Morioka, M., Otsuki, N., 2010. Carbon mineralization of γ-Ca2SiO4 and the Mechanism of Vaterite Formation. *Journal of Advanced Concrete Technology* 8(3), 273–280.
Seo, J.H., Amr, I.T., Park, S.M., Bamagain, R.A., Fadhel, B.A., Kim, G.M., Hunaidy, A.S., Lee, H.K., 2018. CO2 uptake of carbonation-cured cement blended with ground volcanic ash. *Materials* 11(11), 2187.
Seryotkin, Y.V., Sokol, E.V., Kokh, S.N., 2012. Natural pseudowollastonite: crystal structure, associated minerals, and geological context. *Lithos* 134, 75–90.
Sevelsted, T.F., Skibsted, J., 2015. Carbon mineralization of C–S–H and C–A–S–H samples studied by 13 C, 27 Al and 29 Si MAS NMR spectroscopy. *Cement and Concrete Research* 71, 56–65.
Shi, C., He, F., Wu, Y., 2012a. Effect of pre-conditioning on CO_2 curing of lightweight concrete blocks mixtures. *Construction and Building Materials* 26(1), 257–267.

Shi, C., He, P., Tu, Z., & Cao, Z., 2014. Effect of preconditioning on process and microstructure of carbon dioxide cured concrete. *Journal of the Chinese Ceramic Society* 42(8), 996–1004.

Shi, C., Wang, D., He, F., Liu, M., 2012b. Weathering properties of CO_2-cured concrete blocks. *Resources, Conservation and Recycling* 65, 11–17.

Shi, C., Wu, Y., 2008. Studies on some factors affecting CO2 curing of lightweight concrete products. *Resources, Conservation and Recycling* 52(8–9), 1087–1092.

Shtepenko, O., Hills, C., Brough, A., Thomas, M., 2006. The effect of carbon dioxide on β-dicalcium silicate and Portland cement. *Chemical Engineering Journal* 118(1–2), 107–118.

Siddique, S., Naqi, A., Jang, J.G., 2020. Influence of water to cement ratio on CO_2 uptake capacity of belite-rich cement upon exposure to carbonation curing. *Cement and Concrete Composites* 111, 103616.

Smigelskyte, A., Siauciunas, R., 2019. Parameter influence on the rankinite binder paste and mortar accelerated carbonation curing. *Journal of Thermal Analysis and Calorimetry* 138, 2651–2659.

Smigelskyte, A., Siauciunas, R., Hilbig, H., Decker, M., Urbonas, L., Skripkiunas, G., 2020. Carbonated rankinite binder: effect of curing parameters on microstructure, strength development and durability performance. *Scientific Reports* 10(1), 1–13.

Su, A., Chen, T., Gao, X., Li, Q., Qin, L., 2022. Effect of carbonation curing on durability of cement mortar incorporating carbonated fly ash subjected to Freeze-Thaw and sulfate attack. *Construction and Building Materials* 341, 127920.

Svensson, K., Neumann, A., Feitosa Menezes, F., Lempp, C., Pöllmann, H., 2019. Carbon mineralization of natural wollastonite at non-ambient conditions relevant for CCS—the possible use as cementitious material in wellbores. *Applied Sciences* 9(6), 1259.

Svensson, K., Neumann, A., Menezes, F.F., Lempp, C., Pöllmann, H., 2018. The conversion of wollastonite to CaCO3 considering its use for CCS application as cementitious material. *Applied Sciences* 8(2), 304.

Tan, Y., Liu, Z., Wang, F., 2022. Effect of temperature on the carbonation behavior of γ-C_2S compacts. *Cement and Concrete Composites* 133, 104652.

Tlili, M., Amor, M.B., Gabrielli, C., Joiret, S., Maurin, G., Rousseau, P., 2002. Characterization of CaCO3 hydrates by micro-Raman spectroscopy. *Journal of Raman Spectroscopy* 33(1), 10–16.

Tu, Z., Guo M.-z., Poon, C.S., Shi, C., 2016. Effects of limestone powder on $CaCO_3$ precipitation in CO_2 cured cement pastes. *Cement and Concrete Composites* 72, 9–16.

Tu, Z., Shi, C., Farzadnia, N., 2018. Effect of limestone powder content on the early-age properties of CO $_2$-cured concrete. *Journal of Materials in Civil Engineering* 30(8), 04018164.

Vanoutrive, H., Minne, P., Van de Voorde, I., Cizer, Ö., Gruyaert, E., 2022. Carbon mineralization of cement paste with GGBFS: effect of curing duration, replacement level and CO2 concentration on the reaction products and CO2 buffer capacity. *Cement and Concrete Composites* 129, 104449.

Vinoba, M., Bhagiyalakshmi, M., Choi, S.Y., Park, K.T., Kim, H.J., Jeong, S.K., 2014. Harvesting $CaCO_3$ Polymorphs from In Situ CO_2 Capture Process. *The Journal of Physical Chemistry C* 118(31), 17556–17566.

Visser, J.H.M., 2014. Influence of the carbon dioxide concentration on the resistance to carbon mineralization of concrete. *Construction and Building Materials* 67, 8–13.

Wada, N., Yamashita, K., Umegaki, T., 1995. Effects of divalent cations upon nucleation, growth and transformation of calcium carbonate polymorphs under conditions of double diffusion. *Journal of Crystal Growth* 148(3), 297–304.

Wang, D., Fang, Y., Zhang, Y., Chang, J.J.J.o.C.U., 2019. Changes in mineral composition, growth of calcite crystal, and promotion of physico-chemical properties induced by carbon mineralization of β-C_2S. *Journal of CO2 Utilization* 34, 149–162.
Wang, D., Shi, C., Farzadnia, N., Shi, Z., Jia, H., Ou, Z., 2018. A review on use of limestone powder in cement-based materials: mechanism, hydration and microstructures. *Construction and Building Materials* 181, 659–672.
Wang, D., Xiao, J., Duan, Z., 2022. Strategies to accelerate CO_2 sequestration of cement-based materials and their application prospects. *Construction and Building Materials* 314.
Wang, D., Xiong, C., Li, W., Chang, J., 2020. Growth of Calcium Carbonate Induced by Accelerated Carbon mineralization of Tricalcium Silicate. *ACS Sustainable Chemistry & Engineering* 8(39), 14718–14731.
Wang, H., Gillott, J., 1991. Mechanism of alkali-silica reaction and the significance of calcium hydroxide. *Cement and Concrete Research* 21(4), 647–654.
Wang, K., Ren, L., Yang, L., 2018. Excellent carbonation behavior of rankinite prepared by calcining the C-S-H: potential recycling of waste concrete powders for prefabricated building products. *Materials (Basel)* 11(8).
Wang, Q.Q., Li, X.D., Shen, X.D., 2017. Crystal structures of silicate minerals. *Journal of Nanjing Tech University* 39(01), 39–45.
Wang, T., Huang, H., Hu, X., Fang, M., Luo, Z., Guo, R., 2017. Accelerated mineral carbonation curing of cement paste for CO_2 sequestration and enhanced properties of blended calcium silicate. *Chemical Engineering Journal* 323, 320–329.
Wang, Y.-Y., Yao, Q.-Z., Li, H., Zhou, G.-T., Sheng, Y.-M., 2015. Formation of vaterite mesocrystals in biomineral-like structures and implication for biomineralization. *Cement and Concrete Composites* 15(4), 1714–1725.
Wang, Y., Lu, B., Hu, X., Liu, J., Zhang, Z., Pan, X., Xie, Z., Chang, J., Zhang, T., Nehdi, M.L., Shi, C., 2021. Effect of CO2 surface treatment on penetrability and microstructure of cement-fly ash–slag ternary concrete. *Cement and Concrete Composites* 123.
Wen, J., Ning, L., Duan, C.-K., Zhan, S., Huang, Y., Zhang, J., Yin, M., 2015. First-principles study on structural, electronic, and spectroscopic properties of γ-Ca2SiO4: Ce3+ phosphors. *The Journal of Physical Chemistry A* 119(29), 8031–8039.
Xian, X., Shao, Y., 2021. Carbonation curing of concretes in a flexible enclosure under ambient pressure. *Journal of Materials in Civil Engineering* 33(4), 04021025.
Xiang, L., Xiang, Y., Wang, Z.G., Jin, Y., 2002. Influence of chemical additives on the formation of super-fine calcium carbonate. *Powder Technology* 126(2), 129–133.
Xiao, J., Gou, C.-f., Jin, Y.-g., Wang, Y.-h., 2010. Effect of $CaCO_3$ on hydration characteristics of C_3A. *Journal of Central South University of Technology* 17(5), 918–923.
Xu, A.W., Antonietti, M., Cölfen, H., Fang, Y.P., 2006. Uniform hexagonal plates of vaterite CaCO3 mesocrystals formed by biomimetic mineralization. *Advanced Functional Materials* 16(7), 903–908.
Xuan, D., Zhan, B., Poon, C.S., 2018. A maturity approach to estimate compressive strength development of CO 2 -cured concrete blocks. *Cement and Concrete Composites* 85, 153–160.
Yahui, Y., Gang, X., Bin, T., 2021. Carbonation characteristics of cement-based materials under the uniform distribution of pore water. *Construction and Building Materials* 275, 121450.
Yang, X., Wang, X.-Y., 2021. Strength and durability improvements of biochar-blended mortar or paste using accelerated carbonation curing. *Journal of CO2 Utilization* 54, 101766.
You, X., Hu, X., Xiao, Z., Bairq, Z.A.S., Chen, W., Shi, C., 2023. Thermodynamic modelling of CaCO3 polymorphs during CO_2 sequestration by cement slurry with the addition of $MgCl_2$. *Journal of Cleaner Production* 410, 137294.

Young, J.F., Berger, R.L., Breese, J., 1974. Accelerated curing of compacted calcium silicate mortars on exposure to CO_2. Journal of the American Ceramic Society 57(9), 394–397.

Zajac, M., Hilbig, H., Bullerjahn, F., 2023. Ben Haha, M., Reactions involved in carbonation hardening of Portland cement: effect of curing temperature. *Journal of Sustainable Cement-Based Materials* 12(9), 1107–1125.

Zhan, B.J., Poon, C.S., Shi, C.J., 2016a. Materials characteristics affecting CO2 curing of concrete blocks containing recycled aggregates. *Cement and Concrete Composites* 67, 50–59.

Zhan, B.J., Xuan, D.X., Poon, C.S., 2018. Enhancement of recycled aggregate properties by accelerated CO2 curing coupled with limewater soaking process. *Cement and Concrete Composites* 89, 230–237.

Zhan, B.J., Xuan, D.X., Poon, C.S., Shi, C.J., 2016b. Effect of curing parameters on CO2 curing of concrete blocks containing recycled aggregates. *Cement and Concrete Composites* 71, 122–130.

Zhang, C., Liu, S., Luo, S., Chang, X., Shen, P., Guan, X., Shi, C., 2022. Effects of sodium doping on carbonation behavior of α-CS. *Cement and Concrete Composites* 131, 104607.

Zhang, D., Cai, X., Jaworska, B., 2020. Effect of pre-carbonation hydration on long-term hydration of carbonation-cured cement-based materials. *Construction and Building Materials* 231, 117122.

Zhang, D., Cai, X., Shao, Y., 2016. Carbonation curing of precast fly ash concrete. *Journal of Materials in Civil Engineering* 28(11), 04016127.

Zhang, D., Shao, Y., 2016. Effect of early carbonation curing on chloride penetration and weathering carbonation in concrete. *Construction and Building Materials* 123, 516–526.

Zhang, H., Chen, J., Shao, R., Zhang, Q., Lu, B., Xia, G., Hou, G., Yun, Z., 2018. The usage of rankinite for carbon capture and storage and carbonation kinetics. *Energy Sources Part A* 40(13), 1629–1646.

Zhang, Q., Feng, P., Shen, X., Lu, J., Ye, S., Wang, H., Ling, T., Ran, Q., 2023. Utilization of solid wastes to sequestrate carbon dioxide in cement-based materials and methods to improve carbonation degree: a review. *Journal of CO2 Utilization* 72, 102502.

Zhang, W., Xu, Y., Wang, Q., 2022. Coupled CO_2 absorption and mineralization with low-concentration monoethanolamine. *Energy* 241, 122524.

Zhang, X., Liu, Z., Wang, F., 2022. Effect of coral powder on the carbonation behavior of γ-C_2S. *Construction and Building Materials* 347, 128251.

Zhao, M., 2012. Quantitative control of C2S crystal transformation. *Applied Mechanics and Materials* 121, 311–315.

Zhao, S., Liu, Z., Mu, Y., Wang, F., He, Y., 2020a. Effect of chitosan on the carbonation behavior of γ-C2S. *Cement and Concrete Composites* 111, 103637.

Zhao, S., Liu, Z., Wang, F., 2020b. Carbonation reactivity enhancement of γ-C_2S through biomineralization. *Journal of CO2 Utilization* 39, 101183.

Zheng, Q., Liang, C., Jiang, J., Li, S., 2023. Unveiling the carbonation dynamics of hydrated alite using electron microscopy. *Chemical Engineering Journal* 8, 143720.

Zhu, J., Qu, Z., Liang, S., Li, B., Du, T., Wang, H., 2022. Macroscopic and microscopic properties of cement paste with carbon dioxide curing. *Materials* 15(4), 1578.

Zhu, T., Dittrich, M., 2016. Carbonate precipitation through microbial activities in natural environment, and their potential in biotechnology: a review. *Frontiers in Bioengineering and Biotechnology* 4, 4.

5 Thermodynamic modelling of CO_2 mineralization and experimental verification

5.1 INTRODUCTION

Thermodynamics plays a crucial role in understanding complex chemical systems. Given known conditions of temperature, pressure, and composition, thermodynamic calculations can be applied to predict whether a reaction will proceed spontaneously and what the final equilibrium state will be. The thermodynamic method was first applied to cement chemistry by Le Chatelier in the late 19th century to describe hydration as the dissolution of clinker phases, leading to supersaturation and subsequent precipitation of hydration products. Thermodynamic approaches were later applied to cement manufacturing processes, focusing on the chemical reconstitution of raw materials, the release of structural water and CO_2 from precursor minerals, and the reorganization of complex minerals during processing. Since the 1940s, advances in fundamental science and microstructural characterization techniques have significantly enhanced our understanding of the highly disordered phases—including glasses, melts, and gels—which have greatly contributed to the development of structural models of these phases, such as C-S-H, and the establishment of more accurate and systematic thermodynamic databases. Subsequently, the development of computer technologies has led to the creation of a series of open-source Gibbs free energy minimization codes and stimulated the advancement of thermodynamic modeling software.

One of the key characteristics of the carbon mineralization process in cement-based materials is the simultaneous occurrence of hydration and carbonation reactions, which makes the analysis of phase assemblages even more complex than in pure hydration systems. Thermodynamic modeling provides an effective approach for predicting the mineralization products, especially in capturing polymorph transformations of calcium carbonate. In addition, process calculations enable stepwise simulations of phase evolution and pore solution chemistry under varying CO_2 uptake levels, thereby reflecting real-time changes in pH and offering insights into the progression of the whole carbonation process. Based on thermodynamic databases and the oxide or clinker composition of cementitious materials, such models can reliably evaluate the possible phase assemblages, ion concentrations, ion activities, pH evolution, and porosity throughout the carbonation process.

 DOI: 10.1201/9781003627531-5

The primary carbonation product, calcium carbonate, exists in three anhydrous polymorphs: cubic calcite, needle-like aragonite, and metastable spherical vaterite. The polymorphic transformation of calcium carbonate has a considerable influence on the properties of the carbonated products. The mineralization of cement-based materials involves coupled hydration and carbonation reactions as well as the formation of gel phases, making it challenging to experimentally determine parameters such as the supersaturation and surface energy of calcium carbonate. These challenges hinder a comprehensive understanding of polymorphic transformations during mineralization. Thermodynamic parameters, such as the change in Gibbs free energy of formation, serve as direct indicators of polymorphic transitions. Based on thermodynamic models, the proportions of different calcium carbonate polymorphs can be calculated, providing a theoretical basis for optimizing carbonation parameters and enabling effective regulation of calcium carbonate polymorphs.

5.2 METHODS TO SOLVE THE THERMODYNAMIC EQUILIBRIUM PROBLEM

Currently, two primary methods are widely used to solve thermodynamic equilibrium problems: the minimization of the system's Gibbs free energy and the solution of mass action law equations. The former is mainly solved by GEM-Selektor (Kulik, 2013; Wagner et al., 2012), and the latter can be solved by PHREEQC (Han et al., 2021; Michel et al., 2021), CHESS (Taylor and Cihon, 2015), and EQ3/6 (Wolery et al., 1990). It is important to note that both approaches are based on the same thermodynamic framework, as the standard Gibbs free energy of reaction ($\Delta_o G_r$), and the equilibrium constant (K) are directly related through the equation $\Delta_r G^0 = \Sigma_i v_i \Delta_f G^0 = -RTlnK$. Therefore, these methods simply represent two different approaches to solving thermodynamic equilibrium problems.

5.2.1 Minimizing the Gibbs free energy

The Gibbs free energy minimization method is based on the fundamental thermodynamic principle that, at equilibrium—when there is no longer a spontaneous trend of change in the reaction—the total Gibbs free energy of the system reaches its minimum. During the calculation process, the approach also strictly adheres to mass and charge balance across the entire system.

Thermodynamic modeling requires the support of thermodynamic databases. At present, thermodynamic databases mainly include Cement 18, PSI-Nagra, MILES, and Thermodem, among which the first one is most widely used. For alkali-activated binders, the solid solution models of alkali-substituted calcium aluminate silicate hydrate gels (CNASH_ss) can be used (R.J. Myers et al., 2013; Rupert J. Myers, Bernal, and Provis, 2014). Phases with uncertain composition such as CSH and hydrotalcite can also be described by solid solution models (Kulik, 2011; Rupert J. Myers, Lothenbach, et al., 2015; Walker et al., 2016) or surface complexation models (Haas and Nonat, 2015). Data for Na-containing and Ca-containing zeolites can be obtained from Lothenbach et al. (Lothenbach, Bernard, and Mäder, 2017) and magnesium silicate hydrates (M-S-H) from Nied et al. (2016).

Before the thermodynamic equilibrium calculation, the degree of cement hydration and the reaction degree of SCMs are also required to be assumed. Portland cement can be assumed to be fully hydrated before mineralization (De Weerdt et al., 2019), while the reaction degree of fly ash and slag can be assumed by some empirical data (Rupert J. Myers, Bernal, et al., 2015; Palacios and Puertas, 2006; De Weerdt et al., 2011) or modified from previous studies (Martin et al., 2017; Yoon et al., 2021). Shah et al. (2018) used the Rietveld analysis to calculate the degree of hydration, which was used as one of the input parameters of the thermodynamic model. This method is more accurate than direct assumptions. In the simulation process of the mineralization, the stepwise CO_2 addition method is generally used. The simulation results of CO_2 increments of 0.1, 0.3, and 1 g per step were compared by Ke et al., and the 0.3 g CO_2 per step has been proved enough to accurately reflect the overall phase evolution (Ke et al., 2020; Park, 2020). The total amount of added CO_2 is generally set to 0–50 g (Shah et al., 2018; Shi et al., 2016) or 0–40 g (De Weerdt et al., 2019) per 100 g of binder.

In the process of thermodynamic equilibrium calculation, the activity of different species is one of the key parameters, as gaseous CO_2 and solid minerals involved in the mineralization all need to dissolve in the pore solution before the reaction. In order to calculate the activity of an ion species, the activity coefficient γ is introduced, and the value is equal to the activity coefficient multiplied by its concentration, i.e., $\{i\} = \gamma_i [i]$. Based on the ionic strength range (I), different calculation methods of the activity coefficient have been proposed, including Davies equation (Lothenbach and Zajac, 2019), extended Debye–Hückel equation (Helgeson, Kirkham, and Flowers, 1981), and the specific ion interaction theory (Elizalde and Aparicio, 1995). Their expressions and scope of application are shown as follows:

Davies equation, valid for 0.3 < I < 0.5 molal:

$$\log \gamma_i = \frac{-A_\gamma Z_i^2 \sqrt{I}}{1+\sqrt{I}} - 0.3I \tag{5.1}$$

Extended Debye–Hückel equation, valid for I < 1 molal:

$$\log \gamma_i = \frac{-A_\gamma Z_i^2 \sqrt{I}}{1+B_\gamma a_i \sqrt{I}} + b_\gamma I \tag{5.2}$$

Specific ion interaction theory, valid for I < 3 molal:

$$\log \gamma_i = \frac{-A_\gamma Z_i^2 \sqrt{I}}{1+B_y a_i \sqrt{I}} + \sum_k \varepsilon(i,k) m_k \tag{5.3}$$

where A_y is the pressure-dependent coefficient, B_y is the temperature-dependent coefficient, Z_i is the charge of species i, a_i is the ion-size parameter, b_y is the semi-empirical parameter, I is the effective molal ionic strength, $\varepsilon(i,k)$ is the interaction coefficient which describes the short-range interaction between the species *i* and *k* and m_k denotes the molality of species *k*.

Thermodynamic models can take all the thermodynamically possible phases that may occur during the mineralization process into consideration, instead of just discussing the main carbonatable phases. Besides, different forms of carbonatable phases can be described in detail. For example, the CSHQ model proposed by Kulik (2011) can be used in the modeling of CSH mineralization, in which the CSH is divided into four categories based on their calcium-to-silicon ratio (Ca–Si): TobH ($C_{2/3}SH_{1.5}$), TobD ($C_{5/6}S_{2/3}H_{1.83}$), JenH ($C_{1.33}SH_{2.17}$), and JenD ($C_{1.5}S_{0.67}H_{2.5}$). The Ca–Si of CSH depends on the proportion of the four end members. Other thermodynamic models of C-S-H include the CSH-II model, ECSH-1 model, ECSH-2 model, and CSH3T model, with their specific parameters available in the Cemdata18 database.

5.2.2 Solving the Law of Mass Action Equations

In the models discussed above, the thermodynamic equilibrium problem is solved according to the principle of minimizing the Gibbs free energy of the system. In addition to this method, thermodynamic equilibrium problems can also be solved by the law of mass action. This method is faster but requires more assumptions, such as the fugacity of gases. In the models based on the law of mass action, the aqueous reactions that may occur in the system usually need to be determined first, and the mass and charge balance equations are solved by Newton–Raphson iterations (Lothenbach and Zajac, 2019). The calculation of phase equilibrium is based on the thermodynamic equilibrium constant K, which can be expressed by the activity of specific species (Appelo and Postma, 2004). The stability of CH and $CaCO_3$ can be characterized by the threshold of ion activity products, and the variable C/S of CSH can also be considered; the detailed simulation method can be found in Shen, Dangla, and Thiery (2013). Except for the modeling of chemical reactions in the mineralization process, thermodynamic models can also simulate the surface complexation reaction with the generalized two-layer surface complexation model built in PHREEQC (Borkovec and Westall, 1983). As the structure of CSH has not been clearly understood, the surface of CSH is usually assumed to be composed of silanol sites based on a widely recognized fact that the structure of CSH is similar to Tobermorite (Labbez et al., 2006; Viallis-Terrisse, Nonat, and Petit, 2001). Through MCD, another module in PHREEQC, the multi-component diffusion caused by the dissolved CO_2 and other ions can be simulated (Kari, Puttonen, and Skantz, 2014). The application of this module in the mineralization process is very meaningful because mineralization is mostly considered a diffusion-controlled process.

5.3 RECENT DEVELOPMENTS IN THERMODYNAMIC MODELLING OF MINERALIZATION

Due to the advantages in handling complex phase assemblages, thermodynamic models have been applied to simulate the mineralization processes of various cement-based materials. De Weerdt et al. (2019) employed the GEMS software in

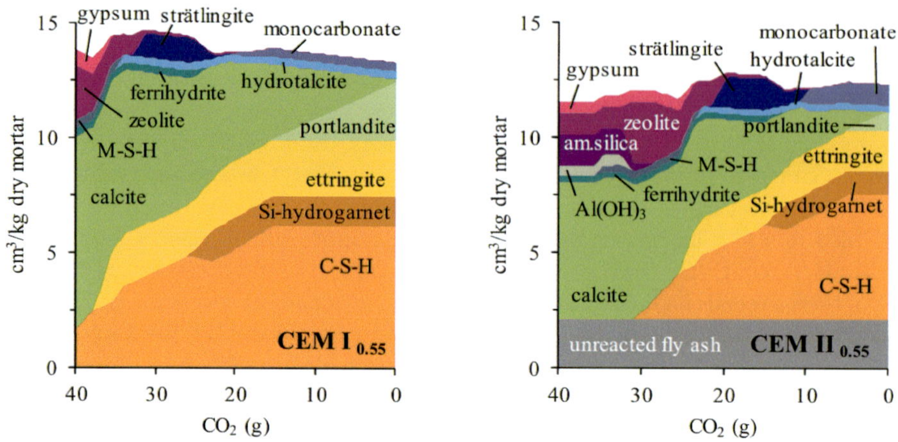

FIGURE 5.1 Phase assemblage evolution during the mineralization of CEM I and CEM II mortars (De Weerdt et al., 2019).

combination with the Cemdata18 thermodynamic database to simulate the carbonation of CEM I and CEM II Portland cements. The predicted phase assemblages are shown in Figure 5.1. It can be observed that portlandite (CH) is the first phase to undergo carbonation. After CH is nearly depleted, C-S-H begins to carbonate and gradually decalcify, accompanied by a continuous increase in solid volume. Calcite forms steadily throughout the carbonation process. In addition to calcite, phases such as M-S-H, zeolites, and gypsum also appear in the later stages of carbonation. While the carbonation products of CEM I and CEM II are generally similar, the latter yields greater amounts of zeolite and gypsum, as well as a small quantity of $Al(OH)_3$.

In addition to phase assemblage analysis, thermodynamic modeling has also been applied to investigate pore solution chemistry. Figure 5.2 illustrates the evolution of pore solution composition during the carbonation of CEM I and CEM II Portland cement. Thermodynamic calculations provide access to key parameters such as solution volume, Ca–Si ratio, the proportion of alkali metal ions, and pH evolution, thereby offering deeper insight into the physicochemical mechanisms underlying phase transformations. In experimental studies, pH changes during mineralization are commonly assessed indirectly using phenolphthalein color change; however, the phase transition behind this change remains unclear. Thermodynamic modeling enables a clearer interpretation of the relationship between phase evolution and pH, highlighting how changes in carbonate phases influence pore solution alkalinity.

In addition to Portland cement, thermodynamic modeling—owing to its distinct advantages in phase assemblage analysis—has also been widely applied to the mineralization of cement-based systems, incorporating supplementary cementitious materials (SCMs), such as fly ash, metakaolin, limestone, and ground granulated blast-furnace slag (GGBFS) (Z. Liu et al., 2023; Shah et al., 2018; Shi et al., 2016), as well as in calcium sulfoaluminate (CSA) cements (Hargis et al., 2017; Park, 2020) and alkali-activated slag binders (Rupert J. Myers, Lothenbach, et al., 2015; Rupert

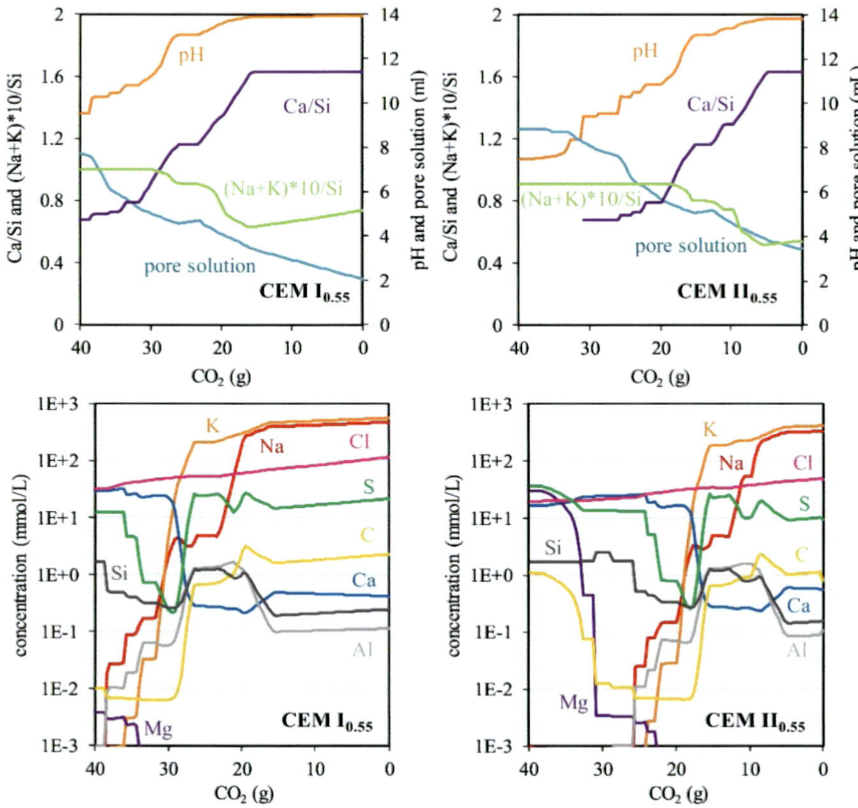

FIGURE 5.2 Evolution of pore solution composition and pH during the mineralization of CEM I and CEM II mortars (De Weerdt et al., 2019).

J. Myers, Bernal, and Provis, 2014). Liu et al. (2023) employed thermodynamic modeling to compare the phase assemblages of OPC and a system containing 50% GGBFS. It was found that, during the early stage of mineralization, the GGBFS-containing system exhibited a significant increase in monocarbonate content and a notable reduction in ettringite and high-calcium C-S-H. As mineralization progressed, the pH in the BFS-containing system declined more rapidly than in the OPC group, accompanied by a marked reduction in calcite and gypsum contents. Hargis et al. (2017) investigated the phase evolution of CSA cements with different mix proportions using thermodynamic modeling, as illustrated in Figure 5.3. The study showed that during mineralization, AH_3 continuously forms, and $CaCO_3$ precipitates following the decomposition of monocarbonate, while gypsum forms as ettringite decomposes. In addition, due to the presence of monosulfate, this phase initially decomposes to form monocarbonate and additional ettringite, resulting in an increase in solid volume and densification of the cement paste in the early stage of mineralization. As the process advances and ettringite decomposes, the solid volume decreases, and pore solution volume increases.

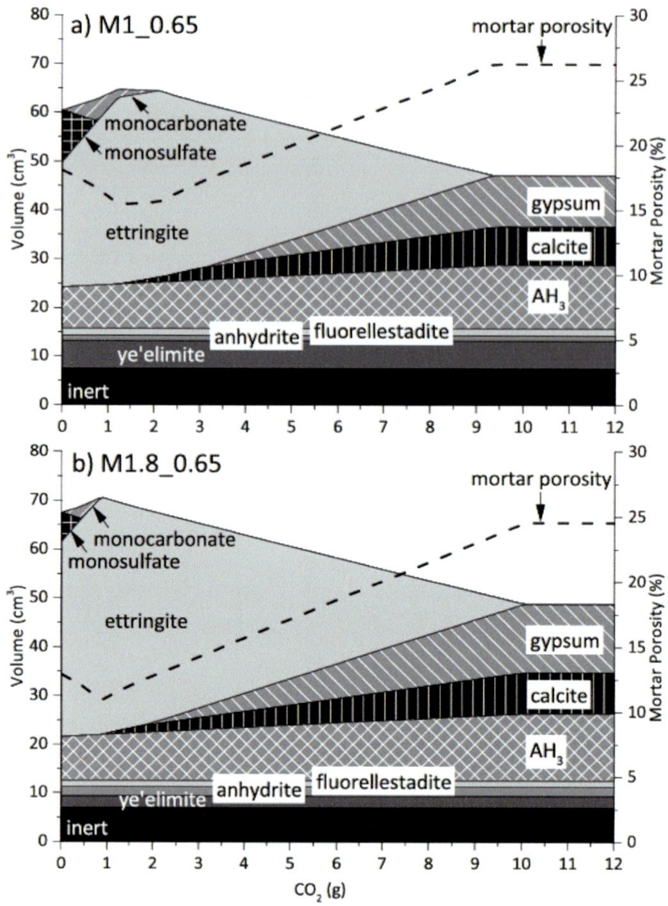

FIGURE 5.3 Phase assemblage evolution during the mineralization of 100 g of (a) M1_0.65 and (b) M1.8_0.65 CSA cement (MX_Y indicates a molar ratio of gypsum to ye'elimite of X and a water-to-cement ratio of Y) (Hargis et al., 2017).

CNASH_ss and hydrotalcite solid solution models (Rupert J. Myers, Lothenbach et al., 2015; Rupert J. Myers, Bernal, and Provis, 2014), in combination with the Cemdata18 and MINES databases, have been applied to simulate the thermodynamic evolution of alkali-activated slag cements during mineralization, as shown in Figure 5.4 (Ke et al., 2020) . In the absence of CO_2 input, the model predicted CNASH_ss as the main reaction product, with strätlingite and hydrotalcite as secondary phases, and minor formation of C_3AH_6. As the mineralization reaction progressed, CNASH_ss and C_4AcH_{11} were rapidly consumed, accompanied by the formation of (Na, Ca)-aluminosilicates and calcite. Subsequently, MgAl-CO-LDH_ss and ettringite were gradually depleted. M-S-H formed as an intermediate phase during the later stages of mineralization but was quickly consumed. The sharp drop in pH corresponded to the disappearance of CNASH_ss.

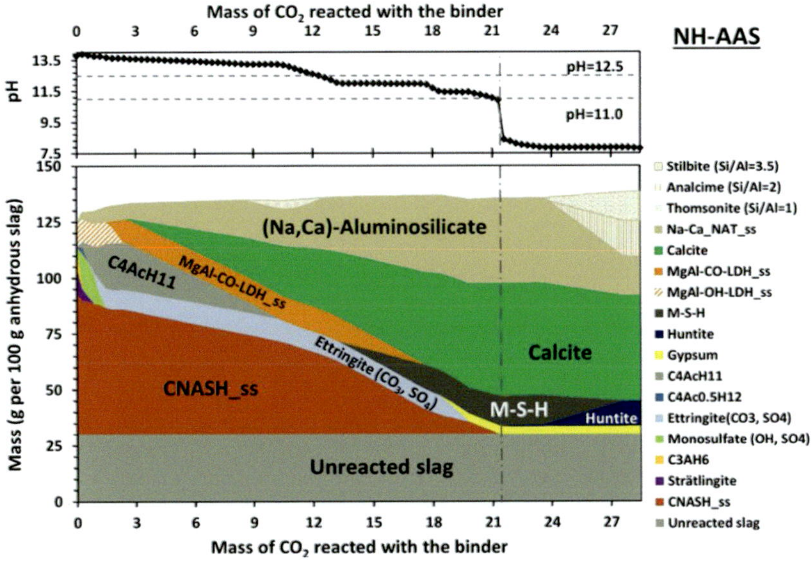

FIGURE 5.4 Phase assemblage and pH evolution of NaOH-activated slag under accelerated mineralization conditions (1% CO_2) (Ke et al., 2020).

Porosity calculation during the mineralization process remains a challenging task due to the complexity of influencing factors. Although several empirical formulas have been proposed (Omikrine Metalssi, Aït-Mokhtar, and Turcry 2020; Papadakis, Vayenas, and Fardis, 1991), the underlying assumptions and the cumulative errors from repeated calculations cannot be neglected. Moreover, such approaches are not able to provide continuous information on porosity evolution throughout the mineralization process. In thermodynamic models, however, porosity can be approximated by subtracting the cumulative volume of all solid phases from the initial system volume, as the phase assemblage is continuously updated over the mineralization. As shown in Figure 5.5, the dashed arrows indicate changes in porosity after mineralization, and the corresponding phase variations can also be observed directly in the figure (Shi et al., 2016).

Thermodynamic modeling provides access to valuable information regarding the reaction sequence of minerals, changes in solid volume, pH evolution, CO_2 uptake, and the composition of mineralization products in various cement-based materials. However, it has also been noted that while software such as GEMS can predict the total amount of calcium carbonate formed, it cannot quantify the distribution among its different polymorphs. Calcium carbonate is one of the predominant products of the mineralization process, and its polymorphic composition significantly influences the properties of the resulting carbonated phases. Therefore, it is necessary to develop a corresponding polymorph model for calcium carbonate based on the thermodynamics of phase transformation to further refine its phase assemblage.

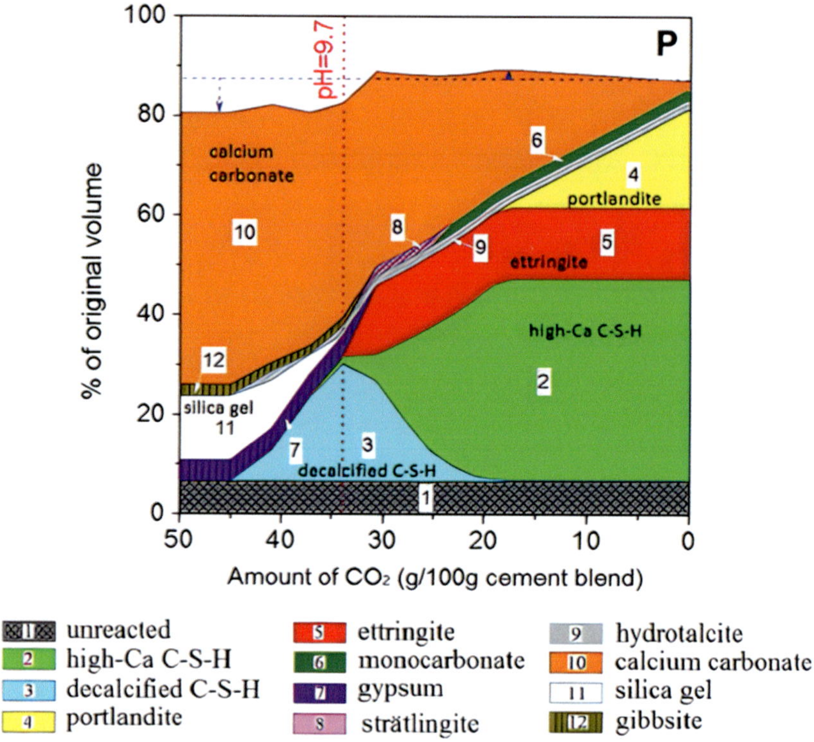

FIGURE 5.5 Phase assemblage and total porosity evolution during the mineralization of Portland cement under 1% CO_2 conditions (Shi et al., 2016).

5.4 MODELLING BACKGROUND OF CALCIUM CARBONATE POLYMORPHS

Generally speaking, calcite is dominant among the three anhydrous polymorphs during mineralization due to its lowest Gibbs free energy of formation. However, crystal stabilizers can inhibit this transformation. Take magnesium ions as an example; part of the calcium ions will be replaced by magnesium ions, leading to the formation of $Ca_{1-x}Mg_xCO_3$. Compared with aragonite, magnesium ions are easier to combine with the structure of calcite. The energy of magnesium ions' entry into aragonite is almost four times higher than that of calcite (Sun et al., 2015). The larger energy consumption comes from the difference in oxygen coordination of divalent cations. In the structure of calcite and aragonite, magnesium ions produce a stress field, making the nearest Mg-O bond about 9% shorter than the Ca-O bond (Sun et al., 2015). However, since divalent cations form six coordination with oxygen atoms in calcite and nine coordination with oxygen atoms in aragonite, the energy consumption required for magnesium ions to replace calcium ions in the aragonite structure is much larger than that in calcite.

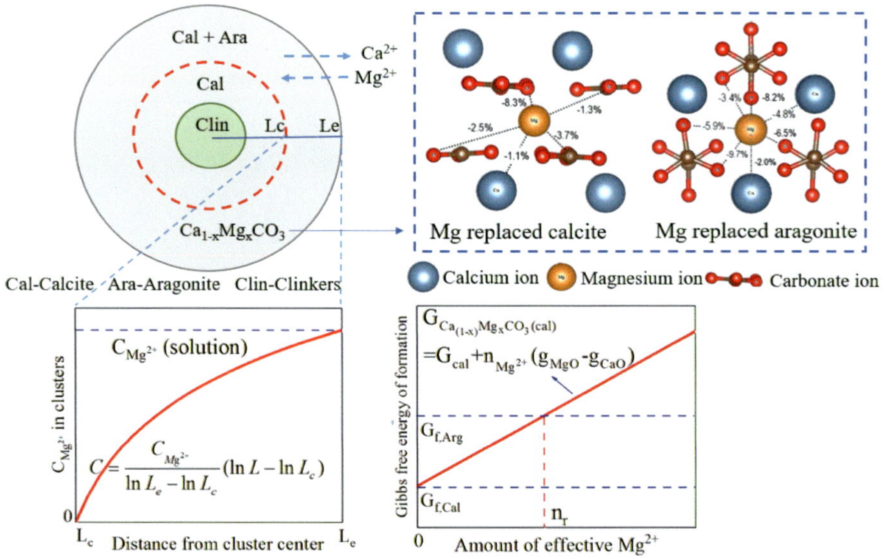

FIGURE 5.6 The changes of different calcium carbonate polymorphs covered on the surface of cement particles, the structure distortion caused by magnesium ions entering calcite and aragonite, the concentration distribution of magnesium ions in calcium carbonate clusters, and the change of Gibbs free energy of formation of calcium carbonate.

As shown in Figure 5.6, the magnesium ions entering the calcium carbonate clusters contribute to the inhibition of the transformation from aragonite to calcite. Although calcium ions in both aragonite and calcite may be replaced by magnesium ions, the structural distortion of magnesium-replaced calcite is smaller than that of aragonite, so magnesium ions tend to enter the calcite structure. This substitution will lead to the increase of Gibbs free energy of formation of calcite, thus making aragonite a relatively stable phase and inhibiting the transformation from aragonite to calcite.

In addition to inorganic regulators, organic additives can also be applied to regulate calcium carbonate polymorphs. The adsorption of organic additives onto calcium carbonate surfaces leads to a reduction in the surface energy of these crystals. However, this reduction effect is relatively limited on the surfaces of calcite and vaterite compared to the notable reduction observed on the aragonite surface. The surface energy of a crystal is pivotal in defining its critical nucleation Gibbs free energy, exhibiting a cubic relationship with the latter. Considering that the system's nucleation free energy inherently seeks minimization, organic additives tend to selectively adsorb on the surface of aragonite. The schematic diagram of organic additive-induced aragonite formation is shown in Figure 5.7.

The carbon mineralization of cement-based materials differs from the pure solution reaction system, as it involves continuous CO_2 input and the concomitant generation of intricate gel phases. Throughout this process, the ion concentration, pH value, and ion activity coefficient of the system undergo constant changes, potentially

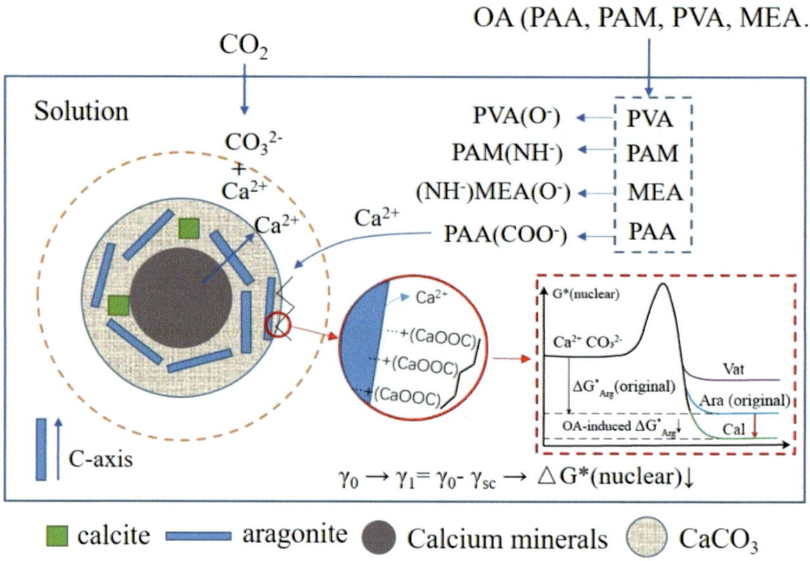

FIGURE 5.7 Schematic diagram of the adsorption of organic additives on the calcium carbonate layer and their effects on nucleation Gibbs free energy (PAA—polyacrylic acid, PAM—polyacrylamide, PVA—polyvinyl alcohol, and MEA—monoethanolamine).

affecting the role of organic additives as crystal stabilizers. To maximize the effects of these organic additives, it is crucial to control the mineralization parameters, especially the input of CO_2, temperature, and the concentration of organic additives. The influence of various mineralization parameters on the polymorphs of calcium carbonate is predominantly achieved through three facets: (1) adjusting the effective concentration of organic additives adsorbed on the surface of calcium carbonate, (2) altering the difference in surface energy and critical nucleation Gibbs free energy between aragonite and calcite, and (3) regulating the reduction in surface energy attributed to per mole organic additive. Figure 5.8 illustrates the mechanistic diagram of various mineralization factors and their influence on the nucleation Gibbs free energy, total amount of calcium carbonate, and the proportion of aragonite.

It is important to highlight that not all initially introduced organic additives can fully manifest their aragonite-inducing effects. The increase in the concentration of organic additives influences the polymorphs of calcium carbonate through the first and third aspects shown above. As shown in Figure 5.8(b), organic additives predominantly distribute in three zones within the system: (I) be adsorbed by the gel phase, (II) penetrate onto the gel phase but do not successfully adsorb onto the calcium carbonate surface, and (III) be adsorbed on the calcium carbonate surface. At low levels of organic additive concentrations, the formation of aragonite is not significant. This is attributed to the limited quantity of organic additives successfully adsorbed on the calcium carbonate surface and the less difference in the effects of these additives in reducing the surface energy of calcite and aragonite at low

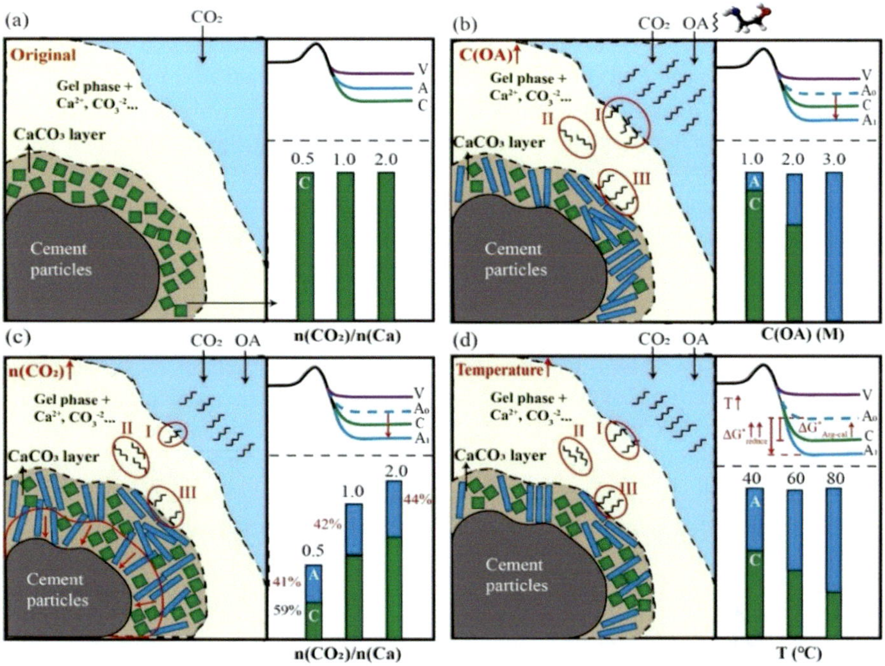

FIGURE 5.8 The mechanistic diagram of various mineralization factors and their impacts on nucleation Gibbs free energy, total $CaCO_3$, and aragonite ratio: (a) original system (without OA), (b) with increasing OA concentrations, (c) with increasing CO_2 input, and (d) with increasing temperature.

concentration range. However, as the concentration of organic additives increases, not only does the amount of additives absorbed on the calcium carbonate surface rise but also the difference in their effects on reducing the surface energies of calcite and aragonite becomes more pronounced. This leads to an increasing proportion of aragonite in carbon mineralization products.

The amount of CO_2 input, or the mineralization degree, primarily impacts the polymorphs of calcium carbonate through the first facet. On the one hand, an increase in CO_2 input results in the gradual transformation of high calcium-silicate ratio CSH gel into low calcium-silicate ratio CSH and silica gel, leading to a reduction in the amount of organic matter adsorbed by the gel phase, as shown in zone I of Figure 5.8(c). On the other hand, an increase in the mineralization degree impacts the ionic strength and activity coefficient, leading to an escalation in the Debye length and a reduction in interfacial charge density. This implies an increased average distance between ions, a weakened ion field, and a reduced probability of complexation between organic additives and calcium ions, ultimately resulting in an increased concentration of organic additives in region II, as shown in Figure 5.8(c). Fortunately, these two effects are not completely conflicting. Controlling the input of CO_2 ($n(CO_2)/n(Ca)>1.2$) can ensure a high interface charge density on the premise of controlling the low adsorption of organic additives by the gel phase. This control promotes the complexation between organic additives and calcium ions. However, despite the fact that an increase in CO_2 input leads to a higher amount of aragonite, the increase in the proportion of aragonite is relatively limited due to the overall growth of total calcium carbonate, as depicted in the bottom right part of Figure 5.8(c).

In comparison, the effect of temperature on the polymorphs of calcium carbonate proceeds mainly through the second and third facets. First, as the temperature rises, an increase in the difference of nucleation Gibbs free energy between aragonite and calcite is observed, leading to a corresponding increase in the difference of critical nucleation Gibbs free energy. Additionally, the increase in temperature also leads to an increase in the extent to which per unit mole organic additives reduce the nucleation Gibbs free energy of calcium carbonate. Thermodynamic calculations indicate that among these two aspects, the latter is predominant. This implies that although the increasing temperature does not significantly affect the quantity of organic additives adsorbed on the calcium carbonate surface, each unit of organic additive plays a more pronounced role in inducing the formation of aragonite. Therefore, in summary, an increase in temperature favors the facilitation of aragonite formation.

5.5 THERMODYNAMIC MODELLING OF $CACO_3$ POLYMORPHS DURING MINERALIZATION

5.5.1 Model framework based on inorganic regulation system

Currently, the most widely used inorganic aragonite regulator is the magnesium-containing additives. This part will take magnesium ions as an example to illustrate the thermodynamic simulation framework. First, the standard Gibbs free energy change, standard enthalpy change, and standard entropy change of magnesium-substituted calcium carbonate are calculated by the group contribution method, the basic principle of which is that the properties of mineral carbonates can be characterized by the combination of basic polyhedral units (Hazen, 1985), as shown in Equation (5.4). Based on this, combined with the standard Gibbs free energy difference between calcite and aragonite, the amount of magnesium ions required to make the Gibbs free energy of magnesium-substituted calcite higher than that of aragonite can be calculated, as shown in Equation (5.5).

$$\Delta G_f^0 = \sum_{i=1}^{N} n_i g_i \tag{5.4}$$

$$\mathrm{n}_{r,Mg^{2+}} = \frac{(\Delta G_{fA}^0 - \Delta G_{fC}^0)}{(g_{MgO} - g_{CaO})} \tag{5.5}$$

where g_i is the contribution coefficient of component i to the standard Gibbs free energy of formation. The physical and chemical properties of different calcium

TABLE 5.1

Thermodynamic parameters of different crystal forms of calcium carbonate

Polymorph	**Calcite**	**Aragonite**	**Vaterite**
Morphology	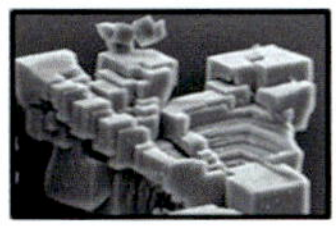	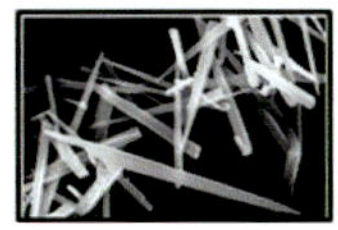	
	Cubic	Needle	Ball
a (nm)	0.524	1.00	10.00
b (nm)	3.142	6.00	42.00
ρ ($kg•m^{-3}$)	2710	2930	2540
V ($m^3•mol^{-1}$)	3.693×10^{-5}	3.415×10^{-5}	3.940×10^{-5}
Ksp ($mol^2•l^{-2}$)	$10^{-8.4}$	$10^{-8.2}$	$10^{-7.9}$
ΔG_f^0 ($kJ•mol^{-1}$)	−1129.2	−1128.4	−1125.5

carbonate polymorphs are shown in Table 5.1. The thermodynamic contribution coefficients of the carbonate polyhedron can be found in the study of Iglesia et al. (La Iglesia and Félix, 1994).

The content of free magnesium ions is the key factor affecting the formation of aragonite. The replacement of some calcium ions by magnesium ions will lead to changes in the thermodynamic properties of calcite. However, the complex chemical reactions occurring simultaneously in the cement system make it hard to obtain the content of free magnesium ions in time. Besides, part of magnesium will not be released into the solution phase, and the magnesium-to-calcium ratio in solution changes with the progression of hydration and mineralization reactions. The effective magnesium ion concentration and volume of solution are also variable. Considering the above problems, using thermodynamic models to calculate the phase assemblages and ion concentration in the system is an efficient method.

Based on GEM SELEKTOR 3.0 software (Kulik et al., 2013; Wagner et al., 2012), PSI/Nagra (Thoenen et al., 2014), and Cemdata 18 (Lothenbach et al., 2019) thermodynamic databases, the evolution of the chemical composition of solution, especially the free magnesium ion concentration, and the phase assemblages can be calculated. The CSHQ model developed by Kulik (2011), which includes four C-S-H end members with different Ca–Si ratios between 0.67 and 1.6 and two end members for K and Na, is used to simulate the C-S-H phase. The decoupled nitrogen is selected to form the model atmosphere, which will be further mixed with pure CO_2 to create the gas environment with different CO_2 concentrations. Compared with the dissolution and precipitation kinetic model, the thermodynamic model can consider all possible phases, including the gel phase, and avoid the assumptions of constant temperature and pressure. The extended Debye–Hückel equation is used for calculating the activity coefficients of the aqueous species, as shown in Equation (5.2).

The free magnesium ion concentration calculated by GEMS cannot be directly used to calculate the proportion of aragonite. Instead, only the magnesium ions entering the calcium carbonate clusters contribute to the inhibition of the transformation from aragonite to calcite. Considering that the density of calcium carbonate in the inner layer is higher than that of newly formed calcium carbonate in the outer layer, it is assumed that the reduction rate of magnesium ion concentration is inversely proportional to its distance from the cluster center, as shown in Equation (5.6).

$$\frac{dc}{dL} = \frac{1}{kL} \tag{5.6}$$

Combined with the boundary condition Equations (5.7)–(5.8), the concentration of magnesium ions in the calcium carbonate clusters can be calculated, as shown in Equation (5.10).

$$f(L_e) = C_{Mg^{2+}} \tag{5.7}$$

$$f(L_c) = 0 \tag{5.8}$$

$$c = f(L) = \frac{C_{Mg^{2+}}}{\ln L_e - \ln L_c}(\ln L - \ln L_c) \tag{5.9}$$

where L_e is the diameter of the calcium carbonate cluster, L_c is the distance between the boundary of aragonite and the pure calcite layer and the center of the carbonate cluster, L is the distance between magnesium ions and the center of carbonate cluster, and $C_{Mg^{2+}}$ is the concentration of free magnesium ions in solution.

Combining the concentration distribution and the volume of calcium carbonate cluster, the amount of effective magnesium ions in the cluster can be calculated, as shown in Equation (5.10).

$$\begin{aligned} n_{e,Mg^{2+}} &= \int_{L_c}^{L_e} \frac{C_{Mg^{2+}}}{\ln L_e - \ln L_c}(\ln L - \ln L_c)dV \\ &= \frac{4\pi}{3} L_e^{\ 3}[C_{Mg^{2+}} - \frac{1}{3}(\ln \frac{L_e}{L_c})^{-1}[1 - (\frac{L_c}{L_e})^3]C_{Mg^{2+}}] \end{aligned} \tag{5.10}$$

From Equation (3.10), it can be observed that the amount of effective magnesium ions is related to the ratio of L_e and L_c, which can be defined as the penetration degree of magnesium ions in calcium carbonate clusters. Choudens-Sánchez (De Choudens-Sánchez and Gonzalez, 2009) and Mucci et al. (Mucci and Morse, 1983) found that the process of magnesium ion entering calcium carbonate was determined by the ratio of magnesium to calcium in solution. Morse (Morse, Wang, and Tsio, 1997) and Burton et al. (Burton and Walter, 1987) proved that higher temperature has a positive effect on increasing the magnesium ion content in calcium carbonate. Based on the above research, the relationship between penetration degree and the ratio of magnesium to calcium in solution and temperature was fitted, as shown in Equation (5.11).

$$\alpha = L_e / L_c = [(-0.0417T + 17.22)[\mathrm{Mg/Ca}] + 0.1671T - 50.144][\mathrm{Mg/Ca}] - 0.0045T + 2.3429 \tag{5.11}$$

Temperature not only affects the effective magnesium ion concentration but also affects the thermodynamic parameters. According to the Harper's equation, the change of Gibbs free energy of aragonite to calcite with temperature can be calculated by the following Equation (5.12):

$$\Delta G_{T,A\to C} = \Delta H_{0,A\to C} - \Delta aT\ln T - \frac{\Delta b}{2}T^2 - \frac{\Delta C}{2}T^{-1} + yT \tag{5.12}$$

where $\Delta G_{T,A\to C}$ is the change of Gibbs free energy of aragonite to calcite at temperature T, Δa, Δb and Δc are the differences between the heat capacity coefficients of aragonite and calcite, respectively.

The relationship between the heat capacity and temperature can be calculated by Equation (5.13). Based on this, the enthalpy change of aragonite to calcite can be calculated by Equation (5.14).

$$C_P = a + bT + cT^{-2} + dT^{-1/2} \tag{5.13}$$

$$\Delta H_{0,A\to C} = \Delta H_{298,A\to C} - \int_0^{298} a + bT + cT^{-2} + dT^{-1/2}dT \tag{5.14}$$

It should be noted that the coefficients of thermodynamic properties in the group contribution method are also affected by temperature. The coefficients at any temperature are derived from the coefficients at the standard state. Assuming that the pressure is constant, they can be calculated by Equations (5.15) and (5.16).

$$(\partial\Delta G / \partial\Delta T)_P = -\Delta S \tag{5.15}$$

$$g_{i,T} - g_{i,298} = -s_{i,298}(T - 298) \tag{5.16}$$

Based on the above equations and the phase assemblages calculated by GEMS, the proportion of aragonite under different temperatures and initial magnesium ion concentrations can be calculated, as shown in Equation (5.17).

$$X_A = \frac{(g_{MgO,T} - g_{CaO,T})M_{CaCO_3}V_S(C_{Mg^{2+}} - \frac{1}{3}(\ln\frac{L_e}{L_c})^{-1}[1-(\frac{L_c}{L_e})^3]C_{Mg^{2+}})}{m_{CaCO_3}(\Delta H_{0,A\to C} - \Delta aT\ln T - \frac{\Delta b}{2}T^2 - \frac{\Delta C}{2}T^{-1} + yT)} \tag{5.17}$$

where $C_{Mg^{2+}}$ is the concentration of free magnesium ion, V_s is the solution volume after mineralization, and m_{CaCO_3} is the mass of calcium carbonate. These three parameters are calculated by GEMS $g_{MgO,T}$ and $g_{CaO,T}$ are thermodynamic property coefficients calculated from Equations (5.15) and (5.16). M_{CaCO_3} is the molar mass of $CaCO_3$. $\alpha = L_e / L_c$ is penetration degree calculated from Equation (5.11). $\Delta H_{0,A \to C}$ is the enthalpy change of aragonite to calcite calculated from Equations (5.12)–(5.14).

5.5.2 Model framework based on organic regulation system

Organic additives will change the critical nucleation energy of calcium carbonate by selectively adsorbing on the crystal surface and changing the interface energy. The selective adsorption preferentially occurs on the aragonite surface parallel to the *c*-axis. Therefore, the growth of crystals in the direction perpendicular to the *c*-axis is inhibited, and the transformation of aragonite to calcite is prevented (Ouhenia et al., 2008; Pan et al., 2007). In this case, aragonite is preferentially generated due to its lower critical nucleation energy than calcite. Therefore, the key point in quantifying the proportion of different calcium carbonate polymorphs is quantifying the effect of organic additives on the interface energy of calcium carbonate.

The classical method for interface energy calculation is to use experimental data to link the induction time of the nucleation of a specific precipitation reaction with the interface energy (Nielsen, 1984). This method is only applicable to the evaluation of the crystal-solution interface energy at the early stage of the reaction, and generally, it is difficult to directly obtain the accurate induction time in some complex systems. In addition, due to the limitations of the measurement method of contact angle, it is quite difficult to test the interface energy of calcium carbonate generated in solution. Therefore, it is significant to establish a thermodynamic model to simulate and determine the interface energy and proportion of different calcium carbonate polymorphs.

The proportion of aragonite is calculated by dividing the effective organic additive content adsorbed on the $CaCO_3$ surface by the amount of organic additives required to inhibit the conversion from a unit mole of aragonite to calcite, as shown in Equation (5.19). Therefore, the model establishment will focus on these two aspects. The solid phase content and ion concentration under different mineralization conditions (such as CO_2 input and temperature) are the basis of model calculations, which are calculated utilizing the thermodynamic simulation software GEM SELEKTOR 3.0. Due to the complexity of cement systems, the gel phase (with varying calcium silicon ratios) adsorption and a constantly thickening layer of calcium carbonate with the increase of mineralization degree on the effective calcium carbonate content will be evaluated. The amount of organic additives required to inhibit the conversion of a unit mole of aragonite to calcite is calculated based on the critical nucleation energy and the reduction of the interfacial energy of aragonite by organic additives. During this process, the influence of the different shapes of various $CaCO_3$ polymorphs and temperature on the interface energy will be analyzed in detail.

The proportion of aragonite refers to the molar ratio of aragonite to the total amount of calcium carbonate. The schematic diagram of the model calculation

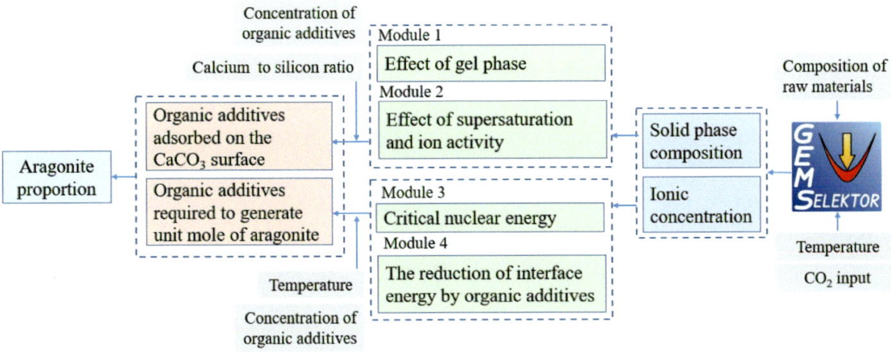

FIGURE 5.9 Schematic diagram of the model calculation flowchart.

flowchart is shown in Figure 5.9. The amount of aragonite depends on the amount of effective organic additives adsorbed on calcium carbonate, divided by the amount of organic additives required to produce per unit mole of aragonite, as shown in Equation (5.18). It is worth noting that in the mineralization system of cement-based materials, the calculation of the former requires considering the changes in surface area of $CaCO_3$, as well as the effect of organic additive adsorption in gel phases with varying Ca–Si ratios. The phase assemblages and ion concentration under different mineralization conditions (such as CO_2 input and temperature) are calculated utilizing the thermodynamic software GEM SELEKTOR 3.0. The amount of organic additives required to generate per unit mole of aragonite is calculated based on the critical nucleation energy and the reduction of the surface energy of aragonite by organic additives. During this process, the influence of the different shapes of various $CaCO_3$ polymorphs and temperature on the interface energy will be analyzed in detail.

$$X_A = \frac{n(\text{Ara})}{n_{\text{tot}}(\text{CaCO}_3)} = \frac{n_c(\text{OA})}{n_r(\text{OA})} \frac{1}{n_{\text{tot}}(\text{CaCO}_3)} \tag{5.18}$$

where $n(\text{Ara})$ is the amount of aragonite, $n_{\text{tot}}(\text{CaCo}_3)$ is the total amount of calcium carbonate, $n_c(\text{OA})$ is the amount of organic additives adsorbed on calcium carbonate layer, $n_r(\text{OA})$ is the amount of organic additives required to generate per unit mole of aragonite. The following parts will focus on the specific calculation procedures of $n_c(\text{OA})$ and $n_r(\text{OA})$, respectively.

Existing studies have investigated the adsorption of organic additives on calcium carbonate surfaces. However, examining this phenomenon within wet mineralization systems of cement-based materials remains challenging. The principal challenge arises from the complexity of varying ion activities and calcium carbonate supersaturation during mineralization—particularly under simultaneous hydration and carbon mineralization reactions—as well as from the influence of gel phases (such as C-S-H with different calcium-to-silicon ratios and distinct adsorption capacity) on organic additive adsorption. In this part, a thermodynamic model will be established using GEMS to calculate phase assemblages, ion activity, and calcium carbonate supersaturation under different mineralization conditions, as well as to quantify the content of C-S-H with varying calcium-to-silicon ratios. Subsequently, the Temkin II model will be utilized to determine the effective adsorption of organic additives on the calcium carbonate surface.

The amount of organic additives adsorbed on the surface of calcium carbonate layer can be calculated using Equation (5.19):

$$n_c(\mathrm{OA}) = \left[n_0(\mathrm{OA}) - n_s(\mathrm{OA})\right] \cdot \theta_c \tag{5.19}$$

where n_0(Ara) is the initial amount of organic additives, n_s(OA) is the adsorption capacity of C-S-H and silica gel on organic additives, and θ_c is the coverage rate of organic additives on the surface of calcium carbonate.

A systematic investigation was conducted on adsorption models applied to cement-based materials, including the Langmuir (Hsu et al., 1999; Marchon et al., 2016), Freundlich (Wu et al., 2019), Fritz-Schlünder (Svilović et al., 2019), Elovich (Wu et al., 2019; Yousef, Farouq, and Hazzaa, 2016), Fowler-Guggenheim (Afonso, Gales, and Mendes, 2016), Temkin, and the modified Temkin II models (Chu, 2021; Svilović et al., 2019). The Temkin II model was ultimately selected because it avoids the monolayer adsorption assumption of the Langmuir model, accounts for the heterogeneity of binding sites, and has been found suitable for adsorption on non-uniform surfaces (Hinz, 2001; Silva Filho et al., 2013; L. Wang et al., 2018). The Freundlich and Elovich models are also applicable to the adsorption on heterogeneous surfaces. However, the Temkin model considers intermolecular interactions between adsorbate molecules, making it more suitable for the adsorption of long-chain molecules like organic additives. In contrast, the Freundlich and Elovich models are more commonly used for the immobilization of heavy metal ions.

The Temkin model describes the functional relationship between the adsorption energy and the surface coverage of surfactant molecules on the adsorbent (MI Temkin, 1940). The Temkin II model is a modified version of the Temkin equation, addressing its limitation of being applicable only to moderate concentration ranges (Chu, 2021). This modification has been successfully applied to both Type I and Type II isotherms (Chu, 2014).

Based on the Temkin II model, the surface coverage rate of organic additives on calcium carbonate can be calculated using the following Equation (5.20).

$$\theta = (RT / b_T)\ln\left(1 + K_T c\right) \tag{5.20}$$

where θ is the surface coverage rate, c is the concentration of additives, R is the gas constant, T is the temperature (K), K_T is the equilibrium binding constant (L/mol), b_T is the variation of adsorption energy (kJ/mol).

Considering the high water-to-solid ratio in aqueous mineralization system, a significant portion of calcium carbonate may precipitate separately, with solution supersaturation serving as a crucial factor in determining its surface area (Jimoh et al., 2018; Shuvo et al., 2024). It has been observed that under certain low supersaturation conditions, the growth rate of calcium carbonate is relatively slow or even incomplete, which significantly affects the surface area (Romanek, Morse, and Grossman, 2011; Zhu et al., 2015). Existing studies have found that relative supersaturation is a key parameter influencing surface area, with a power-law relationship between them. The relationship between the actual surface area and the initial surface area can be approximately characterized using Equation (5.21) (Wiechers, Sturrock, and Marais, 1975; Zhu et al., 2015).

$$A_a = A_i \left[1 + \left(k_b \sqrt{\frac{\left[Ca^{2+}\right]\left[CO_3^{2-}\right] a_D^2}{K_{sp}}} - 1 \right)^2 \right] \tag{5.21}$$

where A_a is the actual surface area, A_i is the initial surface area, $[i]$ is the concentration of species i, k_b is the empirical coefficient, and a_D is the activity coefficient for divalent ions.

The activity coefficient is calculated by the extended Debye–Hückel equation, as shown in Equation (5.22):

$$\log a_i = \frac{-A_y z_i^2 \sqrt{I}}{1 + B_y d_i \sqrt{I}} + b_y I \tag{5.22}$$

where A_y and B_y are coefficients related to pressure and temperature, Z_i is the charge carried by component i, I is the effective molar ionic strength, d_i is the ion size parameter, and b_y is the semi-empirical parameter.

Thermodynamic calculations were performed to determine the phase assemblages, ion concentrations, activity coefficients, and supersaturation under different mineralization conditions. Thermodynamic software GEM SELEKTOR 3.0 (Kulik, 2013; Wagner et al., 2012) and thermodynamic database Cemdata 18 (Lothenbach et al., 2019) and PSI/Nagra (Thoenen et al., 2014) were used for modeling the formation of hydration and mineralization products. The oxide composition of Portland cement used in the experiment, as shown in Table 5.2, was input into the dependent component module of GEMS.

The adsorption of organic additives on C-S-H gel is more complex than on calcium carbonate. Existing literature provides data on the adsorption capacity of C-S-H

TABLE 5.2
Oxide composition of Portland cement (%)

Oxide	CaO	SiO_2	Al_2O_3	Fe_2O_3		f-CaO	Na_2O	SO_3	LOI
Content	62.05	20.53	4.45	3.17	2.81	0.8	0.55	2.1	1.74

TABLE 5.3

Chemical composition and thermodynamic parameters of different gel phase

Phase	Chemical formula	Ca–Si	$\Delta_fG°$(kJ/mol)	$\Delta_fH°$(kJ/mol)	V°(cm^3/mol)
TobH	$C_{2/3}SH_{1.5}$	0.67	−1668.56	−1841.5	55
TobD	$C_{5/6}S_{2/3}H_{1.83}$	1.25	−1570.89	−1742.4	48
JenH	$C_{1.33}SH_{2.17}$	1.33	−2273.99	−2506.3	76
JenD	$C_{1.5}S_{0.67}H_{2.5}$	2.25	−2169.56	−2400.7	81

for organic additives; however, this is often considered an average adsorption value (Androniuk et al., 2017; Nalet and Nonat, 2016). Since the mineralization process is accompanied by continuous C-S-H decalcification, evaluating adsorption capacity based on changes in the calcium-to-silicon ratio (Ca–Si) and the specific composition of C-S-H under different mineralization degrees is more reasonable. To characterize C-S-H gel with different calcium-to-silicon ratios, the CSHQ model (Kulik 2011) was employed, which classifies C-S-H gel into four end-members: CSHQ-JenD ($C_{1.5}S_{0.67}H_{2.5}$), CSHQ-JenH ($C_{1.33}SH_{2.17}$), CSHQ-TobD ($C_{0.83}S_{0.67}H_{1.83}$), and CSHQ-TobH ($C_{0.67}SH_{1.5}$), with their chemical compositions and thermodynamic parameters listed in Table 5.3. As the mineralization progresses, high-calcium phases gradually transform into low-calcium phases.

Using GEMS, the content of different gel phases can be calculated. The Ca–Si ratio significantly influences the adsorption of organic additives on C-S-H, with studies indicating that as the Ca–Si ratio increases, the adsorption of small organic molecules on the C-S-H surface increases approximately linearly (Androniuk et al., 2017; Glaus, Laube, and Van Loon, 2006; Nalet and Nonat, 2016). Therefore, the surface coverage of specific C-S-H end-members can be estimated using Equation (5.23).

$$\theta_{CSH-X} = \text{f}\left(\text{Ca/Si}\right)\theta_{CSH-ave} = \left[2.533\left(\text{Ca/Si}\right) - 2.511\right]\theta_{CSH-ave} \tag{5.23}$$

where θ_{CSH-X} is the surface coverage rate of C-S-H end member X (e.g., C-S-H-TobH), $\theta_{CSH-ave}$ is the average surface coverage rate of C-S-H, $\left(\text{Ca/Si}\right)$ is the calcium to silicon ratio of C-S-H?

Therefore, the effective amount of organic additives adsorbed on the calcium carbonate surface can be calculated using Equation (5.24).

$$n_s\left(\text{OA}\right) = \left[C_0\left(\text{OA}\right)V_{\text{sol}} - \sum_{i=1}^{4} n_i\theta_i\right] \tag{5.24}$$

$$\left(\text{i=1-4 represents TobH, TobD, JenH, JenD, respectively}\right)$$

After calculating the adsorbed capacity of organic additives, the next step is to calculate the amount of organic additives required to form aragonite. The organic additive required to generate one mole of aragonite can be calculated from Equation (5.25).

$$n_r(OA) = \frac{\gamma_0 - \gamma_c}{\gamma_{sc}} \tag{5.25}$$

where γ_0 is the initial interface energy of aragonite, γ_c is the critical interface energy (i.e., the interface energy of aragonite when the critical nucleation of aragonite and calcite is the same), and γ_{sc} is the change of interface energy caused by one mole of organic additives.

The critical nucleation energy ΔG^* is mainly affected by the interface energy, temperature, supersaturation, and molecular volume. Among them, the interface energy γ is the most critical factor, and the relationship between them can be expressed by Equation (5.26) (Ahn et al., 2005).

$$\Delta G^* = \frac{16\pi v^2 \gamma^3}{3{K_B}^2 T^2 (\ln S)^2} \tag{5.26}$$

where γ is interface energy, v is molecular volume, K_B is Boltzmann constant, S is supersaturation, which can be calculated by Equation (5.27):

$$S_{CaCO_3} = \left(\frac{(Ca^{2+})({CO_3}^{2-})}{K_{sp}}\right)^{1/2} = \left(\frac{a_{Ca^{2+}} a_{CO_3^{2-}} [Ca^{2+}][{CO_3}^{2-}]}{K_{sp}}\right)^{1/2} \tag{5.27}$$

where (i) is the activity of ion i,K_{sp} is the solubility product, a_i is the activity coefficients of the ion i, and $[i]$ is the concentration of ion i.

When the interface energy of aragonite is higher than that of calcite, the formed calcium carbonate polymorphs will be transferred from aragonite to calcite. Alternatively, if not, the aragonite will remain in a stable state within the mineralized cement-based materials. In other words:

$$\begin{cases} \text{when } \Delta G^*_{Ara} > \Delta G^*_{Cal} & CaCO_3(\text{Aragonite}) \rightarrow CaCO_3(\text{Calcite}) \\ \text{when } \Delta G^*_{Ara} \leq \Delta G^*_{Cal} & CaCO_3(\text{Aragonite})(\text{stably exist}) \end{cases} \tag{5.28}$$

In this part, the Gibbs free energy minimization method is adopted to calculate the phase assemblages and ionic activity under different conditions. Based on Equations (5.25)–(5.28), the critical interface energy can be calculated using Equation (5.29).

$$\gamma_c = \left(\frac{\gamma_{cal}^3 v_{cal}^2}{v_{ara}^2}\right)^{\frac{1}{3}} \left[\frac{\ln(a_{Ca^{2+}} a_{CO_3^{2-}} [Ca^{2+}][{CO_3}^{2-}]) - \ln ksp_{ara}}{\ln(a_{Ca^{2+}} a_{CO_3^{2-}} [Ca^{2+}][{CO_3}^{2-}]) - \ln ksp_{cal}}\right]^{\frac{2}{3}} \tag{5.29}$$

The effect of ion adsorption on the surface energy of vaterite (sphere), calcite (cube), and aragonite (needle-like) is quantified based on the lattice constants of different calcium carbonate polymorphs. In addition, based on Nernst's law and Debye length calculation, the effect of organic additive concentration on surface energy is evaluated.

The interface energy of the ionic crystal is determined by the interface energy of the crystal at the zero charge point and the surface charge of the ionic crystal, as shown in Equation (5.30).

$$\gamma = \gamma_0 - \gamma_{SC} \tag{5.30}$$

where γ_0 is the interface energy at the zero charge point and can be calculated from Equation (5.31) (Grahame, 1947), γ_{sc} is the change of interface energy caused by the change of surface charge of ionic crystal.

$$\gamma_0 = 0.414 \cdot \mathrm{K_B} T \left(\frac{1000 \cdot \rho}{M_{\mathrm{WC}}} \mathrm{N_{Av}} \right)^{2/3} \ln \left(\frac{1000 \cdot \rho}{M_{\mathrm{W}} \left(\left[\mathrm{Ca}^{2+} \right] \left[\mathrm{CO}_3^{2-} \right] \right)^{1/2}} \right) \tag{5.31}$$

where M_{WC} is the molar mass of calcium carbonate, N_{AV} is the Avogadro constant, and $[i]$ is the concentration of phase i.

The surface charge of ionic crystals γ_{SC} can be determined by the adsorption of ions at the interface and the approach of point charge to the interface.

$$\gamma_{SC} = \frac{\mathrm{R}T}{z\mathrm{F}} \{ f(ad) + f(ap) \} \tag{5.32}$$

where Z is the charge carried by electrolyte, for calcium carbonate Z = 2, F is the Faraday constant, the factors f(ad) and f(ap) indicate the influence of ion adsorption and the approach of point charge to the interface on surface energy, respectively.

The addition of organic additives has a significant impact on the adsorption of ions at the interface. Specifically, the concentration and dissociation constant of additives will affect the change of surface charge, as shown in Equation (5.33). Also, due to the different shapes of calcium carbonate polymorphs, the influence of the adsorption of organic additives on their interface energy is different.

$$\begin{aligned} f(ad) &= f(ads) f(shp) \sigma_{\max} \ln \left(\frac{\sigma_{\max}}{\sigma_{\max} - |\sigma|} \right) \\ &= \frac{k_{ads} c_i}{k_{ads} c_i + 1} \frac{k_i}{k_c} \beta \sigma_{\max} \ln \left(\frac{\sigma_{\max}}{\sigma_{\max} - |\sigma|} \right) \end{aligned} \tag{5.33}$$

where σ_{max} is the maximum charge density at the interface, $f(ads)$ and $f(shp)$ are the adsorption factor and shape factor, k_{ads} is the adsorption coefficient, c_i is the concentration of organic additives, K_i is the pKa of the organic additives, and k_c is the pKa of carbonic acid, β is the shape factor. For the three polymorphs of calcium carbonate, vaterite, calcite, and aragonite, their representative shapes are spherical, cubic, and needle-like, respectively. The shape factor β of different calcium carbonate polymorphs is defined by Equation (5.34) (Ahn et al., 2005):

$$\beta = \frac{4b^3}{27a^2} \tag{5.34}$$

The interface charge density σ and maximum charge density σ_{max} are calculated using Equations (5.35) and (5.36), respectively:

$$\sigma = \frac{4FI}{\kappa}\sinh\left(\frac{F\Phi_0}{2RT}\right) \tag{5.35}$$

$$\sigma_{max} = zF\frac{\rho a_0}{M_W}\cdot 10^6 \tag{5.36}$$

where a_0 is the lattice constant, φ_0 is the surface potential, which is calculated based on Nernst's law, as shown in Equation (5.37).

$$\Phi_0 = \frac{RT}{4F}\left\{\ln\left[\frac{(Ca^{2+})}{(Ca^{2+})_{pzc}}\right] - \ln\left[\frac{(CO_3^{2-})}{(CO_3^{2-})_{pzc}}\right]\right\} \tag{5.37}$$

where (i) is the activity of phase i, and the subscript "PZC" represents the state at zero charge, $(i)_{PZC}$ can be calculated by the equilibrium state of the crystal in pure water.

The influence of the approach of point charge to interface on surface energy can be quantified by Equation (5.38) (Mersmann, 1990):

$$f(ap) = \frac{8FI}{\kappa}\left[\sqrt{1+\left(\frac{\kappa\sigma}{4FI}\right)^2} - 1\right] \tag{5.38}$$

where I is the ionic strength, and 1/k is the Debye length calculated by Equation (5.39):

$$\frac{1}{\kappa} = \sqrt{\frac{\varepsilon_0\varepsilon_r RT}{2F^2 I}} \tag{5.39}$$

where ε_0 is the dielectric constant of vacuum (F/m), ε_r is the relative dielectric constant of solvent.

Overall, the model is constructed based on the idea that the content of aragonite is equal to the effective organic additive content adsorbed on the calcium carbonate surface divided by the amount of organic additives required to thermodynamically inhibit the conversion of one mole of aragonite to calcite. Based on Equations

(5.19)–(5.24), the adsorption capacity of the gel phase with a variable calcium to silicon ratio relative to organic additives can be calculated. The purpose of the above part is to calculate the effective organic additive content that affects the polymorphs of calcium carbonate. Based on Equations (5.25)–(5.29), the critical surface energy at which the transformation between aragonite and calcite occurs can be calculated, while based on Equations (5.30)–(5.39), the extent to which organic additives reduce the surface energy can be calculated. In this process, the difference in shape between calcite and aragonite emerges as a key factor influencing the variation in surface energy changes. Therefore, shape factors of different calcium carbonate polymorphs were constructed based on lattice constants to quantify the effect of organic additives adsorbed on cubic calcite and fibrous aragonite surfaces.

5.5.3 Comparison of Modelling Results with Literature and Experimental Results

To verify the accuracy of the model, the results from several published works were compared with the calculated results. It can be seen from Figure 5.10 that in a wide concentration range, the calculated proportions of aragonite are in good agreement with the published results. With the increase of magnesium chloride concentration, the proportion of aragonite gradually increases until all calcium carbonate exists in the form of aragonite.

Considering the fluctuation of literature data, experiments were also conducted to validate the model. Portland cement was used to prepare the cement paste. Anhydrous magnesium chloride was used as an additive for inhibiting the transformation from aragonite to calcite. CO_2 with a concentration of more than 99.5% was used for aqueous mineralization of Portland cement. Deionized water was used in the entire

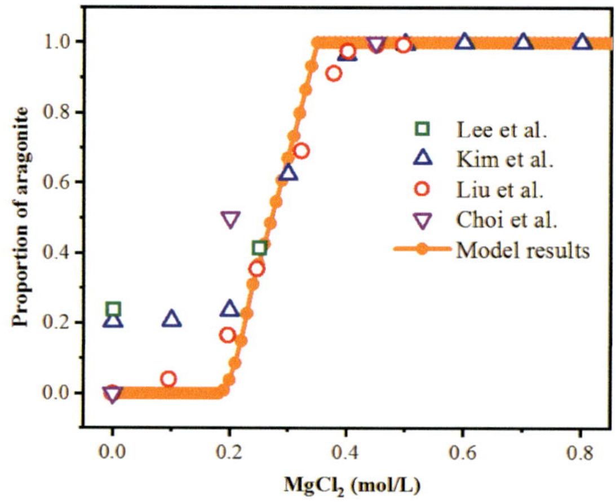

FIGURE 5.10 Proportion of aragonite from the published works and the corresponding calculated results from the model (K.S. Choi, J.K. Park, J.W. Ahn, 2004; J. H. Kim et al., 2006; Lee et al., 2015; Q. Liu et al., 1998).

process of sample preparation. $MgCl_2$ solution (200 ml) with concentrations of 0, 0.05, 0.1, 0.125, 0.15, 0.175, 0.25 mol/L was prepared using deionized water. The reactor containing these solutions were placed in a water bath device with a constant temperature and stirred at a constant rate of 850 rpm. The mineralization experiment will be conducted at different temperatures (25°C and 80°C) and mass of CO_2 input ($M(CO_2)/M(PC)$ = 0.05, 0.25, 0.4, 0.45, 0.65, 1.0). 20 g Portland cement was added to the reactors, and CO_2 gas was introduced into the solution at a constant rate of 0.5 Nl/min. After 60 minutes, stop the reaction, take out the reactor and make them stand for 10 minutes, remove the supernatant to get wet samples. The wet samples were centrifuged at the rate of 8000 rpm for 5 minutes and the process was repeated three times. After centrifugation, the samples were placed in a 40°C vacuum drying oven for 3 days and then ground to obtain mineralized powders. The flowchart of the mineralization experiment and the illustration of the set-up for the mineralization device are shown in Figures 5.11 and 5.12, respectively.

The X-ray diffraction technology was performed using a Cu kα X-ray radiation at 40 kV and 40 mA. The 2θ scanning range was 3°–70° with a step size of 0.02° at the rate of 2 °/min. 20 wt% of rutile (TiO_2) was added to the sample powders as an internal standard. Rietveld quantitative analysis was carried out using X'Pert Highscore Plus software version 5.1 (PANalytical).

The XRD spectra of the mineralized Portland cement prepared at different temperatures and concentrations of $MgCl_2$, and the corresponding proportions of aragonite calculated by XRD Rietveld analysis, and the model are shown in Figures 5.13 and 5.14, respectively. It can be found that after the magnesium chloride concentration reaches the critical value (concentration at which aragonite begins to form), the proportion of aragonite gradually increases with the increase of magnesium chloride concentration. The increase in temperature seems to reduce this critical concentration. The saturation concentration of magnesium chloride also decreases with the increase of temperature, indicating that the concentration of magnesium chloride required to obtain the mineralization products completely composed of aragonite whiskers is reduced. The average error between experimental and calculated results

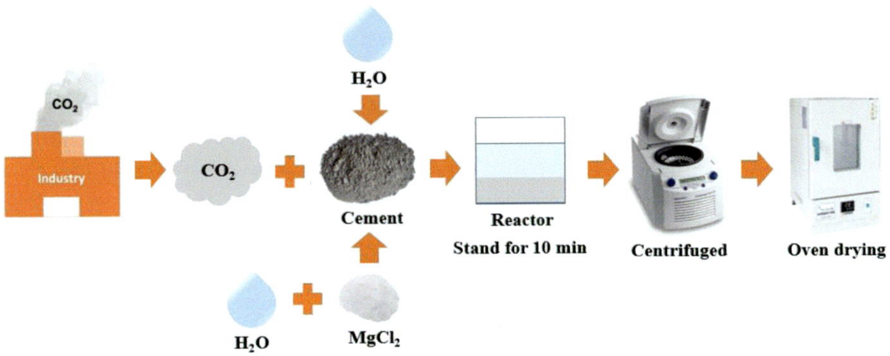

FIGURE 5.11 The flowchart of the mineralization experiment.

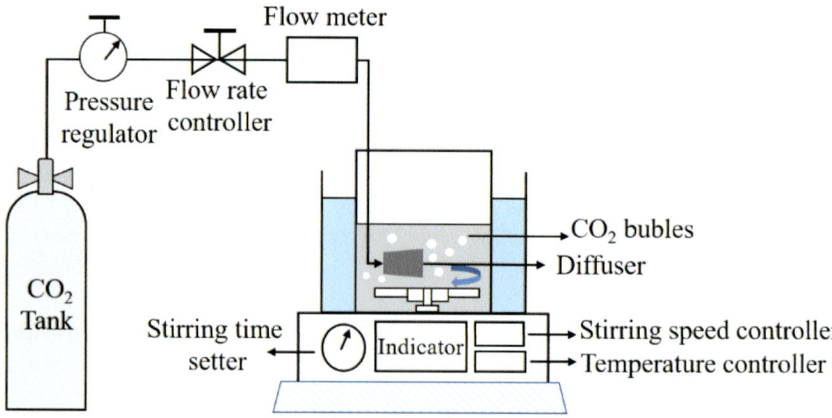

FIGURE 5.12 The illustration of the set-up for the mineralization reactor.

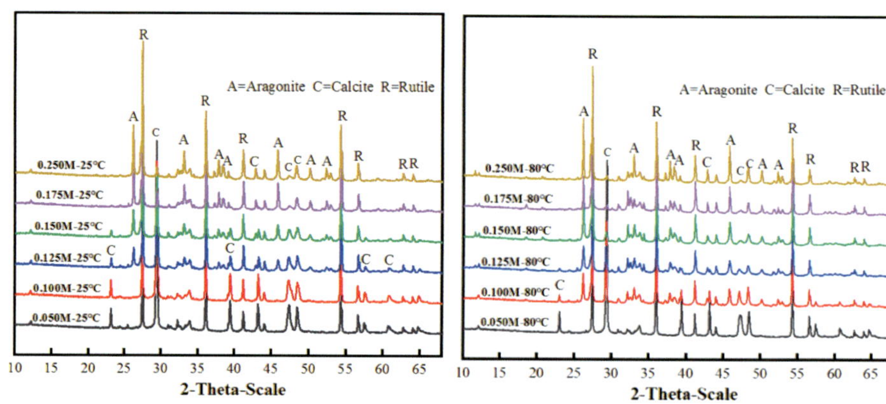

FIGURE 5.13 XRD spectra of mineralized Portland cement prepared at different temperatures and concentration of $MgCl_2$.

from the model is 5.51%, which proves that the model can predict the proportion of different calcium carbonate polymorphs in aqueous mineralization system within a small error range.

In addition to inorganic aragonite inducers, experiments were also conducted using organic additives to induce the formation of aragonite. Considering that there are currently no directly related experimental works available, the developed model was initially used to simulate the most common $Ca(OH)_2$-CO_2-H_2O system with organic additives. In order to verify the accuracy of the model, the literature experimental results of using organic additives with different functional groups (PAA, PAM, and PVA) to prepare calcium carbonate in different polymorphs were compared with the simulation results, as shown in Figure 5.15. Basic information about these three organic additives is shown in Table 5.4. It can be seen from Figure 5.15

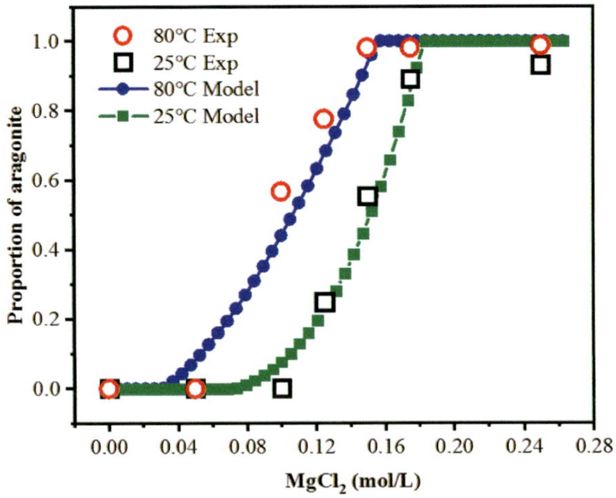

FIGURE 5.14 Proportion of aragonite calculated by XRD Rietveld analysis and the model.

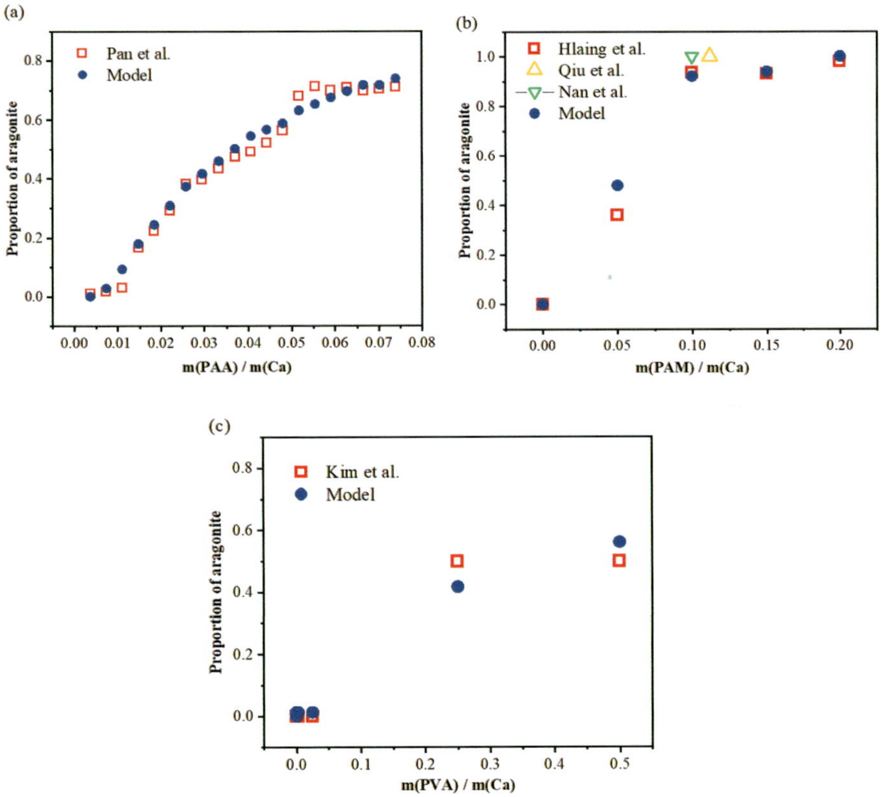

FIGURE 5.15 Comparison between experimental results and corresponding model results in the case of using (a) PAA, (b) PAM, and (c) PVA additives. (Data from Hlaing et al., 2014; I. W. Kim, Robertson, and Zand, 2005; Nan et al., 2008; Pan et al., 2007; Q. Yu et al., 2006).

TABLE 5.4
Properties of the organic additives used as crystalline inhibitors

Organic additives	Abbreviation	Molecular formula	Density ($kg \cdot m^{-3}$)	Molar mass ($g \cdot mol^{-1}$)	Chemical structure
Poly(acrylic acid)	PAA	$(C_3H_4O_2)_n$	1090	72n	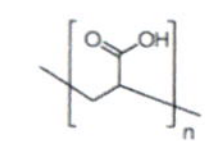
Polyacrylamide	PAM	$(C_3H_5NO)_n$	1302	71n	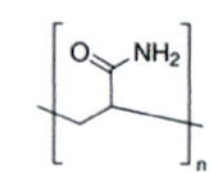
polyvinyl alcohol	PVA	$(C_2H_4O)_n$	1270	44n	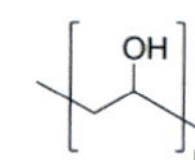

that the model results are close to the experimental results. The average error between the experimental results and the corresponding model results is E(PAA)=2.70%, E(PAM)=4.61% and E(PVA)=3.05%, respectively.

In addition to the concentration of additives, temperature was also considered as an independent variable, with statistical analysis, calculations, and comparisons conducted accordingly. The results are shown in Figure 5.16. Similarly, deviations between the model and literature results fall within an acceptable range. The average error between the literature results and the simulation results is 6.52%.

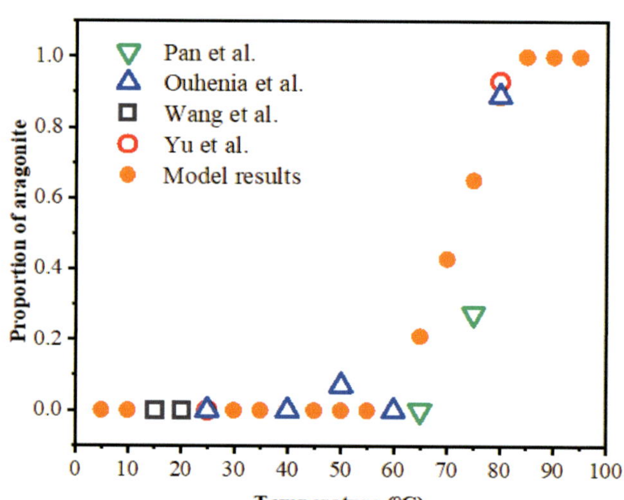

FIGURE 5.16 Variation of experimental and modeling results with temperature (Data from Ouhenia et al., 2008; Pan et al., 2007; C. Wang et al., 2006; J. Yu et al., 2004).

Since no research data has been found on using organic additives to control $CaCO_3$ polymorphs in mineralized cement systems, this work will conduct wet mineralization experiments to validate the accuracy of the developed model within cement systems. Monoethanolamine (MEA) contains both amine and hydroxyl groups as functional groups. It is well known for its exceptional CO_2 capture capability. Additionally, MEA is a highly effective aragonite inducer, although studies on this aspect remain limited. In this study, it is considered a beneficial attempt to promote CO_2 capture during mineralization while regulating the formation of aragonite. In this part, its induction effect on aragonite formation under various mineralization parameters will be examined. Thermodynamic simulations will be performed concurrently, and the simulation results will be compared with the experimental findings.

MEA solution (200 mL) with concentrations of 0, 0.5, 1.0, 1.5, 2.0, 2.5, 3.0, 3.5, 4.0 mol/L was prepared using deionized water. The liquid-to-solid ratio for wet mineralization was set at 10:1, with 20 g of cement added to 200 mL of the MEA solution. The reactors containing these solutions were placed in a water bath device and stirred at a constant rate of 850 rpm. Different temperatures (25°C, 40°C, 50°C, 60°C, 70°C, 80°C) were set to investigate the effect of temperature on aragonite formation. CO_2 with a purity exceeding 99.5% will be injected into the reactor at a flow rate of 0.5 NL/min. The mineralization time was set to 60 minutes. After mineralization was completed, the reactor was removed and left to stand for 10 minutes. The supernatant was then removed to obtain wet samples. The wet samples will be centrifuged at 8000 rpm for 5 minutes, and this process will be repeated three times for each sample. After centrifugation, the samples were placed in a 40°C vacuum drying oven for 3 days and then ground to obtain mineralized powders.

In this part, the input parameters of the thermodynamic model are consistent with the experimental setup: 20 g of cement and 200 mL of water, with the temperature varying within the range of 40–80°C and the concentration of the organic additive ranging from 0-4 M. The cement module in the model was established based on the oxide composition of Portland cement used in experiments. The choice of 60 minutes as the mineralization time was based on the following reason. Due to the high liquid-to-solid ratio, stirring rate, and reaction temperature, experimental results showed that a mineralization time of 60 minutes is sufficient to ensure an almost fully mineralized state in the aqueous mineralization system containing 20 g of cement and 200 mL of water. Since the thermodynamic model is always designed to calculate the final state of the reaction, ensuring full mineralization allows for consistency between the experimental results and the simulation outcomes. The CO_2 input was calculated based on the total amount of $CaCO_3$ measured in the experiments. This builds up the consistency between the total $CaCO_3$ produced in the simulation and the experimental process. It should be acknowledged that this manner may still introduce some discrepancies compared to actual conditions. However, considering that the masses of other reactants—such as cement and water—are controllable, and the experiment is conducted under conditions close to full mineralization, the resulting error in the final amount of mineralization products is considered to be within an acceptable range.

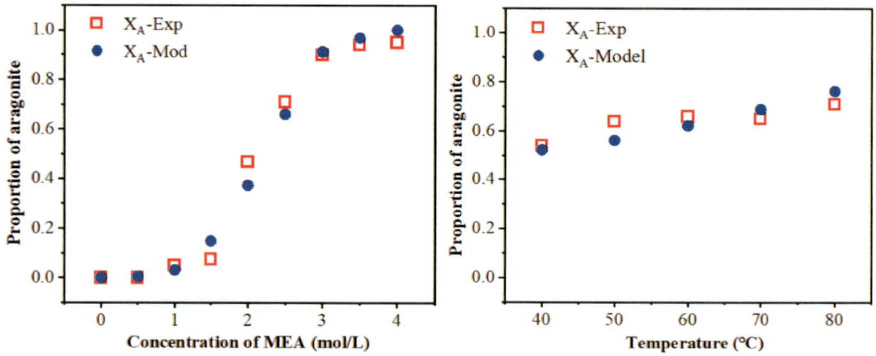

FIGURE 5.17 Comparison between experimental results and the corresponding model results at different (a) concentrations of MEA and (b) temperatures.

The proportion of aragonite measured by XRD Rietveld and the corresponding simulation results are shown in Figure 5.17. The experimental and simulated data are relatively close, with an average error of 3.71% and 4.52% in different concentration and temperature groups, which further verifies the accuracy of the model. The increase in the concentration of organic additives leads to a higher proportion of aragonite. An increase in temperature also promotes a greater proportion of aragonite. When the temperature was raised from 40°C to 80°C, the proportion of aragonite was observed to increase by approximately 30%. The specific mechanisms by which concentration and temperature affect the proportion of aragonite will be analyzed in detail in Section 3.5, in conjunction with thermodynamic simulations. The changes in the proportion of aragonite at different ages are shown in Figure 5.18. The proportion of aragonite fluctuated between 65.46% and 73.94% from day 3 to day 28. Although a maximum fluctuation of 8.47% was observed, the aragonite content can be considered approximately stable within the margin of error inherent to XRD Rietveld refinement. This stability arises from two factors: rapid reaction kinetics of aqueous mineralization and polymorph stabilization by organic additives. The system reached near-complete mineralization within the first few hours, minimizing further changes in $CaCO_3$ content during subsequent aging. The stabilization mechanism is linked to template-controlled epitaxial nucleation. Organic additives adsorbed onto $CaCO_3$ surfaces act as templates, directing oriented growth along specific crystallographic axes (Pan et al., 2007). After aragonite formation, these additives remain embedded within aragonite. Consequently, transforming aragonite to calcite requires energy-intensive lattice rearrangement, which is significantly less favorable than the templated growth of aragonite itself (I. W. Kim, Robertson, and Zand, 2005). This energy barrier inhibits spontaneous phase transformation. Such organically induced stable aragonite structures are actually common in nature, for example in nacreous layers of shells (Yoshimi et al., 2004), pearl oyster (Saruwatari et al., 2009), and red abalone nacre (Coppersmith, Gilbert, and Metzler, 2009). In these systems, biomacromolecules such as polysaccharides function similarly to synthetic organic additives, templating and stabilizing aragonite via interfacial molecular recognition (I. W. Kim, Robertson, and Zand, 2005).

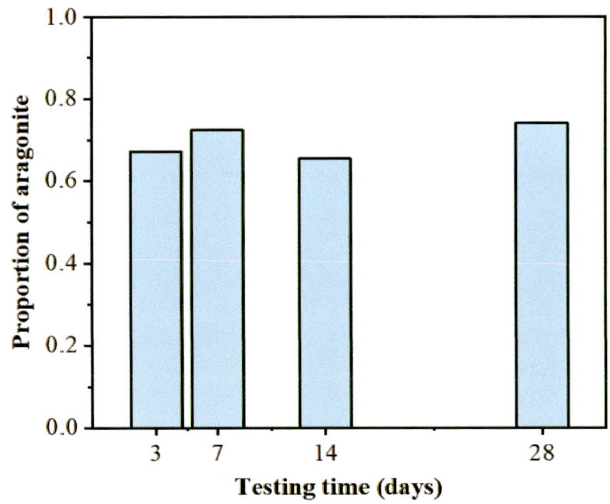

FIGURE 5.18 The proportion of aragonite at different testing times under 2.5 M MEA and 80°C.

5.6 PARAMETERS REGULATION OF $CACO_3$ POLYMORPHS DURING MINERALIZATION

5.6.1 Control of CO_2 Input

The phase assemblages and ion concentration change dramatically with the increasing amount of CO_2 injection. In different mineralization stages, the aqueous environment for the formation of different calcium carbonate polymorphs is also different. Besides, the addition of magnesium ion causes multiple changes to the hydration and mineralization reactions. Therefore, it is important to determine a suitable CO_2 injection amount and aqueous environment for improving the utilization efficiency of magnesium ion as a crystal stabilizer.

Figure 5.19 shows the phase assemblages of the aqueous mineralization system with different mass ratio of CO_2 to Portland cement ($M(CO_2)/M(PC)$) when the concentration of $MgCl_2$ is 0.15 mol/L and the control group without it. It can be found that the direct effect of magnesium chloride on the composition of the solid phase is the increase in the content of hydrotalcite, MSH, and dolomite-ord ($Ca_xMg_{1-x}(CO_3)_2$) and also the advance of their formation time. At the same time, it also leads to the reduction of the total amount of calcium carbonate. Besides, it is noteworthy that in the system with $MgCl_2$ added, when the $M(CO_2)/M(PC)$ is greater than 0.5, the composition of calcium carbonate changes from pure calcite to calcite and aragonite. MSH, as a transition phase, only appears when the $M(CO_2)/M(PC)$ is 0.3–0.55. Although a small amount of brucite appears in the early stage, it dissolves rapidly with the decrease of pH. Magnesium chloride shows a limited impact on hydration, which mainly results in the formation of a smaller amount of portlandite.

In order to evaluate the possible influence of the aqueous environment on the formation of different calcium carbonate polymorphs, the concentrations of different

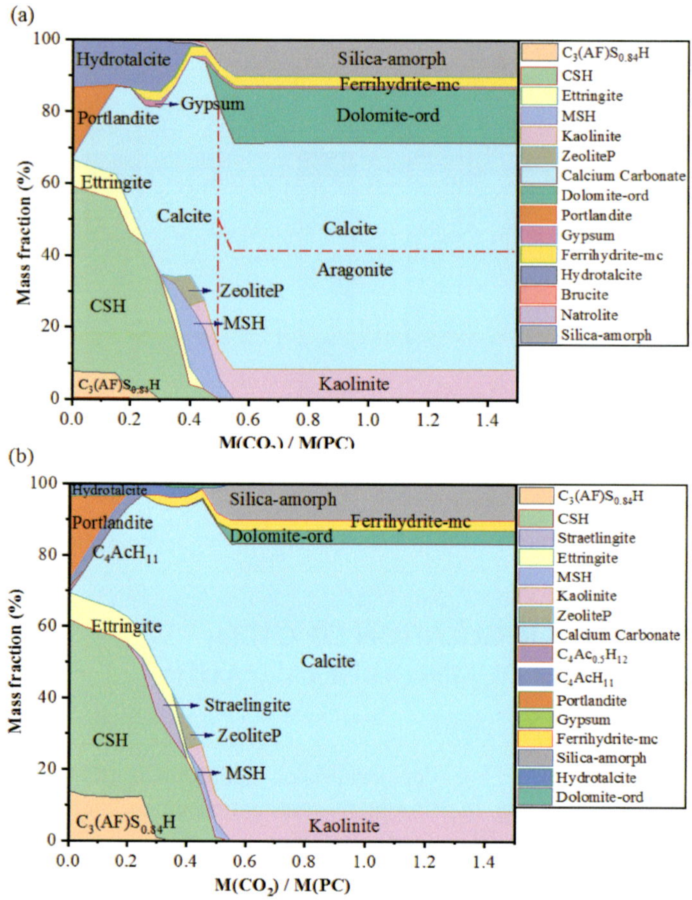

FIGURE 5.19 The change of phase assemblages with different mass ratio of CO_2 to Portland cement when the concentration of $MgCl_2$ is (a) 0.15 mol/L, (b) 0 mol/L.

magnesium species, calcium species, and carbonate species in the solution were calculated, as shown in Figure 5.20. It can be found that the concentration of magnesium ions in the solution is very low when the mass ratio of CO_2 to Portland cement is less than 0.15, only about 10^{-7} mol/L, which means that most of Mg exists in the solid phase dominated by hydrotalcite in the early stage of the mineralization. With the increase of CO_2 amount, the magnesium ion concentration in the solution gradually increases, especially when the M(CO_2)/M(PC) reaches 0.45; it almost increases by 1250 times compared with that of 0.4. This is mainly caused by the dissolution of the solid phases, and it should be noted that this mutation point may be different in different systems. Combining Figures 5.19 and 5.20, it can be found that the free magnesium ions are mainly released during the conversion of MSH to $Ca_xMg_{1-x}(CO_3)_2$. The added magnesium first exists in the form of hydrotalcite and brucite. With the increase of the amount of CO_2, brucite is transformed into hydrotalcite, and

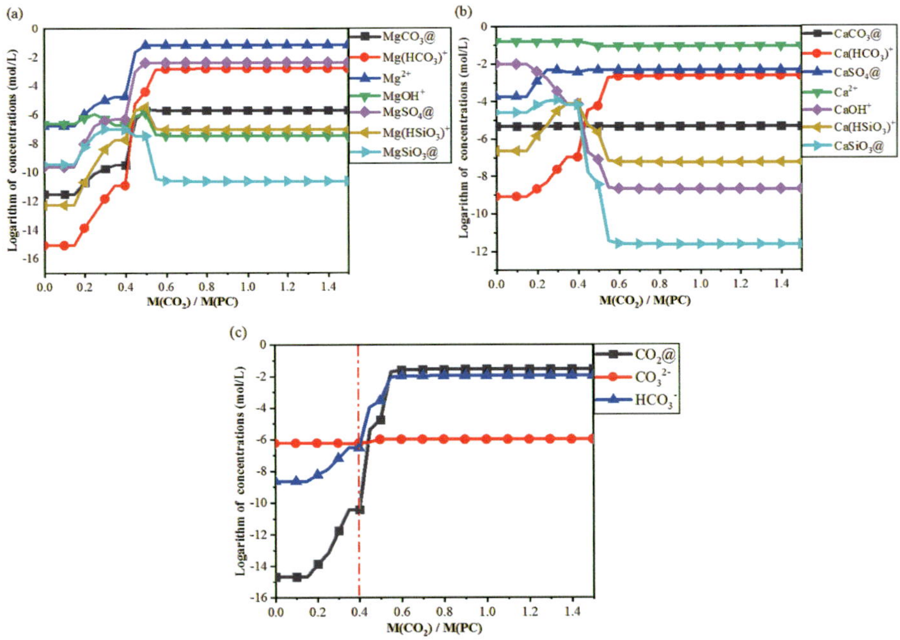

FIGURE 5.20 The concentration of (a) magnesium species, (b) calcium species, and (c) carbonate species in the solution with different mass ratio of CO_2 to Portland cement (@ indicates dissolved species).

then hydrotalcite is converted to MSH. MSH only exists for a short time and will be quickly transformed into $Ca_xMg_{1-x}(CO_3)_2$ and free magnesium ions. This part of magnesium ion has a significant impact on the stability of calcite and indirectly leads to the preferential formation of aragonite.

It can be concluded that the amount of CO_2 injection is one of the most crucial factors affecting the content of Mg in the solution. Therefore, if the magnesium ion is to be used as a crystal stabilizer, a relatively high CO_2 injection amount ($M(CO_2)/M(PC)>0.45$) is better to be ensured. This should be specially noted in the process with a low mineralization rate. When aragonite is not observed in mineralized cement-based material systems with magnesium-containing compounds added, it is helpful to check the mineralization degree and whether the added magnesium ions participate in the reaction of generating solid phases such as hydrotalcite. Unlike magnesium ions, the concentration of calcium ions in the solution remains relatively stable because the calcium in the system is excessive. This is also the reason the concentration of carbonate ions in Figure 5.20 (c) remains stable with the increase of CO_2 amount. In fact, the concentration of calcium ions is controlled by the solubility product of calcium carbonate. In such a system, increasing the amount of CO_2 does not increase the concentration of CO_3^{2-}, but increases the concentration of HCO_3^-. When the ratio of CO_2 to Portland cement reaches 0.4, the concentration of HCO_3^- in the solution is equivalent to that of CO_3^{2-} and will continuously increase with the

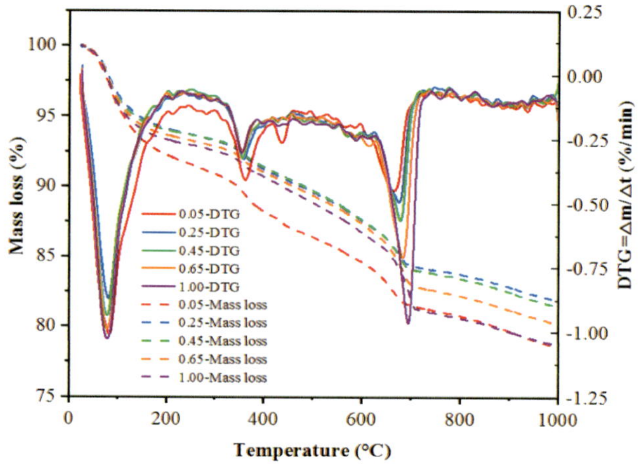

FIGURE 5.21 Thermogravimetric analysis of the mineralized samples at different $M(CO_2)/M(PC)$.

increase of the CO_2 amount. In such cases, $MgCO_3$ will not be formed, although the concentration of magnesium ion is high enough.

The TG-DTG curves of mineralized samples prepared under different $M(CO_2)/M(PC)$ conditions are shown in Figure 5.21. The thermal decomposition of the sample can be clearly divided into three stages, corresponding to three DTG peaks. The first peak is related to the thermal decomposition of amorphous phase CSH and MSH gel from 50°C to 200°C, the second peak corresponds to the decomposition of portlandite at 350–450°C, and the third peak corresponds to the decomposition of calcium carbonate at 600–720°C. Based on the mass loss during the thermal decomposition stage of calcium carbonate, the mass of reacted CO_2 can be obtained. By combining the amount of CO_2 input and CO_2 uptake, the CO_2 sequestration rate can be calculated. The experimental CO_2 sequestration rate has been compared with the results calculated by the model, as shown in Figure 5.22. It can be observed that their trends are consistent, with an average relative error of 6.9%. Although there are some errors, the model predicts the condition ($M(CO_2)/M(PC)=0.5$) under which the CO_2 sequestration rate begins to rapidly decrease within a small deviation range. This turning point may change under different raw material compositions of cement and environmental conditions, and this model can help to determine an appropriate CO_2 input amount, thereby improving the efficiency of CO_2 capture.

The proportion of aragonite under different $M(CO_2)/M(PC)$ conditions was also investigated. Similar trends have been observed in experimental and modeling results, as shown in Figure 5.23. Low $M(CO_2)/M(PC)$ is not conducive to the formation of aragonite. When $M(CO_2)/M(PC)\geq0.45$, magnesium ion can play its role as crystal stabilizer in the cement system. This is because CO_2 input affects the process of magnesium ions being released from the solid phase to the solution. Considering both the CO_2 sequestration rate and the proportion of aragonite, the appropriate CO_2 input range is $M(CO_2)/M(PC)=0.45–0.5$.

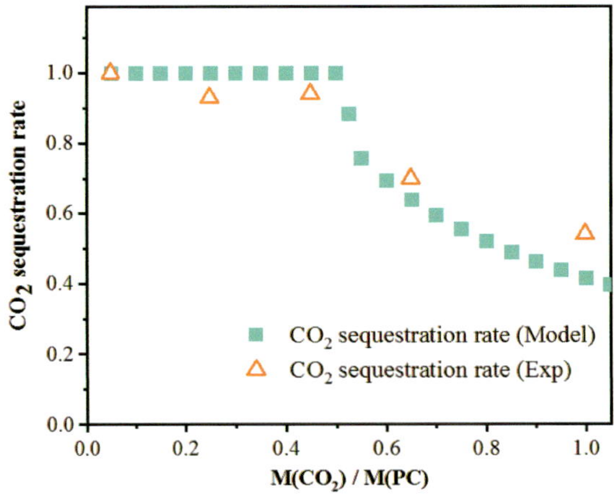

FIGURE 5.22 CO_2 sequestration rate calculated by thermogravimetric analysis and the model.

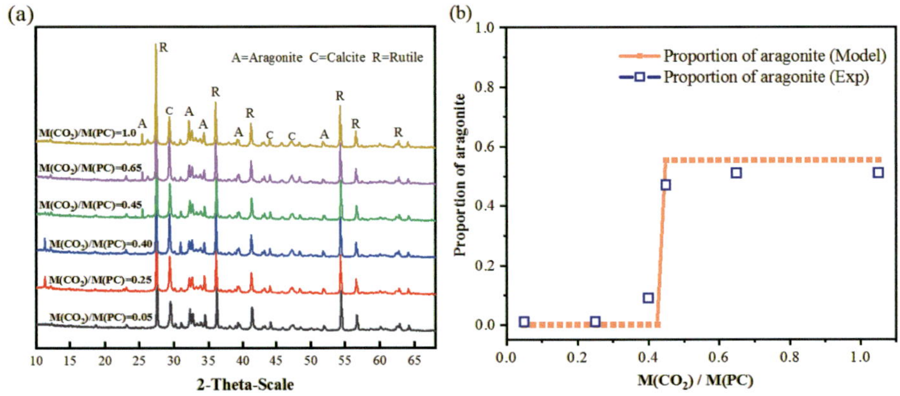

FIGURE 5.23 (a) XRD spectra of mineralized Portland cement prepared at different $M(CO_2)/M(PC)$ and (b) the comparison between results from XRD Rietveld analysis and the model.

Figure 5.24 shows the change of phase assemblages in the system with organic additives with the increase of the molar ratio of initial CO_2 and calcium ($n(CO_2)/n(Ca)$), which can also be regarded as a change in mineralization degree to a certain extent. In the early stage of mineralization, the first phase consumed is portlandite. After its complete consumption, CSH, C_4AcH_{11} and $C_3(AF)S_{0.84}H$ begin to participate in the mineralization reaction. It can be found that the CO_2 input mainly affects the total amount of $CaCO_3$, while the trend of the quality of calcite and aragonite is similar. The reason may be that the amount of CO_2 input does not have a significant impact on the difference in interfacial energy between calcite and aragonite, which will be discussed in detail later. The limited change in the ratio of calcite to aragonite

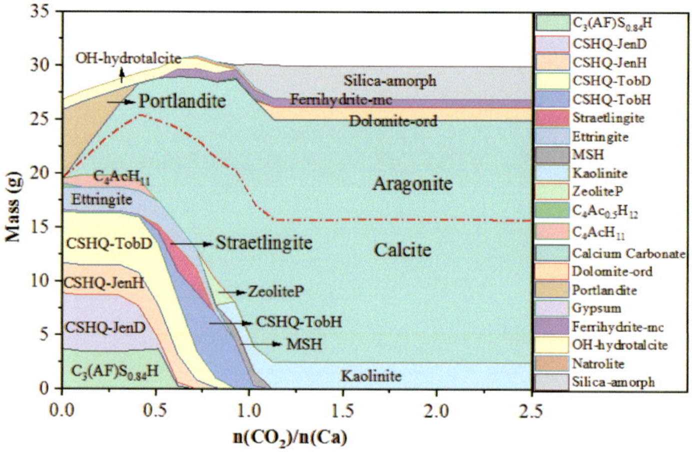

FIGURE 5.24 The change of phase assemblages with the change of $n(CO_2)/n(Ca)$.

may be attributed to the effect of CO_2 input on the content of CSH gel and its Ca–Si ratio, which will further affect the content of effective organic additives. The mineralization of CSH can be divided into two steps. The first step is the mineralization of CSHQ-TobD, CSHQ-JenH, and CSHQ-JenD, followed by CSHQ-TobH. When $n(CO_2)/n(Ca)$ reaches 0.5, part of the first three CSH gels are completely decalcified to form calcium carbonate, and the other part is partially decalcified to transform into CSHQ-TobH with a low calcium silicon ratio. Subsequently, ettringite and CSHQ-TobH reacted with CO_2, resulting in a rapid increase in calcium carbonate content. In this process, straetlingite, zeoliteP, and MSH as intermediate transition phases rapidly formed and then disappeared. After CSHQ-TobH was completely consumed, silica gel began to form and became the main gel phase in the system. This change in gel phase content has an impact on the adsorption of organic additives.

The adsorption of organic additives by the gel phase is one of the key distinctions between the mineralization system of cement-based materials and conventional chemically synthesized calcium carbonate systems. A portion of the initial organic additives will be adsorbed and captured by C-S-H gel rather than effectively binding with calcium ions in calcium carbonate. Therefore, this portion should be subtracted to obtain the effective organic additive concentration. Existing literature provides data on the adsorption capacity of C-S-H for organic compounds, but this is generally considered an average adsorption value (Androniuk et al., 2017; Nalet and Nonat, 2016). Since the mineralization process involves continuous decalcification of C-S-H, it is more appropriate to evaluate adsorption capacity by considering changes in the calcium-to-silicon ratio of C-S-H under different mineralization degrees and the specific composition of different C-S-H species.

Figure 5.25 illustrates the composition change of the gel phase within the system and its adsorption on MEA as the $n(CO_2)/n(Ca)$ ratio increases. According to the different main gel phases, the whole mineralization process can be divided into four stages. In the first stage, the main gel phase consists of three types of C-S-H gel with

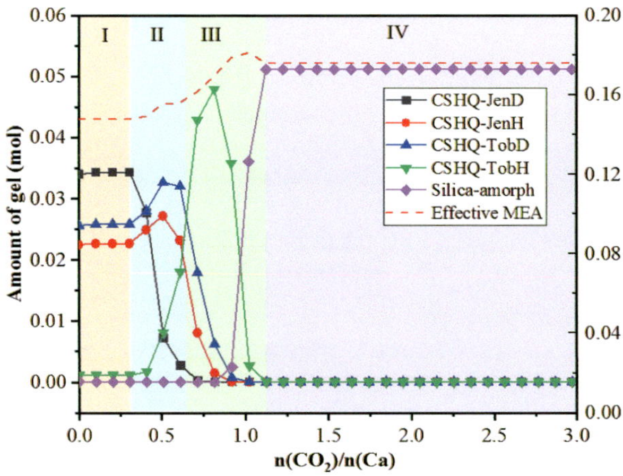

FIGURE 5.25 The changes of gel phase and MEA concentration in different mineralization stages.

a high calcium to silicon ratio: CSHQ-TobD, CSHQ-JenH and CSHQ-JenD. This specific C-S-H phase exhibits the most significant adsorption effect on MEA. When the initial MEA concentration is 0.2 mol/L, the residual concentration after C-S-H adsorption is about 0.14 mol/l, indicating an adsorption rate of about 30%. With the increase in CO_2 input, in the second stage, CSHQ-JenD with the highest calcium-silicon ratio is converted into the other three CSHQ gels. In the subsequent third stage, CSHQ-TobH with the lowest Ca–Si ratio became the main gel phase. In the second and third stages, the residual concentration of MEA continued to increase, but in the fourth stage, when all C-S-H was converted into silica-amorph, the residual concentration of MEA decreased slightly and remained stable. This can be interpreted as an early-stage gelatinous and flocculation structure influencing the interaction of organic additives with calcium carbonate. As mineralization progresses, the calcium carbonate content increases, and high Ca–Si ratio C-S-H gradually transitions into low Ca–Si ratio C-S-H, allowing organic additives to more effectively localize on the calcium carbonate surface, resulting in a higher effective MEA amount. It is worth noting that although the effective amount of MEA is relatively high at the stage IV, this does not mean that the complete mineralization stage is the most conducive to the formation of aragonite, because the amount of CO_2 input will also affect the surface tension and surface energy of $CaCO_3$, which will be discussed below.

The change in different ion concentrations will affect the ionic strength, activity coefficient, Debye length, and surface charge density of the system, thus further affecting the surface tension and interface energy. Figure 5.26 shows the changes in calcium, carbonate, sulfate, potassium, sodium ion concentration, and pH value as the mineralization progresses. The changes in pH value do not affect the amount of $CaCO_3$ but cause significant changes in the ionic composition of the solution (Pan et al., 2007; C. Wang et al., 2006). Understanding this change is important because many mineralization experiments often designate pH=7 as the stopping point for the

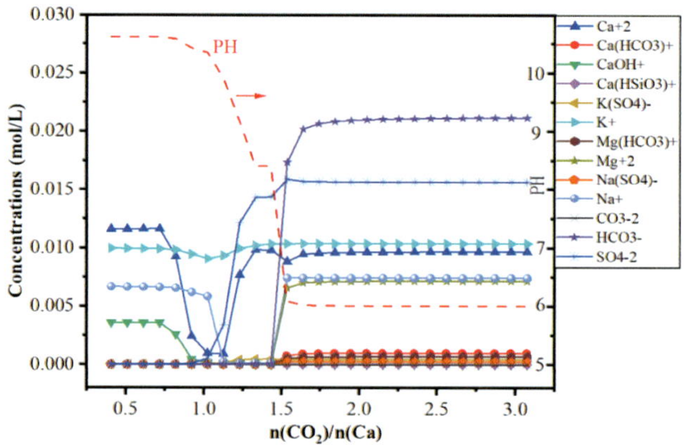

FIGURE 5.26 The concentration of calcium, carbonate, sulfate, potassium, sodium species, and pH value in the solution with the change of $n(CO_2)/n(Ca)$.

reaction. However, numerous carbonatable components remain in the solution at this stage. With the increase of $n(CO_2)/n(Ca)$, the pH of the solution begins to decrease at about $n(CO_2)/n(Ca)=0.75$ after the initial stabilization stage, accompanied by a gradual decrease in the concentration of Ca^{2+} and $CaOH^+$ ions in the solution. The rapid decline of pH is concentrated in the range of $n(CO_2)/n(Ca)=1–1.5$. At this stage, the concentration of Ca^{2+} rises again, with the increase of free sulfate ions, which is caused by the dissolution of ettringite and OH-hydrotalcite. The continuous injection of CO_2 after the consumption of calcium hydroxide will not increase the concentration of CO_3^{2-} in the solution but will lead to an increase in the concentration of HCO_3^- and dissolved CO_2. The bicarbonate ion is the main existing form after the dissolution of CO_2, which controls the final pH of the solution. When the pH value is stable, the concentration of various ions in the solution also tends to be stable, indicating that the excessive CO_2 will not further affect the liquid environment and surface energy of different $CaCO_3$ polymorphs.

Due to the change in ion concentration in the liquid environment, the interface energy of $CaCO_3$ will also change, as shown in Figure 5.27. When $n(CO_2)/n(Ca)$ is around 1.0, the surface energy of calcite and aragonite fluctuates significantly. The effect of the amount of CO_2 on the surface energy of calcite and aragonite is consistent. The surface energy first increases and then decreases, but the final surface energy is slightly higher than the initial value. The difference between the surface energies of calcite and aragonite does not change with the amount of CO_2, which can explain why simply increasing the concentration of CO_2 cannot inhibit the conversion of aragonite to calcite.

Although the change in CO_2 input does not change the initial interfacial energy difference between aragonite and calcite, it changes the ionic strength and activity coefficient of the system and leads to a change in the Debye length, surface charge density, and the reduction of the surface energy of calcium carbonate caused by MEA, as shown in Figure 5.28. Debye length represents the influence scale of electric

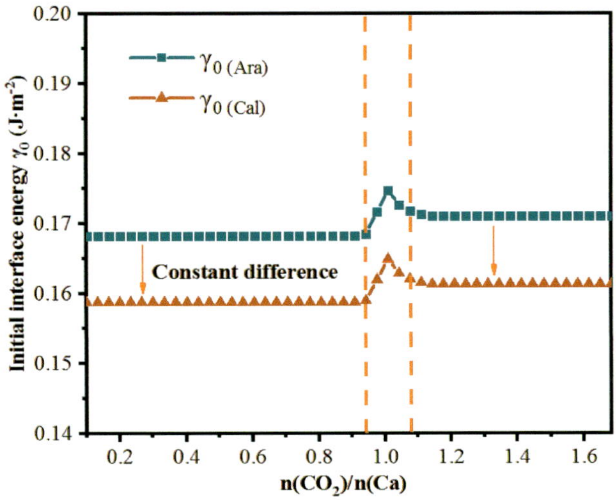

FIGURE 5.27 Variation of initial surface energy of calcite and aragonite with different $n(CO_2)/n(Ca)$.

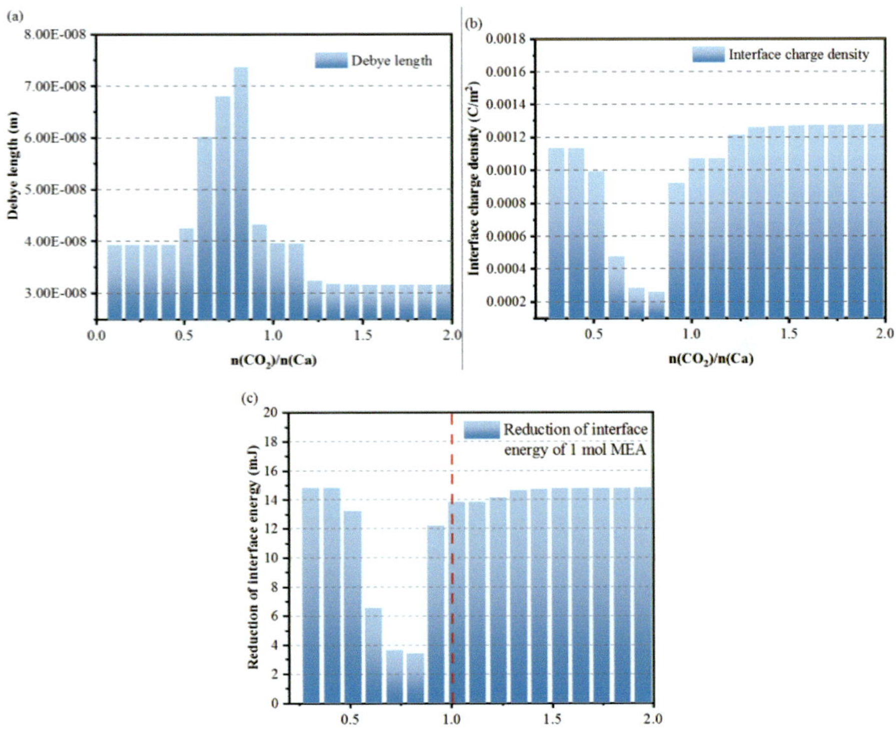

FIGURE 5.28 The change of (a) Debye length, (b) interface charge density, and (c) the reduction of interface energy of unit mol MEA with the change of $n(CO_2)/n(Ca)$.

charge. With the increase in the CO_2 amount input, the Debye length begins to significantly increase at about n(CO_2)/n(Ca)=0.5, which means that the average distance between ions increases, the interface charge density decreases, and the ionic electric field decreases. In this case, the probability of organic additives and calcium ions bonding and complexing becomes lower, thus the effect of adding organic additives on the reduction of calcium carbonate interfacial energy is the weakest at this stage. Therefore, in order to increase the reducing effect of organic additives on interfacial energy and increase the induced aragonite content, it is suggested to avoid adding organic additives in the range of n(CO_2)/n(Ca) between 0.5 and 1.0.

5.6.2 Control of temperature

The influence of temperature on the polymorphs of calcium carbonate in magnesium ion induced aragonite system is complex. It not only affects the concentration of free magnesium ion and the total amount of calcium carbonate, but also affects the permeability of magnesium ions in calcium carbonate clusters. In addition, thermodynamic data such as Gibbs free energy of formation also change with the temperature. Therefore, it is necessary to analyze the influence of temperature on the polymorphs of calcium carbonate and the aqueous mineralization process.

The change of temperature will cause the alteration of thermodynamic properties of different calcium carbonate polymorphs. It can be seen from Figure 5.29 that the difference of Gibbs free energy of formation between calcite and aragonite gradually decreases at 273 K to 295 K and then continues to increase with the increase of temperature. The change of Gibbs free energy caused by the group substitution between magnesium and calcium increases linearly with the increase of temperature. Among these two factors, the change of Gibbs free energy of formation dominates. The trend of magnesium ion concentration required to inhibit the conversion of unit mole

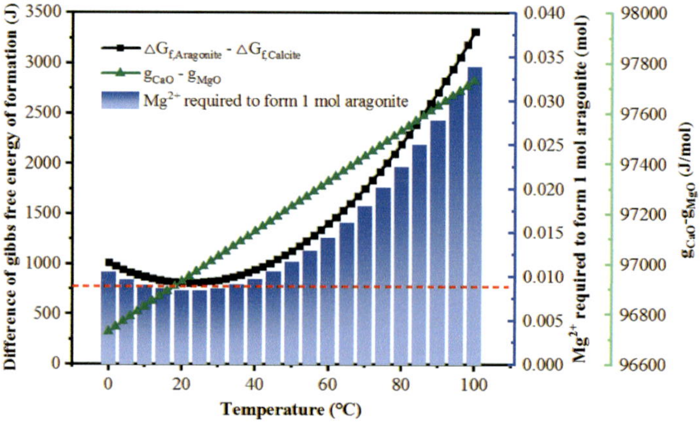

FIGURE 5.29 Difference of $\triangle G_f$ between aragonite and calcite, the change of $\triangle G_f$ caused by 1 mole Mg replacing Ca and the amount of Mg^{2+} required to form 1 mole aragonite at different temperatures.

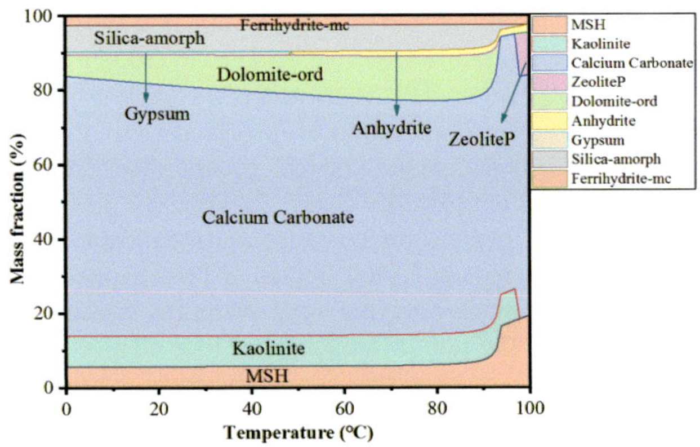

FIGURE 5.30 Phase assemblages with the change of temperature (0.15 mol/L $MgCl_2$).

aragonite to calcite is basically consistent with the difference of Gibbs free energy of formation between calcite and aragonite. Therefore, the effect of temperature on the transformation of aragonite to calcite is mainly the change of Gibbs free energy of formation, rather than the effectiveness of magnesium ions.

As shown in Figure 5.30, when the temperature is lower than 80°C, the increase of temperature will slightly increase the content of $Ca_xMg_{1-x}(CO_3)_2$, but when the temperature is higher than 80°C, the content of $Ca_xMg_{1-x}(CO_3)_2$ decreases rapidly. Similarly, silica gel will gradually decrease when the temperature is greater than 80°C, while the amount of MSH gradually increases. The change of temperature has a significant impact on magnesium ions, calcium ions, and carbonate ions. Combined with Figure 5.31(a), it can be found that the concentration of magnesium ions decreases with the increase of temperature, especially when the temperature is greater than 92°C, which is opposite to the change of calcium ions concentration. Therefore, although $Ca_xMg_{1-x}(CO_3)_2$ dissolves and releases some magnesium ions,

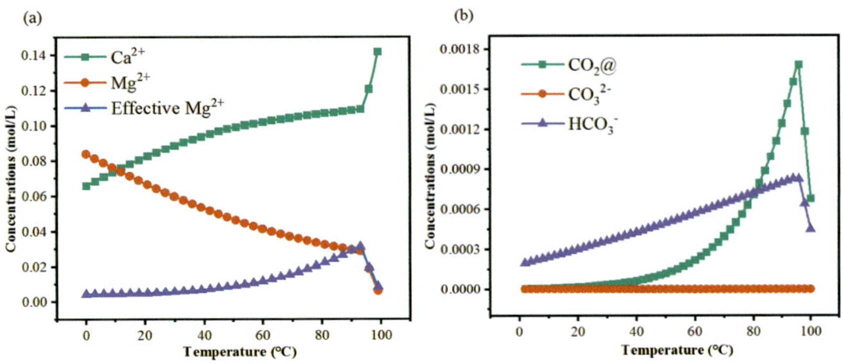

FIGURE 5.31 The concentration of (a) magnesium and calcium ions and (b) carbonate species with the change of temperature.

the amount of the dissolved magnesium ions is less than that consumed by the formation of MSH. Too high temperature will also lead to the reduction of the concentration of dissolved CO_2 and HCO_3^-. This reduction is not caused by the conversion of HCO_3^- to CO_3^{2-}, because the concentration of dissolved CO_3^{2-} does not change significantly. The most likely reason is that the CO_2 escapes back into the atmosphere.

Although the increase of temperature causes the decrease in free magnesium ion concentration, the effective magnesium ion concentration increases with the increase of temperature, as shown in Figure 5.31(a). Too high a temperature is unfavorable to both the free magnesium ion and effective magnesium ion concentration. When the temperature exceeds 92°C, the rate of reduction of free magnesium ion concentration suddenly increases, and the trend of effective magnesium ion concentration also changes from increase to decrease. Therefore, considering the dissolution of CO_2, the concentration of effective magnesium ions and the decomposition of silica gel, 80–90 °C may be the most suitable temperature range for the formation of aragonite in an aqueous mineralization system.

For organic regulator system, the change in temperature will first lead to a change in phase composition (especially the gel phase), thus affecting the effective adsorption of MEA in the system. As shown in Figure 5.32, for a partial mineralization system (n (CO_2)/n (Ca)=0.5), with the increase in temperature, the total amount of main phase calcium carbonate and CSH gel remains relatively stable, accompanied by a slight decrease in the content of ettringite. Straetlingite is formed as a transition phase at a temperature below 80°C, while zeolite P is a stable product at a temperature above 80°C. When the temperature exceeds 74 °C, the overall calcium content of CSH increases and the silicon content decreases, mainly caused by the decomposition of CSHQ-TobD, CSHQ-JenH, and CSHQ-TobH, as well as the formation of the phase CSHQ-JenD with the highest calcium silicon ratio, as shown in Figure 5.33. This is not conducive to the adsorption of MEA on the surface of $CaCO_3$, because

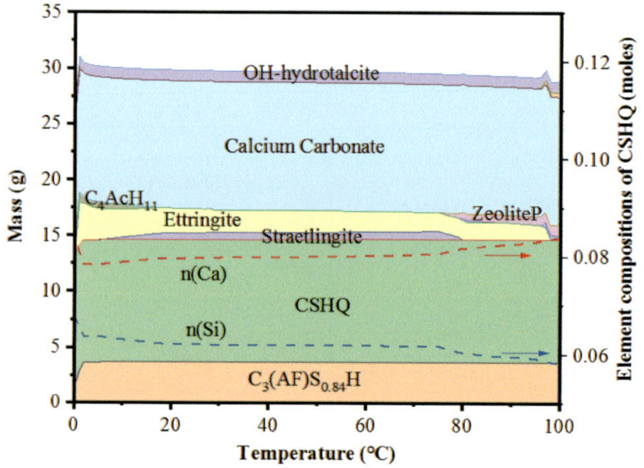

FIGURE 5.32 The change of phase assemblages with the change of temperature ($n(CO_2)/n(Ca)=0.5$).

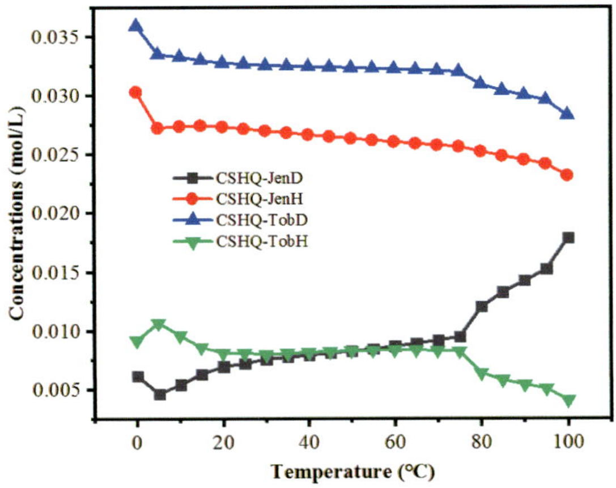

FIGURE 5.33 The change of the concentration of CSH with different Ca–Si ratios when the temperature changes ($n(CO_2)/n(Ca)=0.5$).

CSHQ-JenD has a stronger capture capacity compared with the other three CSH. This indicates that when the CO_2 content is insufficient to ensure the complete mineralization of Portland cement, excessive temperature increase (>74°C) is not conducive to the effectiveness of MEA in inducing aragonite formation. However, in the case of complete mineralization, it is not necessary to consider this temperature limitation because the variation of the content of silica gel with temperature is relatively limited.

Figure 5.34 shows the initial interfacial energy of calcite and aragonite as a function of temperature. The initial interface energies of aragonite and calcite both rise with temperature, and the initial interface energy of aragonite increases more than

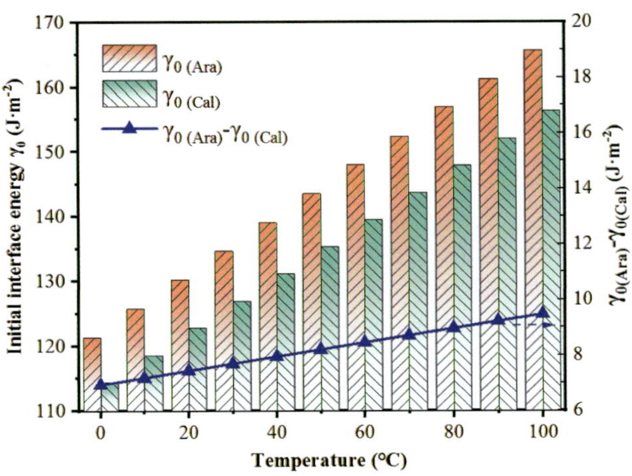

FIGURE 5.34 The change of initial surface energy of calcite and aragonite with different temperatures.

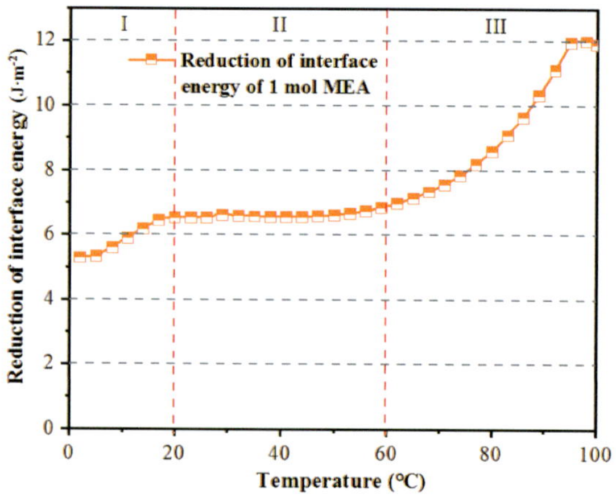

FIGURE 5.35 The change of the reduction of interface energy of unit mol MEA with the change of temperature.

that of calcite. As a result, the difference between the two polymorphs increases slightly (by about 5%) as the temperature rises. However, the change in temperature also affects the effect of MEA on the reduction of interface energy, as shown in Figure 5.35. As the temperature increases, the decrease in interfacial energy caused by MEA also gradually increases, especially in the temperature range above 60°C. According to the calculation results, the latter dominates among these two effects. Therefore, overall, an increase in temperature is beneficial for increasing the proportion of aragonite.

5.6.3 Control of Aragonite Inducer Concentration

Figure 5.36 shows the change of phase assemblages of the system at different concentrations of magnesium chloride. A small amount of magnesium chloride (≤0.0375 mol/L) does not have a significant impact on the phase assemblages, only leading to a slight increase in MSH and kaolinite. After the concentration of magnesium chloride exceeds 0.0375 mol/L, part of calcium carbonate begins to convert to dolomite-ord ($Ca_xMg_{1-x}(CO_3)_2$), and the amount of $Ca_xMg_{1-x}(CO_3)_2$ increases almost linearly with the increase of magnesium chloride concentration. Magnesium mainly exists in three forms in the system: MSH, free magnesium ions, and $Ca_xMg_{1-x}(CO_3)_2$. MSH is obtained by replacing calcium ions in CSH with magnesium ions, while $Ca_xMg_{1-x}(CO_3)_2$ is obtained by the partial substitution of calcium ions in calcium carbonate by magnesium ions. Due to the high CO_2 input amount in the modeling condition, most of the CSH in the system will be mineralized while the remaining amount is very low. Therefore, as the concentration of magnesium chloride increases, CSH will be rapidly consumed, and $Ca_xMg_{1-x}(CO_3)_2$ will become the main reaction product. The amount of $Ca_xMg_{1-x}(CO_3)_2$ is positively proportional to the magnesium to calcium ratio in the solution. Due to the fixed calcium content in the raw materials

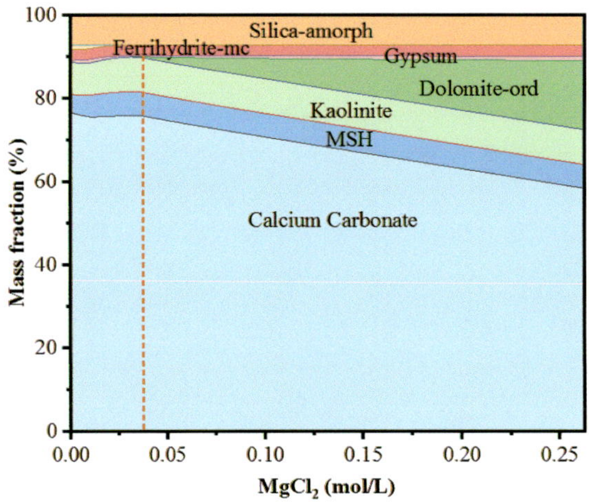

FIGURE 5.36 Phase assemblages of the mineralization system at different $MgCl_2$ concentrations (25°C, 10 g CO_2 participating in the reaction).

and the fixed CO_2 input amount, the magnesium to calcium ratio in the solution increases approximately linearly with the increase of magnesium chloride concentration, resulting in an approximate linear increase in $Ca_xMg_{1-x}(CO_3)_2$ content.

The stable existence of aragonite depends on free magnesium ions, which need to reach a certain critical value to be effective. Therefore, it is necessary to study whether increasing the initial magnesium ion concentration increases the content of magnesium in solid phases or free magnesium ions. As shown in Figure 5.37, when the concentration of magnesium chloride is low (less than 0.015 mol/L), magnesium ions are mainly involved in the reaction of generating MSH. With the increase of

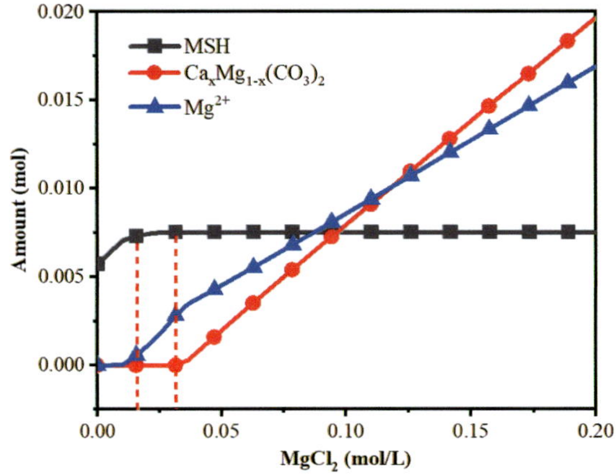

FIGURE 5.37 Amount of Mg species with the change of initial concentration of $MgCl_2$ (25°C, 10 g CO_2 participating in the reaction).

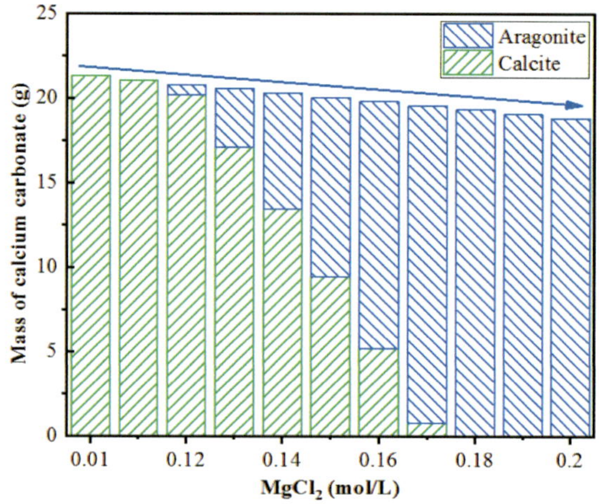

FIGURE 5.38 Modeling results of the total amount of $CaCO_3$ and the proportion of calcite and aragonite with the change of initial $MgCl_2$ concentration (25°C, 10 g CO_2).

magnesium chloride concentration, the amount of MSH will soon stabilize, which makes the free magnesium ions begin to enrich rapidly. When the concentration of magnesium chloride reaches 0.03 mol/L, the formation of dolomite-ord reduces the increasing rate of free magnesium ions. Although the increasing rate of free magnesium ions is lower than that of dolomite-ord, it remains stable.

Different from the sudden increase of aragonite content with the increase of CO_2 injection amount, the increase of initial magnesium chloride concentration results in the gradual increase of aragonite content. As shown in Figure 5.38, with the increase of magnesium chloride concentration, the total amount of calcium carbonate decreases but the proportion of aragonite increases. The minimum initial magnesium chloride concentration that plays a role in inducing aragonite formation is 0.12 mol/L. At about 0.17 mol/L, almost all the calcium carbonate exists in the form of aragonite. After exceeding 0.17 mol/L, excessive magnesium chloride will lead to the continuous reduction of the total amount of calcium carbonate. Therefore, if the purpose is to obtain as much aragonite as possible, a suitable initial magnesium chloride concentration is around 0.18 mol/L. Of course, the threshold of magnesium chloride concentration is different for systems with different compositions of raw materials and temperatures.

FTIR has been applied to characterize the chemical structures of mineralization products with different concentrations of $MgCl_2$, especially the amorphous phase that cannot be monitored by XRD. As shown in Figure 5.39, the band at 874 cm^{-1} corresponds to calcite, while 1460 cm^{-1} corresponds to aragonite. Obviously, the increase in $MgCl_2$ concentration results in stronger aragonite characteristic signals, which is consistent with the results from the model. The band at 970 cm^{-1} corresponds to a typical Si-O-T (T: Si) asymmetric stretching bond by Q^2 units, characterized as the bonding vibration of Si-O-Si in the C-S-H gel. In the sample with 0.25 mol/L $MgCl_2$,

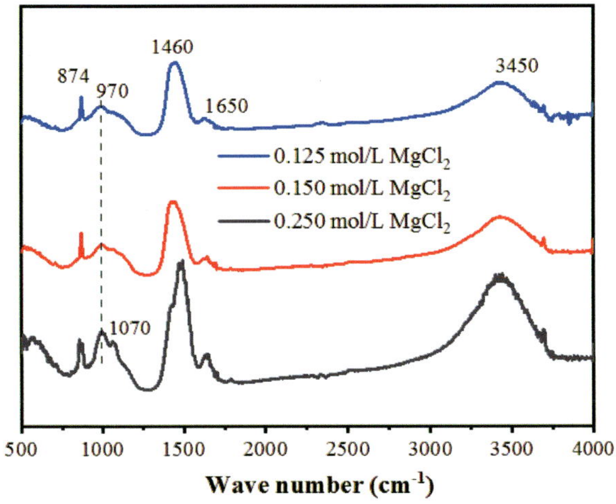

FIGURE 5.39 FTIR spectra of samples prepared with different concentrations of $MgCl_2$.

a 1070 cm^{-1} shoulder was observed, which is attributed to the polymerization of CSH chain (accompanied by a decrease in Ca–Si ratio). This is consistent with the calculation results of the model, as shown in Figure 5.40. The content of four kinds of CSH with different Ca–Si (CSH-JenD, $C_{1.5}S_{0.67}H_{2.5}$; CSH-JenH, $C_{1.33}SH_{2.17}$; CSH-TobD, $C_{5/6}S_{2/3}H_{1.83}$; CSH-TobH, $C_{2/3}SH_{1.5}$) and the overall Ca–Si of CSH were calculated based on the model. It can be found that the overall Ca–Si of CSH shows a decreasing trend, which is due to the conversion of CSH-TobD and CSH-TobH with high Ca–Si to CSH-TobH with low Ca–Si. The narrow band at 1650 cm^{-1} corresponds to the O-H symmetric stretching band in $Ca(OH)_2$. Although the signal observed at this wavelength is weak in all three samples, indicating a low residual $Ca(OH)_2$ content, it can still be distinguished that the highest $Ca(OH)_2$ signal was detected in the 0.25

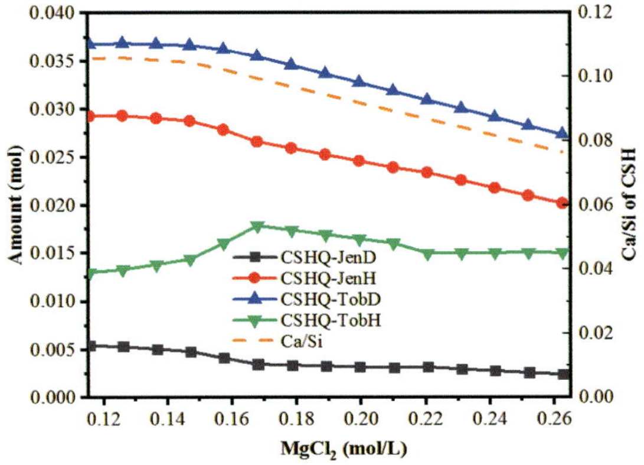

FIGURE 5.40 Changes in the content of different types of CSH and the calcium silicon ratio of CSH.

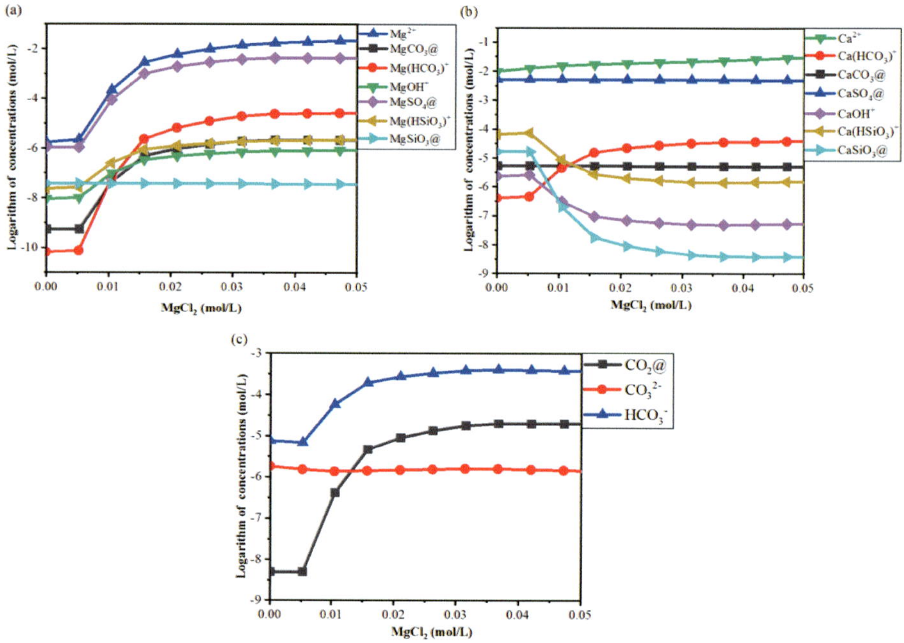

FIGURE 5.41 The concentration of (a) magnesium species, (b) calcium species, and (c) carbonate species in the solution with the change of initial $MgCl_2$ concentration.

mol/L sample. This indicates that the higher $MgCl_2$ concentration leads to a decrease in the degree of mineralization, which is consistent with the calculated decreasing trend of the total amount of calcium carbonate. The broad band with a center of 3450 cm^{-1} is due to the stretching vibrations of O–H groups in H_2O or hydroxyls with a wide range of hydrogen bond strengths.

Figure 5.41 shows the changes in the concentrations of magnesium species, calcium species and carbonate species in solution with different initial magnesium chloride concentrations. With the increase of initial magnesium chloride concentration, not only does the magnesium ion concentration increase but also the calcium ion concentration increases. This part of calcium ion may come from the process of magnesium ion replacing calcium ion in calcium carbonate to form $Ca_xMg_{1-x}(CO_3)_2$. In addition, increasing the concentration of magnesium chloride also leads to a significant increase in the concentration of $Mg(HCO_3)^+$, $Ca(HCO_3)^+$, and $Mg(HSiO_3)^+$, but their final concentration is still relatively low and will not lead to the formation of metastable phases. It is worth noting that with the increase of magnesium chloride concentration, the concentration of dissolved CO_2 and bicarbonate ions in the solution increases while the concentration of carbonate ions remains stable. This means that the addition of magnesium chloride can increase the CO_2 absorption potential of the system, although this part of the absorbed CO_2 will not eventually be converted into calcium carbonate.

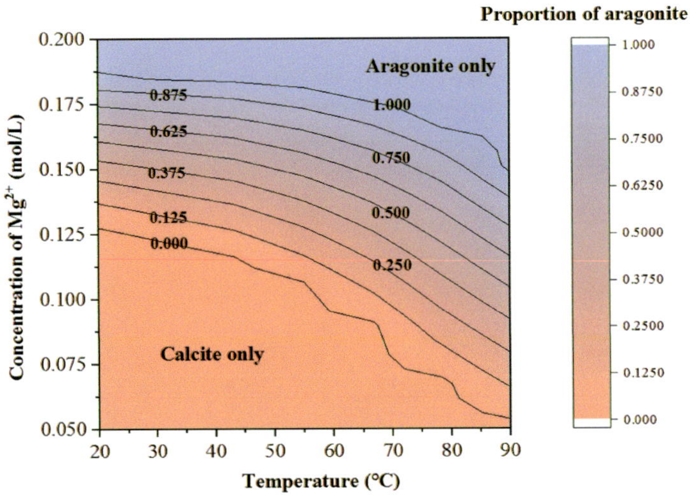

FIGURE 5.42 The modeling proportion of aragonite with the change of temperature and initial magnesium ion concentration.

Considering the influence of initial magnesium ion concentration and temperature, the simulation results of the proportion of aragonite can be obtained, as shown in Figure 5.42. Based on this, conditions for producing calcium carbonate containing aragonite and calcite in a specific proportion can be referenced.

The calculated proportion of aragonite under different concentrations of organic additives MEA is shown in Figure 5.43. It can be observed that the proportion of aragonite increases significantly with the increase in MEA concentration, especially in the range of 1 to 3 M. When the MEA concentration reaches 3 M, almost all calcium

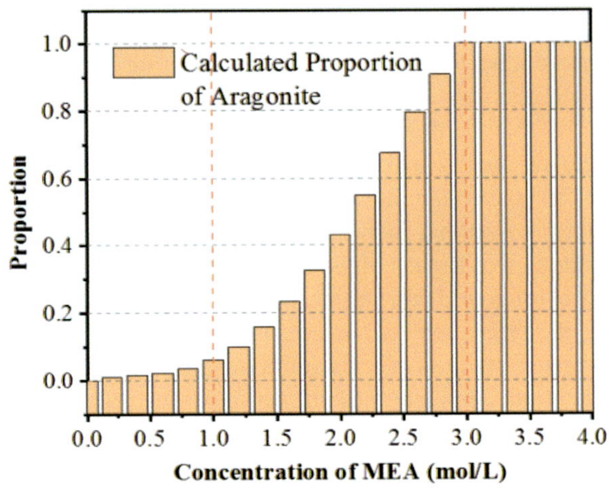

FIGURE 5.43 The proportion of aragonite varies with MEA concentration.

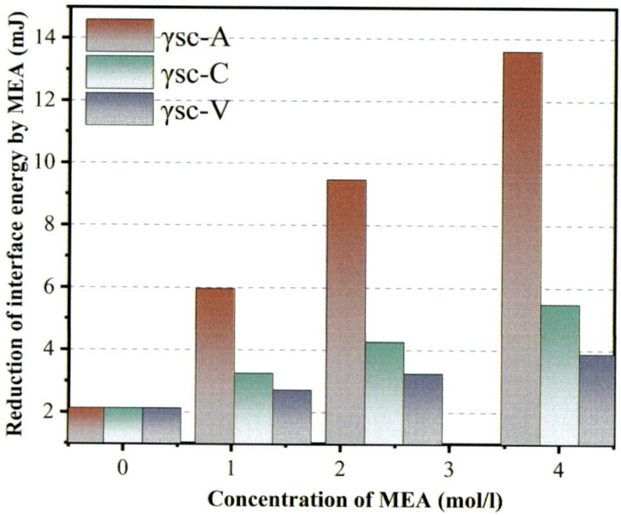

FIGURE 5.44 The reduction of interface energy of different calcium carbonate polymorphs by MEA varies with MEA concentration.

carbonate exists in the form of aragonite. The reason for this selective induction may be that the change in the interfacial energy of aragonite is much more sensitive to the addition of MEA than calcite and aragonite. The increase in MEA concentration leads to an increase in the interfacial energy reduction of the three $CaCO_3$ polymorphs. However, the reduction of interfacial energy of vaterite and calcite caused by the increase in MEA concentration is quite limited, far less than that of aragonite. This gap will continue to expand with the increase in MEA concentration, as shown in Figure 5.44. This may offer an explanation for the selective adsorption mechanism of organic additives on aragonite, as indicated in the literature (Ouhenia et al., 2008; Pan et al., 2007; Sonobe, Watamura, and Hirasawa, 2015), since the total interfacial energy of the system always tends to be the lowest. The reason why the induction effect of MEA on aragonite formation is not significant when the concentration of MEA is low (<1 M) may be that although there is a decrease in interfacial energy, the reduced interfacial energy still cannot reach the critical interfacial energy value corresponding to the conversion of aragonite.

Another primary reason that increasing MEA concentration favors a higher proportion of aragonite is that higher MEA concentrations lead to greater surface coverage of calcium carbonate during mineralization, resulting in a more significant reduction of surface energy, as shown in Figure 5.45. Regardless of the concentration of MEA, the coverage rate at the partial mineralization stage (n $(CO_2)/n(Ca) < 1.0$) is higher than that at the complete mineralization stage, and as the degree of mineralization increases, the coverage rate gradually decreases. Considering the coverage rate, adding MEA in the initial mineralization stage is a more appropriate choice to ensure its effective role. However, it should be noted that this point alone is insufficient to determine the impact of CO_2 input on the proportion of aragonite.

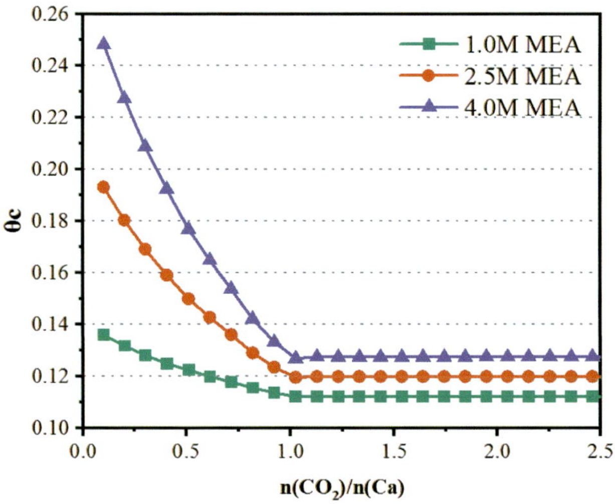

FIGURE 5.45 Coverage of different concentrations of MEA added at different mineralization stages on the surface of calcium carbonate.

5.7 SUMMARY

This chapter introduces the application of thermodynamic simulation in the carbon mineralization process of cement-based materials. It presents two fundamental methods for solving thermodynamic equilibrium problems and discusses the use of thermodynamic approaches to calculate phase assemblages, pore solution ion composition, elemental composition, pH evolution, porosity, and other relevant information. Special emphasis is placed on the transformation mechanisms and control strategies of different calcium carbonate polymorphs during carbon mineralization.

The chapter separately investigates the establishment of thermodynamic model frameworks in inorganic and organic regulation systems, with model accuracy improved through comparisons between literature data, experimental data, and simulation results. By integrating experimental findings and thermodynamic simulations, this chapter also provides specific strategies for controlling different calcium carbonate polymorphs during carbon mineralization by adjusting parameters such as CO_2 input, additives concentration, and temperature. These strategies enable the selective induction of specific calcium carbonate polymorphs, thereby optimizing carbon sequestration efficiency and enhancing the performance of carbon mineralization products.

REFERENCES

Afonso, R., Gales, L., Mendes, A., 2016. Kinetic derivation of common isotherm equations for surface and micropore adsorption. *Adsorption* 22(7), 963–971.

Ahn, J.W., Kim, J.H., Park, J.Y., Kim, W.S., 2005. Synthesis of single phase aragonite precipitated calcium carbonate in $Ca(OH)_2$–Na_2CO_3–NaOH reaction system. *Korean Journal of Chemical Engineering* 22(6), 852–856.

Androniuk, I., Landesman, C., Henocq, P., Kalinichev, A.G., 2017. Adsorption of gluconate and uranyl on C–S–H phases: combination of wet chemistry experiments and molecular dynamics simulations for the binary systems. *Physics and Chemistry of the Earth 99*, 194–203.

Appelo, C.A.J., Postma, D., 2004. Geochemistry, groundwater and pollution, 2nd ed, CRC press.

Borkovec, M., Westall, J., 1983. Solution of the Poisson–Boltzmann equation for surface excesses of ions in the diffuse layer at the oxide–electrolyte interface. *Journal of Electroanalytical Chemistry* 150(1–2), 325–333.

Burton, E.A., Walter, L.M., 1987. Relative precipitation rates of aragonite and Mg calcite from seawater: temperature or carbonate ion control? *Geology* 15(2), 111–114.

Choi, K.S., Park, J.K., Ahn, J.W., Kim, H., 2004. Formation mechanism of aragonite by substitute of Mg^{2+} ions. *Journal of the Korean Ceramic Society* 41(12), 889–892.

De Choudens-Sánchez, V., Gonzalez, L.A., 2009. Calcite and aragonite precipitation under controlled instantaneous supersaturation: elucidating the role of $CaCO_3$ saturation state and Mg/Ca ratio on calcium carbonate polymorphism. *Journal of Sedimentary Research* 79(6), 363–376.

Chu, K.H., 2014. Discrimination of rival isotherm equations for aqueous contaminant removal systems. *Advances in Environmental Research* 3, 131–149.

Chu, K.H., 2021. Revisiting the Temkin isotherm: dimensional inconsistency and approximate forms. *Industrial & Engineering Chemistry Research* 60(35), 13140–13147.

Coppersmith, S.N., Gilbert, P.U.P.A., Metzler, R.A., 2009. Theoretical characterization of a model of aragonite crystal orientation in red abalone nacre. *Journal of Physics A: Mathematical and Theoretical* 42(12), 125101.

Elizalde, M.P., Aparicio, J.L., 1995. Current theories in the calculation of activity coefficients—II. Specific interaction theories applied to some equilibria studies in solution chemistry. *Talanta* 42(3), 395–400.

Glaus, M.A., Laube, A., Van Loon, L.R., 2006. Solid–liquid distribution of selected concrete admixtures in hardened cement pastes. *Waste Management* 26(7), 741–751.

Grahame, D.C., 1947. The electrical double layer and the theory of electrocapillarity. *Chemical Reviews* 41(3), 441–501.

Haas, J., Nonat, A., 2015. From C–S–H to C–A–S–H: experimental study and thermodynamic modelling. *Cement and Concrete Research* 68, 124–138.

Han, S.C., Jo, Y., Yun, J.I., 2021. Chemical degradation of fly ash blended concrete with the seasonal variation of rainwater in a radioactive waste repository: a thermodynamic modeling approach. *Cement and Concrete Research* 141, 106326.

Hargis, C.W., Lothenbach, B., Müller, C.J., Winnefeld, F., 2017. Carbonation of calcium sulfoaluminate mortars. *Cement and Concrete Composites* 80, 123–134. http://dx.doi.org/10.1016/j.cemconcomp.2017.03.003.

Hazen, R.M., 1985. Comparative crystal chemistry and the polyhedral approach. *Reviews in Mineralogy* 14, 317–345.

Helgeson, H.C., Kirkham, D.H., Flowers, G.C., 1981. Theoretical prediction of the thermodynamic behavior of aqueous electrolytes at high pressures and temperatures: IV. Calculation of activity coefficients, osmotic coefficients, and apparent molal and standard and relative partial molal properties to 600 °C. *American Journal of Science* 281(10), 1249–1516.

Hinz, C., 2001. Description of sorption data with isotherm equations. *Geoderma* 99(3–4), 225–243.

Hlaing, N. N., Sreekantan, S., Othman, R., Hinode, H., Kurniawan, W., 2014. Novel aragonite $CaCO_3$ adsorbents: synthesis, characterization and CO_2 adsorption. *Advanced Materials Research* 911, 415–419.

Hsu, K.C., Chiu, J.J., Chen, S.D., Tseng, Y.C., 1999. Effect of addition time of a superplasticizer on cement adsorption and on concrete workability. *Cement and Concrete Composites* 21(5–6), 425–430.

La Iglesia, A., Félix, J.F., 1994. Estimation of thermodynamic properties of mineral carbonates at high and low temperatures from the sum of polyhedral contributions. *Geochimica et Cosmochimica Acta* 58(19), 3983–3991.

Jimoh, O.A., Ariffin, K.S., Hussin, H.B., Temitope, A.E., 2018. Synthesis of precipitated calcium carbonate: a review. *Carbonates and Evaporites* 33(2), 331–346.

Kari, O.P., Puttonen, J., Skantz, E., 2014. Reactive transport modelling of long-term carbonation. *Cement and Concrete Composites* 52, 42–53.

Ke, X., Bernal, S.A., Provis, J.L., Lothenbach, B., 2020. Thermodynamic modelling of phase evolution in alkali-activated slag cements exposed to carbon dioxide. *Cement and Concrete Research* 136(January), 106158. https://doi.org/10.1016/j.cemconres.2020.106158.

Kim, I.W., Robertson, R.E., Zand, R., 2005. Effects of some nonionic polymeric additives on the crystallization of calcium carbonate. *Crystal Growth and Design* 5(2), 513–522.

Kim, J.H., Lee, S.Y., Park, H.S., Kim, W.S., 2006. Inhibition mechanism of magnesium ion on carbonation reaction with $Ca(OH)_2$ and CO_2. *Materials Science Forum* 510–511, 990–93.

Kulik, D.A., 2011. Improving the structural consistency of C–S–H solid solution thermodynamic models. *Cement and Concrete Research* 41(5), 477–95. http://dx.doi.org/10.1016/j.cemconres.2011.01.012.

Kulik, D.A., 2013. GEM-Selektor geochemical modeling package: revised algorithm and GEMS3K numerical kernel for coupled simulation codes. *Computational Geosciences* 17(1), 1–24.

Labbez, C., Nonat, A., Pochard, I., Jönsson, B., Cabane, B., 2006. Surface charge density and electrokinetic potential of highly charged minerals: experiments and Monte Carlo simulations on calcium silicate hydrate. *Journal of Physical Chemistry B* 110(18), 9219–9230.

Lee, J., Kim, H., Park, J., Choi, S., 2015. The effect of chloride additives and pH on direct aqueous carbonation of cement paste. *Journal of the Mineralogical Society of Korea* 28(1), 39–49.

Liu, Q., Zhang, Y., Chen, X., Li, J., 1998. Preparation of aragonite by carbonation process. *Proceedings of the International Symposium on Electrical Insulating Materials* 159–161.

Liu, Z., Van den Heede, P., Zhang, C., Shi, X., Wang, L., 2023. Carbonation of blast furnace slag concrete at different CO_2 concentrations: carbonation rate, phase assemblage, microstructure and thermodynamic modelling. *Cement and Concrete Research* 169, 107161.

Lothenbach, B., Bernard, E., Mäder, U., 2017. Zeolite formation in the presence of cement hydrates and albite. *Physics and Chemistry of the Earth* 99, 77–94.

Lothenbach, B., Durdzinski, P., De Weerdt, K., 2019. Cemdata18: a chemical thermodynamic database for hydrated Portland cements and alkali-activated materials. *Cement and Concrete Research* 115, 472–506.

Lothenbach, B., Zajac, M., 2019. Application of thermodynamic modelling to hydrated cements. *Cement and Concrete Research* 123, 105779.

Marchon, D., Mantellato, S., Eberhardt, A. B., Flatt, R. J., 2016. Adsorption of chemical admixtures, in: *Science and technology of concrete admixtures*, Woodhead Publishing, pp. 219–256.

Martin, L.H.J., Winnefeld, F., Lothenbach, B., Müller, C.J., 2017. Influence of fly ash on the hydration of calcium sulfoaluminate cement. *Cement and Concrete Research* 95, 152–163.

Mersmann, A., 1990. Calculation of interfacial tensions. *Journal of Crystal Growth* 102(4), 841–847.
Michel, A., Marcos-Meson, V., Kunther, W., Geiker, M.R., 2021. Microstructural changes and mass transport in cement-based materials: a modeling approach. *Cement and Concrete Research,* 139, 106285.
Morse, J.W., Wang, Q., Tsio, M.Y., 1997. Influences of temperature and Mg:Ca ratio on $CaCO_3$ precipitates from seawater. *Geology* 25(1), 85–87.
Mucci, A., Morse, J.W., 1983. The incorporation of Mg^{2+} and Sr^{2+} into calcite overgrowths: influences of growth rate and solution composition. *Geochimica et Cosmochimica Acta* 47(2), 217–233.
Myers, R.J., Bernal, S.A., Provis, J.L., 2014. A thermodynamic model for C–(N)–A–S–H gel: CNASH_ss. Derivation and validation. *Cement and Concrete Research* 66, 27–47. http://dx.doi.org/10.1016/j.cemconres.2014.07.005.
Myers, R.J., Bernal, S.A., Provis, J.L., 2015. The role of Al in cross-linking of alkali-activated slag cements. *Journal of the American Ceramic Society* 98(3), 996–1004.
Myers, R.J., Bernal, S.A., San Nicolas, R., Provis, J.L., 2013. Generalized structural description of C–(N)–A–S–H gels: the crosslinked substituted tobermorite model. *Langmuir* 29(17), 5294–5306.
Myers, R.J., Lothenbach, B., Bernal, S.A., Provis, J.L., 2015. Thermodynamic modelling of alkali-activated slag cements. *Applied Geochemistry* 61, 233–247.
Nalet, C., Nonat, A., 2016. Ionic complexation and adsorption of small organic molecules on calcium silicate hydrate: relation with their retarding effect on the hydration of C_3S. *Cement and Concrete Research* 89, 97–108.
Nan, Z., Chang, J., Zhong, S., 2008. A novel morphology of aragonite and an abnormal polymorph transformation from calcite to aragonite with PAM and CTAB as additives. *Journal of Colloid and Interface Science* 317(1), 77–82.
Nied, D., Enemark-Rasmussen, K., L'Hopital, E., Skibsted, J., Lothenbach, B., 2016. Properties of magnesium silicate hydrates (MSH). *Cement and Concrete Research* 79, 323–332.
Nielsen, A.E., 1984. Electrolyte crystal growth mechanisms. *Journal of Crystal Growth* 67(2), 289–310.
Omikrine Metalssi, O., Aït-Mokhtar, A., Turcry, P., 2020. A proposed modelling of coupling carbonation–porosity–moisture transfer in concrete based on mass balance equilibrium. *Construction and Building Materials* 230, 116997. https://doi.org/10.1016/j.conbuildmat.2019.116997.
Ouhenia, S., Chibani, A., Nibou, D., Amrane, A., 2008. Synthesis of calcium carbonate polymorphs in the presence of polyacrylic acid. *Journal of Crystal Growth* 310(11), 2832–2841.
Palacios, M., Puertas, F., 2006. Effect of carbonation on alkali-activated slag paste. *Journal of the American Ceramic Society* 89(10), 3211–3221.
Pan, Y., Xu, H., Chen, J., Tang, R., 2007. Biomimetic synthesis of dendrite-shaped aragonite particles with single-crystal feature by polyacrylic acid. *Colloids and Surfaces A: Physicochemical and Engineering Aspects* 297(1–3), 198–202.
Papadakis, V.G., Vayenas, C.G., Fardis, M.N., 1991. Physical and chemical characteristics affecting the durability of concrete. *ACI Materials Journal* 88(2), 186–196.
Park, S., 2020. Simulating the carbonation of calcium sulfoaluminate cement blended with supplementary cementitious materials. *Journal of CO2 Utilization* 41(August), 101286. https://doi.org/10.1016/j.jcou.2020.101286.
Romanek, C.S., Morse, J.W., Grossman, E.L., 2011. Aragonite kinetics in dilute solutions. *Aquatic Geochemistry* 17(4), 339–356.
Saruwatari, K., Kogure, T., Nishimura, T., Kato, T., Nagasawa, H., 2009. Nucleation and growth of aragonite crystals at the growth front of nacres in pearl oyster, Pinctada Fucata. *Biomaterials* 30(16), 3028–3034.

Shah, V., Scrivener, K., Bhattacharjee, B., Bishnoi, S., 2018. Changes in microstructure characteristics of cement paste on carbonation. *Cement and Concrete Research* 109, 184–197.

Shen, J., Dangla, P., Thiery, M., 2013. Reactive transport modeling of CO_2 through cementitious materials under CO_2 geological storage conditions. *International Journal of Greenhouse Gas Control* 18, 75–87.

Shi, Z., Lothenbach, B., Geiker, M.R., Winnefeld, F., Plank, J., 2016. Experimental studies and thermodynamic modeling of the carbonation of Portland cement, metakaolin and limestone mortars. *Cement and Concrete Research* 88, 60–72. http://dx.doi.org/10.1016/j.cemconres.2016.06.006.

Shuvo, A.K., Sarker, P.K., Shaikh, F.U.A., Rajayogan, V., 2024. Improvement of crushed returned concrete aggregates by wet carbonation. *Construction and Building Materials* 448(September), 138253. https://doi.org/10.1016/j.conbuildmat.2024.138253.

Silva Filho, E.C., Santana, S.A.A., Fonseca, M.G., Airoldi, C., 2013. Immobilization of ethylene sulfide in aminated cellulose for removal of divalent cations. *Carbohydrate Polymers* 92(2), 1203–1210.

Sonobe, Y., Watamura, H., Hirasawa, I., 2015. Polymorphism, size and shape control of calcium carbonate crystals in the presence of a polyelectrolyte. *Chemical Engineering and Technology* 38(6), 1053–1058.

Sun, W., Jayaraman, S., Chen, W., Persson, K.A., Ceder, G., 2015. Nucleation of metastable aragonite $CaCO_3$ in seawater. *Proceedings of the National Academy of Sciences of the United States of America* 112(11), 3199–3204.

Svilović, S., Mužek, M.N., Vučenović, P., Nuić, I., 2019. Taguchi design of optimum process parameters for sorption of copper ions using different sorbents. *Water Science and Technology* 80(1), 98–108.

Chess, P. M., Broomfield, J. P., 2013. *Cathodic protection of steel in concrete and masonry.* CRC Press.

Temkin, M.I., 1940. Kinetics of ammonia synthesis on promoted iron catalysts. *Acta physiochim* 12, 327–356.

Thoenen, T., Hummel, W., Berner, U., Curti, E., 2014. *PSI Bericht Nr. 14-04 The PSI/Nagra Chemical Thermodynamic Database 12/07.*

Viallis-Terrisse, H., Nonat, A., Petit, J.C., 2001. Zeta-potential study of calcium silicate hydrates interacting with alkaline cations. *Journal of Colloid and Interface Science* 244(1), 58–65.

Wagner, T., Kulik, D.A., Hingerl, F.F., Dmytrievava, S.V., 2012. GEM-Selektor geochemical modeling package: TSolMod library and data interface for multicomponent phase models. *Canadian Mineralogist* 50(5), 1173–1195.

Walker, C.S., Lothenbach, B., Bernal, S.A., Provis, J.L., Myers, R.J., 2016. Calcium silicate hydrate (C–S–H) gel solubility data and a discrete solid phase model at 25 °C based on two binary non-ideal solid solutions. *Cement and Concrete Research* 79, 1–30.

Wang, C., Xu, L., Xu, A., Deng, B., 2006. Synthesis of nanosized calcium carbonate (aragonite) via a polyacrylamide inducing process. *Powder Technology* 163(3), 134–138.

Wang, L., Li, S., Zhang, T., Zhang, J., 2018. Microwave-assisted preparation of nitrogen-doped biochars by ammonium acetate activation for adsorption of Acid Red 18. *Applied Surface Science* 433, 222–231.

De Weerdt, K., Kjellsen, K.O., Sellevold, E.J., Justnes, H., 2011. Hydration mechanisms of ternary Portland cements containing limestone powder and fly ash. *Cement and Concrete Research* 41(3), 279–291.

De Weerdt, K., Lothenbach, B., Geiker, M.R., 2019. Effect of carbonation on the pore solution of mortar. *Cement and Concrete Research.*

Wiechers, H.N.S., Sturrock, P., Marais, G.v.R., 1975. Calcium carbonate crystallization kinetics. *Water Research* 118, 38–56.

Wolery, T.J., Jackson, K.J., Delany, J.M., 1990. Current status of the EQ3/6 software package for geochemical modeling. Lawrence Livermore National Laboratory Report.

Wu, L., Zhang, H., Li, T., Liu, C., 2019. Using recycled concrete as an adsorbent to remove phosphate from polluted water. *Journal of Environmental Quality* 48(5), 1489–1497.

Yoon, H. N., Seo, J., Kim, S., Lee, H. K., Park, S., 2021. Hydration of calcium sulfoaluminate cement blended with blast-furnace slag. *Construction and Building Materials* 268, 121214.

Yoshimi, K., Shoji, M., Ogawa, T., Yamauchi, A., Naganuma, T., Muramoto, K., Hanada, S., 2004. Microstructure and orientation distribution of aragonite crystals in nacreous layer of pearl shells. *Materials Transactions* 45(4), 999–1004.

Yousef, N.S., Farouq, R., Hazzaa, R., 2016. Adsorption kinetics and isotherms for the removal of nickel ions from aqueous solutions by an ion-exchange resin: application of two and three parameter isotherm models. *Desalination and Water Treatment* 57(46), 21925–21938.

Yu, J., Lei, M., Cheng, B., Zhao, X., 2004. Facile preparation of calcium carbonate particles with unusual morphologies by precipitation reaction. *Journal of Crystal Growth* 261(4), 566–570.

Yu, Q., Ou, H.D., Song, R.Q., Xu, A.W., 2006. The effect of polyacrylamide on the crystallization of calcium carbonate: synthesis of aragonite single-crystal nanorods and hollow vaterite hexagons. *Journal of Crystal Growth* 286(1), 178–183.

Zhu, G., Li, S., Yang, X., Wang, H., 2015. Crystallization behavior and kinetics of calcium carbonate in highly alkaline and supersaturated system. *Journal of Crystal Growth* 428, 16–23.

6 CO_2 surface treatment of precast concrete products

6.1 INTRODUCTION

Concrete is one of the most widely used construction materials due to its abundant raw material sources and versatile casting properties (Xu et al., 2019). Precast concrete accounts for 20–30% of the total concrete market (Zhang and Shao, 2016a). It is extensively applied in bridges, pavements, sidewalks, and municipal facilities. However, the durability of precast concrete structures is often compromised by issues such as carbonation, steel bar corrosion, and sulfate attack (Chen et al., 2021). Aggressive substances penetrate the concrete surface through water or air, leading to deterioration and reduced service life. Therefore, the permeability of the concrete surface is a critical factor in determining the durability of the entire structure.

Surface protection can enhance the permeability resistance of concrete in harsh environments. Surface treatment has emerged as an economical and effective method for protecting and reinforcing concrete surfaces. It plays a vital role in reducing water absorption, thereby minimizing the penetration of corrosive substances and enhancing the concrete durability (Pan et al., 2017b, 2017c). Commonly used concrete surface treatment agents are categorized into organic and inorganic types. Organic agents typically contain strong hydrophobic components that plug pores and form hydrophobic membranes, mitigating the intrusion of liquids and harmful ions (Pan et al., 2017b). However, due to differences in thermal expansion coefficients, organic agents are prone to cracking, flaking, and degradation, necessitating continuous maintenance (Arabzadeh et al., 2017). In contrast, inorganic agents penetrate the concrete and undergo complex physicochemical reactions with hydration products, forming new compounds that seal surface pores (Pan et al., 2016).

In recent years, CO_2 surface treatment has gained attention as a promising strategy for concrete surface treatment due to its environmental and sustainable nature. CO_2 surface treatment of concrete refers to a technique in which freshly cast or hardened concrete is exposed to a controlled CO_2 environment with specific concentration and pressure. This process utilizes carbonation reactions between CO_2 and calcium-bearing phases in cementitious materials (e.g., $Ca(OH)_2$, C-S-H) to form calcium carbonate ($CaCO_3$) and other modified products, thereby enhancing the surface properties of concrete (Pan et al., 2017a). This process not only optimizes the microstructure of concrete, effectively reducing porosity, but also enhances compressive strength and impermeability (Wang et al., 2021). CO_2 surface treatment significantly improves the durability and extends the service life of concrete, particularly

DOI: 10.1201/9781003627531-6

by preventing the penetration of water, gases, and other deleterious substances (Song et al., 2021, 2022).

Moreover, CO_2 surface treatment is an environmentally friendly process. The capture and utilization of CO_2, a greenhouse gas, is an effective approach, addressing climate change and reducing carbon emissions. Utilizing CO_2 in concrete surface treatment not only enhances concrete performance but also provides an effective method for carbon storage. Thus, CO_2 surface treatment is not only a method to improve concrete durability but also a technology that promotes low-carbon and sustainable development in the construction industry. This chapter provides a comprehensive analysis of the latest advancements in CO_2 surface treatment technology, systematically reviewing its impact on concrete properties and the underlying mechanisms.

6.2 THICKNESS AND MICROSTRUCTURE OF SURFACE TREATED LAYERS

6.2.1 Thickness of surface treated layers

CO_2 surface treatment results in the formation of a dense CO_2 mineralized layer through a series of mineralization reactions. These reactions involve the interaction of CO_2 with hydration products, including $Ca(OH)_2$, ettringite, and C-S-H gel, as well as cement clinkers, such as alite and belite (Shi et al., 2019). When exposed to CO_2, these components undergo chemical transformations to form stable $CaCO_3$, which plays a crucial role in improving the material's durability and mechanical properties (Zajac et al., 2020). The mineralization process begins at the surface and progresses inward, resulting in significant pore-filling and microstructural densification near the exposed areas. This densification reduces porosity, enhances strength, and improves resistance to environmental degradation. However, the extent of the mineralization reaction is inherently limited by the diffusion depth of CO_2 into the material. Factors such as the initial porosity of the concrete, treatment conditions, and CO_2 concentration influence this depth. As a result, the thickness of the mineralized layer typically ranges from a few hundred micrometers to several centimeters, depending on the material and treatment conditions. Beyond this depth, the availability of CO_2 diminishes, leading to a gradual reduction in the mineralization reaction rate. Consequently, the degree of mineralization decreases with increasing distance from the surface, leaving the inner regions of the material relatively unaffected.

The thickness of the surface treatment layer is a critical parameter in CO_2 surface treatment technology, as it directly influences the effectiveness of the treatment in enhancing the material's durability and mechanical properties. This parameter is governed by multiple factors, including mineralization time, CO_2 gas pressure, temperature, and humidity during the mineralization.

Mineralization time plays a particularly significant role. As expected, the mineralization depth of concrete increases progressively with extended exposure to CO_2. Initially, the mineralization reaction proceeds rapidly, as the readily available CO_2 diffuses into the porous structure and reacts with hydration products to form calcium

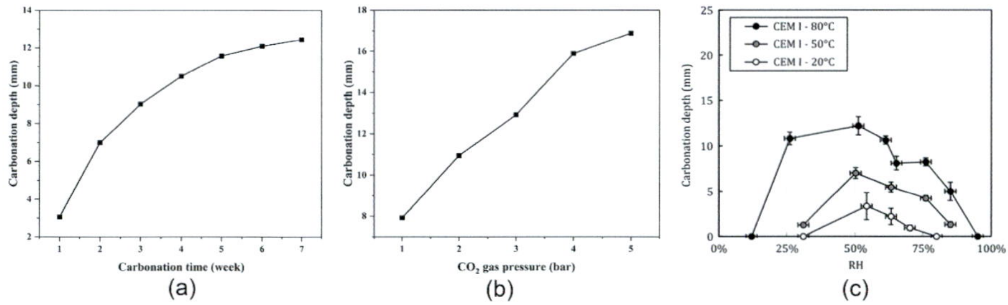

FIGURE 6.1 Relationship between mineralization depth and curing conditions: (a) mineralization time (You et al., 2014); (b) CO_2 gas pressure (Zhang and Shao, 2016a); (c) curing relative humidity (Drouet et al., 2019).

carbonate. However, over time, the rate of mineralization slows down. This deceleration is primarily attributed to the limited diffusion of CO_2 into the concrete and the gradual filling of pores by mineralization products, such as calcite and silica-rich gels. These products act as physical barriers, reducing the permeability of the material and impeding further CO_2 penetration. As illustrated in Figure 6.1(a), You et al. (2014) found that the mineralization depth increased significantly from 3 mm to 7 mm when the surface treatment time by CO_2 mineralization was increased from 7 days to 14 days. However, the mineralization depth increased from 12 mm to 12.5 mm when the treatment time was increased from 42 days to 49 days. These results highlight the diminishing returns of extended mineralization time and emphasize the importance of optimizing treatment conditions to achieve a balance between efficiency and performance. This behavior underscores the importance of optimizing mineralization conditions, such as adjusting CO_2 pressure and environmental humidity, to achieve the desired treatment depth within a practical timeframe.

The CO_2 gas pressure is a key factor influencing the mineralization process, as it directly affects the CO_2 diffusion coefficient, which in turn determines the mineralization depth. As shown in Figure 6.1(b), higher CO_2 gas pressure results in greater mineralization depth. For example, Zhang and Shao (2016a) found that increasing the CO_2 pressure from 1 bar to 5 bar led to a significant rise in mineralization depth from 8 mm to 17 mm. This demonstrates the strong correlation between gas pressure and the extent of mineralization, highlighting the importance of optimizing pressure levels for effective surface treatment.

A critical parameter is the relative humidity (RH) during mineralization treatment, as it plays a dual role in the process. On one hand, higher humidity slows the diffusion of CO_2 into the concrete matrix; on the other hand, it provides the water necessary for mineralization reactions to occur. Experimental results have shown that CO_2 mineralization efficiency peaks at an RH of 50–65% (Liu and Meng, 2021; Zhang and Shao, 2016a). As depicted in Figure 6.1(c), an RH of approximately 50% is optimal for achieving maximum mineralization depth. Both lower and higher humidity levels lead to reduced mineralization efficiency. Low humidity limits the availability of water for reactions, while high humidity hinders CO_2 diffusion.

Temperature also has a profound impact on the mineralization depth. Higher temperature accelerates the reaction kinetics, leading to deeper mineralization. By increasing the mobility of CO_2 molecules and promoting the formation of mineralization products, elevated temperature enhances the overall efficiency of the treatment process.

In summary, mineralization time, CO_2 gas pressure, relative humidity, and temperature are interdependent factors that collectively influence the CO_2 surface treatment thickness. Optimizing these parameters is crucial for maximizing the effectiveness of CO_2 surface treatment, improving the microstructure of cement-based materials and achieving sustainable performance enhancements.

Additionally, the intrinsic properties of the concrete material, including the water–cement ratio, type of mineral admixture, and substitution level, also play critical role in determining the ultimate thickness of the CO_2 mineralized layer, highlighting the need for tailored approaches in CO_2 surface treatment for different concrete compositions and applications. The mineralization depth increases with the water-to-binder ratio. Hussain et al. (2017) observed that increasing the water–cement ratio from 0.35 to 0.65 led to a substantial rise in mineralization depth, from 13 mm to 26 mm, after 28 days of CO_2 treatment. The inclusion of mineral admixtures such as fly ash also affects mineralization rates. While fly ash contributes to a slight increase in mineralization depth, this effect is attributed to the pozzolanic reaction, which consumes calcium hydroxide ($Ca(OH)_2$) and lowers the alkalinity of the pore solution (Branch et al., 2018). This reduction in alkalinity facilitates mineralization reactions but may also compromise the durability of the concrete in aggressive environments. An important observation is the inverse relationship between mineralization depth and the compressive strength of concrete before CO_2 surface treatment (Das et al., 2012), as shown in Figure 6.2. High-strength concrete, characterized by low porosity and denser microstructures, restricts the penetration of CO_2 gas, resulting in shallower mineralization depths. Conversely, lower-strength concrete with higher porosity allows for greater CO_2 diffusion, leading to deeper mineralization layers. This relationship underscores the need to balance material properties and treatment parameters to achieve optimal results.

In summary, the thickness of the CO_2 surface treatment layer is governed by a complex interplay of concrete material properties, mineralization duration, and mineralization conditions. Understanding these factors is essential for tailoring CO_2 surface treatment processes to meet specific performance and sustainability objectives. By optimizing these parameters, it is possible to enhance the durability, resource efficiency, and environmental benefits of CO_2 treated concrete materials.

6.2.2 Phase composition of surface treatment layers

6.2.2.1 Calcium Carbonate

Calcium carbonate ($CaCO_3$) is the primary product formed in the CO_2 surface treatment layers. Its deposition primarily occurs within the surface pores and diminishes gradually with increasing depth, as illustrated in Figure 6.3. Pan et al. (2019) reported

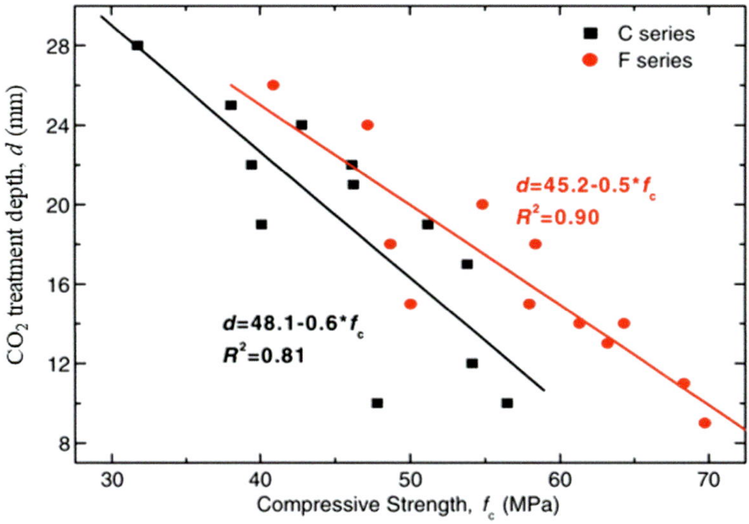

FIGURE 6.2 Relationship between mineralization depth and compressive strength before mineralization (Das et al., 2012).

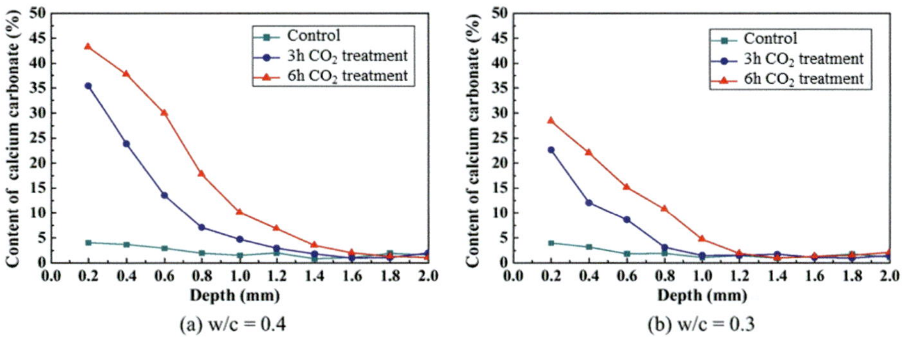

FIGURE 6.3 Effect of CO_2 surface treatment on $CaCO_3$ content at different depths in surface layers (Pan et al., 2019).

28.4% and 43.2% $CaCO_3$ content in the 0.2 mm surface layer for samples with w/c of 0.3 and 0.4, respectively, after 6 hours of CO_2 treatment. The $CaCO_3$ content is affected by the mineralization time and the water–cement ratio. The calcium carbonate content at the same mineralization depth gradually increased with the increase in the mineralization time. Higher $CaCO_3$ content was measured at the same depth for the sample with a 0.4 water–cement ratio compared to the sample with a 0.3 water–cement ratio, which may be attributed to its larger pores, allowing CO_2 to penetrate more easily into the inner parts for mineralization reaction. As the depth increased, the $CaCO_3$ content gradually decreased, reaching approximately 2% at a depth of 2 mm. This gradient is indicative of the rapid mineralization occurring near the surface, where CO_2 diffusion is more efficient, and the slower reaction rates at greater depths, where CO_2 penetration is hindered. The formation of $CaCO_3$ not only enhances the surface hardness and chemical stability of concrete but also plays a crucial role in improving its resistance to environmental degradation.

6.2.2.2 Calcium silicate hydrate gels and silica-rich gels

During the CO_2 treatment on the surface of cement concrete, calcium ions (Ca^{2+}) are leached from the calcium silicate hydrate (C-S-H) gels, leading to a decrease in the Ca–Si ratio in the C-S-H gels. As shown in Figure 6.4, the band (Q^2) of 974 cm^{-1} for the untreated sample was assigned to Si-O stretching vibrations, indicating a wide range of C-S-H for Ca/Si≈2 (Pan et al., 2019). For the CO_2 treated sample, the band was shifted toward higher wave numbers, indicating that the silicate tetrahedra (SiO_4) in the C-S-H structure lost their calcium bonds, leading to the polymerization of the silicate network and the formation of a silica-rich gel. The transformation of C-S-H into a more silica-rich phase is akin to the formation of hydrated silicon dioxide ($SiO_2{\cdot}nH_2O$), which improves the overall microstructure of the CO_2 treated concrete. This process also reduces the volume of free calcium hydroxide in the concrete, contributing to both enhanced mechanical properties and reduced susceptibility to alkali–silica reactions.

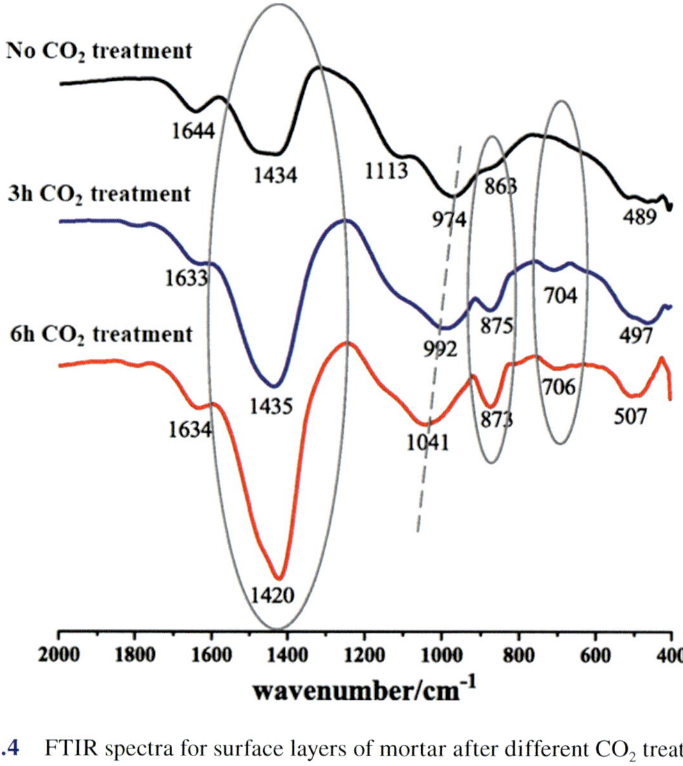

FIGURE 6.4 FTIR spectra for surface layers of mortar after different CO_2 treatments (Pan et al., 2019).

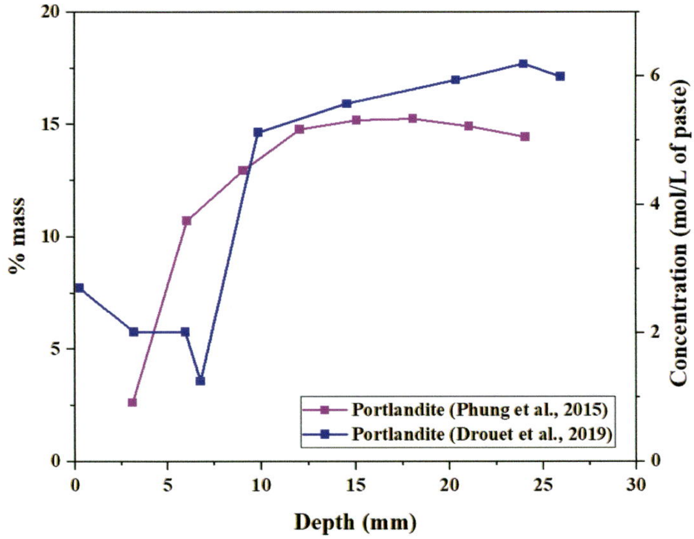

FIGURE 6.5 Changes in portlandite contents as a function of depth (Data from Drouet et al., 2019; Phung et al., 2015).

6.2.2.3 Calcium hydroxide and unhydrated clinker

The calcium hydroxide ($Ca(OH)_2$) content decreases significantly after CO_2 surface treatment, demonstrating the effective reaction of CO_2 with the concrete. As shown in Figure 6.5, the $Ca(OH)_2$ content increases with depth. For instance, Phung et al. (2015) observed that at a depth of 0–3 mm, the $Ca(OH)_2$ content was only about 3%, which increased to about 15% as the depth increased to about 15 mm. This gradient highlights the progressive mineralization of concrete, with the surface layer reacting more readily to CO_2, while deeper regions undergo less mineralization due to restricted CO_2 diffusion.

Additionally, the unhydrated clinker phases present in the concrete, including tricalcium silicate (C_3S) and dicalcium silicate (C_2S), play an important role in the mineralization process. These unreacted clinker phases can participate in reactions with CO_2, contributing to the continued development of mineralization products. The content of unhydrated clinker in the surface treatment layer is influenced by the mixture design, hydration duration, and CO_2 treatment time. The unreacted clinker provides a reservoir for further mineralization reactions, enhancing the long-term durability of the material.

The CO_2 surface treatment of concrete results in a range of chemical and microstructural changes that significantly enhance the material's properties. The formation of $CaCO_3$ strengthens the surface and improves resistance to environmental factors. At the same time, the transformation of calcium silicate hydrate into silica-rich gels enhances the microstructure, contributing to better long-term durability. The reduction in $Ca(OH)_2$ and the potential for secondary reactions involving unhydrated clinker further optimize the performance of CO_2 treated concrete. Understanding these phase transformations is crucial for tailoring CO_2 treatment methods to maximize the material's strength, sustainability, and durability.

6.2.3 Pore structure of surface treatment layers

During CO_2 surface treatment, calcium ions in the concrete matrix react with CO_2 to form $CaCO_3$. This matrix transformation, primarily involving the conversion of $Ca(OH)_2$ into $CaCO_3$, not only contributes to the solidification of the matrix but also plays a pivotal role in the densification and compaction of the CO_2 surface treatment layer. The precipitation of $CaCO_3$ significantly reduces the porosity of the concrete surface, thereby enhancing its density and structural integrity. As shown in Figure 6.6, the porosity of the CO_2 treated samples is consistently lower than that of the untreated samples, indicating the effectiveness of CO_2 treatment in densifying the surface layer. Similar findings were reported by Phung et al. (2015), who observed a 22% reduction in porosity after a 28-day treatment under 6 bar CO_2 pressure. However, as the depth of the concrete increases, the reduction in porosity gradually diminishes. This phenomenon is attributed to the decreasing availability of CO_2 at greater depths, which limits the formation of $CaCO_3$. Consequently, the porosity at deeper levels approaches that of untreated concrete, underscoring the depth-dependent nature of the mineralization process.

Pore size distribution is a critical parameter that reflects the pore structure of concrete and significantly influences its resistance to the transport of water and ions. During CO_2 treatment, a notable shift in pore size distribution occurs. Specifically, there is a marked increase in the proportion of small pores (less than 50 nm), while the proportion of medium and large pores (greater than 50 nm) decreases (Li et al.,

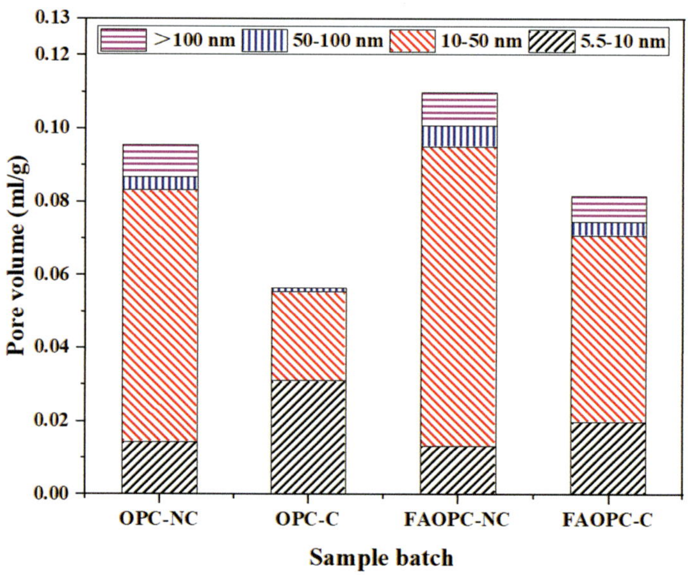

FIGURE 6.6 Pore volume and pore diameter distribution (NC: non-carbonated, C: carbonated) (Zhang and Shao, 2018).

2019). This shift, as illustrated in Figure 6.6, is primarily due to the gradual filling of the medium and large pores by mineralization products, such as $CaCO_3$ and silica-rich gel. As these CO_2 mineralization products accumulate, they block and fill the larger pores, leading to a reorganization of the pore structure. This results in a decrease in the size of the medium and large pores, with an increase in the proportion of smaller pores. In concrete science, pores are classified based on their size, with pores less than 20 nm generally considered harmless, as they do not significantly affect the material's transport properties. Pores in the range of 20–50 nm are regarded as less harmful, while pores between 50 nm and 100 nm are considered harmful due to their role in facilitating the penetration of water and aggressive ions. Pores larger than 100 nm are particularly detrimental, as they can drastically reduce the material's durability by increasing permeability and allowing the ingress of harmful substances (Jiang and Guan, 1999; Yu and Ye, 2013). Therefore, the reduction of larger pores through CO_2 treatment enhances the densification of the concrete surface layer, improving its resistance to water and ion penetration. This increased surface compactness is crucial for enhancing the long-term durability and performance of CO_2 treated concrete, particularly in harsh environments where material degradation due to water and ion ingress is a significant concern.

In summary, CO_2 surface treatment of concrete results in significant improvements in the porosity and pore structure of concrete. The formation of $CaCO_3$ within the surface layer reduces the overall porosity, while the filling of larger pores with mineralization products leads to a more compact and resistant surface. These changes are particularly beneficial for enhancing the durability of concrete by reducing the penetration of water and harmful ions. Additionally, the shift in pore size distribution toward smaller pores enhances the material's long-term performance by minimizing the harmful effects of larger and more permeable pores. As such, CO_2 surface treatment is a promising method for improving the durability and sustainability of concrete, contributing to the development of more resilient and low-carbon building materials.

6.2.4 Micromechanical properties of surface treatment layers

CO_2 surface treatment significantly improves the micromechanical properties of the surface treatment layer through mineralization reaction, primarily manifested in increased hardness and elastic modulus. As shown in Figure 6.7, the microhardness of the CO_2 treated sample increased by about 44% compared to the untreated sample (Li et al., 2019). This substantial enhancement underscores the effectiveness of CO_2 treatment in strengthening the surface layer. During the treatment, $Ca(OH)_2$ and C-S-H gel are transformed into $CaCO_3$. $CaCO_3$ exhibits a higher elastic modulus and hardness compared to the original hydrated products (Li et al., 2019). This transformation significantly contributes to the increased hardness and elastic modulus of the CO_2 treated layer. In addition, the formation of mineralization products, such as $CaCO_3$ and silica-rich gels, fills the pores within the surface layer, resulting in a denser and more compact structure. This densification further enhances the mechanical properties, ensuring improved resistance to external forces and environmental degradation.

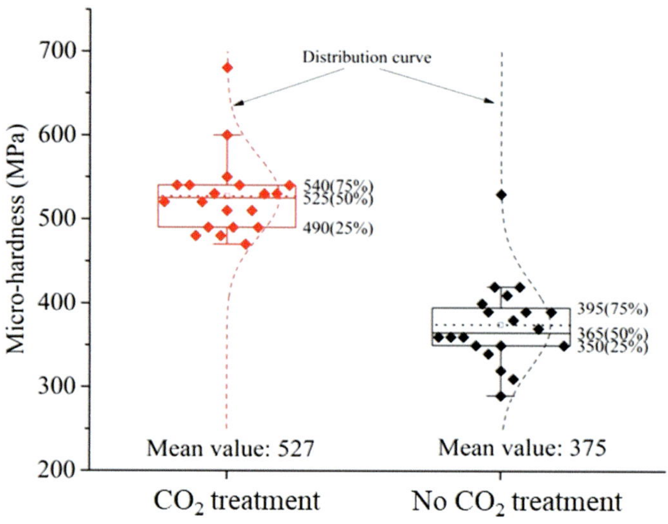

FIGURE 6.7 Micro-hardness of mineralization and hydration paste (Li et al., 2019).

The improvements in hardness and elastic modulus are not uniform throughout the CO_2 treated layer. Instead, they exhibit a gradient variation with depth. Near the surface, where the mineralization reaction is most pronounced, the hardness and elastic modulus reach their peak values. However, as the depth increases, the extent of CO_2 mineralization diminishes due to the limited penetration of CO_2. Consequently, the micromechanical properties gradually decline, approaching the values of untreated material at greater depths. This gradient reflects the depth-dependent nature of the mineralization process and highlights the importance of optimizing CO_2 treatment parameters to achieve the desired performance improvements.

The degree of mineralization plays a pivotal role in determining the micromechanical properties of the treated layer. A higher degree of mineralization corresponds to a more extensive transformation of $Ca(OH)_2$ and C-S-H gels into $CaCO_3$, resulting in greater improvements in hardness and elastic modulus. However, achieving uniform mineralization throughout the surface layer can be challenging due to factors such as pore structure, CO_2 diffusion rate, and treatment duration. Understanding and controlling these factors are essential for maximizing the benefits of CO_2 surface treatment.

The micromechanical enhancements achieved through CO_2 surface treatment make it a promising technique for improving the durability and performance of concrete structures. By increasing the hardness and elastic modulus, the CO_2 surface treatment enhances the material's resistance to mechanical stress and environmental factors. The depth-dependent properties highlight the need for precise control over the mineralization process to ensure consistent performance across the treated layer. As a result, CO_2 surface treatment offers a sustainable and effective solution for enhancing the mechanical properties of concrete, contributing to the development of resilient and long-lasting infrastructure.

6.3 EFFECT OF SURFACE TREATMENT ON STRENGTH

6.3.1 Compressive strength

Concrete compressive strength is a key indicator used to assess the engineering performance of concrete. CO_2 surface treatment technology, through regulating the mineralization process, can effectively improve the concrete microstructure, thereby enhancing its compressive strength. Initially, even a brief mineralization period can lead to significant improvements in the compressive strength of mortar, as demonstrated in Figure 6.8. After 3 hours of CO_2 treatment, the compressive strength of mortar samples with water-to-cement ratios of 0.3 and 0.4 increased by 17.6% and 8.3%, respectively. Furthermore, both mortars showed continued strength gains as the mineralization time was extended. This increase in strength is mainly attributed to the reaction between $Ca(OH)_2$ and CO_2, which forms $CaCO_3$, filling some of the pores within the concrete. Additionally, this reaction lowers the calcium–silicon ratio at the surface, facilitating the formation of calcium silicate gel ($SiO_2 \cdot nH_2O$). However, due to short mineralization duration, the CO_2 penetration is limited, restricting the mechanical enhancement primarily to the surface layer and thus the overall impact on compressive strength remains modest.

In contrast, longer mineralization treatments (7 and 28 days) allow CO_2 to penetrate deeper into the concrete, leading to more profound effects on the overall compressive strength (Chi et al., 2002; Li et al., 2019; Pravalika and Rao, 2018; You et al., 2014). Figure 6.9 illustrates the changes in compressive strength of concrete after 7 and 28 days of CO_2 treatment. The results reveal a substantial increase in compressive strength following CO_2 treatment. After 7 days of mineralization, the compressive strength of treated concrete was 10% to 40% higher than that of untreated concrete. Furthermore, the compressive strength of CO_2 treated concrete reached 70% to 100% of the 28-day strength of untreated concrete. Prolonging the

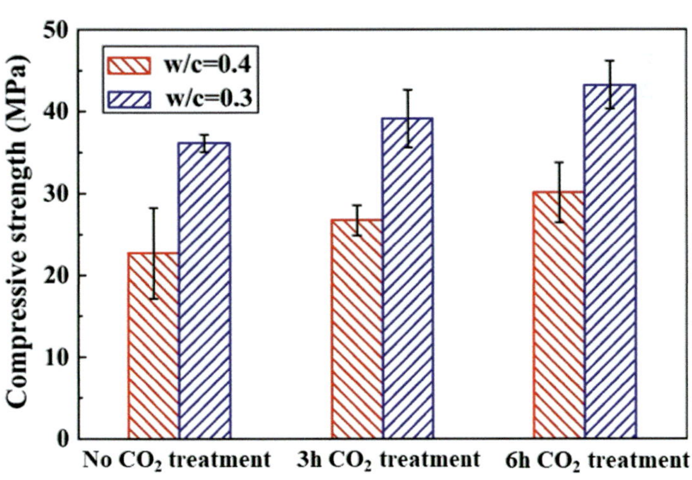

FIGURE 6.8 Effect of CO_2 treatment on compressive strength of cement mortars (Pan et al., 2019).

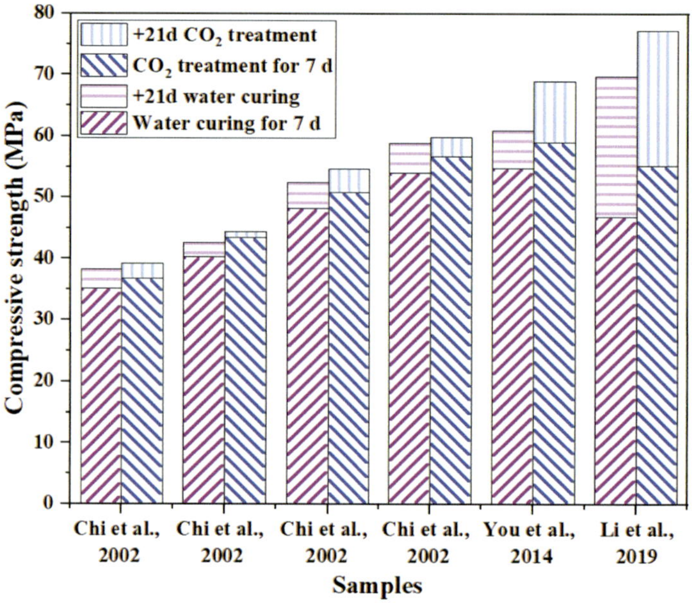

FIGURE 6.9 Compressive strength of CO_2 treated and untreated samples at 7 days and 28 days (Data from Chi et al., 2002; Li et al., 2019; You et al., 2014).

CO_2 treatment duration further enhanced compressive strength, although the rate of improvement gradually slowed. This plateau effect is likely due to the limited penetration depth of CO_2 into the concrete. The dense surface layer formed during early mineralization restricted further CO_2 diffusion, thereby slowing the mineralization reaction. After 28 days of treatment, the compressive strength of CO_2 treated concrete surpassed that of untreated specimens, in some cases reaching up to 120% of the untreated samples' strength. This demonstrates the long-term benefits of CO_2 surface treatment in enhancing the mechanical performance of concrete.

The observed increase in compressive strength of CO_2 treated concrete is primarily due to the presence of the CO_2 treatment layer. This layer, which forms gradually during the mineralization process, creates a dense protective coating on the surface of the concrete, reinforcing its overall structure. The formation of this layer leads to a reduction in porosity, the filling of microcracks, and a substantial increase in strength, thereby significantly improving the concrete's overall performance. As shown in Figure 6.10, there is a distinct positive correlation between compressive strength of concrete and the thickness of the CO_2 treatment layer. As the CO_2 mineralization depth increases, the density of the CO_2 treatment layer also increases, resulting in a corresponding increase in compressive strength. However, despite the observed increase in compressive strength with greater mineralization depth, the rate of this enhancement is relatively slow. This insignificant enhancement is mainly due to the limitation imposed by the CO_2 surface layer thickness, with the compressive strength still predominantly governed by the ongoing hydration reactions within the concrete.

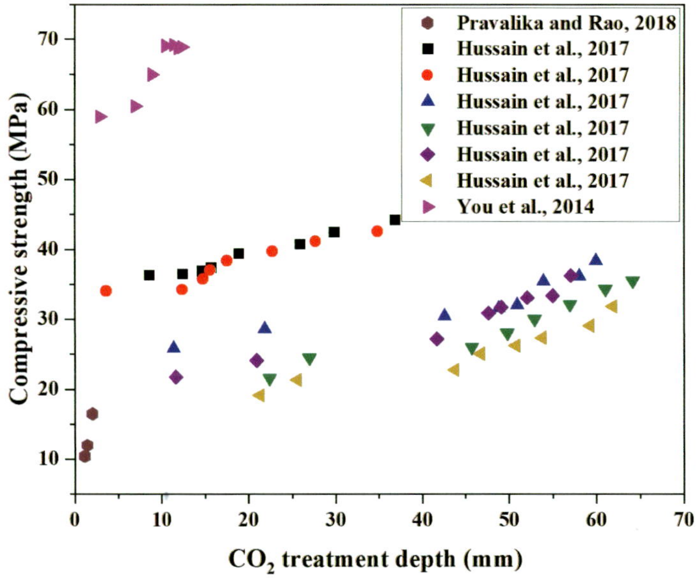

FIGURE 6.10 Relationship between compressive strength and mineralization depth (Data from Hussain et al., 2017; Pravalika and Rao, 2018; You et al., 2014).

6.3.2 Splitting tensile strength

The splitting tensile strength of concrete is a critical mechanical property that reflects its ability to resist tensile stresses, which can lead to cracking and structural failure. CO_2 surface treatment has been shown to significantly enhance this property, as demonstrated in Figure 6.11. The splitting tensile strength of concrete subjected to CO_2 surface treatment is 10–30% higher than that of untreated concrete (Chi et al., 2002; Chun et al., 2007). This improvement is primarily attributed to the microstructural and compositional changes induced by the mineralization process. First, the CO_2 surface treatment effectively reduces the total porosity of the concrete surface layer, particularly targeting large and interconnected pores. This reduction enhances the density and homogeneity of the surface layer, minimizing weak points where tensile stresses could concentrate. Second, the formation of $CaCO_3$ through the mineralization reaction results in a stable structure and is distributed throughout the concrete surface layer. The presence of $CaCO_3$ not only increases the material's stiffness but also enhances its load-bearing capacity, enabling the surface layer to better resist tensile forces. Furthermore, the high strength and stiffness of the CO_2 treated surface layer facilitate the efficient dispersion of external stresses (Hussain et al., 2017). This redistribution of stress delays the onset of tensile failure, allowing the concrete to withstand higher loads before cracking occurs. Consequently, the splitting tensile strength is significantly improved.

The improvement in splitting tensile strength is closely related to the thickness of the CO_2 treated layer. Figure 6.12 illustrates the positive correlation between

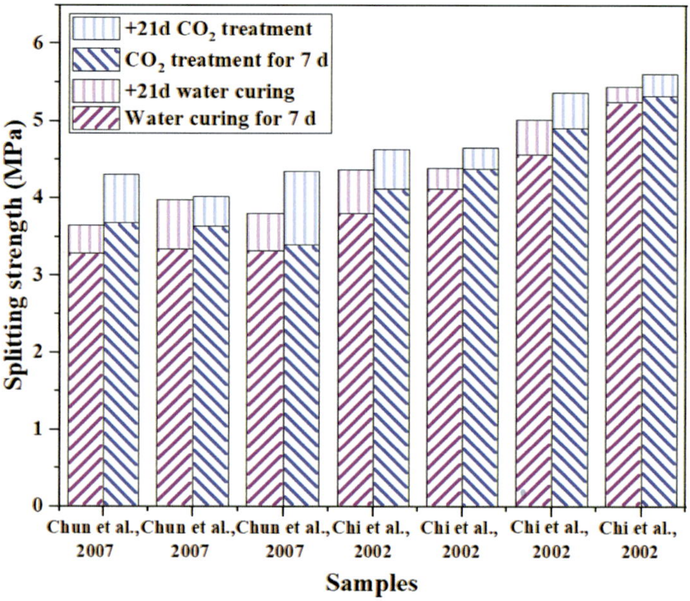

FIGURE 6.11 Splitting tensile strength of uncarbonated and carbonated samples at 7 days and 28 days (Data from Chi et al., 2002; Chun et al., 2007).

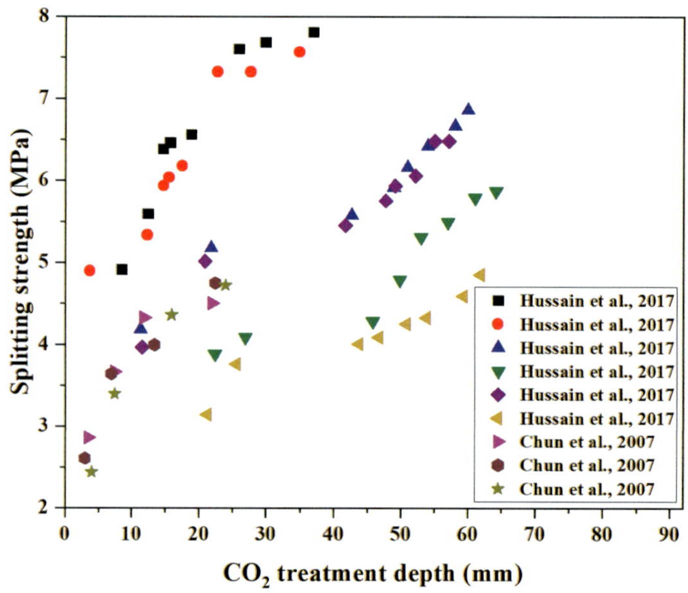

FIGURE 6.12 Relationship between splitting tensile strength and mineralization depth (Data from Chun et al., 2007; Hussain et al., 2017).

splitting tensile strength and CO_2 treated layer thickness. As the thickness of the CO_2 treated layer increases, the densification and reinforcement effects become more pronounced, leading to a gradual increase in splitting tensile strength.

6.4 EFFECT OF SURFACE TREATMENT ON PERMEABILITY

Permeability is a critical parameter in evaluating the durability of concrete, as it directly affects the material's resistance to the ingress of harmful liquids and gases. CO_2 surface treatment induces the formation of substantial quantities of $CaCO_3$ through the reaction of CO_2 with cement clinker and hydration products. The $CaCO_3$ is deposited within the interstitial spaces between the cementitious particles, resulting in the formation of a dense mineralization layer on the surface of the concrete. The mineralization layer exhibits a reduced number of macropores and connecting pores, which effectively impede the penetration of water, gas, and other deleterious substances. The impact of CO_2 surface treatment on the permeability of concrete has been a subject of significant research interest among scholars, particularly regarding water permeability, gas permeability, and chloride ion permeability.

6.4.1 Water permeability

Water permeability is a critical factor influencing the durability and service life of concrete structures, particularly in wet or aggressive environments where moisture ingress can lead to significant damage. The ingress of water facilitates chemical reactions, such as sulfate attack or chloride-induced corrosion of reinforcement, which can compromise the structural integrity of concrete (Phung et al., 2013). CO_2 surface treatment has been demonstrated to effectively mitigate these risks by reducing water permeability and enhancing the impermeability of concrete surfaces. The mechanism behind this improvement lies in the formation of a dense $CaCO_3$ layer on the concrete surface. This layer is generated through the reaction between CO_2 and cementitious components. The resulting $CaCO_3$ precipitates within the pores, filling macropores and interconnecting voids. This densification significantly reduces the pathways available for water penetration, thereby improving the water resistance of the treated surface. Ren et al. (2022) conducted a comprehensive study on the water permeability of concrete samples subjected to CO_2 surface treatment, as illustrated in Figure 6.13. Their findings revealed that, for untreated mortar, water permeability decreased with a lower water–cement ratio, consistent with the densification of the microstructure at lower water–cement ratios. However, following CO_2 surface treatment, a substantial reduction in water permeability was observed across all samples, regardless of their initial water–cement ratios. The reduction in water permeability was more pronounced in samples with higher water–cement ratios. This suggests that CO_2 surface treatment has a greater impact on mortars with initially higher porosity. The mineralization process effectively targets the larger and more interconnected pores prevalent in high water–cement ratio mortars, filling them with $CaCO_3$ and reducing the overall permeability to a greater extent. These findings underscore the versatility and effectiveness of CO_2 surface treatment in enhancing the impermeability of concrete, particularly for materials with less favorable initial properties.

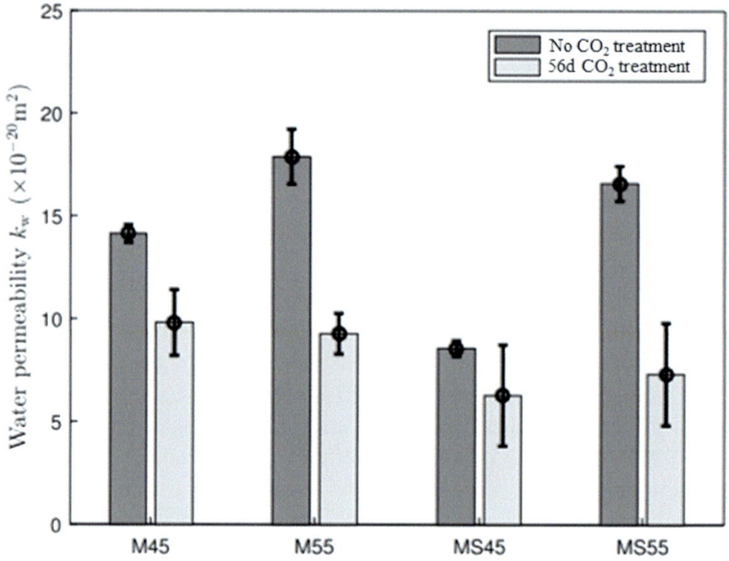

FIGURE 6.13 The evolution of water permeability after mineralization (MS55: mortar, w/c0.55, 10% silica fume) (Ren et al., 2022).

By forming a robust and dense mineralization layer, the treatment not only reduces water permeability but also contributes to the long-term durability of concrete structures exposed to challenging environmental conditions.

6.4.2 Water-vapor transmission

Water-vapor permeability is an important parameter for assessing a concrete's capacity to regulate moisture diffusion, which plays a crucial role in its performance and durability, especially in environments subjected to fluctuating humidity levels. CO_2 surface treatment has been shown to effectively reduce the water-vapor permeability of concrete, thereby enhancing its resistance to water-vapor ingress. Studies have shown that the water-vapor transmission of concrete samples decreases after CO_2 surface treatment (Pan et al., 2019; Tracz and Zdeb, 2019). Pan et al. (2019) observed that, regardless of the w/c, CO_2 treated samples exhibited lower water-vapor transmission rates compared to their corresponding control mortars. As shown in Figure 6.14, after 576 hours of exposure, the mass loss of the control sample with a w/c of 0.4 was 6.4 g, while the mass loss of samples after 3 hours and 6 hours of CO_2 treatment decreased by 16.9% and 32.8%, respectively. Similar findings were observed in the samples with a w/c of 0.3. The migration of water vapor in cementitious materials can facilitate the diffusion of aggressive gases and the formation of moisture gradients, which may compromise the material's durability. CO_2 surface treatment can effectively mitigate these issues by reducing the permeability to water vapor, thus preventing moisture-induced shrinkage or expansion. Additionally, the mineralization process lowers the alkalinity of the cement mortar, further enhancing the material's stability under varying environmental conditions.

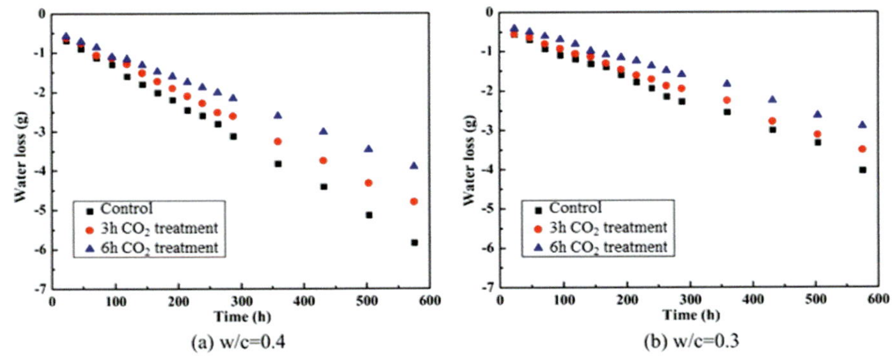

FIGURE 6.14 Effects of CO_2 surface treatment on the water-vapor transmission of cement mortars (Pan et al., 2019).

6.4.3 Chloride permeability

Chloride migration is an important factor affecting the durability of concrete, especially in marine or salty environments, where chloride intrusion can lead to the corrosion of reinforcing steel, thus reducing the service life of the structure. Chloride ions typically penetrate cement-based materials through the pore structure, and it can be hypothesized that the denser mineralization layer formed by CO_2 surface treatment may, at least theoretically, hinder the migration of chloride ions. Pan et al. (Pan et al., 2017a) observed a notable decline in the chloride migration coefficient of mortar following CO_2 surface treatment. As shown in Figure 6.15, a 3-hour CO_2 surface treatment resulted in an approximately 30% reduction in the chloride migration coefficient compared to the control group, while specimens with 6 hours of CO_2 surface treatment demonstrated a reduction of approximately 40%. This indicates

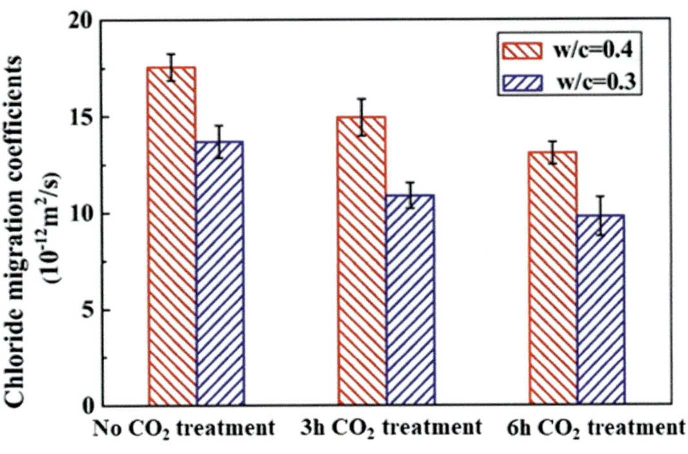

FIGURE 6.15 Effects of CO_2 surface treatment on the chloride migration coefficient of cement mortars (Pan et al., 2017a).

that the effectiveness of CO_2 surface treatment in mitigating chloride ion penetration improves with longer exposure times, likely due to the formation of a thicker and denser mineralization layer.

The effect of supplementary cementitious materials (SCMs), such as fly ash (FA) and ground granulated blast furnace slag (GGBS), on chloride ion permeability in CO_2 treated concrete has been extensively studied. Boa and Topu (2012) reported that an optimal combination of FA and GGBS in cementitious systems can enhance resistance to chloride ion penetration. This improvement is attributed to the formation of additional hydrated calcium silicate phases resulting from the secondary hydration of FA and GGBS. These phases refine the pore structure, reduce pore connectivity, and increase the physical adsorption and chemical binding capacity for chloride ions. However, contrasting findings have been observed regarding the role of FA and GGBS in chloride ion permeability resistance after CO_2 surface treatment. Song et al. (2022) investigated the rapid chloride migration coefficient of Portland cement-FA-GGBS mortar at 28 and 90 days, as shown in Figure 6.16. Their results revealed that pure cement samples exhibited the lowest chloride migration coefficients among ternary systems. As the replacement levels of FA and GGBS increased, the rapid chloride migration coefficient also increased, peaking at the FA maximum. This trend suggests that the chloride permeability resistance of CO_2 treated concrete decreases with higher FA and GGBS content. The underlying mechanism may involve the reaction of CO_2 with $Ca(OH)_2$, a byproduct of cement hydration. The mineralization reaction consumes $Ca(OH)_2$, particularly within the surface layer, reducing its availability for the pozzolanic reactions of FA and GGBS. This limitation hinders the formation of a dense, well-bonded mineralization layer, thereby diminishing the barrier effect against chloride ions. Moreover, the impact of SCMs on the mineralization process is time-dependent. While GGBS exhibits high early activity, contributing to pore refinement during initial hydration stages, FA's pozzolanic effect becomes more pronounced at later stages. Consequently, the interplay between mineralization and SCMs activity influences the overall effectiveness of CO_2 surface treatment in resisting chloride ion migration.

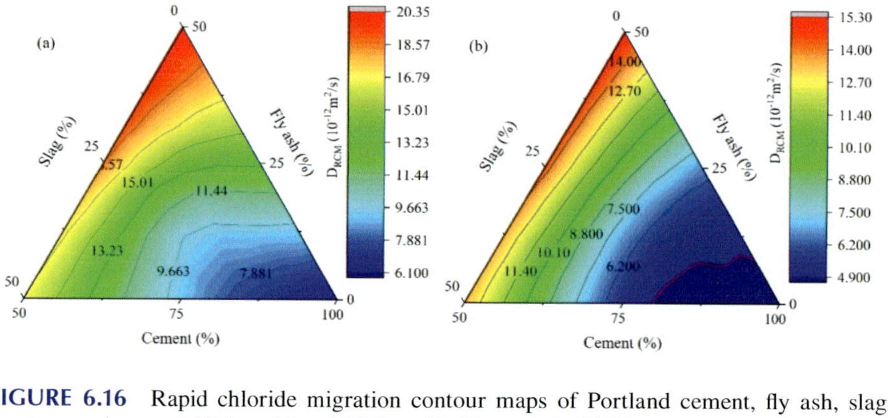

FIGURE 6.16 Rapid chloride migration contour maps of Portland cement, fly ash, slag mortar specimens at 28 days (a) and 90 days (b) (Song et al., 2022).

In summary, CO_2 surface treatment can significantly reduce chloride ion permeability in concrete, particularly when applied for longer durations. However, the incorporation of FA and GGBS introduces complexities that may either enhance or weaken the treatment's effectiveness, depending on the proportions and hydration stage. Optimizing the SCMs composition and tailoring CO_2 treatment protocols are crucial for achieving superior chloride ion resistance in concrete exposed to aggressive environments.

6.5 EFFECT OF SUBSEQUENT HYDRATION ON MECHANICAL PROPERTIES AND MICROSTRUCTURE OF CO_2 SURFACE TREATED CONCRETE

6.5.1 Mechanical properties

After CO_2 surface treatment, concrete undergoes further hydration when exposed to humid curing, which may further improve its mechanical properties. As shown in Figure 6.17, during subsequent water curing, the compressive strength of CO_2 treated concrete increases as curing time extends. This is primarily due to the ongoing hydration of unreacted cement components, such as C_2S and C_3S, which were not fully mineralized during the initial curing stages. Under continued water curing conditions, these unreacted particles further hydrate, leading to the formation of additional hydration products, including C-S-H gel and $Ca(OH)_2$. These hydration products not only enhance the overall mechanical properties of the concrete but also contribute to a denser and more stable microstructure over time. The formed $Ca(OH)_2$ reacts with the mineralization products silica gel to form additional C-S-H gel, further improving the concrete's strength and durability (He et al., 2016).

It is important to note that the duration of the initial CO_2 treatment period can influence the extent of strength gain during the subsequent hydration phase. Concrete

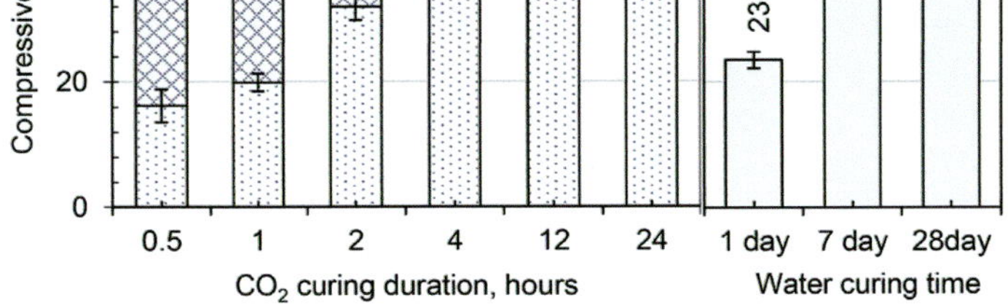

FIGURE 6.17 Strength development of cement paste subjected to coupled CO_2-water curing and water curing (Zhan et al., 2019).

subjected to a shorter CO_2 treatment period may experience a more significant increase in compressive strength during subsequent water curing. This is because a shorter CO_2 treatment period leaves a larger proportion of cement particles without mineralization, which remain reactive and available for continued hydration. The availability of these reactive particles enables further hydration, resulting in a more substantial strength development.

However, regardless of the CO_2 treatment time, the ultimate compressive strengths attained after the subsequent water curing are almost at the same level. This suggests that, although a shorter CO_2 treatment period may result in a more pronounced strength increase during subsequent curing, the long-term ultimate strength reaches similar levels across different CO_2 treatment times. This outcome is likely due to the barrier effect of the mineralization products formed during the CO_2 surface treatment. As the CO_2 treatment time increases, a greater quantity of mineralization products forms within the mineralization layer, reducing surface permeability and limiting the diffusion of water and calcium ions. The mineralization products create a physical barrier that restricts further hydration, making it more difficult for additional C-S-H gel to form. Conversely, when fewer mineralization products are present, there are fewer barriers to water and calcium ion diffusion, facilitating more extensive hydration and leading to enhanced mechanical properties.

The impact of FA and GGBS on the compressive strength of CO_2 treated concrete has been extensively studied. Wang et al. (2021) investigated the development of compressive strength during subsequent water curing after one day of CO_2 surface treatment, as shown in Figure 6.18. The results indicated that the 7-day compressive strength of almost all samples increased slightly following CO_2 surface treatment. This suggests that CO_2 surface treatment increases the surface density of concrete, thereby enhancing its compressive strength. However, since CO_2 surface treatment only affects the surface layer, the overall improvement in compressive strength remains limited.

Regarding the 28-day compressive strength, no significant difference was observed between CO_2 surface treated and untreated samples. However, when the fly-ash content exceeded 30%, a reduction in the 28-day compressive strength was noted after CO_2 surface treatment. This can be attributed to the fact that the pozzolanic reaction in the fly-ash Portland cement system is hindered by the early mineralization reaction. The higher the degree of early mineralization, the lower the pozzolanic reaction of the fly ash. At 90 days, the compressive strength of CO_2 treated samples decreased for all mixtures, with the degradation becoming more pronounced as the SCM content increased. This is because mineralization products such as calcium carbonate and silica gel cover the surface of unreacted clinker, hindering subsequent hydration reactions and suppressing the development of compressive strength (Huang et al., 2025).

6.5.2 Microstructural changes of surface treatment layers

Continued curing of CO_2 surface treatment concrete in a wet environment results in subsequent hydration reactions that significantly alter its microstructure. This process involves the further reaction of unhydrated cement particles, the formation of

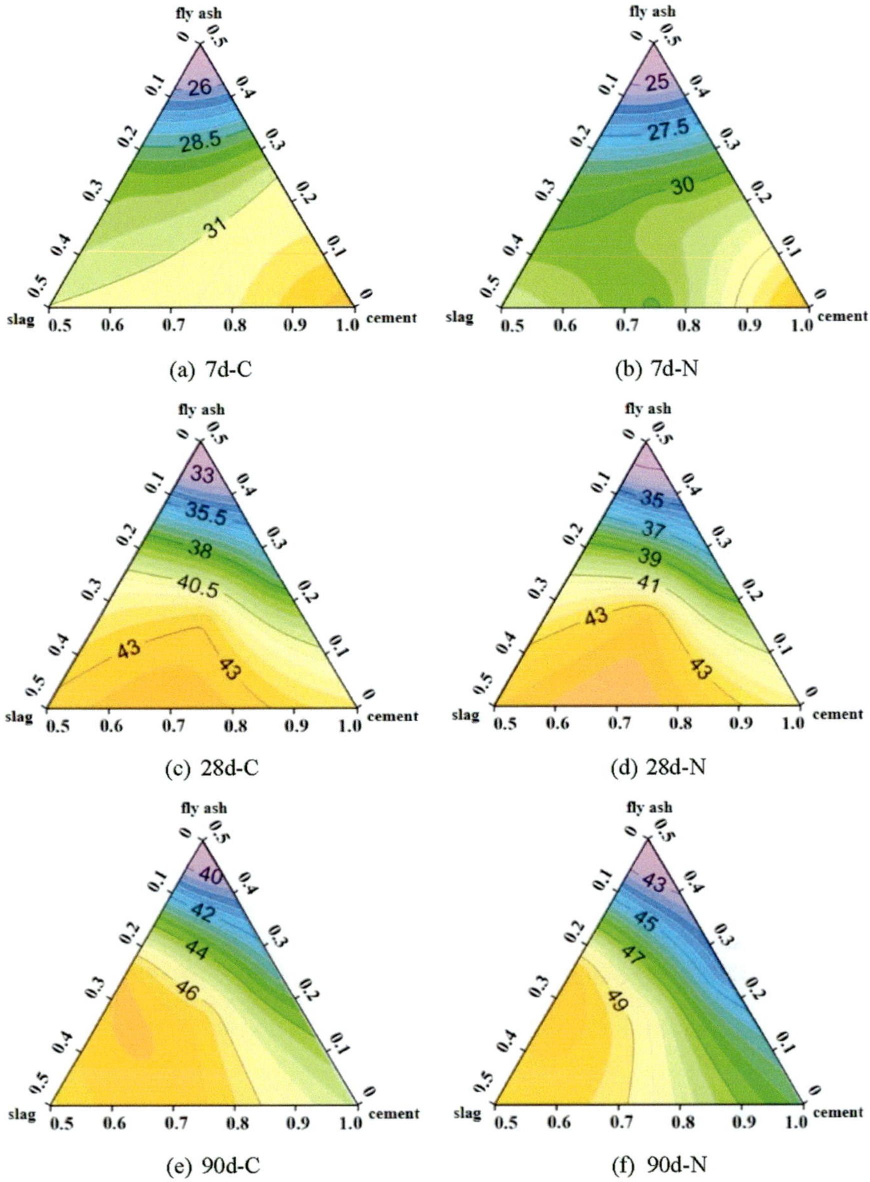

FIGURE 6.18 Compressive strength development in concrete treated with 1-hour CO_2 surface treatment and subsequent water curing (C presents specimens after CO_2 surface treatment, N presents specimens without CO_2 surface treatment) (Wang et al., 2021).

new hydration products, and changes in the pore structure, which are critical for the concrete's overall performance and durability. During CO_2 surface treatment, mineralization primarily affects the surface layer of concrete, consuming hydration products and cement minerals. These phases react with CO_2 to form $CaCO_3$

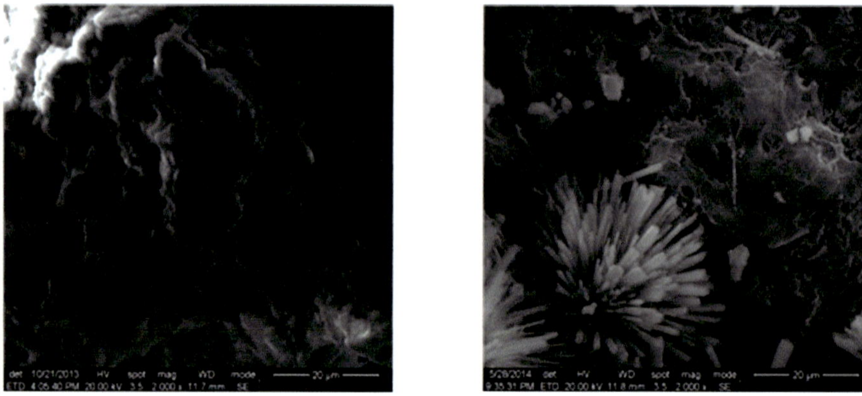

(a) 3h CO_2 treatment

(b) 3h CO_2 treatment + 90d further watercuring

FIGURE 6.19 SEM images of samples (He et al., 2016).

and silica-rich gels, creating a dense mineralization layer at the concrete surface. However, the inner regions of the concrete retain unreacted cement clinker particles, including C_2S and C_3S. These particles undergo hydration during subsequent water curing, forming additional C-S-H, ettringite, and $Ca(OH)_2$, which contribute to the continued strength development of the concrete. As shown in Figure 6.19, it can be seen that calcite grains were formed around the cement particles (Figure 6.19a) after 3 hours of CO_2 treatment. After 90 days of further water curing, large amounts of needle-shaped ettringite crystals and gels were observed (Figure 6.19b). A pozzolanic reaction also occurs between the calcium hydroxide produced during hydration and the silica-rich gels generated during the CO_2 surface treatment. This reaction forms additional hydrated calcium silicate gel, further enhancing the microstructure and mechanical properties of the concrete. Pan et al. (2019) observed an increase in $Ca(OH)_2$ and C-S-H gel content in the CO_2 treated layer after subsequent water curing, indicating ongoing hydration and pozzolanic reactions within the concrete. Furthermore, it was also observed that the decomposition temperature of the $CaCO_3$ produced by the CO_2 surface treatment increased during subsequent water curing. This suggests that the degree of crystallization of the calcium carbonate is improved, indicating that the mineralization products are becoming more stable and less reactive, thus reinforcing the surface layer. Interestingly, it was also observed that the $CaCO_3$ content in the concrete decreased slightly during subsequent water curing. This reduction can be attributed to the hydration of residual calcium aluminates (C_3A) in the untreated cement clinkers. In the presence of $CaCO_3$, these aluminates react to form hydrated monocarbonate calcium aluminates, thereby consuming some of the $CaCO_3$ and leading to a slight decrease in its content (He et al., 2016).

The hydrated calcium silicate gel and other products produced during the subsequent water curing process fill the pores, thereby reducing the total porosity of the concrete. Zhan et al. (2019) found that the porosity of all CO_2 surface treatment samples decreased with the duration of subsequent water curing. As shown in

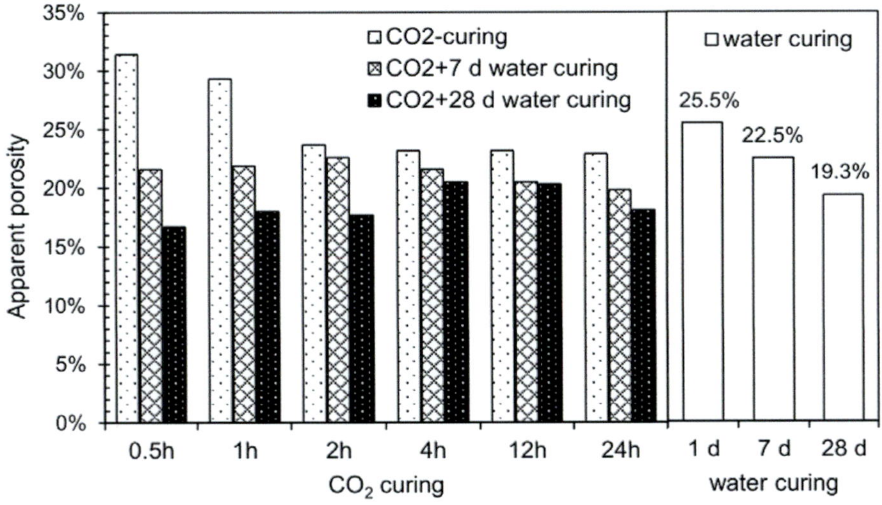

FIGURE 6.20 Comparison of total porosity for cement pastes experienced different curing regimes (Zhan et al., 2019).

Figure 6.20, the reduction in porosity was significant, but the decrease in porosity for CO_2 treated samples was less pronounced compared to that of untreated samples during the first 2 hours of curing. After 28 days of subsequent water curing, the total porosity of all samples, including the ones without CO_2 surface treatment, remained relatively stable, indicating that the primary reduction in porosity occurred in the early stages of curing.

6.6 EFFECT OF SUBSEQUENT HYDRATION ON THE PERMEABILITY OF CO_2 SURFACE TREATED CONCRETE

CO_2 surface treatment significantly reduces the initial permeability of concrete by forming a mineralization layer on the surface. This mineralization layer, primarily composed of $CaCO_3$, increases the density of the concrete surface, thereby acting as an effective barrier to water and ion penetration. However, the permeability of concrete is not solely influenced by the mineralization layer; the subsequent hydration process also plays a crucial role in optimizing the concrete's microstructure, which further impacts its water permeability.

During subsequent water curing, unhydrated cement particles continue to react, leading to the formation of additional hydration products, such as C-S-H gel and $Ca(OH)_2$. These hydration products enhance the density of the mineralization layer by filling voids and refining the pore structure. Particularly, the formation of C-S-H gel serves as a filler, reducing the residual porosity in the mineralization layer. This process refines the pore size distribution, converting larger capillary pores into smaller gel pores. The reduction in pore size and the refinement of the pore structure are key factors in improving the impermeability of the concrete, as they significantly reduce the capacity of water, gas, and other harmful ions to diffuse through the material. The diffusion of water, gases, or other deleterious substances is strongly influenced by the connectivity and size of the pores in the concrete. Smaller pores and more complex pore structures result in a reduced diffusion rate. Studies have shown that reduced pore connectivity and the creation of more tortuous diffusion paths can delay the migration of harmful substances, such as chloride ions or aggressive gases, into the concrete. Therefore, CO_2 surface treatment followed by water curing reduces pore connectivity, leading to a more intricate pore structure that hinders the transportation of water and other contaminants, ultimately improving the durability of the concrete in aggressive environments.

6.6.1 Water permeability

The initial water absorption coefficient is a significant parameter that reflects the water transport properties and pore structure characteristics of cementitious materials. Pan et al. (2019) observed that subsequent water curing significantly reduced the initial water absorption of CO_2 treated mortar. As shown in Figure 6.21, the water absorption rate of CO_2 surface treated mortar decreased with lime water curing time.

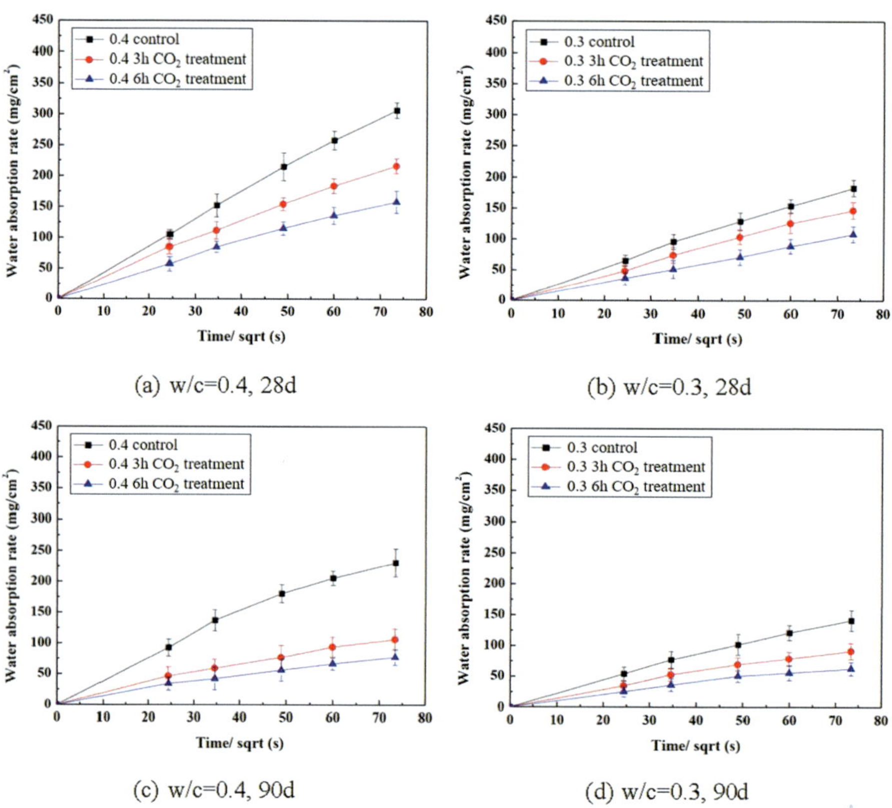

FIGURE 6.21 Effect of CO_2 surface treatment on the water absorption rate of cement mortars after 28 or 90 days of subsequent lime water curing (Pan et al., 2019).

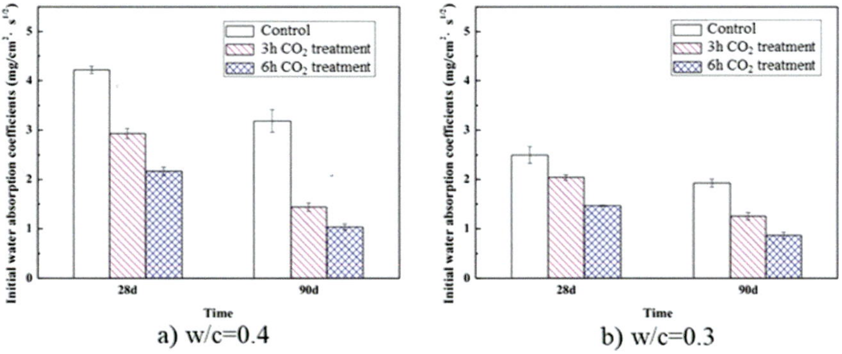

FIGURE 6.22 Initial water absorption coefficients of CO_2 surface treated cement mortars after subsequent lime water curing (Pan et al., 2019).

After 90 days of lime water curing, the water absorption rates of mortar samples with a water-to-cement ratio of 0.4 were reduced by 53.9% and 66.5% after 3 hours and 6 hours of CO_2 treatment, respectively. As shown in Figure 6.22, for mortar with a water–cement ratio of 0.4, subsequent water curing for 28 and 90 days resulted in a reduction of approximately 31% and 50%, respectively, in the initial water absorption coefficient for samples treated with CO_2 for 3 hours. In contrast, samples treated with CO_2 for 6 hours showed a more significant reduction of approximately 55% and 72% at 28 and 90 days, respectively. Similarly, for mortar with a lower w/c of 0.3, the initial water absorption coefficients of CO_2 treated samples decreased by 19.9% and 17%, respectively, after 28 and 90 days of subsequent water curing.

These results suggest that both the w/c and the duration of the CO_2 surface treatment significantly influence the degree of reduction in water absorption during subsequent water curing. A higher water–cement ratio and longer CO_2 treatment duration lead to a more pronounced decrease in water absorption, which suggests more effective densification of the concrete structure. The observed reduction in water absorption can be attributed to two main factors: (1) the continued hydration of un-carbon mineralized cement particles during subsequent curing and (2) the secondary reactions between the silica gel and $Ca(OH)_2$, which form additional C-S-H gel. These processes further enhance the microstructure, reduce porosity, and improve the concrete's ability to resist water penetration.

The incorporation of SCMs, such as FA and GGBS, significantly alters the pore structure of concrete, thereby influencing the effectiveness of CO_2 surface treatment. These SCMs, through their pozzolanic reaction, modify the microstructure of concrete at different stages of hydration, directly impacting its permeability characteristics. GGBS, known for its high early reactivity, induces a pronounced pozzolanic effect during the initial hydration stages. This early reaction densifies the pore structure of concrete, contributing to reduced permeability. In contrast, FA exhibits delayed pozzolanic activity, with its benefits in reducing porosity becoming more evident at later stages of hydration. During the initial stages, FA can increase the overall porosity of concrete, potentially compromising its early impermeability.

These differences in pore structure evolution affect how SCMs-modified concrete interacts with CO_2 surface treatment. Figure 6.23 illustrates the water permeability index of cement, fly-ash, slag concrete at various ages (Wang et al., 2021). In untreated concrete, the ternary system's water permeability index gradually

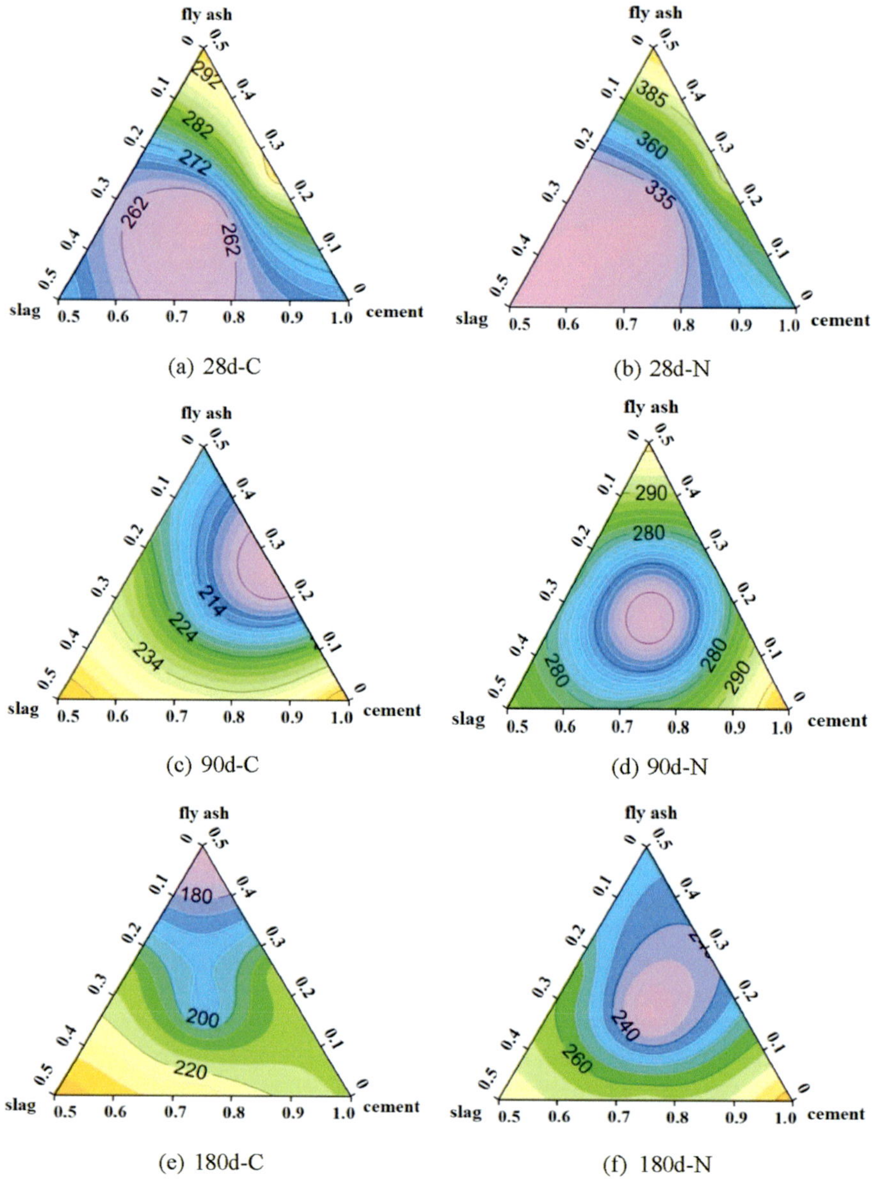

FIGURE 6.23 Effect of subsequent water curing on water permeability index of 1-hour CO_2 surface treated concrete (μL/min$^{1/2}$) (C presents specimens after CO_2 surface treatment, N presents specimens without CO_2 surface treatment) (Wang et al., 2021).

transitions from being predominantly influenced by GGBS content to being more affected by FA content. This shift reflects the growing influence of FA's pozzolanic effect in reducing porosity and hindering water ingress over time.

After CO_2 surface treatment, the permeability characteristics of SCMs-modified concrete exhibit a distinct concave pattern in the ternary diagram. This pattern indicates that the effectiveness of CO_2 surface treatment diminishes with higher SCM content. At 90 days, the concave region shifts away from the GGBS axis, suggesting that GGBS's influence on CO_2 treatment becomes less favorable over time. This shift may be attributed to the reduced availability of reactive components in GGBS-rich systems for mineralization. At 180 days, the minimum water permeability index of CO_2 treated concrete is observed near the FA apex. This finding highlights the enhanced effectiveness of CO_2 surface treatment in FA-rich systems during later hydration stages. The delayed pozzolanic reaction of FA not only complements the mineralization process but also synergistically reduces porosity, further improving the impermeability of the treated concrete. These results emphasize the complex interplay between SCMs, hydration age, and CO_2 surface treatment in determining the permeability performance of concrete.

In conclusion, subsequent water curing significantly improves the water permeability of CO_2 treated concrete by enhancing the density and refining the pore structure of the mineralization layer. This process reduces both the pore size and connectivity, creating more tortuous paths for water and harmful ions, thus enhancing the concrete's impermeability and overall durability. The findings highlight the importance of considering both the CO_2 treatment duration and the subsequent curing conditions when evaluating the long-term performance of CO_2 treated concrete, particularly in terms of water resistance.

6.6.2 Water-vapor transmission

Water-vapor penetration is an important index for assessing the ability of concrete to resist water-vapor diffusion, directly impacting its durability and long-term performance. The subsequent hydration process plays a pivotal role in enhancing the water-vapor resistance of CO_2 surface treated concrete by refining its pore structure and blocking vapor infiltration pathways.

After CO_2 surface treatment, the initial reduction in water-vapor diffusion capacity is attributed to the formation of a dense CO_2 mineralization layer. This layer significantly reduces the connectivity and size of pores near the surface, forming a barrier to water-vapor penetration. The effectiveness of this barrier is further enhanced during subsequent water curing. As unhydrated and un-mineralized cement particles in the inner regions of the concrete continue to react, additional hydration products are formed, further densifying the pore structure. This process fills voids and refines the pore size distribution, particularly within the mineralization layer. The gradual reduction in pore size and connectivity due to subsequent hydration is key in limiting water-vapor permeability. Smaller pores and more tortuous diffusion pathways impede the movement of water vapor, reducing its permeability. Over time, the prolonged water curing results in a cumulative densification effect, further blocking

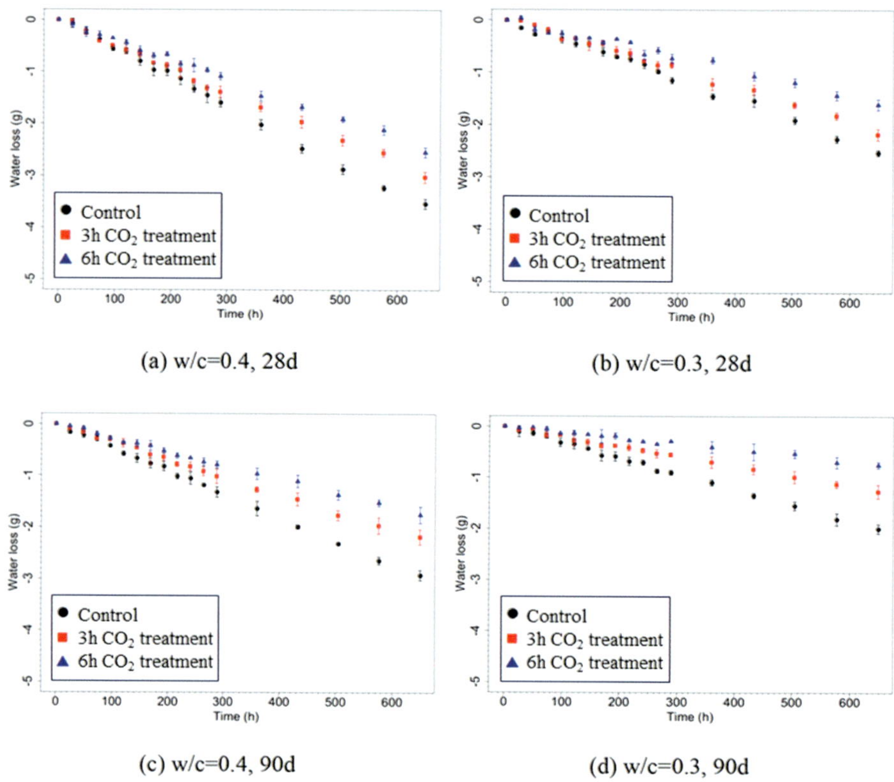

FIGURE 6.24 Effects of CO_2 surface treatment on the water-vapor transmission of cement mortars after subsequent lime water curing (Pan et al., 2019).

vapor diffusion channels. The water-vapor permeability coefficient is a quantitative measure of water-vapor diffusion through concrete. Pan et al. (2019) demonstrated the effectiveness of subsequent water curing in reducing this coefficient for CO_2 treated concrete. As shown in Figure 6.24, compared to untreated mortar, the water loss rate of the mortar treated for 3 hours and 6 hours with CO_2 was reduced by 25.2% and 39.8%, respectively, after 90 days of subsequent lime water curing. In addition, the water-vapor permeability coefficient of the 6-hour CO_2 surface treated specimen decreased by 33.3% after 28 days of subsequent water curing and 40.9% after 90 days of subsequent water curing, as shown in Figure 6.25. This progressive reduction highlights the significant impact of extended water curing on enhancing the impermeability of CO_2 treated concrete. These findings underscore the synergistic effects of CO_2 surface treatment and subsequent water curing in improving the water-vapor resistance of concrete. While the initial CO_2 treatment creates a dense and impermeable surface layer, subsequent hydration reactions further optimize the microstructure, reducing pore connectivity and refining pore size distribution. This dual mechanism not only limits water-vapor diffusion but also enhances the overall durability and service life of the concrete.

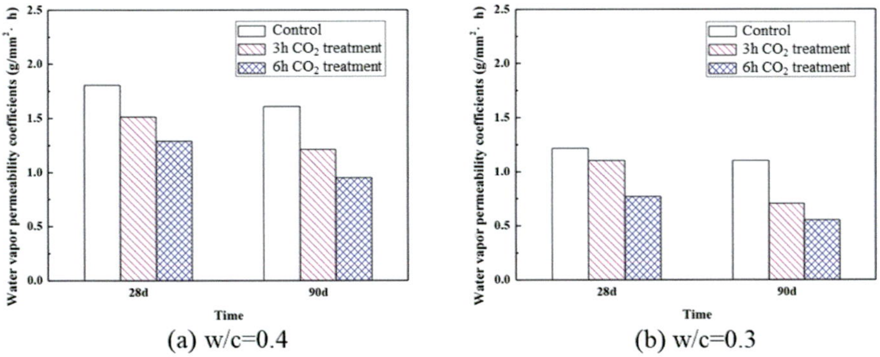

FIGURE 6.25 Effects of CO_2 surface treatment on the water-vapor transmission of cement mortars after subsequent lime water curing (Pan et al., 2019).

The incorporation of SCMs further influences the effectiveness of CO_2 surface treatment and subsequent water curing on gas permeability. Figure 6.26 illustrates the relationship between the air permeability index and SCMs content in CO_2 treated and subsequent water cured concrete. Among the SCMs, FA exhibited a pronounced effect in enhancing the resistance to gas permeability. As the proportion of FA substitution increased, a corresponding reduction in air permeability was observed, particularly in concrete treated with CO_2 for 28 days. This behavior can be explained by the delayed pozzolanic reaction of FA. During the early stages of hydration, FA participates minimally in the chemical reactions, resulting in a relatively loose and porous surface layer. This porous structure facilitates deeper penetration of CO_2 into the concrete, leading to a thicker mineralization layer and a higher degree of mineralization. Consequently, the increased thickness and density of the mineralization layer contribute to a more substantial reduction in gas permeability over time. In contrast, the inclusion of GGBS showed a contrasting effect. While GGBS improved the gas permeability of untreated concrete due to its high reactivity and pozzolanic effect, it appeared to weaken the impact of CO_2 surface treatment. This may be due to GGBS's high reactivity, which consumes $Ca(OH)_2$ during the pozzolanic reaction, reducing the availability of $Ca(OH)_2$ for CO_2 mineralization. As a result, the formation of the mineralization layer is less pronounced in GGBS-rich systems, limiting the reduction in gas permeability achieved through CO_2 surface treatment.

6.6.3 Chloride permeability

The formation of a dense mineralization layer due to CO_2 surface treatment significantly enhances the chloride ion resistance of concrete. This mineralization layer serves as a physical barrier, restricting the movement of chloride ions. However, the effectiveness of this barrier is further improved through subsequent hydration, which optimizes the microstructure of the concrete and enhances its protective properties. During subsequent hydration, unhydrated cement particles react, forming additional hydration products that fill the remaining pores. This process refines the pore

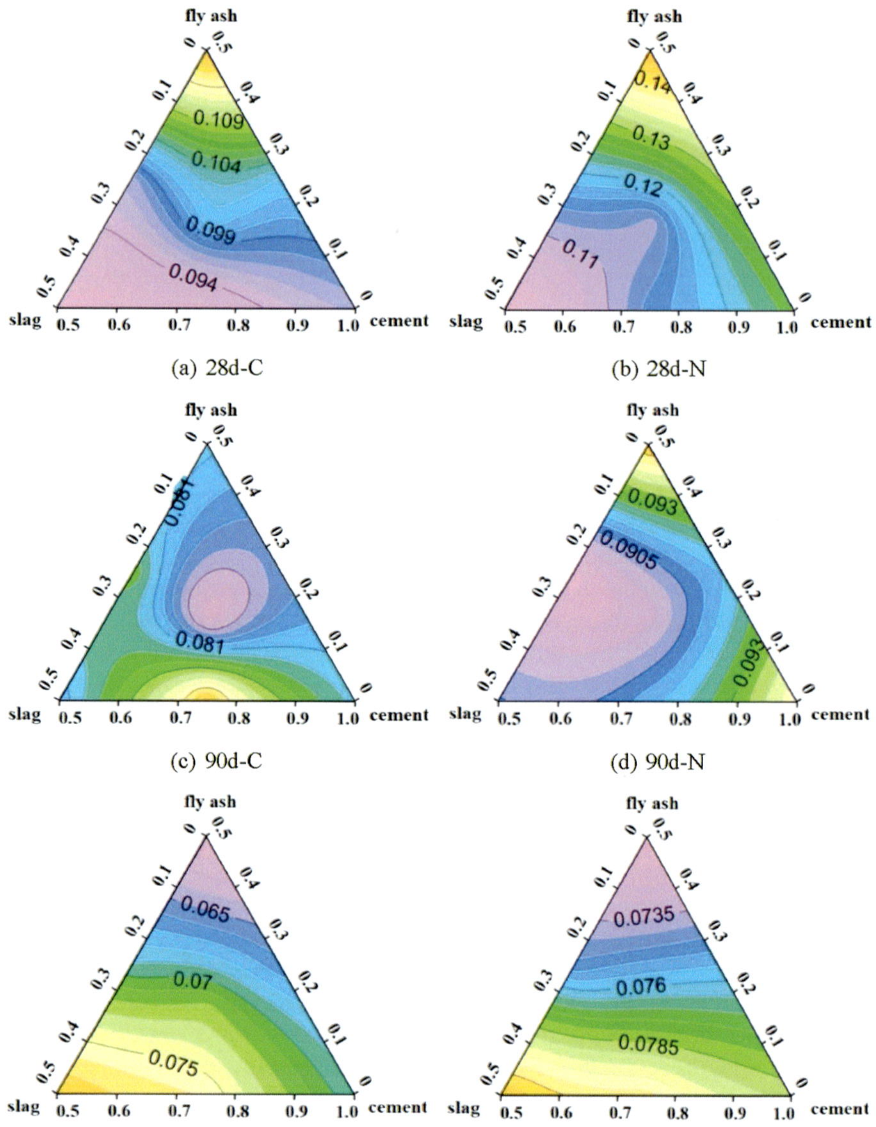

FIGURE 6.26 Effect of subsequent water curing on air permeability index of 1-hour CO_2 surface treated concrete (in (kPa)/min) (C presents specimens after CO_2 surface treatment, N presents specimens without CO_2 surface treatment) (Wang et al., 2021).

structure, reducing the overall porosity of the concrete. As a result, the chloride ion migration pathway becomes more tortuous and obstructed, further reducing permeability to chloride ions. Pan et al. (2019) observed that after 28 days of subsequent water curing, concrete specimens treated with CO_2 for 3 and 6 hours exhibited a

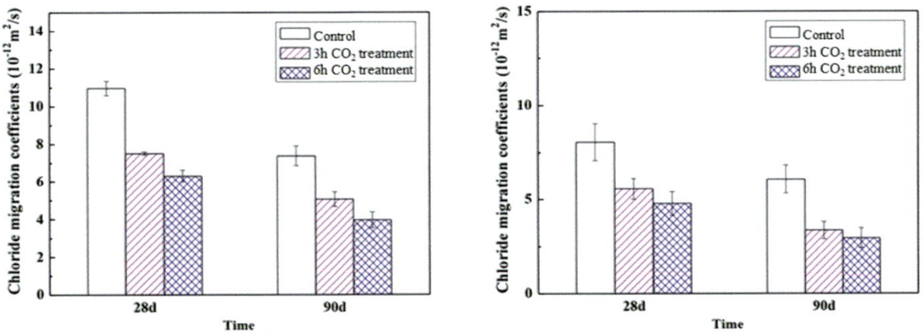

(a) w/c=0.4 (b) w/c=0.3

FIGURE 6.27 Chloride migration coefficients of CO_2 treated mortars after 28 and 90 days of lime water curing (Wang et al., 2021).

reduction of 47% and 51.7%, respectively, in the chloride ion migration coefficient, as shown in Figure 6.27. Specifically, after 28 days of water curing, the chloride ion migration coefficients for untreated, 3-hour CO_2 treated, and 6-hour CO_2 treated concrete were 10.9×10^{-12} m^2/s, 7.5×10^{-12} m2/s, and 6.3×10^{-12} m^2/s, respectively. These results highlight the crucial role of CO_2 treatment followed by subsequent water curing in reducing chloride permeability and improving concrete durability.

In addition to reducing chloride ion permeability, subsequent water curing also aids in the partial recovery of alkalinity in concrete. CO_2 surface treatment typically results in a reduction in the concrete's alkalinity due to the consumption of $Ca(OH)_2$ during the mineralization process. However, subsequent water curing helps to reform hydration products, partially restoring the lost alkalinity. This recovery of alkalinity further enhances the concrete's resistance to chloride ion penetration by improving the durability of the mineralization layer. Zhang and Shao (2016b) observed that CO_2 treated concrete, when subjected to a one-year chloride immersion/air drying cycle, exhibited significantly higher alkalinity and a more than 50% reduction in total chloride content compared to samples subjected to hydration control. This result underscores the synergistic effects of CO_2 surface treatment and subsequent water curing, which significantly enhance the concrete's ability to resist the penetration of harmful ions like chlorides. By reducing porosity, refining pore size distribution, and limiting pore connectivity, this combined treatment optimizes the microstructure of concrete and improves its overall impermeability. Consequently, the permeability resistance of CO_2 treated concrete is substantially increased, thereby enhancing its durability and service life in harsh and challenging environments.

6.7 SUMMARY

This chapter presents a comprehensive summary of the current state of research on CO_2 surface treatment technology. It discusses the effects of CO_2 treatment on the microstructure of the CO_2 mineralized layer, the mechanical properties, and permeability of concrete, as well as the influence of subsequent water curing on structural properties post-treatment. The following conclusions can be drawn:

(1) CO_2 surface treatment technology facilitates the formation of a carbonated layer through a CO_2 mineralization reaction in the concrete surface layer. The calcium carbonate produced by mineralization fills the pore structure of the concrete, thereby reducing the proportion of large pores, increasing the densification and enhancing micromechanical properties of the surface layer.
(2) The thickness of the CO_2 surface treatment layer is affected by mineralization conditions and the intrinsic properties of the concrete. As the CO_2 concentration and mineralization duration increase, the thickness of the carbonated layer increases gradually. However, the inherent strength of the concrete members may limit the efficacy of the CO_2 surface treatment.
(3) CO_2 surface treatment enhances the compressive strength and splitting tensile strength of concrete. Both strengths increase as the mineralization layer thickens.
(4) CO_2 surface treatment significantly reduces the permeability of concrete, including air permeability, water permeability, and chloride ion migration, thereby enhancing the durability of the concrete.
(5) After CO_2 surface treatment, concrete continues to hydrate during subsequent water curing. The additional hydration products further densify the pore structure, improving both the mechanical properties and the resistance to penetration.

REFERENCES

Arabzadeh, A., Ceylan, H., Kim, S., Gopalakrishnan, K., Sassani, A., Sundararajan, S., Taylor, P.C., 2017. Superhydrophobic coatings on Portland cement concrete surfaces. *Construction and Building Materials* 141, 393–401. https://doi.org/10.1016/j.conbuildmat.2017.03.012

Boa, A.R., Topu, L.B., 2012. Influence of fly ash on corrosion resistance and chloride ion permeability of concrete. *Construction and Building Materials* 31, 258–264. https://doi.org/10.1016/j.conbuildmat.2011.12.106

Branch, J.L., Epps, R., Kosson, D.S., 2018. The impact of carbonation on bulk and ITZ porosity in microconcrete materials with fly ash replacement. *Cement and Concrete Research* 103, 170–178. https://doi.org/10.1016/j.cemconres.2017.10.012

Chen, Y., Qian, C., Hao, Z., Zhou, H., 2021. Effect of bio-mineralization on concrete performance: Carbonation, microhardness, gas permeability and Cl- migration. *Biochemical Engineering Journal* 171, 108024. https://doi.org/10.1016/j.bej.2021.108024

Chi, J.M., Huang, R., Yang, C.C., 2002. Effects of carbonation on mechanical properties and durability of concrete using accelerated testing method. *Journal of Marine Science and Technology* 10, 14–20. https://doi.org/10.51400/2709-6998.2296

Chun, Y., Naik, T.R., Kraus, R.N., 2007. *Carbon dioxide sequestration in concrete in different curing environments*. Proceedings of the Coventry University International Conference on Sustainable Construction Materials and Technologies, Coventry 18–24.

Das, B.B., Singh, D.N., Pandey, S.P., 2012. Rapid chloride ion permeability of OPC- and PPC-based carbonated concrete. *Journal of Materials in Civil Engineering* 24, 606–611. https://doi.org/10.1061/(asce)mt.1943-5533.0000415

Drouet, E., Poyet, S., Le Bescop, P., Torrenti, J.M., Bourbon, X., 2019. Carbonation of hardened cement pastes: Influence of temperature. *Cement and Concrete Research* 115, 445–459. https://doi.org/10.1016/j.cemconres.2018.09.019

He, P., Shi, C., Tu, Z., Poon, C.S., Zhang, J., 2016. Effect of further water curing on compressive strength and microstructure of CO2-cured concrete. *Cement and Concrete Composites* 72, 80–88. https://doi.org/10.1016/j.cemconcomp.2016.05.026

Huang, Y., Zhang, J., Hu, X., Wang, Y., Drissi, S., 2025. Effect of early CO 2 curing and subsequent water curing on the interfacial transition zone between cement paste and aggregates. *Construction and Building Materials* 460, 139857. https://doi.org/10.1016/j.conbuildmat.2025.139857

Hussain, S., Bhunia, D., Singh, S.B., 2017. Comparative study of accelerated carbonation of plain cement and fly-ash concrete. *Journal of Building Engineering* 10, 26–31. https://doi.org/10.1016/j.jobe.2017.02.001

Jiang, L., Guan, Y., 1999. Pore structure and its effect on strength of high-volume fly ash paste. *Cement and Concrete Research* 29, 631–633. https://doi.org/10.1016/S0008-8846(99)00034-4

Li, Z., He, Z., Chen, X., 2019. The performance of carbonation-cured concrete. *Materials* (Basel) 12, 1–14. https://doi.org/10.3390/ma12223729

Liu, Z., Meng, W., 2021. Fundamental understanding of carbonation curing and durability of carbonation-cured cement-based composites: A review. *Journal of CO2 Utilization* 44, 101428. https://doi.org/10.1016/j.jcou.2020.101428

Pan, X., Shi, C., Farzadnia, N., Hu, X., Zheng, J., 2019. Properties and microstructure of CO 2 surface treated cement mortars with subsequent lime-saturated water curing. *Cement and Concrete Composites* 99, 89–99. https://doi.org/10.1016/j.cemconcomp.2019.03.006

Pan, X., Shi, C., Hu, X., Ou, Z., 2017a. Effects of CO2 surface treatment on strength and permeability of one-day-aged cement mortar. *Construction and Building Materials* 154, 1087–1095. https://doi.org/10.1016/j.conbuildmat.2017.07.216

Pan, X., Shi, Z., Shi, C., Hu, X., Wu, L., 2016. Interactions between inorganic surface treatment agents and matrix of Portland cement-based materials. *Construction and Building Materials* 113, 721–731. https://doi.org/10.1016/j.conbuildmat.2016.03.091

Pan, X., Shi, Z., Shi, C., Ling, T.C., Li, N., 2017b. A review on concrete surface treatment Part I: Types and mechanisms. *Construction and Building Materials* 132, 578–590. https://doi.org/10.1016/j.conbuildmat.2016.12.025

Pan, X., Shi, Z., Shi, C., Ling, T.C., Li, N., 2017. A review on surface treatment for concrete – Part 2: Performance. *Construction and Building Materials* 133, 81–90. https://doi.org/10.1016/j.conbuildmat.2016.11.128

Phung, Q.T., Maes, N., Jacques, D., Bruneel, E., Van Driessche, I., Ye, G., De Schutter, G., 2015. Effect of limestone fillers on microstructure and permeability due to carbonation of cement pastes under controlled CO2 pressure conditions. *Construction and Building Materials* 82, 376–390. https://doi.org/10.1016/j.conbuildmat.2015.02.093

Phung, Q.T., Maes, N., De Schutter, G., Jacques, D., Ye, G., 2013. Determination of water permeability of cementitious materials using a controlled constant flow method. *Construction and Building Materials* 47, 1488–1496. https://doi.org/10.1016/j.conbuildmat.2013.06.074

Pravalika, A., Rao, N.V., 2018. Effect of carbonation on the properties of concrete. *International Journal of Civil Engineering and Technology* 9, 1605–1611. https://doi.org/10.32861/sr.512.205.214

Ren, F., Chen, X., Zeng, Q., Zhou, C., 2022. Effects of pure carbonation on pore structure and water permeability of white cement mortars. *Cement* 9, 100040. https://doi.org/10.1016/j.cement.2022.100040

Shi, C., Qu, B., Provis, J.L., 2019. Recent progress in low-carbon binders. *Cement and Concrete Research* 122, 227–250. https://doi.org/10.1016/j.cemconres.2019.05.009

Song, B., Liu, S., Hu, X., Ouyang, K., Li, G., Shi, C., 2022. Compressive strength, water and chloride transport properties of early CO2-cured Portland cement-fly ash-slag ternary mortars. *Cement and Concrete Composites* 134, 104786. https://doi.org/10.1016/j.cemconcomp.2022.104786

Song, B., Shi, C., Hu, X., Ouyang, K., Ding, Y., Ke, G., 2021. Effect of early CO2 curing on the chloride transport and binding behaviors of fly ash-blended Portland cement. *Construction and Building Materials* 288, 123113. https://doi.org/10.1016/j.conbuildmat.2021.123113

Tracz, T., Zdeb, T., 2019. Effect of hydration and carbonation progress on the porosity and permeability of cement pastes. *Materials* (Basel) 12. https://doi.org/10.3390/ma12010192

Wang, Y., Lu, B., Hu, X., Liu, J., Zhang, Z., Pan, X., Xie, Z., Chang, J., Zhang, T., Nehdi, M.L., Shi, C., 2021. Effect of CO2 surface treatment on penetrability and microstructure of cement-fly ash–slag ternary concrete. *Cement and Concrete Composites* 123, 104194. https://doi.org/10.1016/j.cemconcomp.2021.104194

Xu, J., Wang, X., Yao, W., 2019. Coupled effects of carbonation and bio-deposition in concrete surface treatment. *Cement and Concrete Composites* 104, 103358. https://doi.org/10.1016/j.cemconcomp.2019.103358

You, K., Jeong, H., Hyung, W., 2014. Effects of accelerated carbonation on physical properties of mortar. *Journal of Asian Architecture and Building Engineering* 13, 217–221. https://doi.org/10.3130/jaabe.13.217

Yu, Z., Ye, G., 2013. The pore structure of cement paste blended with fly ash. *Construction and Building Materials* 45, 30–35. https://doi.org/10.1016/j.conbuildmat.2013.04.012

Zajac, M., Skocek, J., Durdzinski, P., Bullerjahn, F., Skibsted, J., Ben Haha, M., 2020. Effect of carbonated cement paste on composite cement hydration and performance. *Cement and Concrete Research* 134, 106090. https://doi.org/10.1016/j.cemconres.2020.106090

Zhan, B.J., Xuan, D.X., Poon, C.S., Shi, C.J., 2019. Mechanism for rapid hardening of cement pastes under coupled CO2-water curing regime. *Cement and Concrete Composites*. 97, 78–88. https://doi.org/10.1016/j.cemconcomp.2018.12.021

Zhang, D., Shao, Y., 2016a. Early age carbonation curing for precast reinforced concretes. *Construction and Building Materials* 113, 134–143. https://doi.org/10.1016/j.conbuildmat.2016.03.048

Zhang, D., Shao, Y., 2016b. Effect of early carbonation curing on chloride penetration and weathering carbonation in concrete. *Construction and Building Materials* 123, 516–526. https://doi.org/10.1016/j.conbuildmat.2016.07.041

Zhang, D., Shao, Y., 2018. Surface scaling of CO2-cured concrete exposed to freeze-thaw cycles. *Journal of CO2 Utilization* 27, 137–144. https://doi.org/10.1016/j.jcou.2018.07.012

7 CO_2 mineralization by recycled concrete aggregate

7.1 INTRODUCTION

With increasing urbanization in developing countries, the demand for high-strength, durable, and low-cost concrete is increasing. This is accompanied by an increased demand for coarse and fine aggregates, which account for about 70% of the total concrete volume. This will result in over-exploitation of stone and river sand and an increased burden on the natural environment globally. At the same time, the construction of high-rise buildings will result in the massive demolition of old and low-rise buildings, which will lead to a large number of discarded concrete components and serious environmental problems. The use of C&D waste as a substitute for NA can reduce the exploitation of natural resources and contribute to sustainable social and economic development. In the past 20 years, many researches on recycled concrete aggregates (RCA) have been conducted, and RCA are recognized to play an important role in reducing the use of NA and reducing environmental pollution. However, RCA usually consists of virgin aggregate, adherent mortar, and an interfacial transition zone (TTZ) between the adherent mortar and virgin aggregate (Figure 7.1). The presence of adherent mortar and more interfacial transition zones causes RCA to have higher porosity and water absorption and more microcracks than NA, which in turn leads to more fragile mechanical properties and durability of RCA (Guo et al., 2018; Shi et al., 2016).

One method of improving the properties of RCA, as described above, is CO_2 mineralization. This is a recognized, environmentally friendly method. On the one hand, the high alkalinity of recycled concrete aggregates gives them a high potential for CO_2 sequestration, which is a good practice of the global low-carbon strategy of the construction industry. On the other hand, the calcium carbonate and silica gel produced by accelerated mineralization fill the pore structure and increase the solid volume, thus improving the various properties of RCA.

In this section, the mechanism of CO_2 sequestration and the mineralization characteristics of RCA are described. On this basis, the enhancement of CO_2 mineralization for RCA is reviewed and discussed. In addition, the effect of CO_2 mineralization of RCA on concrete properties is also presented.

DOI: 10.1201/9781003627531-7

FIGURE 7.1 Composition of RCA (Mao et al., 2021).

7.2 PROPERTIES OF RCA

RCA is mainly derived from crushing and processing of previously used concrete structural elements. RCA is different from virgin aggregate mostly owing to the adhered old cement mortar. The fraction of adhered mortar (by volume of the sample) decreases with the size of the aggregate increasing. Hansen et al. (Torben and Henrik, 1983) observed that for different sizes of RCA i.e., for 4–8 mm size, 8–16 mm size and 16–32 mm size, the mortar content (percentage of volume) is 60%, 40% and 35%, respectively. Research on RCA shows that the water absorption capacity of RCA is higher than that of virgin aggregates, and this is because of the larger porosity of adhered mortar. Some researchers observed that the water absorption of RCA is 3-12% higher than that of natural coarse and fine aggregates. Water absorption was observed to decrease with the size of aggregate increasing. This behavior is due to the higher absorption rate of the adhered old cement mortar present in larger RCA particles. The research indicates that aggregates with excessive water absorption have a great effect on the mechanical and durability properties development of concrete. Therefore, the higher water absorption capacity of RCA significantly restricts its utilization in structural concrete(Fang et al., 2021).

Generally, properties of concrete depend on the properties of aggregates used. Thus, properties of aggregate are a crucial aspect of the mechanical and chemical properties of concrete made with RCA. Studies have exhibited that the mechanical strength of RCA is lower than that of virgin aggregates, primarily due to the adhered old mortar (Fang et al., 2020). A previous study (Parthiban and Saravana Raja Mohan, 2017) showed that the crushing value of RCA was higher than that of virgin aggregates. The RCA crushing values were 45% and 33% higher than those of virgin aggregates for 5–10 mm and 10–20 mm, respectively. Durability of RCA depends on its chemical properties which are quantified by sulfate soundness and chloride content. Nagataki et al. (2004) reported sulfate soundness losses to be 29.1–49% for RCA depending on the source, while the value was 9.1% and 2.6% for coarse aggregate and fine aggregate, respectively. This study also observed that when subjected to another level of crushing, the performance of RCA improved drastically.

Excessive chloride content can lead to rapid corrosion of steel reinforcement; thus, it is important to control chloride content in concrete structures. RCA has a higher chloride content, but it is generally in the acceptable range. RCA should be immersed in water to remove the excessive chloride, which restrains the excessive chloride content in concrete (Guo et al., 2023). The presences of adhered mortar and processing-induced cracks have a detrimental effect on the durability of RCA. However, from all the literature, it has been observed that the properties of RCA depended largely on the source of aggregate, age, and quality of the parent concrete. Thus, the design of mixture proportions should be done carefully since RCA exhibits lower performance in tests as compared to NA.

7.3 CO_2 MINERALIZATION OF RCA

In a CO_2 atmosphere, the hydration products in the mortar attached to the surface of the RCA are subjected to a series of complex physicochemical reactions with CO_2. As one of the most important hydration products of cement, $Ca(OH)_2$ has a very high mineralization activity. The mineralization process of $Ca(OH)_2$ can also be divided into two stages. In the first stage, the Ca^{2+} concentration leached out from $Ca(OH)_2$ surged to about 700 mg/L at the beginning of the reaction. Throughout the CO_2 mineralization process, the leached Ca^{2+} stabilized at a high value. This ensures that the dissolution of Ca^{2+} does not become a limiting factor for the extent of CO_2 mineralization throughout the process, and in the second stage, when the CO_2 source is sufficient, the mineralization can be maintained at a high rate. In the second stage, according to the morphology of $Ca(OH)_2$ after CO_2 mineralization, the unmineralized $Ca(OH)_2$ is hardly covered by the formed mineralization products. This makes the mineralization product layer less of an obstacle to the CO_2 mineralization process.

The other major hydration product of Ordinary Portland Cement is C-S-H, which accounts for about 50% of the volume of the cement paste. The mineralization process of C-S-H gel can be divided into two stages. In the first stage, it is also the leaching properties of calcium ions that control the C-S-H mineralization process. As the CO_2 mineralization proceeds, the mineralization of C-S-H enters the diffusion stage. In this period, C-S-H begins to transform into silica gel, and the outward diffusion of Ca^{2+} and the inward diffusion of CO_2 become more and more difficult. When the mineralization product of silica gel is thick enough, the mineralization of C-S-H will enter the sustained reaction period.

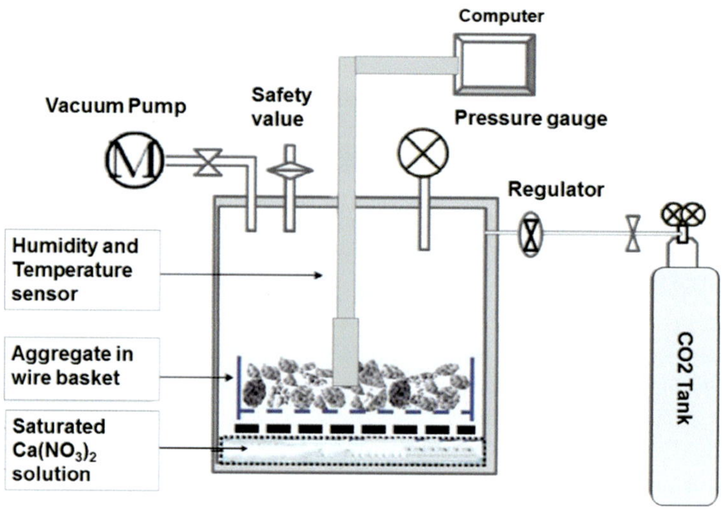

FIGURE 7.2 Typical experimental setup of CO_2 mineralization of RCA.

$CaCO_3$ and C-S-H product from pozzolanic reaction of silica gel can effectively fill the pores and microcracks of adherent mortar and ITZ. This reduces the porosity of the RCA and improves the properties of the RCA. Based on the above principles, CO_2 mineralization of RCA was proposed, and the setup of CO_2 mineralization for RCA is shown in Figure 7.2.

7.4 PHYSICAL CHANGES OF RCA DURING CO_2 MINERALIZATION

Figure 7.3 shows the XRD results of unmineralized and mineralized RCA. The $Ca(OH)_2$, C_2S, and C_3S peaks in the mineralized RCA basically disappeared. This suggested that $Ca(OH)_2$, C_2S, and C_3S in the attached cement paste were transformed into $CaCO_3$ during mineralization. By comparing the quantity of $Ca(OH)_2$, C_2S, and C_3S in the unmineralized recycled gravel aggregates, only a few C_2S and C_3S were found. Thus, the quantity of $CaCO_3$ transformed from C_2S and C_3S during mineralization was very small compared with that transformed from $Ca(OH)_2$.

Figure 7.4 shows FTIR spectra of mineralized and unmineralized RCA. The particle size of G-RCA (recycled gravel aggregate) is 1.25–2.5 mm, and the particle size of C-RCA (recycled crushed aggregate) is 0.16–0.32mm. The strong IR frequencies at 1430/cm were typical bands assigned to v_3-CO_2^{-3}, a sharp band at 875/cm assigned to v_2-CO_2^{-3} and a strong band at 1080/cm assigned to v_3-Si-O-Si. The areas of bands in 433 and 1084/cm of mineralized samples were larger than those of unmineralized samples. This indicated that the quantity of $CaCO_3$ in attached cement paste grew with the degree of mineralization. The band of the silica gel generated from the

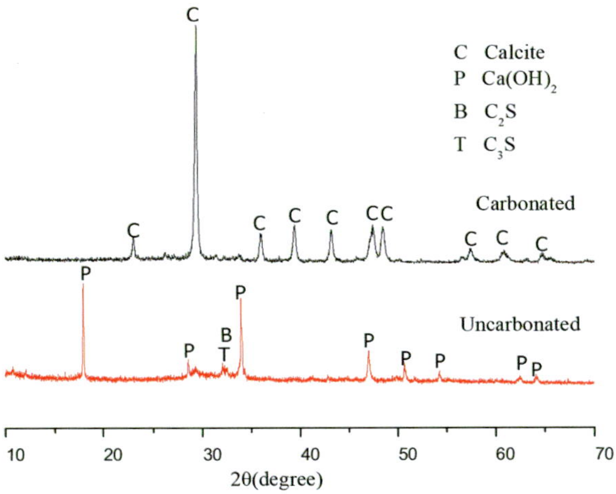

FIGURE 7.3 X-ray diffractograms of cement paste attached to aggregates.

mineralization of C-S-H occurred at about 1084/cm (Rostami et al., 2012). Though the band of vibration of Si-O-Si in quartz may affect the identification of silica gel, the increased band area at 1080/cm also indicated that silica gel formed during mineralization.

The frequency of the band assigned to v_3-Si-O-Si associated with the Ca–Si ratio, and the band of v_3-Si-O-Si is at 970/cm for ordinary cement (Thiery et al., 2007). However, the frequency increases with the reduction of the Ca–Si ratio, which reached 1080/cm as silica gel appeared in mineralized C-S-H.

Thermogravimetric analysis (TGA) was conducted to measure the quantity of $CaCO_3$ and $Ca(OH)_2$ in cement paste attached to the RCA after mineralization. On the differential scanning calorimetry (DSC) curve, the peaks are about 430–520°C and 530–950°C corresponding to the dehydration of $Ca(OH)_2$ and decarbonization of $CaCO_3$. In the TGA curve, the mass percentage of $Ca(OH)_2$ (*CH*) and CaCO3 (C_C) can be calculated from the mass loss at 430–520°C (d_h) and 530–950°C (d_c), as shown in Equations (7.1) and (7.2):

$$\mathrm{CH}(\%)=\left(\frac{\mathrm{d}_{\mathrm{H}^{\times}}\left(\frac{74}{18}\right)}{\mathrm{M}_{950^{\circ}\mathrm{C}}}\right)\cdot 100\% \tag{7.1}$$

$$Cc(\%)=\left(\frac{d_C\times\left(\frac{100}{44}\right)}{M_{950^{\circ}\mathrm{C}}}\right)\cdot 100\% \tag{7.2}$$

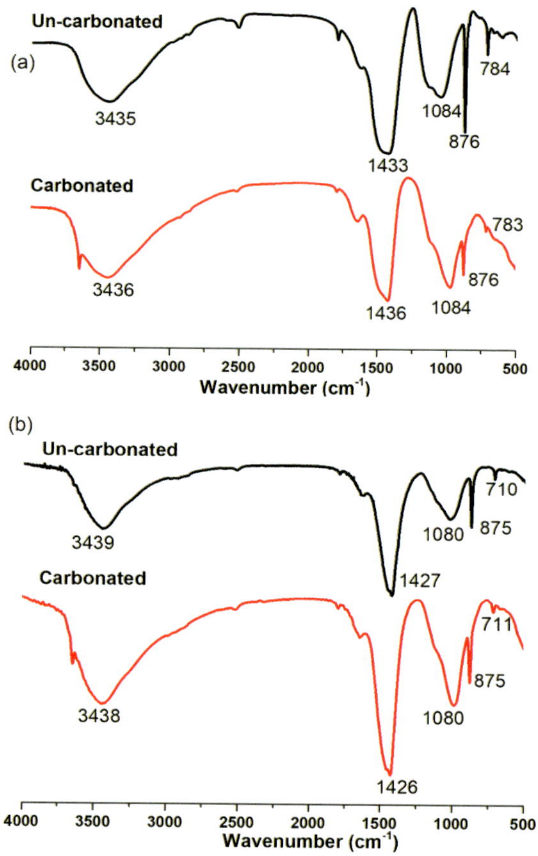

FIGURE 7.4 FTIR spectrum of recycled concrete aggregate.

The quantity of $CaCO_3$ generated during the mineralization of $Ca(OH)_2$ can be calculated based on the percentage of $Ca(OH)_2$ in the cement paste after mineralization. The total amount of $CaCO_3$ includes the $CaCO_3$ produced from the mineralization of $Ca(OH)_2$ and C-S-H. Table 7.1 shows the composition of the mineralized and unmineralized RCA with different particle sizes. The $Ca(OH)_2$ content was reduced after mineralization, and the $CaCO_3$ content increased significantly. RCA with a smaller particle size had more $CaCO_3$ generated from C-S-H. $Ca(OH)_2$ in hydrated cement paste appeared as large hexagonal plate crystals; mineralization consumed and refined the $Ca(OH)_2$ crystal to form $CaCO_3$. The formed $CaCO_3$ filled the pores in hardened cement paste, which increased the density of hardened cement paste. However, the mineralization products covered the surface of $Ca(OH)_2$, which impedes further reactions. Thus, solid $Ca(OH)_2$ cannot be fully mineralized, which was also evidenced by the weak peak around 450–500°C in the DSC curve (Figure 7.5(a)). The peak at b indicated that the $CaCO_3$ with low crystallinity was decomposed. The peak at around 450–500°C corresponds to both the dehydration of $Ca(OH)_2$ and the

TABLE 7.1
Effect of mineralization on the component of G-RCA

	Unmineralized (%)		Mineralized (%)					
	CH	CC	CH	CC	Δ_{CH}	CC_{CH}	CC_{CSH}	$(CC_{CSH})/(\Delta_{CC})$
Particle size (mm)	**A**	**B**	**C**	**D**	**E**	**F**	**G**	**H**
0.16–0.315	3.86	17.30	2.22	25.24	1.64	2.22	5.72	72.0
0.315–0.63	4.52	16.85	2.63	24.45	1.89	2.55	5.05	66.5
0.63–1.25	4.32	15.75	1.81	20.95	2.51	3.39	1.81	34.8
1.25–2.5	4.23	16.55	2.38	19.43	1.85	2.50	0.38	13.2

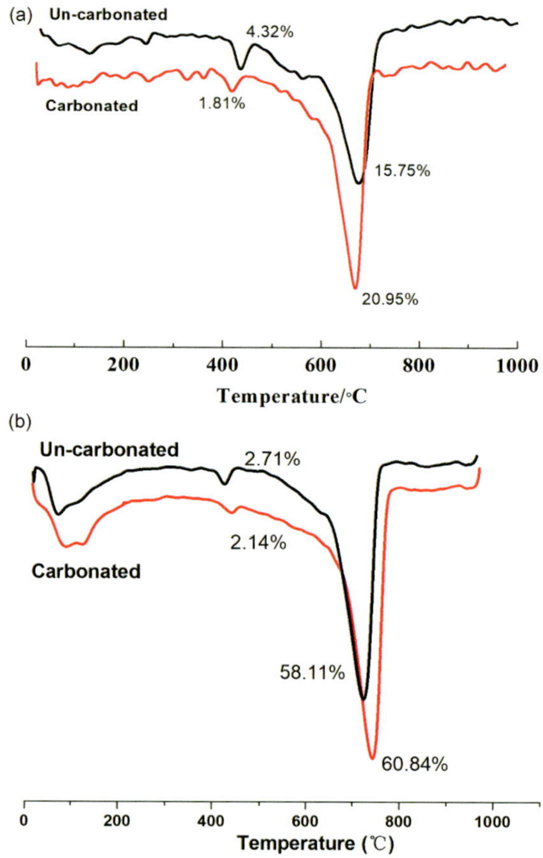

FIGURE 7.5 Thermal gravimetric curves of the attached cement on RCA.

decomposition of $CaCO_3$. Therefore, the calculated quantity of $Ca(OH)_2$ was higher than the real measurement. Thus, the mineralization degree was underestimated as well. Table 7.2 summarizes the composition of C-RCA before and after mineralization. It was found that the quantity of $Ca(OH)_2$ decreased after mineralization, while the quantity of $CaCO_3$ increased significantly. Aggregates with smaller particle sizes generated more $CaCO_3$ during mineralization, and more $CaCO_3$ was transformed from C-S-H.

Figure 7.5(b) shows the DTG curve of mineralized cement paste attached to C-RCA. The mass change of C-RCA was smaller than that of G-RCA by comparing Figures 7.5(a) and 7.5(b). This is because the C-RCA was very dense due to a low initial W/C ratio, which makes it more difficult for CO_2 to infiltrate into the pastes on C-RCA.

Tables 7.1 and 7.2 show that $CaCO_3$ generated from C-S-H in C-RCA was higher than that in G-RCA. This may be due to the different composition of initial cement paste in the two concretes. Concrete with lower strength usually contains more

TABLE 7.2
Effect of mineralization on the component of C-RCA

Particle size (mm)	Unmineralized (%)		Mineralized (%)					
	CH	CC	CH	CC	Δ_{CH}	CC_{CH}	CC_{CSH}	$(CC_{CSH})/(\Delta_{CC})$
	A	B	C	D	E	F	G	H
0.16–0.315	2.14	64.42	1.93	67.69	0.21	0.28	2.99	91.4
0.315–0.63	2.55	59.68	2.26	61.07	0.29	0.39	1.00	71.9
0.63–1.25	2.71	58.11	2.14	60.84	0.57	0.77	1.96	71.7
1.25–2.5	3.33	55.71	2.80	56.75	0.53	0.71	0.33	31.7

pozzolanic materials, which would consume some $Ca(OH)_2$ and generate more C-S-H. C-S-H generated from pozzolanic reactions had a lower Ca–Si ratio and was easier to be mineralized.

7.5 PERFORMANCE ENHANCEMENT OF RCA AFTER CO_2 MINERALIZATION

7.5.1 Porosity

CO_2 mineralization can reduce the porosity of RCA by 19–49.89%, which correlates with the original strength grade of the concrete (Zhang et al., 2023), the water content of the aggregate (Pan et al., 2017), and the external calcium source. CO_2 mineralization for 24 hours at a pressure of 10 kPa was able to reduce the porosity of RCA (particle sizes of 5–20 mm, from C30–C80 strength classes of waste concrete) by 18.5–21.4% (Zhan et al., 2014). Excessively high or low moisture content can limit CO_2 diffusion (Pan et al., 2017; Zhan et al., 2016). The porosity of the treated recycled fine aggregate decreased to 3.66% as the moisture content of the aggregate increased from 2.5% to 5%. Further increase in moisture content (7.5%) failed to reduce its porosity (Pan et al., 2017). Notably, CO_2 mineralization combined with limewater curing was able to reduce the total porosity by 31.7% and significantly reduced the mesoporosity with diameters of 100–1000 nm (Zhan et al., 2018). In addition, CO_2 mineralization reduced the cumulative pore surface area, cumulative pore volume, and average pore size of RCA (Liang et al., 2019).

On the other hand, CO_2 mineralization has been demonstrated to significantly reduce the porosity of the ITZ in RCA. Zhan et al. (Zhan et al., 2020) reported a substantial decrease in ITZ porosity from 49.1% to approximately 25% in plain RCA concrete as the distance from the RCA surface increased from 10 μm to 60 μm, irrespective of the parent material composition of RCAs (Figure 7.6) (Gao et al., 2022). Comparative analysis revealed that while CO_2 mineralized RCA concrete exhibited

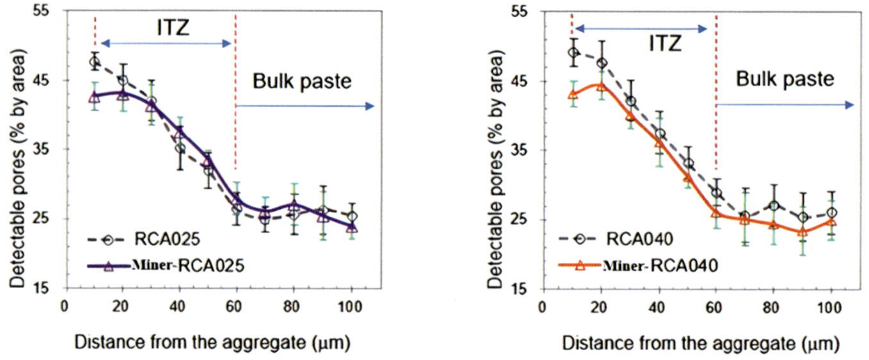

FIGURE 7.6 Porosity of interfacial transition zone of concrete with plain RCA and carbonated RCA from parent materials with w/c ratios of (a) 0.25 and (b) 0.40 (Gao et al., 2022).

similar porosity characteristics to non-treated RCA concrete in terms of average porosity and pore gradient distribution, a notable difference was observed within the innermost 10 μm region, where CO_2 mineralized RCAs showed a 6.1% reduction in porosity compared to their counterparts (Zhan et al., 2020). Furthermore, experimental data indicated consistent porosity levels in the new bulk paste, maintaining approximately 25% in both RCA and CO_2 mineralized RCA concrete systems. These findings suggest that while the CO_2 mineralization process does not significantly alter the total pore volume of either the ITZ or the new mortar matrix, it effectively enhances the density and microstructure of the immediate interfacial region surrounding the RCA surface within the concrete composite.

7.5.2 Water Absorption

The decrease in water absorption of RCA after mineralization is shown in Figure 7.7. The CO_2 mineralization efficiency of RCA is mainly determined by material properties and mineralization conditions. Material properties mainly include RCA reactivity, open porosity, water content, and RCA particle size. The reactivity of RCA strongly depends on the strength, storage time, and calcium content of the original concrete. Zhan et al. (2014) and Kou et al. (2014) concluded that the increase in the strength of the original concrete caused the RCA to contain more mineralized fractions without the capacity to be mineralized during the CO_2 mineralization treatment. For example, CO_2-treated RCA with the original strength of C30 (particle size 0.16–2.5 mm) showed a 28.29% reduction in water absorption, while those with C50 (particle size 0.16–2.5 mm) showed a 22.64% reduction in water absorption (Zhang et al., 2015b). CO_2-treated RCA at 6 months of age showed a decrease in water absorption of 18.03%, whereas RCA stored in the laboratory for

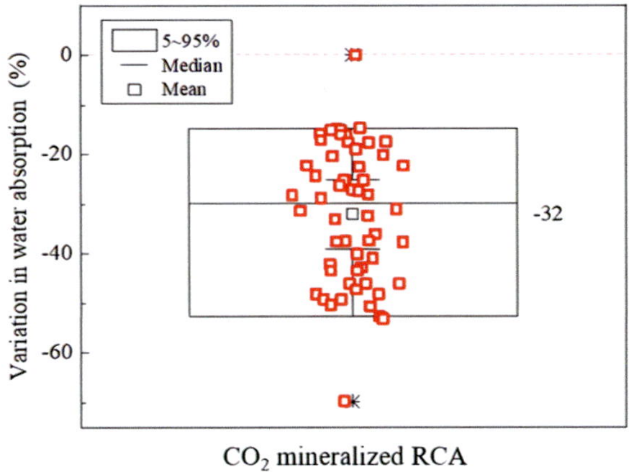

FIGURE 7.7 Variation in water absorption of recycled concrete aggregates after CO_2 mineralization.

1.5 years showed a decrease in water absorption of only 5.45% (Xuan et al., 2016). Furthermore, to increase the content of additional compounds to be able to be mineralized, Pan et al. (2017) immersed fine RCA (particle size less than 4.75 mm) in calcium hydroxide solutions at different concentrations (0.01 to 0.25 mol) and found that the water absorption of fine RCA immersed in 0.01 mol calcium hydroxide solution was reduced by up to 62.07% after CO_2 mineralization. Zhan et al. (2018) found that the RCA absorbed about 4.5% to 5.8% more CO_2 after several immersions in saturated limestone water. As a result, the water content of CO_2 mineralized RCA decreased by 55.31% and the apparent density increased by 5.83%. It is well known that RCA's open porosity and moisture content significantly affect the diffusion rate of CO_2 gas. Higher open porosity promotes deeper penetration of CO_2 in RCA. However, either too low or too high moisture content is detrimental to the mineralization reaction. Low moisture content limited the dissolution of calcium hydroxide, while high moisture content hindered the effective diffusion of CO_2. If most open pores inside the RCA are filled with free water, CO_2 cannot easily penetrate the RCA (Zhan et al., 2014). Previous studies (Lu et al., 2018) have concluded that RCA with smaller particle sizes and larger specific surface areas can capture more CO_2 during the accelerated mineralization process. About 55.2% of $CaCO_3$ was present in fully mineralized hardened cement paste powder (particle size less than 75 mm).

CO_2 mineralization conditions mainly include CO_2 concentration, ambient pressure, ambient humidity, and mineralization time. The water absorption of RCA decreased by 17.35%, and the apparent density increased by 1.49% after 24 hours of CO_2 mineralization at a pressure of 5.0 bar (Xuan et al., 2016). As the CO_2 concentration increased from 20% to 70%, the water absorption of RCA first decreased from 2.00% to 1.65% and then increased to 2.00% at a CO_2 concentration of 100% (Pan et al., 2017). The mineralization degree of RAC reached 25% after 2 hours of mineralization and increased to only 28.47% after 4 hours (Zhan et al., 2014). Kou et al. (2014) and Lou et al. (2018) suggested that the optimum CO_2 mineralization time is 24 hours. This may be due to the fact that the performance of RCA mineralized for 48 hours and 72 hours was slightly better than that of the aggregate maintained for 24 hours. The results indicate that higher calcium hydroxide content, larger open porosity, and suitable mineralization conditions are favorable for improving the mineralization degree of RCA. It should be noted that the efficiency of the mineralization is also affected by the source and particle size of the aggregate.

7.5.3 Density

As shown in Figure 7.8, both aragonite and calcite can effectively fill the pores of attached old mortar on recycled concrete aggregates, resulting in a denser internal structure of the RCA and helps to increase the density of the RCA. Figure 7.9 shows the variation in the density of the RCA after CO_2 mineralization treatment. It can be seen that there is basically an increase in the density of the RCA after CO_2 mineralization, with a maximum increase of 4.92%, which is attributed to the fact that the mineralization reaction produces a higher density of the mineralization products that will fill the pores within the concrete matrix. In addition, a previous study (Liang

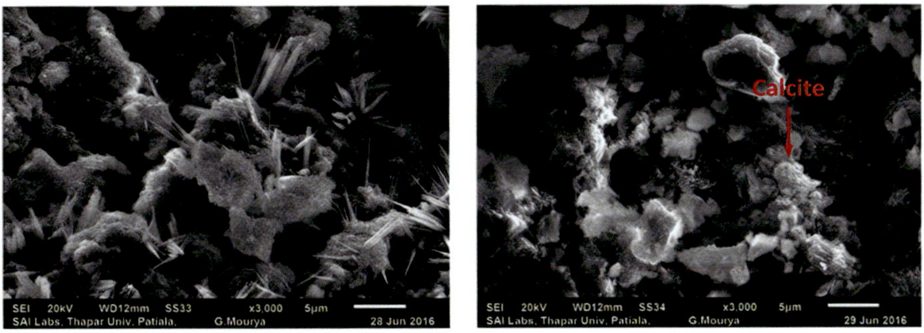

(a) Unmineralized RCA (b) CO_2 mineralized RCA

FIGURE 7.8 Morphology of plain and mineralized RCAs (Singh et al., 2021).

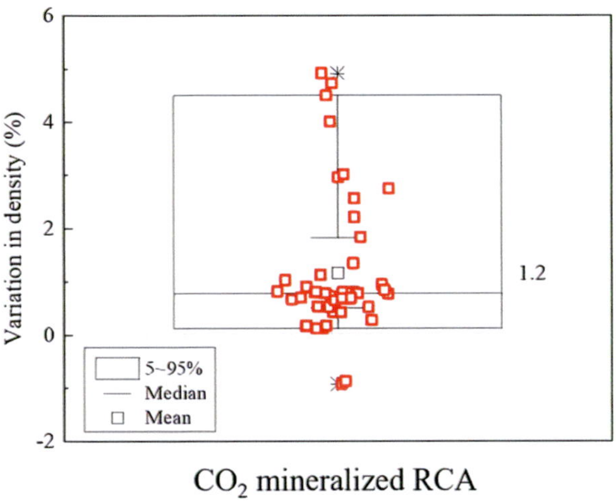

FIGURE 7.9 Variation in density of recycled concrete aggregates after CO_2 mineralization.

et al., 2019) indicated that the density of fine RCA increased significantly more than that of coarse RCA through CO_2 mineralization. This is because the mineralization products fill the larger pore spaces between the fine RCA.

7.5.4 Crushing Value

The decrease in crushing value of RCA after CO_2 mineralization (at 95% confidence interval) is shown in Figure 7.10. CO_2 mineralization resulted in a 17.4% increase in the strength of the adhered mortar in the RCA (Li et al., 2019). The increase in strength by CO_2 mineralization mainly depends on the original concrete and mineralization conditions. First, the crushing value of mineralization-treated recycled

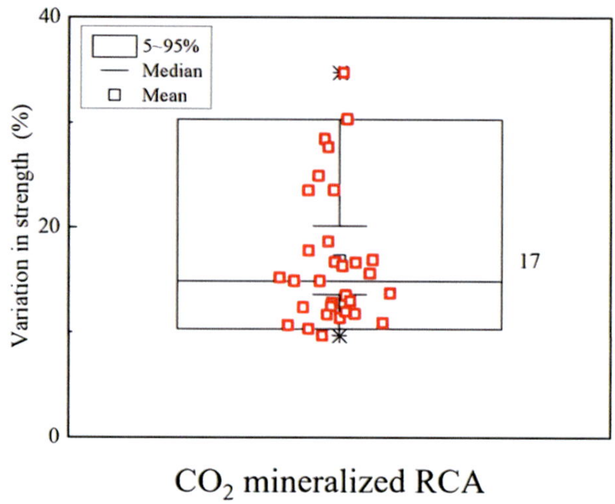

FIGURE 7.10 Variation in crushing value of recycled concrete aggregates after CO_2 mineralization.

concrete fine aggregates of strength C30 (particle size 0.16–2.5 mm) decreased from 18.6% to 16.9%, a reduction of 9.1%, which was superior to the crushing value of those recycled concrete fine aggregates of strength C50 of 7.6%. Second, increasing the mineralization time in the range of 6–72 hours decreased the crushing value of RCA (Luo et al., 2018) and increased the fine-grained value (Kou et al., 2014). The strength of the adhered mortar increased by about 17.4% after CO_2 mineralization (Li et al., 2019). However, CO_2 mineralization at 5.0 bar and 0.1 bar CO_2 pressure for 24 hours decreased the aggregate crushing value of RCA by 25.9% and 21.2%, respectively (Xuan et al., 2016). Pan et al. (2017) conducted a comprehensive

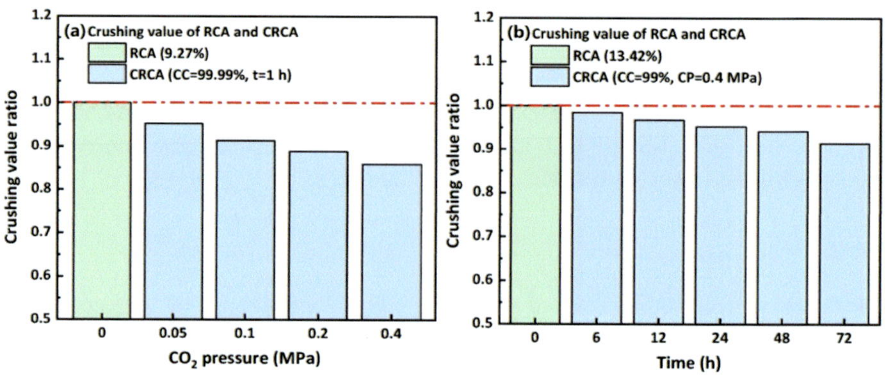

FIGURE 7.11 Effect of mineralization conditions on crushing value of mineralized RCAs (Pan et al., 2017; Pu et al., 2021).

investigation on the relationship between CO_2 concentration and the crushing value of fine RCA. Their experimental results demonstrated a progressive increase in crushing value from 10% to 15.8% as the CO_2 concentration was elevated from 20% to 100%. Notably, the optimal reduction in crushing value was achieved at a CO_2 concentration of 20%, as illustrated in Figure 7.10(a). Furthermore, the study revealed a significant correlation between mineralization duration and crushing value reduction, with extended mineralization periods (0–72 h) resulting in substantially lower crushing values, as depicted in Figure 7.10(b). This phenomenon can be attributed to the enhanced formation of mineralization products under conditions of elevated CO_2 pressure and prolonged mineralization duration, which effectively fill the pores within the attached old mortar of RCAs, thereby creating a more compact internal microstructure and consequently reducing the crushing value (Xuan et al., 2017). However, it is noteworthy that while both CO_2 concentration and pressure demonstrate substantial influence on crushing value reduction, the parameter exhibits relatively lower sensitivity to mineralization duration, particularly during the initial 6-hour period of the CO_2 mineralization process.

7.6 MICROSTRUCTURE AND PROPERTIES OF CO_2 MINERALIZED RECYCLE AGGREGATE CONCRETE (RAC)

7.6.1 Microstructure

The microstructure of RAC is different from that of normal concrete as RCA consists of adhered mortar from the old cement matrix, thus leading to the formation of two Interfacial Transition Zones (ITZ), one between NA and the old cement matrix and another between the old cement matrix and the new cement matrix (Kisku et al., 2017). The microstructure of concrete has been studied using techniques such as nano-indentation, atomic force microscopy, scanning electron microscopy, and Energy Dispersive X-ray Analyser. The ITZ consists of loose particles, and the density of particles increases as the distance from the interface increases, thus acting as the weakest link in concrete. The old ITZ makes the concrete microstructure more fragile due to higher porosity and cracks, thus acting as the weakest link. It was observed that the new cement paste is more compact and closed than the old paste matrix (Lee and Choi, 2013). The characteristics of ITZ depend on the quality of the mortar around it; however, its quantity does not have any influence on ITZ characteristics. The properties of ITZ also depend on the type of NA, as ITZ thickness and indentation modulus were found to be higher for limestone than for gravel. This was attributed to the higher porosity of limestone than gravel. The processing method of RCA also influences the ITZ, as the crushing of source concrete has been observed to reduce the density of cracks in the ITZ and introduce negligible new cracks.

SEM images of the RCA surface are shown in Figure 7.12. It can be seen that a denser ITZ was observed around the mineralized RCA, as shown in Figure 7.19b, and the surface of the unmineralized RCA was more porous (Figure 7.12(a)). Also, the needle-like ettringite was found in Figure 7.12(b), and such crystals were not observed in the mineralized RCA. Both ettringite (AFt) and monosulfate (AFm)

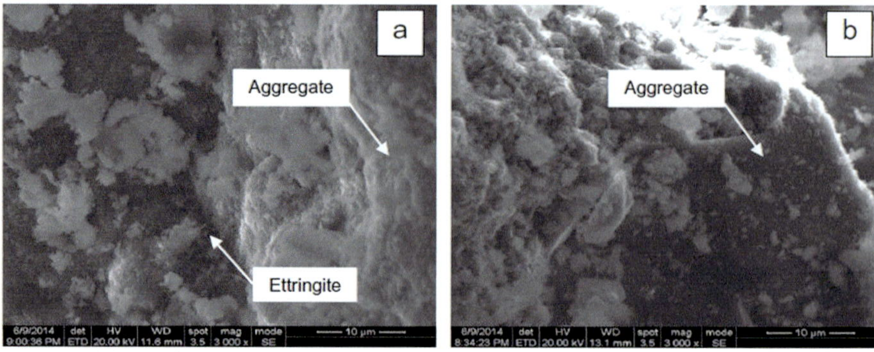

(a) un-mineralized RCA (b) mineralized RCA

FIGURE 7.12 SEM images of recycled aggregate surfaces.

crystals are the hydration products of typical Portland cement. AFm was converted into AFt and aluminum oxide at a pH of 11.6, and AFt was decomposed into sulfate ions and aluminum hydroxide ions at a pH of 10.6 during mineralization (Visser, 2014). The reaction products were less porous than the reactants after mineralization, which might contribute to the reduced water absorption of the mortars. The ITZ between cement paste and aggregate is an important region in concrete. A typical ITZ microstructure in natural fine aggregate mortar is shown in Figure 7.13. The microstructures of the ITZ and cement paste were dense; there were only a few large holes in both the ITZ and the paste (Figure 7.13(a)). Needle-like ettringite crystals were found in Figure 7.13(b).

The SEM images of mineralized and unmineralized RCA mortar, which was mineralized for 3 days, are shown in Figure 7.14. The interface of unmineralized RCA mainly consists of loose granular C-S-H and needle-like ettringite crystals. The region around the ITZ was more porous than that around the natural sand–cement interface. The absorbed water by the attached cement paste resulted in a water film around the aggregate surface, which increased the local w/c. Thus, the

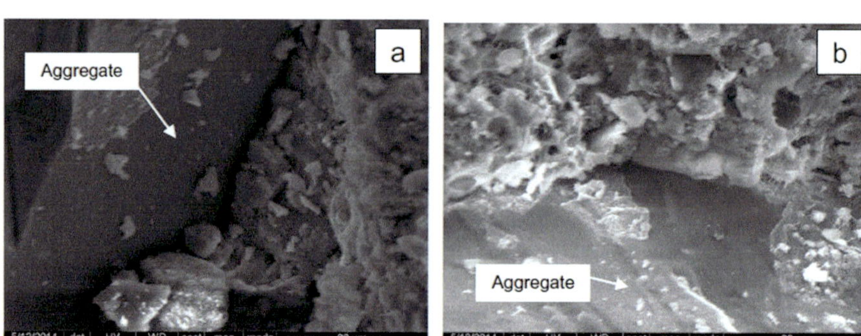

FIGURE 7.13 Typical interfacial morphology of natural sand mortar.

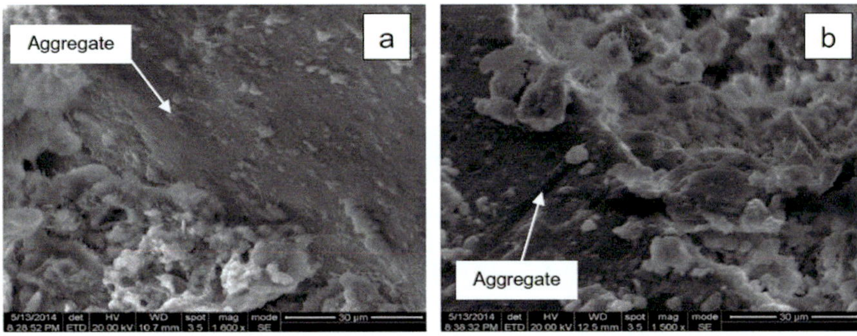

(a) un-mineralized mortar (b) mineralized mortar

FIGURE 7.14 Interfacial morphology of RCA mortars at 3 days.

porosity of the ITZ in RCA mortar was higher and less ettringite would form compared with the natural sand mortar. Since mineralization reduced the water absorption of the RCA, the thickness of the water film was decreased, and the ITZ of the mineralized RCA mortar was improved. The interfaces of un-mineralized and mineralized G-RCA mortars at 28 days are shown in Figure 7.15. It could be found that the ITZ of the mineralized mortar was significantly denser than the unmineralized mortar. The ITZ of the un-mineralized G-RCA mortar exhibited a large amount of loose hydrate products, and connected pores can be seen in the interface. The bonding between cement paste and aggregate was weak. The ITZ of the mineralized G-RCA was much denser (Figure 7.15(b)), and large pores between the cement paste and aggregate can hardly be found. The water absorption of RCA was reduced, and the water film around the aggregate surface became thinner after mineralization. Therefore, the local w/c of the interface was decreased, and the improved ITZ would increase the compressive strength and other mechanical properties of the concrete. Several reasons can contribute to the strengthened bonding between cement

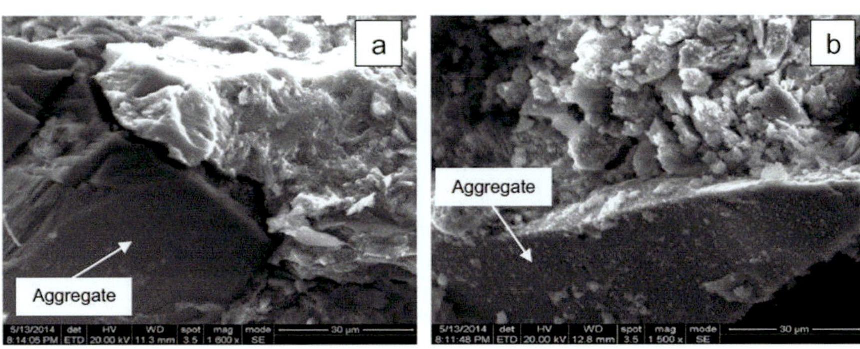

(a) un-mineralized mortar (b) mineralized mortar

FIGURE 7.15 Interfacial morphology of recycled gravel aggregate mortars at 28 days.

paste and RCA: (1) the surfaces of RCA were rougher and more angular than the natural aggregate, (2) the elastic modulus of RCA was closer to cement paste due to the original attached cement paste, and (3) the microcracks in the attached cement paste absorbed cement particles from fresh cement paste, which filled the cracks and reduced the porosity of RCA.

7.6.2 Flowability

Shi et al. [27] studied the flowability of the fresh mortars with different sands and sand–cement ratios. The flowability of natural sand mortar was better than that of the recycled aggregates mortar. This is because the attached cement pastes in RCA had high absorption, which increased the friction between the aggregate and the fresh paste. The flowability of mineralized recycled aggregate mortar was better than that of the unmineralized RCA mortar. This is because mineralization reduced the water absorption of the RCA. In Figure 7.16, the flowability of the C-RCA mortar was slightly higher than that of the G-RCA mortar, which might be due to the lower original water-cement ratio and less porous original attached cement paste. It is also noted that the flowability of the fresh mortars decreased with the sand–cement ratio because sand has a higher water absorption.

7.6.3 Compressive strength

As shown in Figure 7.17, the variation in compressive strength of concrete containing RCA after CO_2 mineralization ranged from –5.8% to 35% (at 95% confidence level).

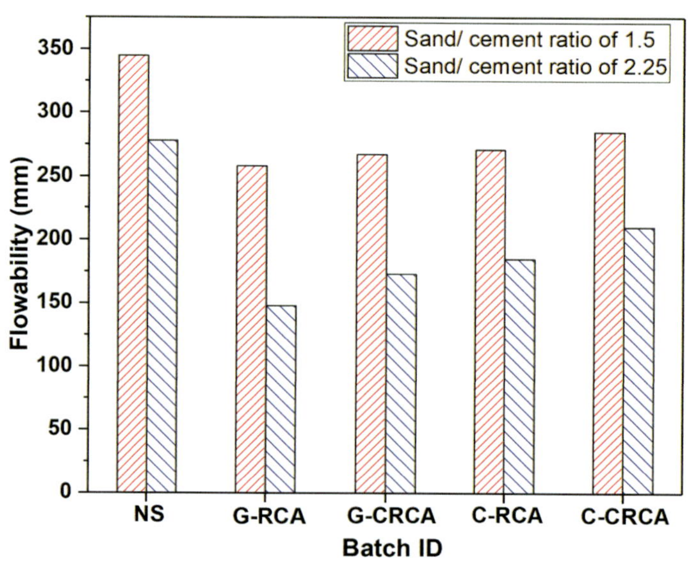

FIGURE 7.16 Flowability of fresh recycled aggregate mortars.

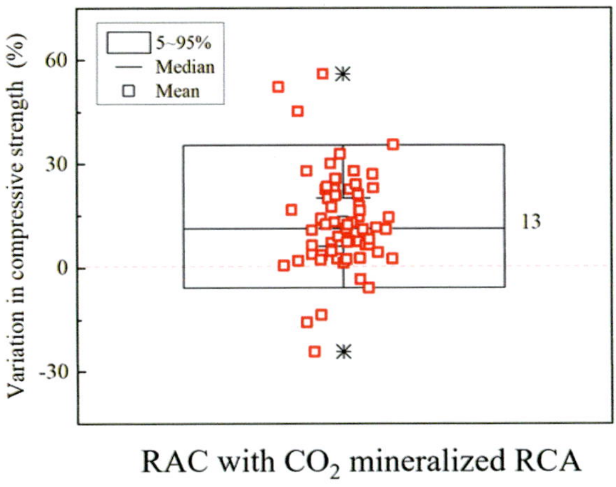

FIGURE 7.17 Variation in compressive strength of recycled aggregate concrete containing recycled concrete aggregates after CO_2 mineralization.

The change in compressive strength was mainly related to material properties, CO_2 mineralization conditions, and replacement ratio of the RCA.

The strength of the parent concrete of RCA had a significant effect on the compressive strength of the RAC. The increase in compressive strength of RAC prepared with CO_2 mineralized RCA produced from higher strength parent concrete was lower than that using CO_2 mineralized RCA produced from lower strength parent concrete. This indicates that CO_2 mineralization is beneficial in improving the properties of low-quality RCA (Kou et al., 2014; Li et al., 2018; Zhang et al., 2015a). This is attributed to the high permeability of the low strength parent concrete which favors CO_2 mineralization. On the other hand, the amount of carbonatable material in the RCA, e.g. the amount of adhered mortar, can influence the mineralization efficiency. Unfortunately, there has been no relevant research in this area.

CO_2 mineralization conditions, which mainly include pressure, temperature, and mineralization period of time, affect the strength of RCA and also the produced RAC. As the mineralization pressure of RCA increased from 0.01 MPa to 0.5 MPa, the compressive strength of the RAC increased slightly by 2.3% to 6.6% (Xuan et al., 2016). However, the higher pressure of 0.15 MPa had a more negative effect compared to the compressive strength of RCA with 0.075 MPa (Tam et al., 2016). Zhan et al. (2017) reported that increasing the CO_2 pressure from 0.5 MPa to 1.0 MPa only resulted in an increase in compressive strength from 22.58 to 27.96%, compared to the reference sample prepared with RCA without CO_2 mineralization. When the CO_2 pressure was increased from 1.0 MPa to 1.5 MPa, the compressive strength decreased. As far as the temperature is concerned, the strength of recycled concrete decreases from 11.9 MPa to 11.4 MPa when the mineralization temperature of the RCA is increased from 25°C to 50°C and increases to 12.1 MPa when the temperature is increased to 75°C (Zhan et al., 2017). This may be associated with the inhomogeneity of the mineralization reaction between CO_2 and RCA. In summary, extending the CO_2 mineralization time of RCA can increase the compressive strength of RAC, and the compressive strength increases faster in the first period of CO_2 mineralization and slower in the later period. For example, when the mineralization time of RCA was increased from 7 days to the 28th day, the compressive strength of RAC containing RCA with mineralization increased from 14.6% to 23.4% (Li et al., 2020). A longer mineralization time (90 minutes) favors the strength of the RAC. In addition to optimizing the CO_2 mineralization conditions, it is interesting to note that by combining the lime soaking process, the RAC will achieve higher compressive strength as the additional calcium hydroxide will form more calcium carbonate in the old mortar to fill more pores and lead to a denser microstructure of RCA. After three cycles of CO_2 mineralization and saturated lime water, the strength of RAC increased by 22.8% relative to concrete prepared with untreated RCA (Zhan et al., 2018). On the other hand, the type of calcium source is also an important indicator of modification efficiency. The compressive strength of mortar with fine RCA pre-soaked with different calcium source solutions was greater for agglomerates than for reference samples (Zhan et al., 2017). Pre-soaking with calcium chloride resulted in the highest mortar strength, superior to calcium hydroxide and calcium nitrate, regardless of the negative consequences of introducing chloride ions into the treated RCA.

There are some contradictions regarding the effect of CO_2 mineralization of RCA on the strength growth of RAC at different ages. Some researchers have reported that CO_2 mineralization improves the strength of RAC at the 28th day or 90th day compared to RAC containing untreated RCA (Kou et al., 2014; Li et al., 2017; Zhang et al., 2015a). On the contrary, other researchers claimed that CO_2 mineralization treatment of RCA facilitates the enhancement of early strength (Kou et al., 2014; Lu et al., 2019; Shi et al., 2018; Zhang et al., 2015a). This may be related to the different adsorption/desorption behavior of RCA toward cement-based material hydration after CO_2 mineralization. In addition, the increase rate in compressive strength decreased with the increase in the replacement ratio of RCA with CO_2 mineralization, which indicated that RCA with CO_2 mineralization reduced the negative effect of RCA on RAC strength (Lu et al., 2019; Luo et al., 2018; Xuan et al., 2016).

From a microstructural point of view, the macroscopic mechanical properties of concrete are related to the properties of the interfacial transition zone (Kazmi et al., 2019). Similarly, the mechanical properties of RAC containing CO_2 mineralized RCA are closely related to the average microcrack width of the interfacial transition zone in RCA. A low average microcrack width indicates a denser microstructure, which favors the strength of RAC (Li et al., 2020).

7.6.4 Flexible strength

Figure 7.18 shows the flexural strengths of the RAC after 28 days of standard curing. Incorporating the mineralized RCA enhanced the flexural strength of the RAC (Xuan et al., 2016). Compared to RAC with 100% non-mineralized RCA, the flexural strength of the concrete with 100% mineralized RCA was increased by 28.7%.

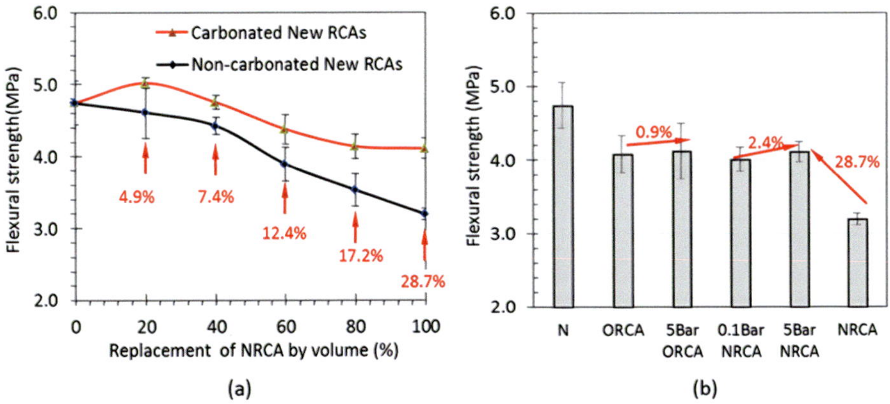

FIGURE 7.18 Flexural strength of concrete with non-mineralized and mineralized RCAs at 28 days.

Moreover, up to 60% of the mineralized RCA may be used in the new concrete to attain a flexural strength comparable to the control concrete. The mineralization pressure had a lesser influence on the properties of concrete, and mineralization pressure had a lesser influence on the increase in the flexural strength of the concrete with the mineralized RCA.

Figure 7.19 illustrates the relationship between CO_2 mineralized fine RCA replacement levels and the flexural strength evolution of mortar specimens, with particular emphasis on mineralization duration of cement mortar (7 and 28 days) and spatial anisotropy. Notably, while increasing replacement ratios of untreated fine RCA progressively diminished flexural strength (–12.3% to –18.7% at 50–100% replacement), CO_2 mineralization pretreatment consistently enhanced this parameter across all mineralization stages ($p < 0.05$). The strength differential between control and mineralized RCA groups exhibited progressive amplification with both replacement magnitude and mineralization time, culminating in a 21.4% strength superiority for 100% mineralized RCA specimens at 28 days (Figure 7.17(c)). Compared to the direction-dependent flexural strength, the highest and lowest strength occurred along the *y*-direction and *x*-direction, respectively, while the enhancement in strength by CO_2 mineralized fine RCA was maintained in all directions. However, the replacement of mineralized fine RCA was more effective in enhancing the compressive strength of mortar (i.e., up to 68.5% improvement in 28 days compressive strength (Sun et al., 2022)) than that in flexural strength, which can be explained by the densified pore structure of mortar. This divergent behavior originates from microstructural alterations characterized through X-ray computed tomography, which demonstrated pore structure refinement via two mechanisms: (1) 38% increase in sub-0.2-mm micropores and (2) 27% reduction in 0.2–1-mm mesopores (Sun et al., 2022). The CO_2 mineralization precipitation-induced pore densification preferentially resists axial compression loads rather than flexural stresses, as evidenced by the disproportionate strength enhancement ratios (Sun et al., 2022).

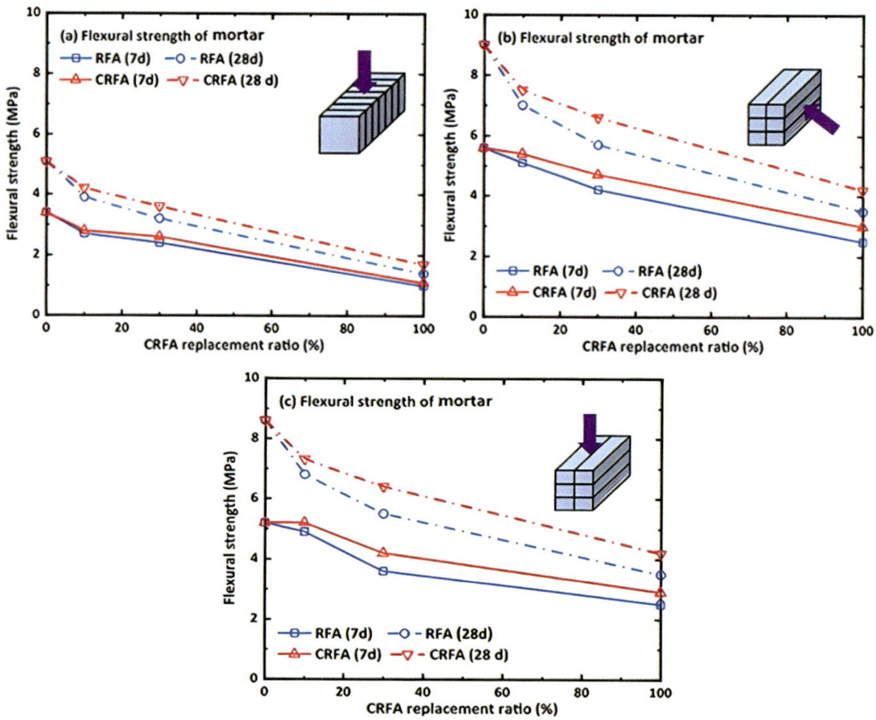

FIGURE 7.19 Effect of carbonated fine RCA replacement on flexural strength of mortar along (a) *x*-, (b) *y*-, and (c) *z* direction (Sun et al., 2022).

7.6.5 Tensile splitting strength

Figure 7.20 demonstrates that the RAC containing 30% CO_2 mineralized RCA performs better than the natural concrete. The RAC with 100% mineralized RCA replacement percentage performs poorly due to excess water in the mixture. The 30-0-0 and 100-0-0 mixes (specimen titles are designed as "A-B-C," "A" is the replacement ratio of CO_2 mineralized RCA, "B" is the duration of CO_2 mineralization, and "C" is the pressure of CO_2 mineralization) achieve a middle-tier tensile strength result of about 3.43 MPa and 2.96 MPa, respectively. The concrete made with 30% mineralized RCA realized an increased tensile strength performance when compared to the concrete with 30% RCA replacement level. The 30-30-75 mix measures the greatest tensile strength performance. The tensile strength results indicate that a lower CO_2 pressure overrides a longer CO_2 mineralization duration of RCA for the recycled concrete. The concrete with 100% mineralized RCA replacement percentage does not procure quality performance that equals the 100-0-0 mix. It indicates that the surplus water left in the cement deteriorates the concrete's tensile strength quality.

Figure 7.21 further displays the evolution of splitting tensile strength of mortar with CO_2 mineralized fine RCA replacement ratio ranging from 0% to 100%. It is

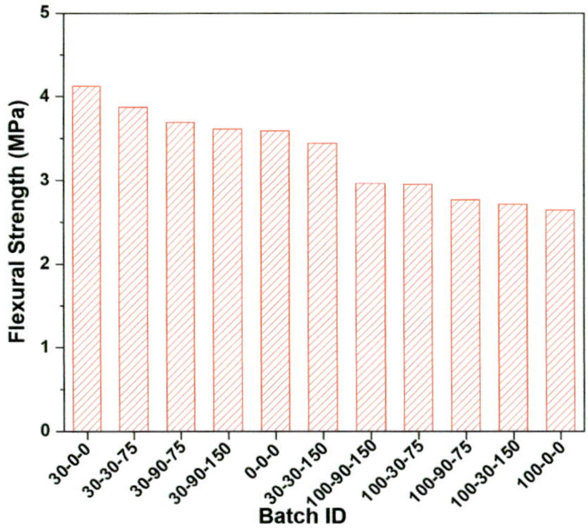

FIGURE 7.20 Experimental results on tensile strength.

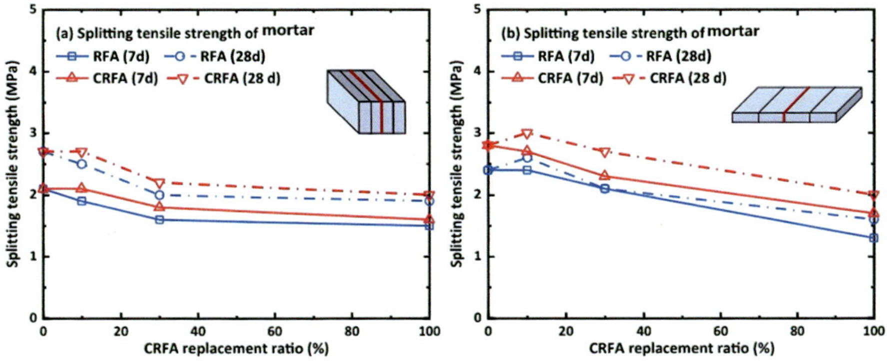

FIGURE 7.21 Effect of carbonated fine RCA replacement on splitting tensile strength of mortar (Sun et al., 2022).

worth mentioning that the splitting tensile strength of mortar refers to the interfacial bonding strength between mortar layers, which plays a critical role in determining the overall mechanical properties. In general, the splitting tensile strength of mortar prepared with mineralized fine RCA was higher than that with plain fine RCA under the same replacement ratio, which was maintained throughout the measured replacement ratio and mineralization time. Compared to mortar with equal content of plain fine RCA, the splitting tensile strength along the horizontal and vertical directions increased by up to 17.4% and 10.0%, respectively, due to the 30% replacement of mineralized fine RCA (Sun et al., 2022). The enhanced splitting tensile strength of mortar was ascribed to the formation of calcite precipitate in the adhered old mortar during CO_2 mineralization, which not only densifies the microstructure of fine RCA but also generates more ettringite crystals in the new cement paste of mortar (Gao et al., 2022). Moreover, the improved efficiency of mortar along the horizontal direction (Figure 7.21(a)) was higher than that along the vertical direction (Figure 7.21(b)), which can be ascribed to the closer contact interface along the horizontal direction as a result of gravity, as well as the larger contact area of the horizontal interlayer than the vertical interlayer.

7.6.6 Modulus of Elasticity

As shown in Figure 7.22, the variation in modulus of elasticity of concrete containing CO_2 mineralized RCA (at the 95% confidence level) was –3.4 to 17%. The change in the modulus of elasticity is closely related to the strength of the parent concrete, mineralization conditions, the replacement ratio of the RCA, and the mineralization time of RAC.

The modulus of elasticity of concrete containing CO_2 mineralization RCA mainly depends on the strength of the parent concrete. The higher the quality of the RCA, the greater the observed increase in the modulus of elasticity of the RAC at 28 days (Kou et al., 2014; Li et al., 2018). The increase in the modulus of elasticity of RAC may be attributed to the improvement of CO_2 mineralization on the microstructure of RCA (Zhan et al., 2018).

From the previous studies, CO_2 mineralization pressure and mineralization duration significantly influence the modulus of elasticity of RAC. Xuan et al. (2016) suggested that the modulus of elasticity of RAC prepared with RCA mineralized by 0.01 and 0.5 MPa of CO_2 pressures increased by 8.3–13.2% compared to the control samples. The elastic modulus value of RAC prepared with RCA further increased by 8.3% when mineralized at 0.5 MPa for 24 hours (Xuan et al., 2016). On the contrary,

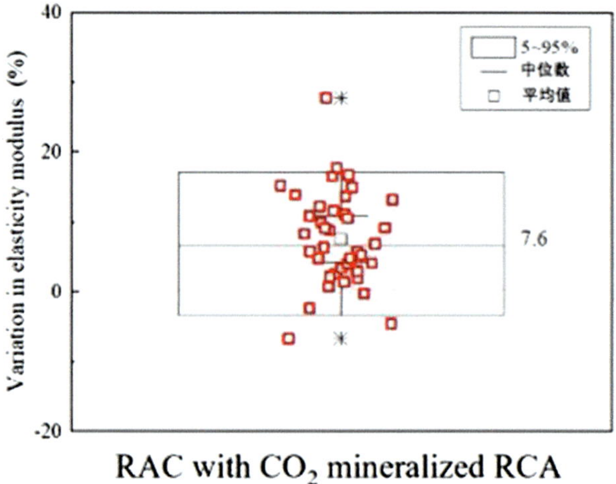

FIGURE 7.22 Variation in modulus of elasticity of recycled aggregates concrete containing recycled concrete aggregates after CO_2 mineralization.

the modulus of elasticity of concrete containing CO_2 mineralized RCA increased by 5.59% when RCA was treated at a pressure of 75 kPa, which was superior to RAC containing RCA with CO_2 mineralization at a CO_2 pressure of 150 kPa (an average increase of 1.29%) (Tam et al., 2016). Second, the modulus of elasticity of concrete containing RCA increased by 14.9% when the CO_2 mineralization time of RCA was increased from 0 days to 14 days, whereas the modulus of elasticity of concrete containing CO_2 mineralized RCA increased by just 1.3% when the RCA was further mineralized from 14 days to 28 days. This may be attributed to the reduction in microcrack width in the post-conservation after CO_2 mineralization (Li et al., 2020). On the contrary, the average modulus of elasticity of RAC containing 100% RCA mineralized in a shorter mineralization duration (30 minutes) had a higher improvement efficiency than a mineralization duration of 90 minutes (Tam et al., 2016).

Although the modulus of elasticity of the pretreated RAC was still lower than that of the control samples, the mineralization increased the modulus of elasticity of the concrete mixtures to some extent at all ages. Previous study (Kazmi et al., 2019) reported that the modulus of elasticity of RAC containing combined mineralization with lime-impregnated RCA increased by 6.4% after 90 days of mineralization compared to that at 28 days. It is noteworthy that the increase in the modulus of elasticity of RAC specimens tested at 28 days was higher than that at 90 days (Kazmi et al., 2019; Kou et al., 2014). The modulus of elasticity values gradually decreased with the increase in the content of the RCA replacement ratio. However, Xuan et al. (2016) and Luo et al. (2018) found that higher content of CO_2 mineralized RCA (80–100%) was beneficial in increasing the elastic modulus of RAC as compared to concretes containing untreated RCA. On the contrary, with changes in mineralization pressure and duration of mineralization, the modulus of elasticity of RAC containing 30% RCA mineralized at room temperature increased compared to that of RAC containing 100% RCA mineralized at room temperature (Tam et al., 2016).

7.6.7 Shrinkage

The autogenous shrinkage strain of mortar with different aggregate sand is shown in Figure 7.23. The autogenous shrinkage strain of the recycled aggregate mortar almost reached a constant value after 24 hours. The autogenous shrinkage strain of the C-RCA mortar before mineralization was 774 $\mu\varepsilon$ at 3 days, which increased by 126% after mineralization. A similar observation was noticed in the G-RCA mortar whose autogenous shrinkage strain was raised from 310 $\mu\varepsilon$ to 1940 $\mu\varepsilon$ (by 526%) after mineralization. Higher water absorption of the original attached paste resulted in a lower autogenous shrinkage strain of the unmineralized RCA. Because the autogenous shrinkage was caused by the chemical reaction during cement hydration, it was measured in a sealed condition where water could hardly evaporate. The absorbed water in the recycled aggregates can be released during cement hydration, which would reduce the effect on the internal humidity. Also, the pore diameter of the aggregates and cement paste affected the release of absorbed water. Therefore, denser mineralized recycled aggregates made the absorbed water more difficult to transport.

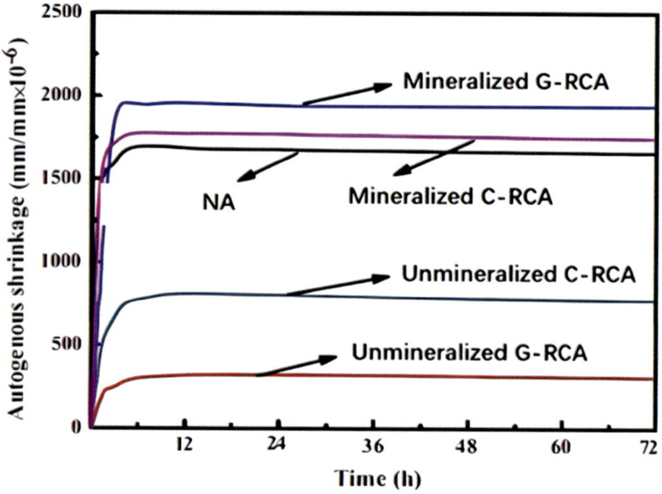

FIGURE 7.23 Autogenous shrinkage of mortars with different sands.

As shown in Figure 7.24, the variation of concrete drying shrinkage (at 95% confidence level) of CO_2 mineralized RCA was -34–2.5%. The main factors influencing drying shrinkage of RAC include the strength of the parent concrete, the replacement ratio of RCA, and the CO_2 mineralization conditions of RCA.

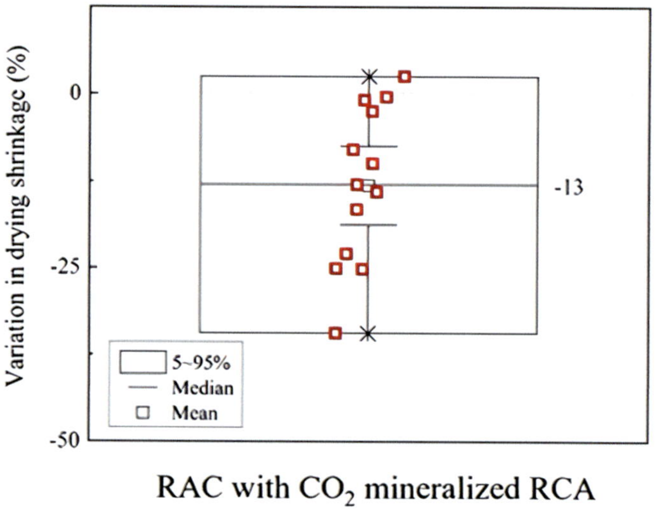

FIGURE 7.24 Variation in drying shrinkage of recycled aggregates concrete containing recycled concrete aggregates after CO_2 mineralization.

Generally, high quality RCA helps to reduce the drying shrinkage of RAC (Kou et al., 2014; Zhang et al., 2015a). Lu et al (2019) investigated the effect of CO_2 mineralized RCA (5–20 mm) on the drying shrinkage of RAC. The results indicated that the 28-day drying shrinkage of RAC containing CO_2 mineralized RCA was reduced by 16.6% and 25.1%, respectively, compared to RAC containing untreated RCA when 50% and 100% replacement ratios were used. At day 112, the drying shrinkage of RAC containing RCA mineralized in a lime solution soaked environment was reduced by 13.5%, which was superior to that of RAC containing conventional CO_2 mineralized RCA (Wang et al., 2020). Theoretically, these factors influenced the properties of RAC, e.g., mineralization temperature and relative humidity, gas pressure and CO_2 concentration, also influence the drying shrinkage of RAC (Fang et al., 2017; Liang et al., 2020; Zhan et al., 2017). However, previous study (Xuan et al., 2017) found that increasing the mineralization pressure had an insignificant effect on reducing the drying shrinkage of RAC. When the mineralization pressure was increased from 0.01 MPa to 0.5 MPa, the drying shrinkage of RAC at 112 days decreased by just 2.2%. Unfortunately, there are few studies regarding the drying shrinkage of mineralized recycled concrete at various temperatures and humidity. More research is needed to explain the mechanisms behind these influences.

7.7 DURABILITY OF MINERALIZED RECYCLED AGGREGATE CONCRETE

According to previous study (Guo et al., 2018), the durability of RAC is mainly determined and affected by the adhered mortar of RCA, the higher the amount of adhered mortar of RCA, the higher the porosity and water absorption leading to poorer durability performance of RAC. Higher RCA content and w/c ratio can result in poor durability of RAC, and the negative effect is more obvious for fine RCA than for coarse RA. Modified RCA with CO_2 can also react with hydration products to produce $CaCO_3$ for micro-pore structure strengthening. Therefore, in order to produce a more durable RAC, it is suggested to use mineralization prior to use in concrete.

7.7.1 Water absorption

As shown in Figure 7.25, the variation of water absorption (at 95% confidence level) of concrete containing CO_2 mineralized RCA was –29% to –3.4%. The results of Zhang et al. (2015a) indicated that the strength of the parent concrete had an insignificant effect on the water absorption of the recycled mortar. Kazmi et al. (2019) reported that the water absorption of RAC (pretreated with RCA pretreated with CO_2 mineralization after lime impregnation) was reduced by 8.3%–10.9%, which was superior to that of the RAC containing conventional CO_2 mineralized RCA. Increasing the CO_2 mineralization pressure is beneficial in reducing the water absorption of RAC. Xuan et al. (2017) reported that the water absorption of RAC containing RCA mineralized at a pressure of 0.5 MPa CO_2 was reduced by 10.5%

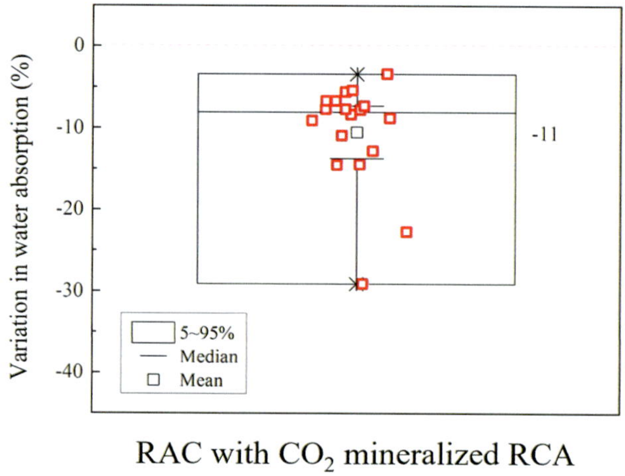

FIGURE 7.25 Variation in water absorption and permeability of recycled aggregates concrete containing recycled concrete aggregates after CO_2 mineralization.

compared to that of RAC containing RCA mineralized at a pressure of 0.01 MPa. In addition, the use of CO_2 mineralized RCA reduces the water absorption of RAC. When 100% replacement of CO_2 mineralized RCA was used, the water absorption was reduced by 43.6% compared to that of RAC containing untreated RCA.

Figure 7.26 illustrates the influence of carbonation conditions on the water absorption of carbonated RCA. For carbonated fine RCA, the water absorption initially decreases and then increases with rising CO_2 concentration, as depicted in Figure 7.26(a). Fine RCA carbonated at a CO_2 concentration of 50% exhibits the lowest water absorption value of 1.23% (Pan et al., 2017). This phenomenon can be attributed to the rapid formation of carbonation products at higher CO_2 concentrations, which fill the micropores and inhibit further CO_2 penetration into the RCAs (You et al., 2022). In contrast, the water absorption of carbonated RCAs demonstrates a continuous decline with increasing CO_2 concentration, reaching an approximate 25% reduction at 100% CO_2 concentration, as shown in Figure 7.26(b). Since the pore size of RCAs is larger than that of RFAs, the carbonation products generated at higher CO_2 concentrations effectively fill the macropores without significantly hindering further CO_2 diffusion (You et al., 2022). Additionally, the inhibitory effect on water absorption becomes less pronounced at CO_2 concentrations exceeding 20% (Pu et al., 2021), likely due to the denser surface layer of carbonated RCAs, which restricts CO_2 diffusion into the internal pores (Cui et al., 2015).

The impact of CO_2 pressure on the water absorption of carbonated RCAs is further demonstrated in Figure 7.26(c). Unlike the trend observed for apparent density, the water absorption of carbonated RCAs decreases significantly with increasing CO_2 pressure; however, this reduction becomes less pronounced when the CO_2 pressure exceeds 0.2 MPa (Wu et al., 2022). Therefore, a CO_2 pressure equivalent to twice atmospheric pressure can be considered optimal, as it benefits both water

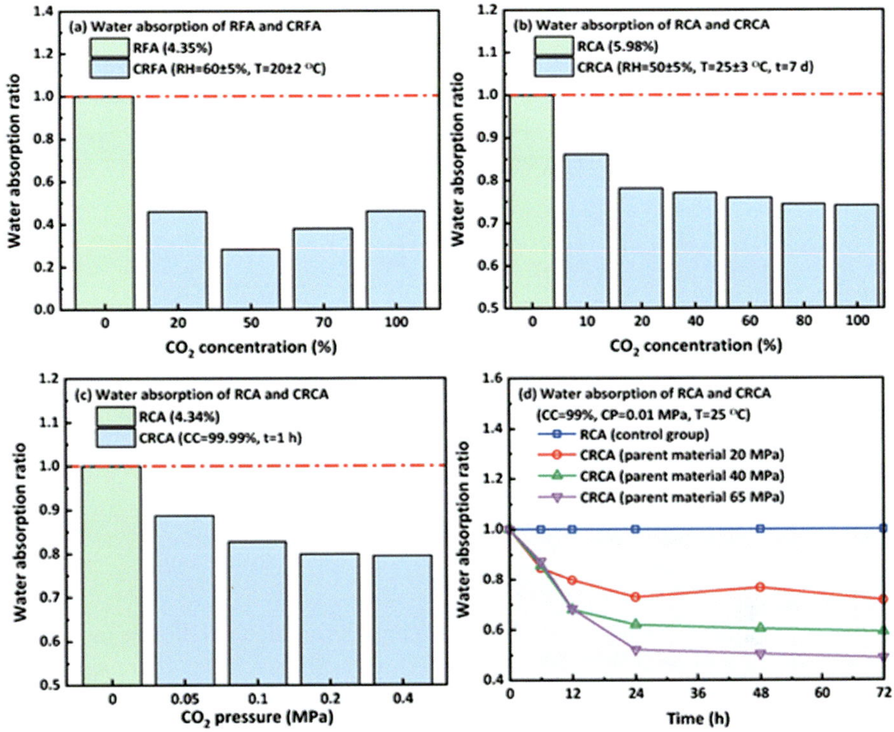

FIGURE 7.26 Effect of carbonation condition on water absorption of carbonated RCAs.

absorption and apparent density of carbonated RCAs. Figure 7.26(d) presents the effect of carbonation duration on the water absorption of carbonated RCAs. A rapid decline in water absorption occurs primarily within the first 24 hours of carbonation, coinciding with the notable increase in apparent density. Furthermore, the reduction in water absorption for carbonated RCAs with a parent material strength of 20 MPa is approximately half that of RCAs with a parent material strength of 65 MPa. This suggests that smaller pores are more effectively filled with carbonation products during the carbonation process (Cui et al., 2015).

7.7.2 Chloride penetration resistance

As shown in Figure 7.27, the variation (at the 95% confidence level) in the chloride permeability of RAC containing CO_2 mineralized RCA was -86 to 16%. The most important factors influencing chloride permeability of RAC include the strength of the parent concrete, mineralization conditions, replacement ratio of RCA, and concrete age.

According to previous studies (Kou et al., 2014; Liang et al., 2019; Zhang et al., 2015a), the strength of the parent concrete and the added calcium source have an

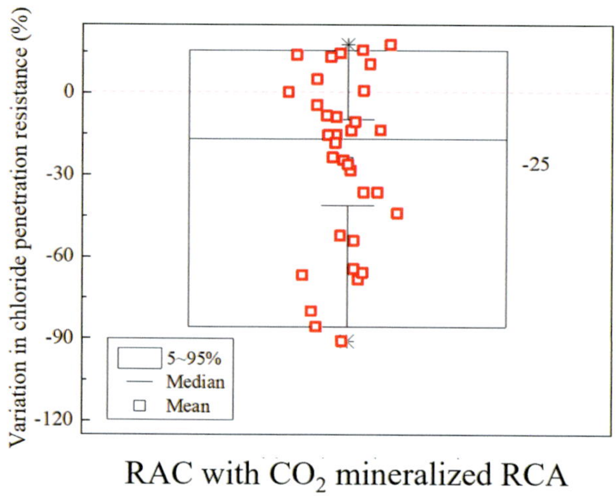

FIGURE 7.27 Variation in chloride penetration resistance of recycled aggregates concrete containing recycled concrete aggregates after CO_2 mineralization.

effect on the chloride permeability of RAC. The deterioration of chloride permeability of concrete prepared with low quality CO_2 mineralized RA is more significant than that of RAC containing high quality CO_2 mineralized RCA. In addition, Kazmi et al. and Wang et al (Kazmi et al., 2020; Wang et al., 2020) found that the chloride permeability of RAC containing lime-impregnated CO_2 mineralized RCA decreased by 19.53–23.5%, which was superior to RAC containing conventional CO_2 mineralized RCA. This is attributed to the superior performance of lime-impregnated CO_2 mineralized RCA over conventional CO_2 mineralized RCA.

CO_2 mineralization pressure and mineralization time influence the degree of CO_2 mineralization, which consequently influences the chloride permeability of RAC. However, previous study (Xuan et al., 2017) concluded that the chloride permeability of RAC containing RCA mineralized by a CO_2 pressure of 0.5 MPa was reduced by 11.2% compared to that of 0.01 MPa. The results indicate that increasing the CO_2 mineralization pressure has a significant effect on improving the chloride permeability of RAC. Unfortunately, there is no relevant study on the chloride permeability of RAC containing RCA CO_2 mineralized at various temperatures and humidity. Figure 7.28 illustrates the effect of the replacement level of carbonated RCA on the chloride migration coefficient of concrete (Liu et al., 2022). The chloride migration coefficient of RCA concrete increases with the use of repeatedly recycled RCAs, primarily due to the presence of numerous cracks and pores in the bonding mortar on the surface of repeatedly RCAs (Liu et al., 2022). Even with the addition of 25% carbonated repeatedly RCAs, the chloride migration coefficient of RCA concrete rises, indicating that the porosity reduction achieved by a small proportion of carbonated repeatedly RCAs is insufficient to offset the weakening

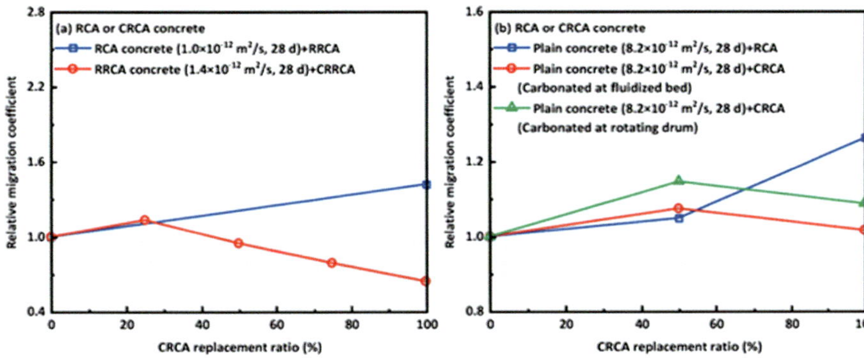

FIGURE 7.28 Chloride migration coefficient of concrete with different replacement levels of carbonated RCAs (Liu et al., 2022).

effect of RCAs. However, an opposite trend is observed at higher replacement levels of carbonated repeatedly RCAs (up to 100%), where the chloride migration coefficient decreases and even falls below that of conventional RCA concrete (Figure 7.34(a)). This can be attributed to the denser internal structure of carbonated repeatedly RCAs at higher substitution levels (i.e., above 40% by mass), which effectively reduces the overall porosity of the concrete. Liang et al. (2019) and Zhang et al. (2019) investigated the effect of RCA on the corrosion behavior of reinforcing bars, which is closely related to chloride permeability in reinforced concrete. They claimed that CO_2 mineralization improves the pore structure of the adhered mortar in RCA and reduces the chloride concentration around the steel reinforcement, and therefore reduces the corrosion of reinforced concrete. In addition, the bonding properties of concrete containing CO_2 mineralized RCA to corroded steel reinforcement need to be further investigated.

7.7.3 Resistance to carbonation

Figure 7.29 depicts the change in carbonation depth with time for cement mortar prepared with unmineralized and mineralized RCA. The carbonation depth increases with the increase in time for both samples. Compared with the concrete prepared with G-RCA, the carbonation depth of concrete prepared with G-CRCA was 22.4% and 33.0% lower than that of concrete prepared with G-RCA at 7 and 28 days, respectively. This indicates that the mineralization of RCA is an effective method to reduce carbonation depth. The higher carbonation depth of concrete prepared with G-RCA is due to the higher water absorption and more porous adherent cement mortar. However, the water absorption and porosity of G-RCA are reduced by mineralization, and the ITZ and microstructure of concrete prepared with G-CRCA are denser and improved. Therefore, the concrete prepared with G-CRCA has a lower carbonation depth.

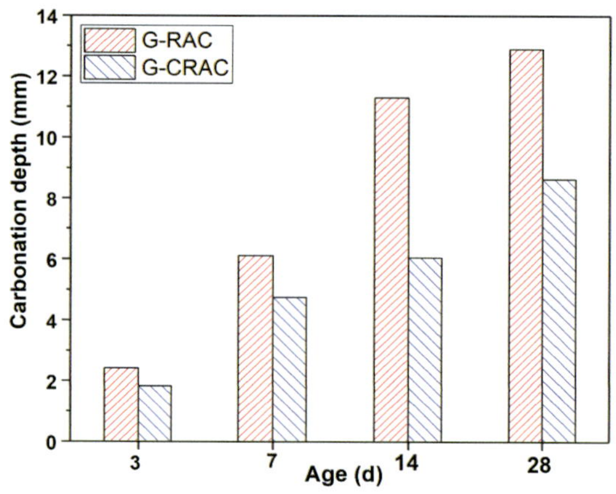

FIGURE 7.29 carbonation depth of concrete prepared with G-RCA and G-CRCA.

7.8 SUMMARY

A CO_2 mineralization as a method to improve the quality of RCAs has been reviewed in this study. The following conclusions can be drawn:

(1) CO_2 mineralization not only improved the original ITZ in RCA but also improved the newly formed ITZ in the RAC. When subjected to CO_2 mineralization, the physical and mechanical properties of RCAs were improved. The apparent density was increased, and water absorption and crushing value were decreased.
(2) Mechanical properties of RAC were lower than those of the natural concrete. But CO_2 mineralization increased the flowability and compressive strength, flexural strengths, and tensile splitting strength of the RAC.
(3) Compared to RAC containing unmineralized RCA, RAC containing CO_2 mineralized RCA has higher autogenous shrinkage, lower drying shrinkage, water absorption, mineralization depth, and chloride migration coefficient.

REFERENCES

Cui, H., Tang, W., Liu, W., Dong, Z., Xing, F., 2015. Experimental study on effects of CO2 concentrations on concrete carbonation and diffusion mechanisms. *Construction and Building Materials* 93, 522–527.

Fang, X., Xuan, D., Shen, P., Poon, C.S., 2021. Fast enhancement of recycled fine aggregates properties by wet carbonation. *Journal of Cleaner Production* 313, 127867.

Fang, X., Zhan, B., Poon, C.S., 2020. Enhancing the accelerated carbonation of recycled concrete aggregates by using reclaimed wastewater from concrete batching plants. *Construction and Building Materials* 239, 117810.

Fang, X.L., Xuan, D.X., Poon, C.S., 2017. Empirical modelling of CO_2 uptake by recycled concrete aggregates under accelerated carbonation conditions. *Materials and Structures* 50(4), 200.

Gao, S., Guo, J., Gong, Y., Ban, S., Liu, A., 2022. Study on the penetration and diffusion of chloride ions in interface transition zone of recycled concrete prepared by modified recycled coarse aggregates. *Case Studies in Construction Materials* 16, e01034.

Guo, H., Shi, C., Guan, X., Zhu, J., Ding, Y., Ling, T.-C., Zhang, H., Wang, Y., 2018. Durability of recycled aggregate concrete – A review. *Cement and Concrete Composites* 89, 251–259.

Guo, Y.-C., Li, X.-M., Zhang, J., Lin, J.-X., 2023. A review on the influence of recycled plastic aggregate on the engineering properties of concrete. *Journal of Building Engineering* 79, 107787.

Kazmi, S.M.S., Munir, M.J., Wu, Y.-F., Patnaikuni, I., Zhou, Y., Xing, F., 2019. Influence of different treatment methods on the mechanical behavior of recycled aggregate concrete: A comparative study. *Cement and Concrete Composites* 104, 103398.

Kazmi, S.M.S., Munir, M.J., Wu, Y.-F., Patnaikuni, I., Zhou, Y., Xing, F., 2020. Effect of recycled aggregate treatment techniques on the durability of concrete: A comparative evaluation. *Construction and Building Materials* 264, 120284.

Kisku, N., Joshi, H., Ansari, M., Panda, S.K., Nayak, S., Dutta, S.C., 2017. A critical review and assessment for usage of recycled aggregate as sustainable construction material. *Construction and Building Materials* 131, 721–740.

Kou, S.-C., Zhan, B.-j., Poon, C.-S., 2014. Use of a CO_2 curing step to improve the properties of concrete prepared with recycled aggregates. *Cement and Concrete Composites* 45, 22–28.

Lee, G.C., Choi, H.B., 2013. Study on interfacial transition zone properties of recycled aggregate by micro-hardness test. *Construction and Building Materials* 40, 455–460.

Li, L., Poon, C.S., Xiao, J., Xuan, D., 2017. Effect of carbonated recycled coarse aggregate on the dynamic compressive behavior of recycled aggregate concrete. *Construction and Building Materials* 151, 52–62.

Li, L., Xiao, J., Xuan, D., Poon, C.S., 2018. Effect of carbonation of modeled recycled coarse aggregate on the mechanical properties of modeled recycled aggregate concrete. *Cement and Concrete Composites* 89, 169–180.

Li, Y., Fu, T., Wang, R., Li, Y., 2020. An assessment of microcracks in the interfacial transition zone of recycled concrete aggregates cured by CO_2. *Construction and Building Materials* 236, 117543.

Li, Y., Zhang, S., Wang, R., Zhao, Y., Men, C., 2019. Effects of carbonation treatment on the crushing characteristics of recycled coarse aggregates. *Construction and Building Materials* 201, 408–420.

Liang, C., Ma, H., Pan, Y., Ma, Z., Duan, Z., He, Z., 2019. Chloride permeability and the caused steel corrosion in the concrete with carbonated recycled aggregate. *Construction and Building Materials* 218, 506–518.

Liang, C., Pan, B., Ma, Z., He, Z., Duan, Z., 2020. Utilization of CO_2 curing to enhance the properties of recycled aggregate and prepared concrete: A review. *Cement and Concrete Composites* 105, 103446.

Liu, H., Zhu, X., Zhu, P., Chen, C., Wang, X., Yang, W., Zong, M., 2022. Carbonation treatment to repair the damage of repeatedly recycled coarse aggregate from recycled concrete suffering from coupling action of high stress and freeze-thaw cycles. *Construction and Building Materials* 349, 128688.

Lu, B., Shi, C., Cao, Z., Guo, M., Zheng, J., 2019. Effect of carbonated coarse recycled concrete aggregate on the properties and microstructure of recycled concrete. *Journal of Cleaner Production* 233, 421–428.

Lu, B., Shi, C., Zhang, J., Wang, J., 2018. Effects of carbonated hardened cement paste powder on hydration and microstructure of Portland cement. *Construction and Building Materials* 186, 699–708.

Luo, S., Ye, S., Xiao, J., Zhen, J., Zhu, Y., 2018. Carbonated recycled coarse aggregate and uniaxial compressive stress-strain relation of recycled aggregate concrete. *Construction and Building Materials* 188(2018), 956–965.

Mao, Y., Liu, J., Shi, C., 2021. Autogenous shrinkage and drying shrinkage of recycled aggregate concrete: A review. *Journal of Cleaner Production* 295, 126435.

Nagataki, S., Gokce, A., Saeki, T., Hisada, M., 2004. Assessment of recycling process induced damage sensitivity of recycled concrete aggregates. *Cement and Concrete Research* 34(6), 965–971.

Pan, G., Zhan, M., Fu, M., Wang, Y., Lu, X., 2017. Effect of CO2 curing on demolition recycled fine aggregates enhanced by calcium hydroxide pre-soaking. *Construction and Building Materials* 154, 810–818.

Parthiban, K., Saravana Raja Mohan, K., 2017. Influence of recycled concrete aggregates on the engineering and durability properties of alkali activated slag concrete. *Construction and Building Materials* 133, 65–72.

Pu, Y., Li, L., Wang, Q., Shi, X., Fu, L., Zhang, G., Luan, C., Abomohra, A.E.-F., 2021. Accelerated carbonation treatment of recycled concrete aggregates using flue gas: A comparative study towards performance improvement. *Journal of CO2 Utilization* 43, 101362.

Rostami, V., Shao, Y., Boyd, A.J., He, Z., 2012. Microstructure of cement paste subject to early carbonation curing. *Cement and Concrete Research* 42(1), 186–193.

Shi, C., Li, Y., Zhang, J., Li, W., Chong, L., Xie, Z., 2016. Performance enhancement of recycled concrete aggregate – A review. *Journal of Cleaner Production* 112, 466–472.

Shi, C., Wu, Z., Cao, Z., Ling, T.C., Zheng, J., 2018. Performance of mortar prepared with recycled concrete aggregate enhanced by CO2 and pozzolan slurry. *Cement and Concrete Composites* 86, 130–138.

Singh, M., Danie Roy, A.B., Waseem, S., Singh, H., 2021. Feasibility and performance analysis of carbonated recycled aggregate concrete. *International Journal of Sustainable Engineering* 14(4), 761–775.

Sun, B., Zeng, Q., Wang, D., Zhao, W., 2022. Sustainable 3D printed mortar with CO2 pretreated recycled fine aggregates. *Cement and Concrete Composites* 134, 104800.

Tam, V.W.Y., Butera, A., Le, K.N., 2016. Carbon-conditioned recycled aggregate in concrete production. *Journal of Cleaner Production* 133, 672–680.

Thiery, M., Villain, G., Dangla, P., Platret, G., 2007. Investigation of the carbonation front shape on cementitious materials: Effects of the chemical kinetics. *Cement and Concrete Research* 37(7), 1047–1058.

Torben, C.H., Henrik, N., 1983. Strength of recycled concrete made from crushed concrete coarse aggregate. *Concrete International* 5(1), 79–83.

Visser, J.H.M., 2014. Influence of the carbon dioxide concentration on the resistance to carbonation of concrete. *Construction and Building Materials* 67, 8–13.

Wang, J., Zhang, J., Cao, D., Dang, H., Ding, B., 2020. Comparison of recycled aggregate treatment methods on the performance for recycled concrete. *Construction and Building Materials* 234, 117366.

Wu, K., Luo, S., Zheng, J., Yan, J., Xiao, J., 2022. Influence of carbonation treatment on the properties of multiple interface transition zones and recycled aggregate concrete. *Cement and Concrete Composites* 127, 104402.

Xuan, D., Zhan, B., Poon, C.S., 2016. Assessment of mechanical properties of concrete incorporating carbonated recycled concrete aggregates. *Cement and Concrete Composites* 65, 67–74.

Xuan, D., Zhan, B., Poon, C.S., 2017. Durability of recycled aggregate concrete prepared with carbonated recycled concrete aggregates. *Cement and Concrete Composites* 84, 214–221.

You, X., Hu, X., He, P., Liu, J., Shi, C., 2022. A review on the modelling of carbonation of hardened and fresh cement-based materials. *Cement and Concrete Composites* 125, 104315.

Zhan, B., Poon, C.S., Liu, Q., Kou, S., Shi, C., 2014. Experimental study on CO2 curing for enhancement of recycled aggregate properties. *Construction and Building Materials* 67, 3–7.

Zhan, B.J., Poon, C.S., Shi, C.J., 2016. Materials characteristics affecting CO2 curing of concrete blocks containing recycled aggregates. *Cement and Concrete Composites* 67, 50–59.

Zhan, B.J., Xuan, D.X., Poon, C.S., 2018. Enhancement of recycled aggregate properties by accelerated CO2 curing coupled with limewater soaking process. *Cement and Concrete Composites* 89, 230–237.

Zhan, B.J., Xuan, D.X., Poon, C.S., Scrivener, K.L., 2020. Characterization of interfacial transition zone in concrete prepared with carbonated modeled recycled concrete aggregates. *Cement and Concrete Research* 136, 106175.

Zhan, B.J., Xuan, D.X., Zeng, W., Poon, C.S., 2019. Carbonation treatment of recycled concrete aggregate: Effect on transport properties and steel corrosion of recycled aggregate concrete. *Cement and Concrete Composites* 104.

Zhan, M., Pan, G., Wang, Y., Fu, M., Lu, X., 2017. Effect of presoak-accelerated carbonation factors on enhancing recycled aggregate mortars. *Magazine of Concrete Research* 69(16), 838–849.

Zhang, J., Shi, C., Li, Y., Pan, X., Poon, C.-S., Xie, Z., 2015a. Influence of carbonated recycled concrete aggregate on properties of cement mortar. *Construction and Building Materials* 98, 1–7.

Zhang, J., Shi, C., Li, Y., Pan, X., Poon, C.-S., Xie, Z., 2015b. Performance enhancement of recycled concrete aggregates through carbonation. *Journal of Materials in Civil Engineering* 27(11), 04015029.

Zhang, T., Chen, M., Wang, Y., Zhang, M., 2023. Roles of carbonated recycled fines and aggregates in hydration, microstructure and mechanical properties of concrete: A critical review. *Cement and Concrete Composites* 138, 104994.

8 CO_2 mineralization by recycled concrete fines

8.1 INTRODUCTION

It is estimated that over 10 billion tons of construction waste are generated globally on an annual basis (Akhtar and Sarmah, 2018). Of this quantity, concrete waste constitutes approximately 60–70% in volume (K. Ouyang et al., 2020). The recycling of concrete waste into recycled concrete aggregate (RCA) represents a highly effective technique for addressing the scarcity of aggregate resources and reducing carbon emissions. Dismantled concrete is pulverized and processed into fine powders with particle sizes less than 0.075 mm (Lu et al., 2018), which is known as recycled concrete fines (RCF). Its principal composition comprises hydration products such as C-S-H, $Ca(OH)_2$, AFm, and inert quartz sand (Lu et al., 2018; L. Li et al., 2022). Its main uses are in the preparation of recycled cement from burnt matrices or as a supplementary cementitious material (SCM) to produce concrete; the latter has been extensively studied. However, the use of RCF as SCM is limited by its rough surface, high number of micropores and microcracks, and low reactivity, which result in deterioration of the matrix microstructure and reduction of mechanical properties (Zhang et al., 2015; Wu, Yang et al., 2021).

In recent years, enhancement of RCF through accelerated CO_2 mineralization has emerged as a highly efficient carbon capture, utilization, and storage (CCUS) technology aligning with global strategies for carbon reduction (Thonemann et al., 2022; Zhang et al., 2023). The mineralization between CO_2 and RCF can produce CO_2 mineralization products, including calcium carbonate ($CaCO_3$) and amorphous silica gel with significant pozzolanic activity, which enhances the granular matrix of RCF and increases its potential reactivity (Lu et al., 2018, 2023; Mao et al., 2023). When utilizing CO_2 mineralization recycled concrete fines (cRCF) as SCM, $CaCO_3$ can generate filling, nucleation, and chemical effects, whereas amorphous silica gel has a high pozzolanic activity. Thus, cRCF shows great potential as a new SCM (Kwon et al., 2015; Likes et al., 2022).

This chapter outlines the CO_2 mineralization of RCF and the decoupling analysis to examine the effects of cRCF on cement paste and its underlying mechanism, drawing from existing literature. Based on this foundation, the study reviews the effects of cRCF as an SCM on the hydration, rheology, microstructure, mechanical properties, and durability of cement-based materials. The decoupling analysis aims to elucidate the critical factors of the cRCF that influence the properties of cement-based materials and provide insights into controlling these properties in cement-based materials containing cRCF. Through reviewing, analyzing, and discussing existing knowledge, this chapter aims to promote cRCF as a sustainable and environmentally friendly

DOI: 10.1201/9781003627531-8

SCM in practical engineering applications. Furthermore, it clarifies the direction for future research and application of cRCF as an SCM.

8.2 PROPERTIES OF RCF

8.2.1 Component properties

The chemical and mineral composition of RCF depends on the composition of the cementitious materials in the matrix concrete and the aggregates used, which are usually high in impurities and complex in composition (Mao et al., 2023). Its mineral composition mainly consists of hydration product phases (calcium hydroxide (CH), AFm, and C-S-H gel), a small amount of clinker (C_3S, C_2S, etc.), and a certain portion of inert sand remaining during mechanical processing of stripping. In this instance, the primary phases to be mineralized are alkaline hydrates (CH, C-S-H gels), whereas inert sands are not. In order to avoid the influence of impurity components, nearly fully hydrated cement paste powder is typically used in studies to simulate RCF. Consequently, the data presented in the literature are often regarded as representing the upper limit of the CO_2 mineralization reactivity of RCF, while the lower limit of the reactivity is determined from real samples of waste concrete. However, Jiang et al. (2023) have indicated that the CO_2 mineralization efficiency of real RCF and its effect on mortar properties is only slightly lower than that of simulated RCF. Furthermore, they have proposed that the simulated RCF can be considered as real RCF with a large amount of residual cement paste in actual production.

The chemical composition of RCF is illustrated in Figure 8.1, indicating that the major chemical components are SiO_2, CaO, Al_2O_3, and Fe_2O_3, accounting

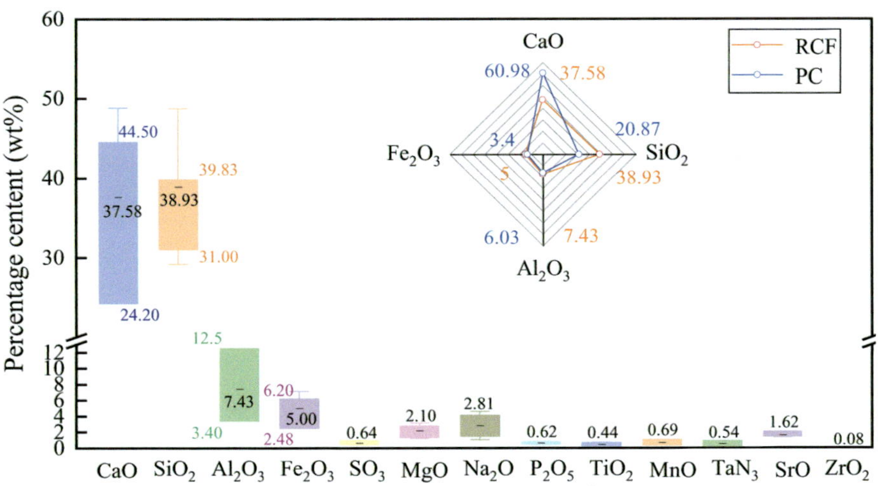

FIGURE 8.1 Chemical composition of RCF and ordinary Portland cement (PC). (Data from Horsakulthai, 2021; Liang et al., 2021; Sun et al., 2021, 2022; Gao et al., 2022; He et al., 2022; X. Li et al., 2022; Sun et al., 2022).

for 31.00–39.83%, 24.20–44.50%, 3.40–12.50%, and 2.48–6.20%, respectively (Horsakulthai, 2021; Liang et al., 2021; Sun et al., 2021; Gao et al., 2022; He et al., 2022; X. Li et al., 2022; Sun et al., 2022). According to Steinour's (Steinour, 1959) formula (Eq. 8.1), CO_2 uptake is positively correlated with calcium content. Then, the higher CaO content in RCF demonstrated its excellent carbon sequestration ability.

$$X_{CO_2,max} = 0.785(CaO - 0.56CaCO_3 - 0.7SO_3) + 1.091MgO + 1.42Na_2O + 0.93K_2O \quad (8.1)$$

8.2.2 Activity potential

The activity index is an important indicator for evaluating the activity potential of SCMs (Horsakulthai, 2021). Table 8.1 lists the average particle size, replacement, and activity index requirement of RCF in different standards. For RCF with a particle size less than 20 μm, the 28-day activity index generally meets the requirement of the Chinese national standard GB/T 1596–2017 of not less than 70%, and a higher activity index is observed with decreasing particle size (Wu, Yang et al., 2021). It is well known that the activity of RCF can be enhanced when doped with nano-SiO_2 (Liu et al., 2022) and thermal treatment (Wu, Xu et al., 2021) or CO_2 mineralization treatment (Kaliyavaradhan et al., 2022). Among them, the CO_2 mineralization treatment of RCF is more significant for the enhancement of the activity index (Kaliyavaradhan et al., 2022), which can make it easier for RCF with regular particle size to meet the requirement of the activity index.

8.3 CO_2 MINERALIZATION OF RCF

8.3.1 CO_2 mineralization methods

8.3.1.1 Dry CO_2 mineralization

During dry CO_2 mineralization of RCF, the reaction between CO_2 and RCF generally takes place at multiple localized reaction fronts. As the mineralization front advances toward the interior of RCF, CO_2 is rapidly extracted by a limited number of reaction sites, hindering further CO_2 diffusion. At this point, the ionic saturation in the pore solution of RCF becomes a key factor in controlling the rate of the CO_2 mineralization and significantly affects the properties of the CO_2 mineralization products.

The main reactant phases in RCF that can be CO_2 mineralized are calcium-containing phases (e.g., CH, C-S-H gels). As shown in Figure 8.2, literature (Zajac, Skibsted, Bullerjahn et al., 2022) investigated the relationship between the contents of formed calcium carbonate and CH consumption during RCF CO_2 mineralization and found that in dry CO_2 mineralization with relative humidity (RH) lower than 90% (the red part in Figure 8.2), CO_2 was observed to react with all calcium-containing phases in the first few minutes of mineralization. The CO_2 mineralization rate of CH is higher than that of C-S-H gel. In contrast, in the dry CO_2 mineralization with 100% RH (blue part in Figure 8.2), it was found that the CH consumption reached almost half before it started to deviate from the theoretical line, i.e., at this time, the

TABLE 8.1
Average particle size, replacement, and activity index of RCF

Standard	Activity index requirement (%)	Replacement (%)	Category	D50 (μm)	Treatment	28-days activity index (%)	Ref.
GB/T 1596	70	30	WCP	9.0	–	70.0	(Zhang et al., 2022)
GB/T 1596	70	30	Coarse WCP	104.0	–	54.0 (Dissatisfy)	(Wu et al., 2021)
BS EN 450-1			Fine WCP	12.0	–	72.7	
			Finer WCP	9.0	–	81.0	
GB/T 1596	70	30	WCP	52.3	Nano SiO_2	71.0	(Liu et al., 2022)
GB/T 1596	70	30	WCP	10.8	Thermal	73.7	(Wu et al., 2021)
ASTM C311	75	20	WCP	26.5	–	76.0	(Kaliyavaradhan et al., 2022)
			cWCP	38.4	CO_2 mineralization	77.0	

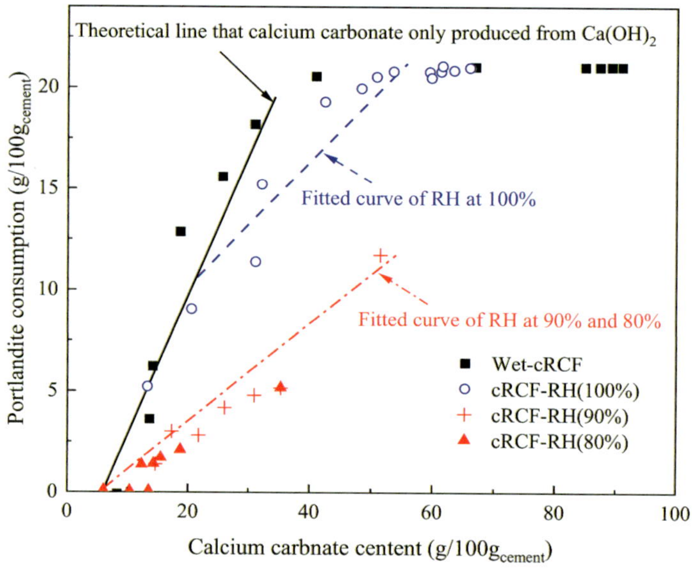

FIGURE 8.2 The relation between calcium carbonate formation (carbonation degree) and portlandite consumption (Zajac, Skibsted, Lothenbach et al. 2022, 其他地方也要修改).

mineralization of RCF preferentially showed a rapid dissolution of CH prior to starting the decalcification of the C-S-H gel (Zajac, Skibsted, Bullerjahn et al., 2022). The difference in this sequence of physical phase reactions lies in the difference in the water content of the RCF pore solution. After the rapid depletion of CH, a dense $CaCO_3$ layer forms in the mineralization zone. This limits the dissolution of the remaining CH in the RCF, and the mineralization rate decreases sharply (Thiery et al., 2007). As the CO_2 mineralization front advances toward the C-S-H gel inside the RCF, Ca^{2+} is mainly decalcified through the C-S-H gel and diffuses through the microporous and tortuous $CaCO_3$ layer.

Theoretically, the water released in the pre-carbonation stage promotes the dissolution dynamics of RCF, but the exothermic CO_2 mineralization reaction accelerates the loss of internal water, which further reduces the pore water content in the late stage of CO_2 mineralization. The water released before CO_2 mineralization helps RCF dissolve, but the CO_2 mineralization reaction also loses water. This makes it harder for ions to move, limiting the mineralization process (Lu et al., 2018). Therefore, the CO_2 mineralization kinetics of RCF is significantly affected by the water content during dry CO_2 mineralization, resulting in a lower degree of mineralization (as shown in Table 8.2) and relatively few mineralization products.

8.3.1.2 Wet CO_2 mineralization

Wet CO_2 mineralization of RCF occurs in a solution medium, where CO_2 is no longer highly dependent on gaseous diffusion, but is strongly dissolved in solution. At the same time, the calcium-containing hydrates are undersaturated, which constantly

TABLE 8.2
CO_2 mineralization degree and conditions of RCF

CO_2 mineralization degree	Dry/wet	Particle size (μm)	CO_2 mineralization condition: Liquid–solid ratio	Temperature (°C)	CO_2 concentration (%) / CO_2 flow rate (L/h/per g_{WCP})	Duration (h)	Relative humidity (%) / Aqueous	Ref.
48.6	Dry	2–75	–	20±1	99/-	24	60±5	(Lu et al., 2018)
30.83	Dry	D90<100	–	20	25/-	6	80	(Zajac, Skibsted, Bullerjahn et al., 2022)
50.84	Dry	D90<100	–	20	25/-	6	90	
66.15	Dry	D90<100	–	20	25/-	6	100	
59.52	Dry	D90<253	–	20	25/-	6	100	
4.0	Dry	D90<150	0.3	20	20/-	0.5	5–10	(Wu et al., 2022)
16.3	Dry	D90<150	0.3	60	20/-	0.5	5–10	
30.6	Dry	D90<150	0.3	100	20/-	0.5	5–10	
18.2	Dry	D90<150	0.3	140	20/-	0.5	5–10	
60.19	Wet	D90<100	10g/100 mL	25	100/0.06	1	Aqueous	(Mao et al., 2023)
61.35	Wet	D90<100	10g/100 mL	80	100/0.06	1	Aqueous	
90.74	Wet	D90<100	10g/100 mL	20	25/0.16	6	Aqueous	(Zajac, Skibsted, Bullerjahn et al., 2022)
81.20	Wet	D90<100	15g/100 ml	20±1	10 (with N2)/0.16	6	deionized water	(Zajac et al., 2021)
77.42	Wet	D90<100	15g/100 mL	20±1	10 (with N2)/0.16	6	0.1M NaOH	
82.78	Wet	D90<100	15g/100 mL	20±1	10 (with N2)/0.16	6	0.4M NaOH	
61.56	Wet	D90<100	15g/100 mL	20±1	10 (with N2)/0.16	6	0.25M Na_2SO_4	(Zajac, Skibsted, Lothenbach et al., 2022)
60.13	Wet	D90<100	15g/100 mL	20±1	10 (with N2)/0.16	6	0.5M Na_2SO_4	

promotes the continuous leaching of Ca^{2+}, thus increasing the precipitation of the mineralization products. Therefore, compared to dry CO_2 mineralization, wet CO_2 mineralization significantly improves the mineralization efficiency (see Table 8.2) (Zajac, Skibsted, Durdzinski et al., 2020; Zajac, Skibsted, Skocek et al., 2020; Mao et al., 2023), and RCF can reach more than 80% mineralization level in a few hours (Zajac, Skibsted, Bullerjahn et al., 2022).

The sequence of physical phase reactions in the wet CO_2 mineralization of RCF is not identical to dry mineralization (see black scattered dots in Figure 8.2), CH has a high solubility in solution, and the C-S-H gel starts to decalcify only when CH is almost depleted (Zajac, Skibsted, Bullerjahn et al., 2022). Generally, the wet mineralization process of RCF was divided into four stages and plotted in Figure 8.3 according to Shen et al. (Shen, Lu et al., 2022). In the first stage, CO_2 is dissolved in water and ionizes to CO_3^{2-} and H^+, while Ca^{2+} is released from CH. A slight increase in pH during this phase indicates that the dissolution rate of RCF exceeds the dissolution rate of CO_2 (Shen, Zhang et al., 2022). It is worth noting that mineralization in an alkaline environment increases the concentration of Al and Si in the first stage (Zajac, Skibsted, Durdzinski et al., 2020), which may affect the precipitation of $CaCO_3$. The dissolution of CO_2 or the precipitation of $CaCO_3$ at this time may

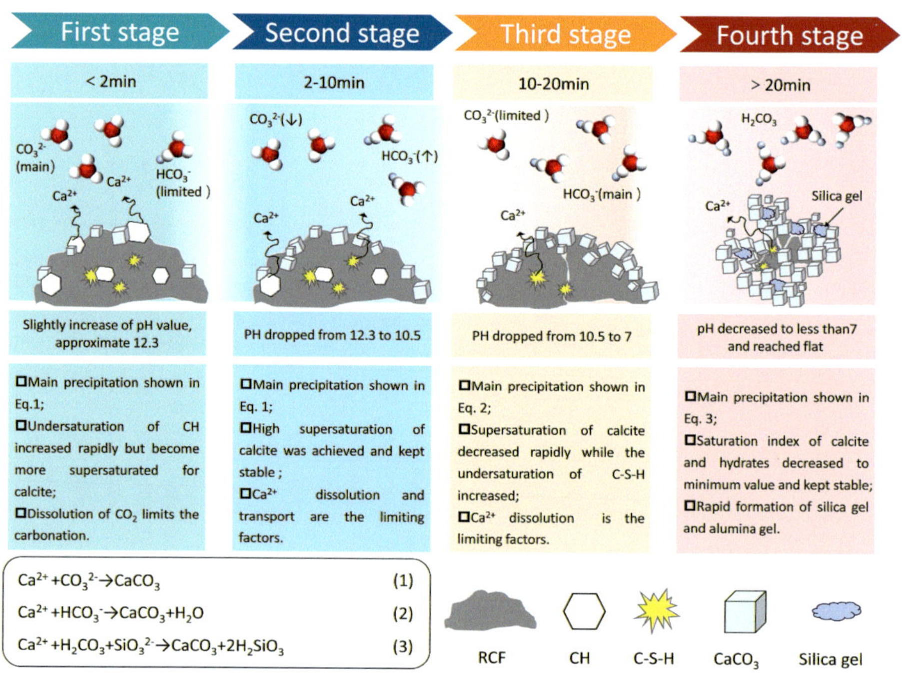

FIGURE 8.3 The four stages of wet carbonation are illustrated based on Shen, Zhang et al. (2022).

constrain the mineralization process. In the second stage, Ca^{2+} comes from CH dissolution and C-S-H gel decalcified. The gradual accumulation of $CaCO_3$ on the surface of RCF impedes the transport of Ca^{2+} from the interior of the particles to the surface, resulting in a rapid decrease in pH and a lower leaching rate, which restricts the process of the mineralization reaction. Entering the third stage, the mineralization rate gradually slows down. The $CaCO_3$ layer on the RCF surface continues to thicken, which hinders the C-S-H gel decalcification inside the RCF, and the reaction rate depends on the dissolution of the hydride. A recent study by Qian et al. (Qian et al., 2023) demonstrated that the use of pH cycling mineralization method could always ensure a higher degree of undersaturation of ions in the solution during the interval of the mineralization cycle, which promoted the dissolution dynamics of RCF and could effectively alleviate the second and third stages. In the fourth stage, amorphous silica and alumina gels are formed rapidly. Meanwhile, the $CaCO_3$ layer turns homogeneous, the core shrinks and becomes porous, and decomposition eventually occurs. Although the $CaCO_3$ layer no longer hinders Ca^{2+} diffusion at this time, Ca^{2+} diffusion is slow due to the difficulty of C-S-H decalcification with a low Ca–Si ratio (Tine F. Sevelsted et al., 2015). Then the carbonation reaction tends to stabilize.

8.3.2 Influencing factors on CO_2 mineralization of RCF

8.3.2.1 Dry CO_2 mineralization

In order to improve the efficiency of the dry CO_2 mineralization process of RCF and extend it to industrial-scale applications, researchers have extensively studied ways to regulate the degree of mineralization (López-Arce et al., 2011; Drouet et al., 2019; Wu et al., 2022; Zajac, Skibsted, Bullerjahn et al., 2022), including the optimization of several influencing factors, such as RH, mineralization temperature, mineralization pressure, CO_2 concentration, and mineralization time.

RH value is a key factor in the dry CO_2 mineralization of RCF; choosing an appropriate RH value is crucial to ensure an efficient mineralization reaction. Higher RH values may hinder CO_2 diffusion and inhibit the mineralization reaction. Galan et al. (Galan et al., 2013) suggested that the diffusion coefficient of CO_2 can be reduced by seven orders of magnitude as the water saturation increases from dry to near saturation. In view of this, dry CO_2 mineralization usually requires pre-treatment to ensure that the RH value of RCF is within the optimal interval of 50–70% (López-Arce et al., 2011; Galan et al., 2013). However, Zajac et al. (2022) found that in the range of RH values from 80% to 100%, the degree of CO_2 mineralization increased dramatically in response to increasing RH values and far exceeded the degree of that in the range of 50% to 70% (see bar graph in Figure 8.4(a)). This is attributed to the promotion of calcium dissolution by the higher water content. This facilitation can even cause the initial mineralization rate of RCF to exceed that of wet CO_2 mineralization at RH=100% (Zajac, Skibsted, Bullerjahn et al., 2022) (Figure 8.4(a)).

Elevated temperature can promote CO_2 diffusion (Drouet et al. 2015), leading to an overall accelerated trend in the mineralization rate, thus increasing the

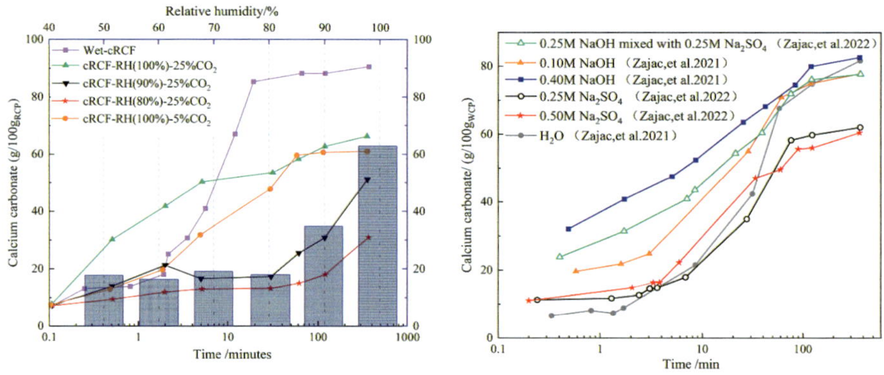

FIGURE 8.4 Variation of the calcium carbonate content with mineralization duration when RCF is CO_2 mineralized at (a) different relative humidity and different CO_2 concentrations (Zajac, Skibsted, Bullerjahn et al., 2022); (b) different solution environments (Zajac, Skibsted and Ben Haha, 2021; Zajac, Skibsted, Lothenbach et al., 2022).

mineralization degree. However, powdered RCF has a large specific surface area and CO_2 mineralization at high temperatures tends to cause water evaporation (Drouet et al., 2015). Once the temperature threshold is exceeded, the mineralization rate decreases (Drouet et al., 2019; Wu et al., 2022). Wu et al. (2022) suggested that the mineralization temperature threshold of RCF may be around 100 °C. When the temperature is between the interval of 20°C and 100°C, the mineralization rate of RCF is accelerated with the elevated temperature, whereas a further increase in temperature exacerbated water evaporation and the mineralization rate showed a decreasing trend.

In addition, higher CO_2 concentration and pressure drive the diffusion of CO_2 into the deeper portion of RCF through a larger gradient of concentration or pressure (Morandeau et al., 2014; Wang et al., 2017), which facilitates the second stage of dry CO_2 mineralization and significantly improves the degree and efficiency of mineralization. However, the effect of CO_2 concentration on the degree of mineralization is time-dependent; when the mineralization duration is sufficient, the difference will gradually decrease (Figure 8.4(a)). This is due to the fact that the mineralized layer keeps accumulating during mineralization, which hinders the diffusion of CO_2 gas, and then leads to a slow growth or even stagnation of the mineralization rate. According to Algourdin et al. (Algourdin et al., 2023), the mineralization reaction of RCF only lasted up to 17 hours before it stabilized, after which the change in the mineralization degree was no longer significant. He also showed that a CO_2 concentration of 18% also achieves the desired CO_2 uptake after 17 hours of mineralization. Based on this, the CO_2-rich flue gas (concentration around 15–30%) emitted from cement plants with residual heat has great potential for industrial application (He et al., 2020), not only to avoid separating the CO_2 from the captured gas but to obtain the heat energy required for CO_2 mineralization. It should be noted that the flue gas contains small amounts of NO_2 and SO_2, whose influence mechanism on mineralization is not yet clear. Hence, an in-depth study of them can lay the foundation for promoting the utilization of flue gas.

8.3.2.2 Wet CO_2 mineralization

The wet CO_2 mineralization of RCF is controlled by dissolution-precipitation. The influencing factors such as mineralization temperature, CO_2 concentration, and mineralization duration can be optimized to regulate the dissolution rate of CO_2 and Ca^{2+} as well as the precipitation of mineralization products, so as to improve the degree and efficiency. As in Section 8.3.2.1, the effects of temperature and duration on the mineralization rate and degree are basically consistent with the dry CO_2 mineralization. The difference is that the excessively high temperature constrains the rate by reducing the solubility of CO_2 and ions (Zhan et al., 2016). In addition, the CO_2 concentration is controlled by the CO_2 flow rate in wet CO_2 mineralization. Mao et al. (Mao et al., 2023) noted that changing the CO_2 flow rate from 3 L/min to 5 L/min significantly increased the $CaCO_3$ content during the first 10 minutes, and this enhancement was more pronounced than elevated temperature.

As wet CO_2 mineralization occurs in solution, the degree of mineralization can be increased by controlling the solution environment (Zając, Skibsted and Ben Haha, 2021; Zajac, Skibsted, Durdzinski et al., 2021; Zajac, Skibsted, Lothenbach et al., 2022; Jiang et al., 2023). Figure 8.4(b) synthesizes the findings of Zajac et al. (Zajac, Skibsted, Durdzinski et al., 2021; Zajac, Skibsted, Lothenbach et al., 2022) which investigated the variation in $CaCO_3$ content with mineralization time in different solution environments. The findings revealed that, although alkaline solutions (NaOH concentrations of 0.1 mol/L and 0.4 mol/L) accelerated the initial stage of the mineralization reaction, they delayed or had minimal impact on the subsequent stages and the final mineralization degree. However, the study of Jiang et al. (2023), which also used alkali solution (NaOH concentrations of 0.5 M and 1.0 M), concluded that alkali solution promoted the rate of mineralization and increased the extent of mineralization, leading to an increase in the depth of $CaCO_3$ deposition by approximately three times. Jiang et al. (2023) explained that NaOH acts in a catalyst-like role in high-concentration alkali solutions, which can continue to promote the dissolution of CO_2 while driving $CaCO_3$ deposition. Although the formation of a reactive layer in the later stages will limit the advancement of the $CaCO_3$ deposition front (Castellote and Andrade 2008; Galan et al., 2015), the excess presence of carbonate will generate a high concentration gradient. Such high concentration gradients enhance the diffusion driving force of the substrate inside the RCF and break through the limitation of the $CaCO_3$ layer to establish a mineralization equilibrium in a deeper region inside the RCF (Lee 2004). In addition, due to the presence of sulfate ions in the RCF base material, these ions will be rapidly leached from the RCF during wet CO_2 mineralization. Recycling of the mineralization solution during industrial production may lead to an increase in sulfate concentration (Zajac, Skibsted, Durdzinski et al., 2020), which consequently produces calcium sulfate precipitation affecting the degree of mineralization. The results of the study (Zajac, Skibsted, Lothenbach et al., 2022) showed that the sulfate solution enhanced the mineralization degree at the initial stage but reduced the mineralization rate after 30 minutes, and the final degree was reduced by about 25% compared to the aqueous environment (Figure 8.4(b)). In contrast, in a solution of sulfate mixed with alkali, the mineralization process is expedited by preventing calcium sulfate precipitation, and the degree is significantly enhanced, which can be equivalent to the aqueous environment (Zajac, Skibsted, Lothenbach et al., 2022).

8.4 CO_2 MINERALIZATION PRODUCTS

The activity of mineralized RCF is primarily derived from $CaCO_3$ and amorphous silica (Zajac, Skibsted, Skocek et al., 2020; Shen, Zhang et al., 2022). In particular, amorphous silica gel has high pozzolanic activity (see Section 9.4.3 for further details), and its content is typically low, with a maximum of only 0.1 g per gram of RCF (Mao, Drissi, He et al., 2024). In contrast, $CaCO_3$ has a relatively high content, but its reactivity is low, and it mainly performs a filler effect and nucleation effect when added to the cement.

8.4.1 Distribution of CO_2 mineralization products

Herterich et al. (2022) suggested that the distribution of $CaCO_3$ depends on the origin of the C-S-H: inner C-S-H or outer C-S-H (inner C-S-H more dense). Generally, in RCF obtained from dry mineralization, it can be observed that the location of $CaCO_3$ precipitation is mainly in the RCF particles. Further, $CaCO_3$ and decalcified C-S-H grow intermixed in the outer C-S-H region. This is due to the fact that the dense morphology of the inner C-S-H layer inhibits $CaCO_3$ crystal growth, leading to Ca^{2+} migration to the outer product region, which is easier to form. Differently, in wet mineralization, $CaCO_3$ is mainly precipitated in the solution-occupied spaces, i.e., among the RCF particles. Also, it has been shown (Zajac, Skibsted, Skocek et al., 2020) that $CaCO_3$ precipitation is also present in the inner C-S-H region of RCF obtained by wet mineralization. This is attributed to the higher concentration in the wet mineralization solution promoting diffusion of CO_3^{2-} into the inner space. In addition, the distribution of alumina-silica gel varies depending on the mineralization conditions (Zając, Skibsted and Ben Haha, 2021): the silica gel is uniformly distributed in wet mineralization without alkali, whereas in alkaline solutions, the silica gel is non-uniformly distributed and less dense. Especially, the relative distribution of $CaCO_3$ and silica gel also differed in the dry and wet mineralization process. In dry mineralization, $CaCO_3$ is covered underneath by the silica gel layer (X. Ouyang et al., 2020; Mao et al., 2023), with obvious layering. While in wet mineralization, silica gel and $CaCO_3$ are mixed and distributed without obvious layering (Mao et al., 2023).

8.4.2 Calcium carbonate

$CaCO_3$ is mainly obtained by the combination of Ca^{2+} with CO_3^{2-} and precipitation. The grain size of $CaCO_3$ is typically 90–120 nm (McDonald et al., 2022), and the smaller nanocrystal size allows it to fill in the tiny pores of RCF to improve matrix packing. Studies (Zając, Skibsted and Ben Haha, 2021; Zajac, Skibsted, Lothenbach et al., 2022) have shown that the crystal size of $CaCO_3$ in RCF is affected by the solution environment. In high alkali solution or sodium sulfate solution, $CaCO_3$ exists in the form of large crystals, while the crystal size is smaller in low alkali solution or aqueous solution.

In addition, $CaCO_3$ is polymorphic, including amorphous calcium carbonate and a variety of crystal types: spherical vaterite, fibrous aragonite, and cubic calcite. Each crystal type has its own particular advantages in cementitious materials. Due to its porous internal structure, vaterite has the potential to act as an internal curing agent in concrete to delay the release of water and reduce the negative effects of self-shrinkage during the curing process (Chen et al., 2017). Aragonite, owing to its unique fibrous shape, on the one hand, can prevent the generation and extension of microcracks in brittle concrete and effectively improve the toughness of cementitious materials (Cao et al., 2014). On the other hand, it will also significantly affect the rheological properties of the cement matrix, which will be discussed in detail in Section 9.5.1. In contrast, the microstructure of the common cleavage surface of calcite has a higher similarity to the C-S-H structure than aragonite, which is more likely to serve as a site for C-S-H nucleation and growth (Li et al., 2019). When having the same specific surface area, calcite shows more significant nucleation effects (Zhao, Williams, Li et al., 2023).

Conventional concrete is considered brittle, making it susceptible to cracking under thermal, shrinkage stresses, and other external loads. Replacing high-cost fibers with aragonite to enhance the toughness and tensile strength of cementitious materials shows promise (Cao et al., 2014). Consequently, scholars worldwide have devoted considerable attention to aragonite crystalline modulation in RCF (Lu et al., 2022; Shen, Lu et al., 2022; Wu et al., 2022; Shen et al., 2023; You et al., 2023). In the study conducted by Zajac et al. (2023), $CaCO_3$ was observed to exist in the form of calcite, aragonite, and vaterite when subjected to dry mineralization conditions with an RH of 100%. The aragonite content was found to be higher in comparison to the other observed forms. This phenomenon can be attributed to the lower moisture content compared to that of wet mineralization, which inhibits the dissolution and recrystallization of aragonite to calcite conversion. Based on the same principle, aragonite content is typically lower in wet mineralization. Accordingly, the wet mineralization of RCF necessitates the regulation of aragonite crystal morphology through the modification of the mineralization conditions, including the control of the CO_2 flow rate, the mineralization temperature, and the introduction of agents that influence the formation of the crystals. Mao et al. (Mao et al., 2023) observed that the CO_2 flow rate has a negligible impact on the formation of aragonite. However, an increase in temperature facilitates aragonite nucleation and growth by reducing the critical aragonite nucleation rate and decelerating the transition of aragonite to calcite (Montes-Hernandez et al., 2007; Pan et al., 2007), which results in the rate of aragonite production exceeding the rate of aragonite transformation to calcite. It is also important to note that crystal control agents play a significant role in regulating the different forms of $CaCO_3$. Magnesium admixtures are the most prevalent type of crystal regulator. Given that the energy required for Mg^{2+} to displace Ca^{2+} in aragonite is significantly greater than that of calcite, it is more readily displaced in calcite.

The presence of Mg^{2+} in calcite has been observed to result in an increase in surface energy and solubility, accompanied by a notable decline in nucleation rate, which ultimately leads to an enhanced formation of aragonite (You et al., 2023). In accordance with the aforementioned principle, Shen et al. (Shen, Lu et al., 2022) expeditiously synthesized aragonite with single crystal lengths of 10–30 μm and diameters of 0.5–2 μm under 80 °C and 0.05 M $MgCl_2$. It is noteworthy that, although the proportion of aragonite increases with the growth of $MgCl_2$ concentration, the total amount of $CaCO_3$ gradually decreases, indicating a reduction in the degree of mineralization (Shen, Lu et al., 2022; You et al., 2023). It is therefore important to ensure that the concentration of $MgCl_2$ is not excessive in order to maintain the desired mineralization degree of RCF. It is unfortunate that, although temperature has a facilitating effect on aragonite generation, when temperature is coupled with $MgCl_2$, excessive temperature will lead to a decrease in effective Mg^{2+} concentration, which is unfavorable for massive aragonite synthesis. You et al. (You et al., 2023) found that the optimal temperature at this point is approximately 92°C.

8.4.3 Silica gel

Amorphous silica gel plays a significant role in the properties of mineralized RCF, with its formation largely dependent on the continuous decalcification of C-S-H gels. Dissolved CO_2 gradually penetrates the silica-rich zone, facilitating the removal of Ca^{2+} from the structure. The decalcified C-S-H undergoes a transformation whereby it becomes [SiO_4], and protonates to $Si(OH)_4$. Due to its reduced stability, this rapidly polymerizes to form silica gel.

Silica gel can rapidly undergo hydration with CH to form secondary C-S-H gel, which is the main source of the RCF pozzolanic effect. According to Zajac et al. (2024), a thermodynamic model was employed to calculate the reactivity of the pozzolanic effect through CH consumption. It was found that the alumina-silica gel reacted rapidly, approaching complete reaction in just seven days, with higher pozzolanic activity than slag (Skibsted and Snellings, 2019; Zajac, Skocek, Durdzinski et al., 2020) (e.g., Figure 8.5). The higher pozzolanic activity leads to an important contribution of silica gel to the early strength of cementitious materials. A number of scholars (Mao et al., 2023; Mao, Drissi, He et al., 2024) have investigated the effect of mineralization conditions on the silica gel content and polymerization degree on RCF. The finding (Zajac et al., 2023) indicated that the silica gel content of wet mineralization RCF was greater than that of dry mineralization for the same mineralization time. Furthermore, an increase in temperature, an extension of mineralization time, and the setting of the solution environment will have a significant impact on the silica gel content (Jiang et al., 2023). As demonstrated in Reference (Mao, Drissi, He et al., 2024), elevating the temperature from 25°C to 80°C in the context of wet mineralization resulted in an increase of 11.4% in silica content and a 68.2% enhancement in polymerization degree. Jiang et al. (Jiang et al., 2023) also enhanced the efficiency of silica formation by employing NaOH alkaline solution to facilitate CO_2 dissolution. At 45°C in a 0.5 M alkali solution, 22.0% silica gel was obtained, of which 11.5% was fully polymerized.

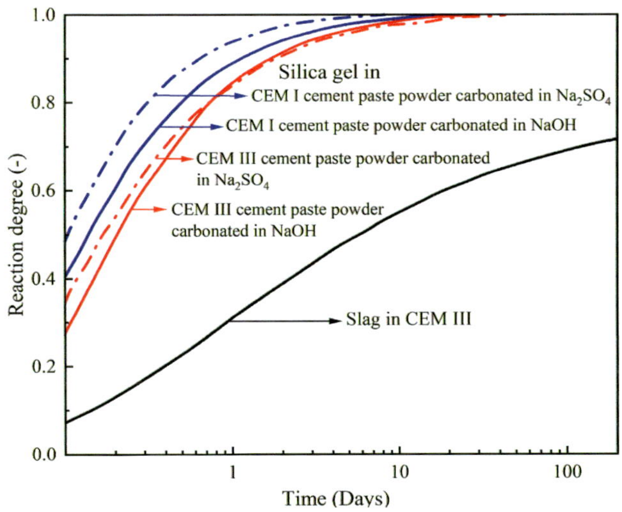

FIGURE 8.5 Degrees of reaction for the alumina-silica gel and slag in the cRCF-portlandite mixes calculated by thermodynamic modeling (Zajac et al., 2024).

Furthermore, silica gel exhibits a pronounced affinity for water, which may exert a considerable detrimental effect on the rheological properties (X. Ouyang et al., 2020). The high water absorption capacity of silica gel is primarily determined by the pore structure and surface morphology. The molecular structure of silica gel contains a substantial number of hydrophilic hydroxyl groups, which are capable of adsorbing water molecules through the formation of hydrogen bonds. It is important to note, however, that the structural hydroxyl groups within the silica gel have no effect on its water absorption (eds. Vansant et al., 1995). Therefore, the determination of the number of hydroxyl groups on the surface of the silica gel alone is the most accurate method for indicating the silica gel's hydrophilicity. Additionally, the inner walls of the minute pores on the silica gel surface exhibit a slight negative charge, which can facilitate the electrostatic attraction and subsequent adsorption of water molecules from the surrounding environment. During the mineralization process, an increase in temperature or CO_2 flow rate has been observed to promote the transformation of capillary pores in silica gel to finer gel pores (Mao et al., 2023), thereby increasing the number of minute pores and further enhancing the hydrophilicity.

8.5 EFFECT OF CO_2 MINERALIZED RCF AS SCM ON THE PROPERTIES OF CEMENTITIOUS MATERIALS

8.5.1 Rheology of Fresh Paste

The impact of RCF on the rheological characteristics of the paste is contingent upon five physicochemical mechanisms attributes of its surface: elevated specific surface area, hydrophilicity of the surface silica gel, geometrical properties and distribution of the surface $CaCO_3$, and surface electrochemical characteristics. The incorporation of CO_2 mineralized RCF (cRCF) typically results in a higher yield stress and plastic viscosity of cement paste compared to unmineralized recycled concrete fine (uRCF) (X. Ouyang et al., 2020; Peng et al., 2023). Furthermore, the yield stress and plastic viscosity increase significantly with increasing cRCF incorporation (Peng et al., 2023). This phenomenon is primarily attributed to the elevated specific surface area of cRCF, in conjunction with its enhanced hydrophilic characteristics of silica gel. On the one hand, the higher specific surface area results in cRCF requiring a larger wetting surface and exhibiting higher water absorption (Drouet et al., 2015), which increases as particle size decreases (Peng et al., 2023). Conversely, the enhanced hydrophilic properties of silica gel exert a considerable influence. Upon mixing cRCF with the cement paste, a considerable quantity of water is adsorbed by silica gel, resulting in a notable reduction in the paste's rheological properties (X. Ouyang et al., 2020). However, Mehdizadeh et al. (2020, 2021) proposed that although the flowability of cement pastes doped with both cRCF and uRCF was inferior to that of pure pastes, the flowability of the cRCF-doped pastes was markedly superior to that of the uRCF-doped pastes (an average increase of 58.7%). This is attributed to the filling effect of $CaCO_3$ on the surface of cRCF and the electrochemical properties of the surface. The fine-size $CaCO_3$ on the cRCF surface not only enhances the water absorption performance of RCF by filling the internal pores of the paste but also releases the free water. The $CaCO_3$ present in RCF is predominantly in the form of calcite, which has a higher absolute value of the surface electric potential. This disrupts the flocculation of the cement paste structure, thereby further promoting the release of free water and consequently improving the rheological properties of the cRCF-doped paste. However, as demonstrated by Mao et al. (2024), the impact of $CaCO_3$ on rheological characteristics was found to be insignificant in comparison to the combined influence of RCF. It is noteworthy that Mehdizadeh et al. (Mehdizadeh et al., 2020, 2021) employed dry mineralization in all of their studies, with a lower degree of mineralization, and did not demonstrate a higher silica gel content, which may account for the inconsistency of their results with those of other studies. In contrast, uRCF also exhibits high specific surface area characteristics. From this, it can be postulated that the hydrophilicity of silica gel may be the dominant mechanism by which RCF causes deterioration of rheological properties. Furthermore, Peng et al. (2023) proposed that the pozzolanic effect of silica gel influences the rheological properties of cement pastes. This is attributed to the fact that the pozzolanic effect accelerates the setting and hardening process of the paste, thereby increasing the frictional resistance between RCF and the paste particles. However, previous studies have indicated that the pozzolanic effect of silica gel occurs at 1–6 hours of hydration, which is later than the time at which rheological properties are typically tested. The effect on the rheology is slight and is therefore usually not considered.

The significant influence of aragonite morphology on the rheological properties of cement pastes has also been the subject of considerable attention. The special fibrous geometrical properties of aragonite exert a profound influence on the surrounding fluid flow lines. In comparison to calcite and vaterite, the incorporation of aragonite has been observed to not only enhance the initial static yield stress, dynamic yield stress, and energy storage modulus of the cement paste, but also accelerate the structural buildup rate and induce a higher degree of thixotropy, characterized by a higher static yield stress and a lower dynamic yield stress, in the cement paste. Furthermore, the surface electrochemical properties of aragonite are positively charged in contrast to calcite and vaterite. This will affect the attraction of aragonite to ions in the pore solution and the mutual charge interaction between aragonite and the phases in the cement system. However, the effect is more complex and subtle, and thus far, has not been studied in detail.

The preceding analysis indicates that the hydrophilic properties of the mineralization products, particularly silica gel, and the polymorphic form of $CaCO_3$, exert a considerable detrimental impact on the rheological properties. Reducing the content and the specific surface area of silica gel, or abandoning the polycrystalline modulation of $CaCO_3$ by controlling the mineralization conditions will have a significant negative impact on the positive contribution to the mechanical properties. This approach is therefore not a sensible one. Currently, the rheological properties of cRCF incorporated cement-based materials are improved through the addition of a high-efficiency water reducing agent, viscosity modifier, or by compounding with the "ball effect" of fly ash and other conventional means. The positive effects of cRCF on mechanical and durability properties are ensured while improving rheological properties.

8.5.2 Hydration

The incorporation of cRCF has been demonstrated to accelerate the early hydration of cementitious materials and to increase the cumulative heat generation. Additionally, it affects the three main hydration peaks (X. Ouyang et al., 2020; Zajac, Skocek, Durdzinski et al., 2020; Peng et al., 2023; Zajac et al., 2023): The first hydration peak, associated with ion dissolution, is enhanced by the incorporation of cRCF. Secondly, the addition of cRCF results in an earlier onset and increased magnitude of the second peak linked to C_3S hydration, which forms C-S-H and CH. Additionally, the inclusion of cRCF advances and enhances the peak representing the hydration reaction of the aluminate phase (Peng et al., 2023). This reaction is associated with the interaction of $CaCO_3$ in cRCF with C_3A (Peng et al., 2023; Zajac et al., 2023).

The advancement or enhancement of the hydration peak, as previously mentioned, can be attributed to the coupling of the pozzolanic effects of alumina-silica gel and the nucleation effect or chemical reaction of $CaCO_3$ in cRCF. Among these factors, the high pozzolanic activity of alumina-silica gel is the primary driver of the notable acceleration in the early hydration of cement systems doped with cRCF. The acceleration effect on hydration is more pronounced when the silica gel content is higher and the degree of polymerization is greater. The notable elevation of the initial peak may be attributed to the pozzolanic effect exerted by alumina-silica gel, which consumes a considerable quantity of CH and accelerates ion dissolution (Peng et al., 2023). The advancement and enhancement of the second peak can be attributed to the nucleation effects of $CaCO_3$ and secondary C-S-H produced by the pozzolanic effect. On the one hand, the secondary C-S-H can serve as an ideal nucleation site for the C-S-H gel formed at the later stage of hydration, which is conducive to the nucleation and growth of C-S-H gel (Zhao, Williams, Park et al., 2023). On the other hand, the adsorption of Ca^{2+} by $CaCO_3$ is driven by strong acid–base interactions (X. Ouyang et al., 2020). In cement systems incorporating cRCF, $CaCO_3$ may facilitate C-S-H nucleation and reinforce the C-S-H phase through a robust chemical interaction, thereby accelerating early cement hydration (X. Ouyang et al., 2020). The nucleation effect is enhanced when the particle size of $CaCO_3$ decreases and the specific surface area increases, thereby increasing the number of nucleation sites. Furthermore, the chemistry of $CaCO_3$ corresponds to the advancement of the third peak, which is associated with the interaction of $CaCO_3$ with C_3A, resulting in the formation of monocarbonates and hemicarbonates.

8.5.3 Microstructure

The microstructure of cementitious materials undergoes alterations during the hydration process of cement, exerting a significant impact on the mechanical and durability properties. In the initial stages of hydration, the cRCF-doped samples exhibit the formation of needle-like C-S-H on the surface of the particles, resulting in the development of a dense structure. In contrast, only a limited quantity of hydration products was observed in the uRCF-doped samples, and the pores were clearly discernible (Mao, Drissi, He et al., 2024). The results of some studies (Chen et al., 2021; Peng et al., 2023) indicated that the pore structure of the cement system doped with cRCF was refined. Specifically, the proportion of capillary pores larger than 10 nm decreased, while the percentage of gel pores smaller than 10 nm increased (Peng et al., 2023). Additionally, the total porosity was reduced by up to 41.30–48.81% (Pan et al., 2023). Another study (X. Ouyang et al., 2020) observed that the interface between the cRCF particles and the matrix was denser and exhibited higher interfacial strength (Figure 8.6). In the uRCF system, cracks were consistently observed to propagate along the interface between uRCF and the matrix. Further hydration did not result in enhanced interfacial strength. In contrast, the interface between cRCF and the matrix exhibited significantly enhanced strength, with crack extension along the interface becoming difficult to observe at the age of 28 days. As the age of the cement paste increased, the cracks gradually expanded within the cRCF.

The microstructural differences between cement systems doped with uRCF and cRCF can be attributed to the combined influence of the pozzolanic effect, nucleation effect, chemical reaction, and filling effect. The decoupling analysis is as follows: (i) The nucleation effect of $CaCO_3$ induces the C-S-H gel to grow closely, uniformly, and vertically on the surface of cRCF, thereby bridging the cRCF and the cement particles and refining the pores. (ii) The pozzolanic effect creates secondary C-S-H gel with a longer average chain length and higher degree of polymerization, which results in excellent micromechanical properties (higher modulus of elasticity and hardness) of the cRCF system (X. Ouyang et al., 2020). (iii) The reaction of $CaCO_3$ with C_3A stabilized the early-formed AFt and inhibited the formation of AFm. Furthermore, the molar volume of AFt ($707 cm^3 / mol$) was markedly larger than that of MS ($309 cm^3 / mol$), which augmented the volume of the solid phase

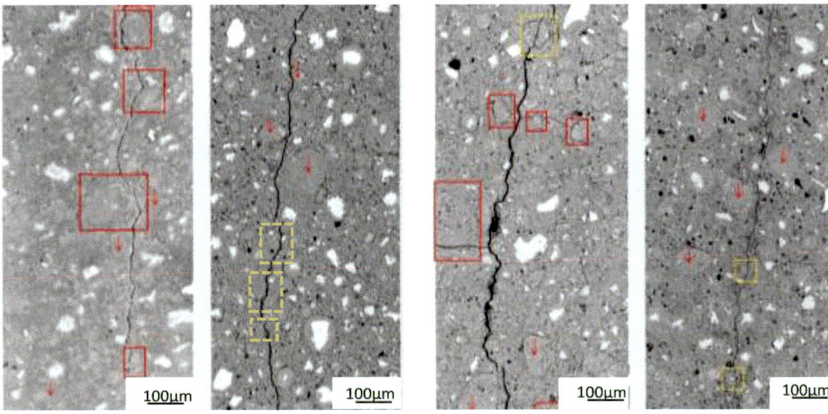

uRCF-28d cRCF-28d uRCF-96d cRCF-96d

Crack propagated along the interface between the matrix and the uRCF or cRCF particles.

Crack passed through the uRCF or cRCF particles.

uRCF or cRCF particles.

FIGURE 8.6 The crack propagation in cement paste after 28 and 96 days of hydration with 30% uRCF or cRCF (X. Ouyang et al., 2020).

of the hydration product and diminished the porosity. (iv) $CaCO_3$ has a nanoscale, which can be incorporated into minute pores in cementitious materials to enhance the matrix packing density and obstruct pore connectivity.

8.5.4 Mechanical property

The impact of cRCF on the compressive strength of cementitious materials is well substantiated in the literature. In comparison to uRCF, cRCF has been demonstrated to exert a considerable influence on the enhancement of compressive strength (Mehdizadeh et al., 2020; Peng et al., 2023). This enhancement can be attributed to the coupling of the four effects previously outlined. Figure 8.7 illustrates the normalized effects of uRCF and cRCF on compressive strength at varying incorporation ranges and carbonation techniques. At lower dosages, both dry and wet mineralization are capable of compensating for the loss of mechanical properties, thereby enabling the compressive strength to be comparable or even higher than that of pure cement samples. At higher dosages (30%, 40%), the negative dilution effect is predominant (Mao, Drissi, He et al., 2024), resulting in a significant reduction in the compressive strength of the cementitious material. At this dosage, the compressive strength is more variable, and the enhancement effect is largely dependent on the degree of mineralization.

The mechanism of the coupling of four effects has been described in detail in the previous section. This section focuses on the identification of the dominant effect contributing to the compressive strength and the quantification of the contribution of each effect at different stages. Lu et al. (Lu et al., 2018) asserted that

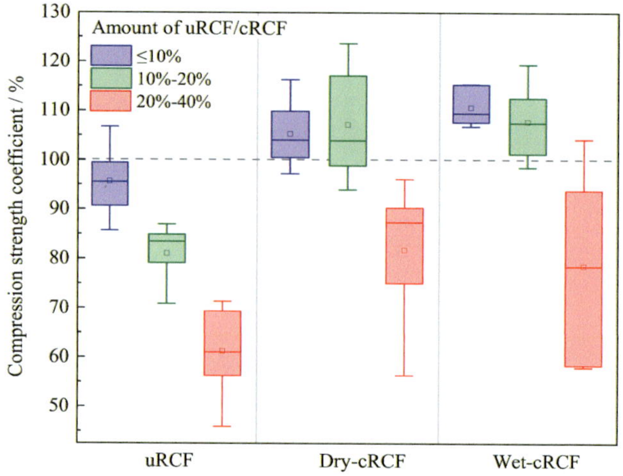

FIGURE 8.7 The influence of uRCF and cRCF on compressive strength of cement paste (Lu et al., 2018; Mehdizadeh et al., 2020; Peng et al., 2023; Mao, Drissi, He et al., 2024).

the enhancement in compressive strength resulting from the incorporation of cRCF was predominantly attributable to a synergistic interaction between the nucleation effect, chemical reaction, and filling effect of $CaCO_3$. In contrast, Zajac et al. (2020) proposed that the pronounced pozzolanic effect of silica gel was the primary factor responsible for the accelerated development of early strength. Ouyang et al. (2020) posited that the nucleation effect is the primary reason for the significantly higher compressive strength of cementitious materials doped with cRCF compared to uRCF. Mao et al. (2024) decoupled and quantified the contribution values of the four effects to the compressive strength at different hydration stages by porosity. Their findings suggest that the filling effect is the most contributing factor to the increase in compressive strength at any hydration stage. Furthermore, it is noted that the lower silica content may restrict the contribution of the pozzolanic effect. Additionally, it can be observed that the nucleation effect of dry mineralization at one day of hydration is less pronounced than that of wet mineralization. This outcome is reflected in the discrepancy in the distribution of the relative positions of the mineralization products after dry and wet mineralization, as illustrated in Section 8.4.1. Furthermore, the contribution of nucleation effects is exclusive to the initial stages of hydration (Wang et al., 2019; Bai et al., 2022) and has a negligible impact on the subsequent stages. Conversely, the contribution of chemical reactions increases significantly, reaching 10.9% at 28 days of hydration (Mao, Drissi, He et al., 2024).

Notably, there is no consensus regarding the optimal incorporation value of cRCF for enhancing compressive strength. This is because the incorporation level is affected by the degree of mineralization of cRCF and the morphology of the mineralization products, which cannot be generalized. The substitution rate is constrained when the objective is to ensure compressive strength. To increase the substitution rate of cRCF, it is necessary to enhance the degree of mineralization to counteract the negative effects of dilution by promoting chemical reactions, pozzolanic effects, and nucleation effects. However, there is a contradiction between the energy consumption required to enhance the degree of mineralization and the reduction of energy consumption by replacing cement with large dosages. Therefore, a systematic and comprehensive link between the degree of mineralization, cRCF dosage, and energy loss and mechanical properties should be established. Furthermore, a balance between economic and environmental benefits should be considered, which will help to promote high-quality resource utilization of cRCF.

8.5.5 Water permeability

Moisture penetration plays a significant role in the deterioration of concrete. Section 8.5.3 illustrates that the incorporation of cRCF densifies the pore structure of cementitious materials, which in turn reduces the occurrence of cracks. This effectively blocks the penetration of moisture and erosive substances, thus enhancing the durability of concrete structures. Figure 8.8 provides the water absorption curve and penetration depth of uRCF (or uRCP in the figure) incorporated cement mortar, cement mortar with different content of cRCP (30% and 33%) and different substrate strength (72.6 Mpa, 45.9 Mpa, 61.4 Mpa, 69.5 Mpa) and ordinary mortar (ref) in referenced papers (Wu, Liang et al., 2021; Pan et al., 2023). In comparison to uRCF, the water absorption of mortar containing cRCF was markedly diminished, and the depth of water intrusion also decreased to a level comparable to that of ordinary mortar. The maximum water absorption of cRCF mortar samples was reduced by approximately 50% (Pan et al., 2023). The influence of incorporating cRCF into mortar with varying base material strengths on the water absorption improvement effect is not significant. This suggests that the quality of the RCF base material can be disregarded and treated similarly in practical projects, which considerably simplifies the recycling of RCF and enhances its utilization efficiency.

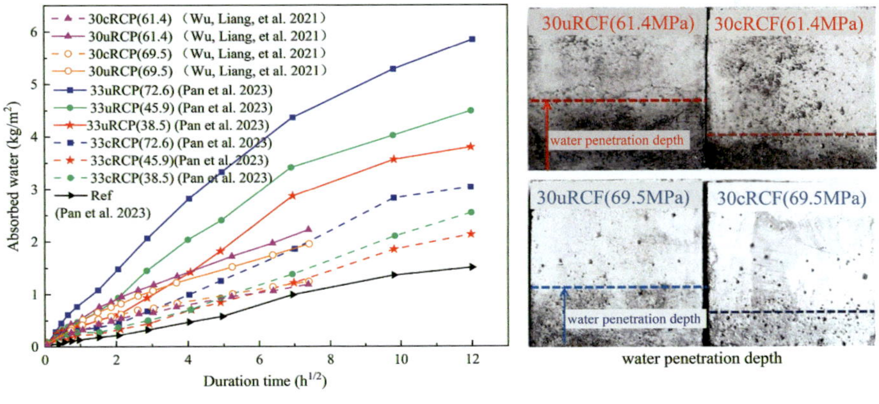

FIGURE 8.8 Water absorption curve and depth of water penetration of mortar with cRCF and uRCF (Wu, Liang et al., 2021; Pan et al., 2023).

8.6 SUMMARY

RCF possesses the characteristics of small particle size, high specific surface area, and high CaO content. CO_2 mineralization treatment can not only stimulate its reaction activity but also sequester a large amount of CO_2. Characteristics of cRCF and its effect on the properties of cementitious materials as a supplementary cementitious material have been reviewed in this chapter; the following conclusions can be drawn:

(1) $CaCO_3$ and amorphous silica are the main mineralization products of RCF. The content, morphological characteristics, chemical properties, and relative distribution of the mineralization products are related to the mineralization method and conditions. Optimization of the mineralization conditions can improve the degree of mineralization of RCF and regulate the morphology of $CaCO_3$ or the degree of polymerization of silica gel.

(2) When cRCF replaces cement, $CaCO_3$ has a nucleation effect, filling effect, and can react chemically with C_3A; silica gel has a high pozzolanic effect, which is the main reason that cRCF can be used as an efficient SCM. When the degree of mineralization is high, the coupled positive effects of the above four effects exceed the dilution effect; the incorporation of cRCF can accelerate the hydration of cement, optimize the microstructure, improve the compressive strength, and achieve a large dosage.

(3) The five surface physicochemical properties of cRCF: high specific surface area, hydrophilicity of silica gel, geometrical properties, filling effect of $CaCO_3$, and surface electrochemical properties are the inherent mechanisms influencing the rheological properties of the paste. Due to the fact that the filling effect of $CaCO_3$ and the electrochemical properties of the surface have a less significant effect than the hydrophilicity of silica gel, the high specific surface area and the geometrical properties of $CaCO_3$ on the rheological properties of freshly pastes, eventually, the rheological properties deteriorate significantly. In this case, the hydrophilicity of silica gel dominated the deterioration mechanism.

(4) The incorporation of cRCF enhanced the impermeability of the mortar and inhibited ionic migration.

(5) There is still a lack of research on the durability of cRCF-incorporated cement systems, especially on the corrosion resistance of reinforcing steel, which does not address the relationship between the amount of cRCF, the degree of mineralization, and the corrosion of reinforcing steel. Decoupling and quantifying the contribution of each effect to reinforcement corrosion will help to improve the knowledge of reinforcement corrosion in cRCF cement systems and lay the foundation for the industrial use of cRCF.

REFERENCES

Akhtar, A., Sarmah, A.K., 2018. Construction and demolition waste generation and properties of recycled aggregate concrete: a global perspective. *Journal of Cleaner Production* 186, 262–281.

Algourdin, N., Larbi, K.S., Santos, I., Mesticou, Z., Pimienta, P., Pinoteau, N., Larbi, A.S., 2023. Experimental and numerical investigation of accelerated carbonation of recycled fines. *Construction and Building Materials* 382, 131286.

Bai, S., Guan, X., Li, G., 2022. Early-age hydration heat evolution and kinetics of Portland cement containing nano-silica at different temperatures. *Construction and Building Materials* 334, 127363.

Cao, M., Zhang, C., Lv, H., Xu, L., 2014. Characterization of mechanical behavior and mechanism of calcium carbonate whisker-reinforced cement mortar. *Construction and Building Materials* 66, 89–97.

Castellote, M., Andrade, C., 2008. Modelling the carbonation of cementitious matrixes by means of the unreacted-core model, UR-CORE. *Cement and Concrete Research* 38(12), 1374–1384.

Chen, P., Wang, J., Wang, L., Xu, Y., Qian, X., Ma, H., 2017. Producing vaterite by CO2 sequestration in the waste solution of chemical treatment of recycled concrete aggregates. *Journal of Cleaner Production* 149, 735–742.

Chen, X., Li, Y., Bai, H., Ma, L., 2021. Utilization of recycled concrete powder in cement composite: Strength, microstructure and hydration characteristics. *Journal of Renewable Materials* 9(12), 2189–2208.

Drouet, E., Poyet, S., Bescop, P.L., 2019. Carbonation of hardened cement pastes: Influence of temperature. *Cement and Concrete Research* 115, 445–459.

Drouet, E., Poyet, S., Torrenti, J.-M., 2015. Temperature influence on water transport in hardened cement pastes. *Cement and Concrete Research* 76, 37–50.

Galan, I., Andrade, C., Castellote, M., 2013. Natural and accelerated CO 2 binding kinetics in cement paste at different relative humidities. *Cement and Concrete Research* 49, 21–28.

Galan, I., Glasser, F.P., Baza, D., Andrade, C., 2015. Assessment of the protective effect of carbonation on portlandite crystals. *Cement and Concrete Research* 74, 68–77.

Gao, Y., Cui, X., Lu, N., Hou, S., He, Z., Liang, C., 2022. Effect of recycled powders on the mechanical properties and durability of fully recycled fiber-reinforced mortar. *Journal of Building Engineering* 45, 103574.

He, Z., Han, X., Zhang, M., Yuan, Q., Shi, J., Zhan, P., 2022. A novel development of green UHPC containing waste concrete powder derived from construction and demolition waste. *Powder Technology* 398, 117075.

He, Z., Wang, S., Mahoutian, M., Shao, Y., 2020. Flue gas carbonation of cement-based building products. *Journal of CO2 Utilization* 37, 309–319.

Herterich, J., Richardson, I., Moro, F., Marchi, M., Black, L., 2022. Microstructure and phase assemblage of low-clinker cements during the early stages of carbonation. *Cement and Concrete Research* 152, 106643.

Horsakulthai, V., 2021. Effect of recycled concrete powder on strength, electrical resistivity, and water absorption of self-compacting mortars. *Case Studies in Construction Materials* 15, e00725.

Jiang, Y., Peng, L., Ma, Z., Lu, J., Shen, P., Poon, C.S., 2023. Enhancing the treatment efficiency of recycled concrete fines with aqueous carbonation. *Cement and Concrete Research* 174, 107338.

Kaliyavaradhan, S.K., Li, L., Ling, T.-C., 2022. Response surface methodology for the optimization of CO2 uptake using waste concrete powder. *Construction and Building Materials* 340, 127758.

Kwon, E., Ahn, J., Cho, B., Park, D., 2015. A study on development of recycled cement made from waste cementitious powder. *Construction and Building Materials* 83, 174–180.

Lee, D.K., 2004. An apparent kinetic model for the carbonation of calcium oxide by carbon dioxide. *Chemical Engineering Journal* 100(1), 71–77.

Li, L., Cao, M., Yin, H., 2019. Comparative roles between aragonite and calcite calcium carbonate whiskers in the hydration and strength of cement paste. *Cement and Concrete Composites* 104, 103350.

Li, L., Liu, Q., Huang, T., Peng, W., 2022. Mineralization and utilization of CO2 in construction and demolition wastes recycling for building materials: A systematic review of recycled concrete aggregate and recycled hardened cement powder. *Separation and Purification Technology* 298, 121512.

Li, X., Lv, X., Zhou, X., Meng, W., Bao, Y., 2022. Upcycling of waste concrete in eco-friendly strain-hardening cementitious composites: Mixture design, structural performance, and life-cycle assessment. *Journal of Cleaner Production* 330, 129911.

Liang, C., Zhang, Y., Wu, R., Yang, D., Ma, Z., 2021. The utilization of active recycled powder from various construction wastes in preparing ductile fiber-reinforced cementitious composites: A case study. *Case Studies in Construction Materials* 15, e00650.

Likes, L., Markandeya, A., Haider, M.M., Bollinger, D., McCloy, J.S., Nassiri, S., 2022. Recycled concrete and brick powders as supplements to Portland cement for more sustainable concrete. *Journal of Cleaner Production* 364, 132651.

Liu, X., Liu, L., Lyu, K., Li, T., Zhao, P., Liu, R., Zuo, J., Fu, F., Shah, S.P., 2022. Enhanced early hydration and mechanical properties of cement-based materials with recycled concrete powder modified by nano-silica. *Journal of Building Engineering* 50, 104175.

López-Arce, P., Gómez-Villalba, L.S., Martínez-Ramírez, S., Álvarez de Buergo, M., Fort, R., 2011. Influence of relative humidity on the carbonation of calcium hydroxide nanoparticles and the formation of calcium carbonate polymorphs. *Powder Technology* 205(1), 263–269.

Lu, B., Drissi, S., Liu, J., Hu, X., 2022. Effect of temperature on CO2 curing, compressive strength and microstructure of cement paste. *Cement and Concrete Research* 157, 106827.

Lu, B., Huo, Z., Xu, Q., Hou, G., Wang, X., Liu, J., Hu, X., 2023. Characteristics of CSH under carbonation and its effects on the hydration and microstructure of cement paste. *Construction and Building Materials* 364, 129952.

Lu, B., Shi, C., Zhang, J., Wang, J., 2018. Effects of carbonated hardened cement paste powder on hydration and microstructure of Portland cement. *Construction and Building Materials* 186, 699–708.

Mao, Y., Drissi, S., He, P., Hu, X., Zhang, J., Shi, C., 2024. Quantifying the effects of wet carbonated recycled cement paste powder on the properties of cement paste. *Cement and Concrete Research* 175, 107381.

Mao, Y., Drissi, S., Hu, X., Zhang, J., Shi, C., 2024. Effect of wet carbonated recycled cement paste powder on the rheology of cement paste. *Cement and Concrete Research* 181, 107553.

Mao, Y., He, P., Drissi, S., Zhang, J., Hu, X., Shi, C., 2023. Effect of conditions on wet carbonation products of recycled cement paste powder. *Cement and Concrete Composites* 144, 105307.

McDonald, L.J., Carballo-Meilan, M.A., Chacartegui, R., Afzal, W., 2022. The physicochemical properties of Portland cement blended with calcium carbonate with different morphologies as a supplementary cementitious material. *Journal of Cleaner Production* 338, 130309.

Mehdizadeh, H., Ling, T.-C., Cheng, X., Mo, K.H., 2020. Effect of particle size and CO2 treatment of waste cement powder on properties of cement paste. *Canadian Journal of Civil Engineering* 48(5), 522–531.

Mehdizadeh, H., Ling, T.-C., Cheng, X., Pan, S.-Y., Hung Mo, K., 2021. CO2 Treatment of hydrated cement powder: Characterization and application consideration. *Journal of Materials in Civil Engineering* 33(4), 04021041.

Montes-Hernandez, G., Renard, F., Geoffroy, N., Charlet, L., Pironon, J., 2007. Calcite precipitation from CO2–H2O–Ca(OH)2 slurry under high pressure of CO2. *Journal of Crystal Growth* 308(1), 228–236.

Morandeau, A., Thiéry, M., Dangla, P., 2014. Investigation of the carbonation mechanism of CH and C-S-H in terms of kinetics, microstructure changes and moisture properties. *Cement and Concrete Research* 56, 153–170.

Ouyang, K., Shi, C., Chu, H., Guo, H., Song, B., Ding, Y., Guan, X., Zhu, J., Zhang, H., Wang, Y., Zheng, J., 2020. An overview on the efficiency of different pretreatment techniques for recycled concrete aggregate. *Journal of Cleaner Production* 263, 121264.

Ouyang, X., Wang, L., Xu, S., Ma, Y., Ye, G., 2020. Surface characterization of carbonated recycled concrete fines and its effect on the rheology, hydration and strength development of cement paste. *Cement and Concrete Composites* 114, 103809.

Pan, B., Mao, Y., Hou, S., Liang, C., Gou, Y., 2023. *Corrosion mechanism of recycled mortar prepared from CO2-treated hardened cement paste powder. Construction and Building Materials* 384, 22.

Pan, Y., Zhao, X., Sheng, Y., Wang, C., Deng, Y., Ma, X., Liu, Y., Wang, Z., 2007. Biomimetic synthesis of dendrite-shaped aragonite particles with single-crystal feature by polyacrylic acid. *Colloids and Surfaces, A: Physicochemical and Engineering Aspects* 297(1), 198–202.

Peng, L., Jiang, Y., Ban, J., Shen, Y., Ma, Z., Zhao, Y., Shen, P., Poon, C.-S., 2023. Mechanism underlying early hydration kinetics of carbonated recycled concrete fines-ordinary portland cement (CRCF-OPC) paste. *Cement and Concrete Composites* 144, 105275.

Qian, X., Xu, W., Wang, Y., Fang, H., Jing, Z., Chen, P., 2023. Revisiting the carbonation of recycled concrete fine: A pH-cycle carbonation method. *Journal of Building Engineering* 77, 107438.

Shen, P., Jiang, Y., Zhang, Y., Liu, S., Xuan, D., Lu, J., Zhang, S., Poon, C.S., 2023. Production of aragonite whiskers by carbonation of fine recycled concrete wastes: An alternative pathway for efficient CO2 sequestration. *Renewable and Sustainable Energy Reviews* 173, 113079.

Shen, P., Lu, J., Zhang, Y., Jiang, Y., Zhang, S., Poon, C.S., 2022. Preparation aragonite whisker-rich materials by wet carbonation of cement: Towards yielding micro-fiber reinforced cement and sequestrating CO2. *Cement and Concrete Research* 159, 106891.

Shen, P., Zhang, Y., Jiang, Y., Zhan, B., Lu, J., Zhang, S., Xuan, D., Poon, C.S., 2022. Phase assemblance evolution during wet carbonation of recycled concrete fines. *Cement and Concrete Research* 154, 106733.

Skibsted, J., Snellings, R., 2019. Reactivity of supplementary cementitious materials (SCMs) in cement blends. *Cement and Concrete Research* 124, 105799.

Sun, C., Chen, L., Xiao, J., Singh, A., Zeng, J., 2021. Compound utilization of construction and industrial waste as cementitious recycled powder in mortar. *Resources, Conservation and Recycling* 170, 105561.

Sun, C., Chen, L., Xiao, J., Zuo, J., Wu, H., 2022. Effects of eco powders from solid waste on freeze-thaw resistance of mortar. *Construction and Building Materials* 333, 127405.

Steinour, H, H,1959. Some effects of carbon dioxide on mortars and concrete-discussion. *Journal of the American Concrete Institute* 30(2), 905–907.

Thiery, M., Villain, G., Dangla, P., Platret, G., 2007. Investigation of the carbonation front shape on cementitious materials: Effects of the chemical kinetics. *Cement and Concrete Research* 37(7), 1047–1058.

Thonemann, N., Zacharopoulos, L., Fromme, F., Nühlen, J., 2022. Environmental impacts of carbon capture and utilization by mineral carbonation: A systematic literature review and meta life cycle assessment. *Journal of Cleaner Production* 332, 130067.

Tine, F.S., Sevelsted, T.F., Skibsted, J., 2015. Carbonation of C–S–H and C–A–S–H samples studied by ^{13}C, ^{27}Al and ^{29}Si MAS NMR spectroscopy. *Cement and Concrete Research* 71(71), 56–65.

Vansant, E.F., Van Der Voort, P., Vrancken, K.C. (eds.), 1995. Chapter 5 the distribution of the silanol types and their desorption energies, in *Studies in surface science and catalysis, characterization and chemical modification of the silica surface*. Elsevier, vol. 93, pp. 93–126.

Wang, D., Shi, C., Farzadnia, N., Jia, H., Zeng, R., Wu, Y., Lao, L., 2019. A quantitative study on physical and chemical effects of limestone powder on properties of cement pastes. *Construction and Building Materials* 204, 58–69.

Wang, T., Huang, H., Hu, X., Fang, M., Luo, Z., Guo, R., 2017. Accelerated mineral carbonation curing of cement paste for CO2 sequestration and enhanced properties of blended calcium silicate. *Chemical Engineering Journal* 323, 320–329.

Wu, H., Liang, C., Xiao, J., Ma, Z., 2021. Properties and CO2-curing enhancement of cement-based materials containing various sources of waste hardened cement paste powder. *Journal of Building Engineering* 44, 102677.

Wu, H., Xu, J., Yang, D., Ma, Z., 2021. Utilizing thermal activation treatment to improve the properties of waste cementitious powder and its newmade cementitious materials. *Journal of Cleaner Production* 322, 129074.

Wu, H., Yang, D., Xu, J., Liang, C., Ma, Z., 2021. Water transport and resistance improvement for the cementitious composites with eco-friendly powder from various concrete wastes. *Construction and Building Materials* 290, 123247.

Wu, Y., Mehdizadeh, H., Mo, K.H., Ling, T.-C., 2022. High-temperature CO2 for accelerating the carbonation of recycled concrete fines. *Journal of Building Engineering* 52, 104526.

You, X., Hu, X., Xiao, Z., Saleh Bairq, Z.A., Chen, W., Shi, C., 2023. Thermodynamic modelling of CaCO3 polymorphs during CO2 sequestration by cement slurry with the addition of MgCl2. *Journal of Cleaner Production* 410, 137294.

Zajac, M., Skibsted, J., Ben Haha, M., 2021. Effect of alkalis on enforced carbonation of cement paste: Mechanism of reaction. *Journal of the American Ceramic Society* 104(2), 1076–1087.

Zajac, M., Skibsted, J., Bullerjahn, F., Skocek, J., 2022. Semi-dry carbonation of recycled concrete paste. *Journal of CO2 Utilization* 63, 102111.

Zajac, M., Skibsted, J., Durdzinski, P., Bullerjahn, F., Skocek, J., Ben Haha, M., 2020a. Kinetics of enforced carbonation of cement paste. *Cement and Concrete Research* 131, 106013.

Zajac, M., Skocek, J., Durdzinski, P., Bullerjahn, F., Skibsted, J., Ben Haha, M., 2020b. Effect of carbonated cement paste on composite cement hydration and performance. *Cement and Concrete Research* 134, 106090.

Zajac, M., Skibsted, J., Lothenbach, B., Bullerjahn, F., Skocek, J., Ben Haha, M., 2022. Effect of sulfate on CO2 binding efficiency of recycled alkaline materials. *Cement and Concrete Research* 157, 106804.

Zajac, M., Skibsted, J., Skocek, J., Durdzinski, P., Bullerjahn, F., Ben Haha, M., 2020. Phase assemblage and microstructure of cement paste subjected to enforced, wet carbonation. *Cement and Concrete Research* 130, 105990.

Zajac, M., Skibsted, J., Durdzinski, P., Haha, M.B., 2021. Effect of alkalis on products of enforced carbonation of cement paste. *Construction and Building Materials* 291, 123203.

Zajac, M., Song, J., Skocek, J., Mahsen, B.H., 2023. Composite cements with aqueous and semi-dry carbonated recycled concrete pastes. *Construction and Building Materials* 407, 133362.

Zajac, M., Song, J., Ullrich, P., Skocek, J., Ben Haha, M., Skibsted, J., 2024. High early pozzolanic reactivity of alumina-silica gel: A study of the hydration of composite cements with carbonated recycled concrete paste. *Cement and Concrete Research* 175, 107345.

Zhan, B.J., Xuan, D.X., Poon, C.S., Shi, C.J., 2016. Effect of curing parameters on CO2 curing of concrete blocks containing recycled aggregates. *Cemen., Concrete Composites* 71, 122–130.

Zhang, H., Xiao, J., Tang, Y., Poon, C.-S., 2022. Long-term shrinkage and mechanical properties of fully recycled aggregate concrete: Testing and modelling. *Cement and Concrete Composites* 130, 104527.

Zhang, J., Shi, C., Li, Y., Pan, X., Poon, C.-S., Xie, Z., 2015. Influence of carbonated recycled concrete aggregate on properties of cement mortar. *Construction and Building Materials* 98, 1–7.

Zhang, Q., Feng, P., Shen, X., Lu, J., Ye, S., Wang, H., Ling, T., Ran, Q., 2023. Utilization of solid wastes to sequestrate carbon dioxide in cement-based materials and methods to improve carbonation degree: A review. *Journal of CO2 Utilization* 72, 102502.

Zhao, D., Williams, J.M., Li, Z., Park, A.-H.A., Radlińska, A., Hou, P., Kawashima, S., 2023. Hydration of cement pastes with calcium carbonate polymorphs. *Cement and Concrete Research* 173, 107270.

Zhao, D., Williams, J.M., Park, A.-H.A., Kawashima, S., 2023. Rheology of cement pastes with calcium carbonate polymorphs. *Cement and Concrete Research* 172, 107214.

9 CO_2 mineralization of steel slag

9.1 INTRODUCTION

Steel slag is an industrial by-product generated during the steelmaking process, accounting for approximately 15–20% of crude steel production by mass. In 2024, the world produced approximately 1.839 billion tons of crude steel and 300 million tons of steel slag (Worldsteel, 2025). This substantial output underscores the importance of addressing the efficient utilization and management of steel slag. Due to its excellent mechanical properties, including high strength and superior wear resistance, steel slag has been widely employed as an aggregate in road infrastructure and highway construction (Shen et al., 2024; Zhao et al., 2024).

Compared with the extensively utilized blast furnace slag, steel slag remains underutilized due to its complex and variable chemical and mineralogical composition (Li et al., 2024; Shi, 2002). The primary mineral phases in steel slag bear similarities to those found in cementitious materials, including tricalcium silicate (C_3S), dicalcium silicate (C_2S), tetra-calcium aluminoferrite (C_4AF), and smaller amounts of dicalcium ferrite (C_2F) (Fan et al., 2024; Fu et al., 2024). Additionally, steel slag contains a unique phase known as the RO phase, a solid solution composed of magnesium, iron, and manganese oxides (Li et al., 2023b; Liu et al., 2024). While these components suggest potential hydraulic activity, the low content of C_3S and amorphous SiO_2, combined with a high proportion of the RO phase, results in poor hydraulic properties (Kriskova et al., 2012; Li et al., 2023a). This limits its reactivity compared to blast furnace slag, which is widely recognized for its effectiveness in cement production. Another critical issue associated with steel slag is its high content of free lime (f-CaO) and free magnesia (f-MgO), which typically ranges from 5% to 10% and causes volume instability, leading to undesirable expansion and cracking during practical applications (Wang et al., 2017, 2024). This problem is particularly significant in construction materials, where long-term durability and dimensional stability are critical considerations (He et al., 2024).

Steel slag exhibits a remarkable CO_2 mineralization reactivity, largely due to its abundant chemical components, including γ-C_2S (gamma dicalcium silicate), free calcium oxide (f-CaO), and free magnesia (f-MgO). These compounds play a crucial role in facilitating mineralization reactions, which makes accelerated mineralization an effective pretreatment method for steel slag. This process serves multiple purposes: it stabilizes the material by eliminating free oxides, sequesters CO_2, and produces high-value-added byproducts that can be used in a variety of industrial applications (Ghouleh et al., 2017; Shuo Yang et al., 2024).

DOI: 10.1201/9781003627531-9

This chapter presents an in-depth review of steel slag, encompassing its classification, chemical composition, thermodynamic analysis, and methods for characterizing the degree of mineralization. It also explores the various methods of mineralization, their influencing factors, and the application of mineralized steel slag in different industries. By addressing these aspects, this review aims to provide a clearer understanding of steel slag and its future utilization, particularly in the context of sustainability and CO_2 reduction technologies.

9.2 CHARACTERISTICS OF STEEL SLAG

Steel slag is the main co-product during the steel manufacturing process, which adds lime, dolomite, and other supplementary materials into the blast furnace to remove carbon, phosphorus, sulfur, and other elements in pig iron (Mo et al., 2020). It can be divided into three categories according to the main production processes of steel. The basic-oxygen furnace (BOF) is employed during the primary steel refinement process, which uses molten iron and steel scraps with lime or dolomite as the main raw materials (Yildirim and Prezzi, 2011). The electric-arc furnace (EAF) has a similar steelmaking process to BOF, except it uses high-power electric arcs to produce high-quality steel mainly from recycled steel scrap. The EAF process requires much less capital than the BOF process and involves using a combination of steel scrap that will be purified in the EAF process (Shi, 2004; Tanaka, 2014). The ladle furnace (LF) process is the final operation of steel refinement, which is generally called the secondary metallurgy process. After the primary steel refinement process in the BOF or EAF, both carbon and stainless steel are poured into an LF and are deoxidized, desulfurized, and suitably alloyed under the protection of a basic slag. The chemical composition and properties of LF slags are quite different from those of BOF and EAF slags because of the use of various fluxes (Setién et al., 2009).

As for the physical properties, steel slag is hard and wear-resistant due to its high Fe content. The grindability index is 0.7, in contrast with the value of 0.96 and 1.0 for blast furnace slag and standard sand, respectively (Yi et al., 2012). The results of morphological tests indicate that steel slags have a rough surface and lots of pores (Ye et al., 2019). Moreover, these honeycomb-structured pores distribute on the internal and external surface of steel slag grains. Further observation exhibits that there are numerous small white spots contained in steel slags which are alkaline hydration products or alkaline oxides (Yang et al., 2020). The density of steel slag is higher than natural aggregate because of the high content of dense oxides. The apparent specific gravity value of steel slag is between 2.73 g/cm^3 and 3.82 g/cm^3, and the bulk density of steel slag is between 1.95 g/cm^3 and 3.37 g/cm^3 (Martins et al., 2021).

The chemical composition of steel slags depends on different raw materials, production processes, and furnace conditions (Dhoble and Ahmed, 2018). The main chemical compositions are iron oxides (FeO and Fe_2O_3), silica (SiO_2), and alumina (Al_2O_3), which derive from the iron ore or scrap metal, and calcium oxide (CaO) and magnesium oxide (MgO), which derive from the lime or dolomite (Yildirim and Prezzi, 2011). Most steel slags are alkaline due to the releasing of calcium and magnesium oxides. These alkaline oxides combine with water to produce calcium hydroxide and magnesium hydroxide solutions so that the pH value is high (Mayes et al., 2018).

The mineral phases contained in steel slag are dependent on its chemical composition and cooling system, as shown in Table 9.1. Common mineral phases include rhodonite ($3CaO{\cdot}MgO{\cdot}2SiO_2$), olivine ($2MgO{\cdot}2FeO{\cdot}SiO_2$), β-dicalcium silicate (β-C_2S), γ-dicalcium silicate (γ-C_2S), tetracalcium aluminoferrite ($4CaO{\cdot}Al_2O_3{\cdot}FeO_3$), dicalcium ferrate ($2CaO{\cdot}Fe_2O_3$), free calcium oxide (f-CaO), periclase (MgO), hematite (Fe_2O_3), magnetite (Fe_3O_4), a solid solution of CaO-FeO-MnO-MgO which is termed as RO phase, and amorphous phases (Roslan et al., 2016). Mineralogy is dominated by the cooling rates. Silicate melts have high viscosity because of their long molecule chains; as a result, it takes a lot of time to rearrange into crystals. Therefore, a rapid cooling rate is beneficial to the formation of glass phases, and the minerals found in slags have complex crystalline and amorphous structures containing silicates and non-silicates (Tossavainen et al., 2007). This is a way to improve the hydraulic property of steel slag by increasing the content of amorphous phases.

9.3 MINERALIZATION THERMODYNAMICS OF STEEL SLAG

The main mineral phases contained in steel slag such as silicates can react with CO_2 and form carbonates that have the potential to be applied as building materials (Salman et al., 2014). The mineralization reaction of silicates can be expressed as follows:

$$(Mg, Ca)\,Si_y\,O_{x+2y} + xCO_2 \rightarrow x(Mg, Ca)CO_3 + ySiO_2 \tag{9.1}$$

Silicate minerals can autonomously sequester CO_2 under natural conditions; however, the reaction rate is quite slow, and the process is relatively complicated (X. Zhang et al., 2021). The main mineralization component is calcium silicate because the content of calcium is much greater than magnesium. The mineralization reaction converts calcium silicate into $CaCO_3$ and near-amorphous C-S-H. The main crystalline phase of calcium carbonate after mineralization is calcite, and γ-C_2S is the major contributor to the formation of calcite (Ghouleh et al., 2015). In the above reaction process, the leaching behavior of Ca^{2+} from the steel slag and the subsequent carbonate precipitation process are essential. The diffusion rate of Ca^{2+} through the solid matrix to the surface is the determining factor of the reaction rate (Huijgen, 2005). According to the above descriptions of the composition of steel slag, this section summarizes the thermodynamic parameters of mineral phases, which may participate in mineralization reactions. The main mineralization components are C_3S, C_2S, CaO, $Ca(OH)_2$, $Ca_3Mg(SiO_4)_2$, Mg_2SiO_4, MgO, $Mg(OH)_2$, and FeO. Hence, the main mineralization equations are presented in Table 9.1. It can be found that the Gibbs free energy changes of all reactions are negative, indicating that the mineralization

TABLE 9.1
Content, mineralization equations, and thermodynamics parameters of different minerals in steel slag

Mineral	Formula	Content			Mineralization	ΔH (kg·mol^{-1})	ΔG_T (kg·mol^{-1})
		BOFs	EAFs	LFs			
Alite (Chang, 2010)	Ca_3SiO_5	trace	trace	minor	$Ca_3SiO_5 + 3CO_2(g) = 3CaCO_3 + SiO_2$	–148	–234.7
Larnite (Chang, 2010)	Ca_2SiO_4	minor	minor	major	$Ca_2SiO_4 + 2CO_2(g) = 2CaCO_3 + SiO_2$	–81	–135.4
Lime (Yildirim and Prezzi, 2011)	CaO	minor	minor	trace	$CaO + CO_2(g) = CaCO_3$	–99	–57.16
Portlandite (Chang, 2010)	$Ca(OH)_2$	major	major	trace	$Ca(OH)_2 + CO_2(g) = CaCO_3 + H_2O$	–68	–73.3
Srebrodolskite (Chang, 2010)	$Ca_2Fe_2O_5$	major	major	trace	$Ca_2Fe_2O_5 + 3CO_2(g) = 3CaCO_3 + Al_2O_3 + 3SiO_2$	–109	–74.7
Forsterite (Chang, 2010)	Mg_2SiO_4	trace	trace	trace	$Mg_2SiO_4 + 2CO_2(g) = 2MgCO_3 + SiO_2$	–139	–351.38
Periclase (Yildirim and Prezzi, 2011)	MgO	trace	minor	trace	$MgO + CO_2(g) = MgCO_3$	–78	–49.4
Brucite (Yildirim and Prezzi, 2011)	$Mg(OH)_2$	trace	trace	trace	$Mg(OH)_2 + CO_2(g) = MgCO_3 + H_2O$	–37	–38.06

Note: ΔH—Reaction enthalpy change; ΔG—Gibbs free energy change. Trace, minor, and major represent the content of minerals.
Source: (Chang, 2010; Martins et al., 2021; Yildirim and Prezzi, 2011).

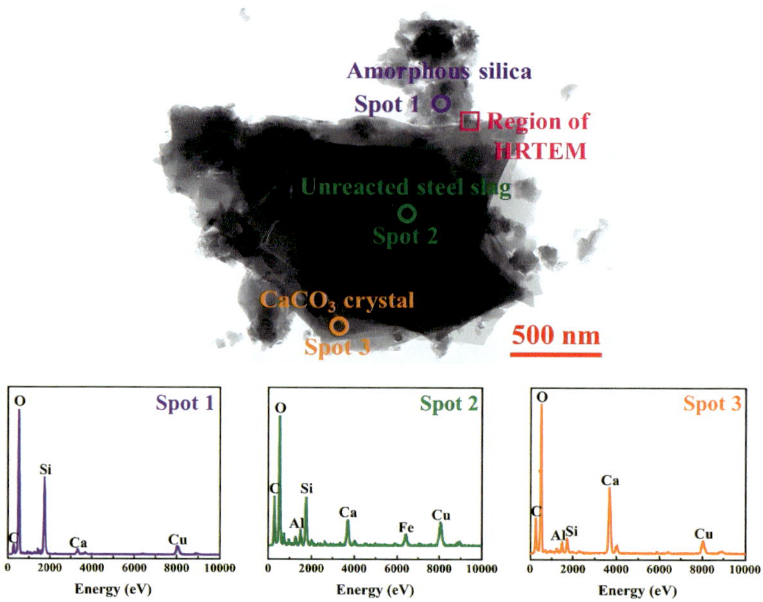

FIGURE 9.1 TEM-EDS image of mineralized steel slag (Zhang et al., 2025).

of steel slag can spontaneously proceed under normal conditions without regard to the kinetic behavior of mineralization (Chang, 2010; Martins et al., 2021; Yildirim and Prezzi, 2011).

TEM-EDS analyses were performed on steel slag after 1 day (l/s = 0.15) of mineralization to examine mineralization products, particularly the amorphous phase (Figure 9.1). The results revealed three distinct morphologies: cloud-like structures (Spot 1), black clumps (Spot 2), and crystalline phases (Spot 3). EDS showed the cloud-like structures had a Si:O atomic ratio of 1:2, indicating amorphous silica gel, as no SiO_2 diffraction peaks were observed in XRD. The black clumps contained Ca, Si, Fe, Al, and O, suggesting they are unreacted steel slag. The crystalline phases were identified as calcite ($CaCO_3$) based on XRD and electron diffraction. Fourier transform of high-resolution images further confirmed amorphous SiO_2 in Region A and calcite in Region B. Calcite contributes to pore filling and improved micromechanical properties, while silica gel enhances bonding strength despite its lower mechanical performance (Zhang et al., 2025).

In order to further understand the reaction kinetics of steel slag at different temperatures, HSC Chemistry was applied to obtain the thermodynamic parameters such as enthalpy changes (ΔH), Gibbs free energy changes (ΔG), and chemical equilibrium constants ($\log K$) (Jeon et al., 2012; Jose-Luis et al., 2019). Figure 9.2(a) represents the mineralization enthalpy changes and Gibbs free energy changes of C_3S, C_2S, CaO, $Ca(OH)_2$, $Ca_3Mg(SiO_4)_2$, Mg_2SiO_4, MgO, $Mg(OH)_2$, and FeO. It can

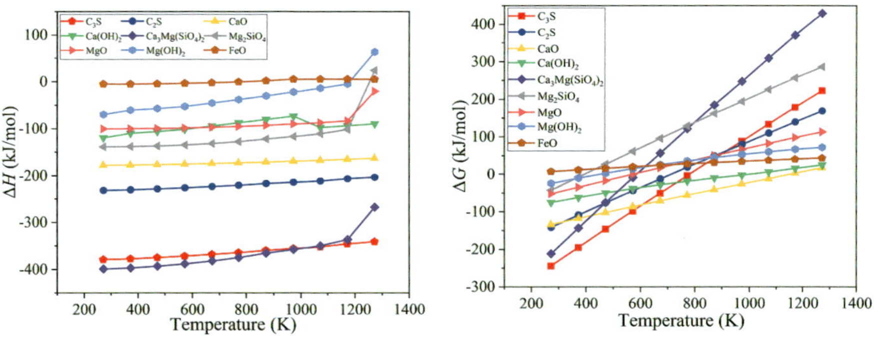

FIGURE 9.2 Mineralization enthalpy changes (a) and Gibbs free energy changes (b) of different mineral phases in steel slag.

be noticed that the mineralization enthalpy changes of C_3S, C_2S, CaO, $Ca(OH)_2$, $Ca_3Mg(SiO_4)_2$, Mg_2SiO_4, MgO, and $Mg(OH)_2$ are all below zero ($\Delta H < 0$) in the temperature range of 273K (0 °C°C) to 1273K (1000 °C) and at normal atmospheric pressure. This indicates that these mineralization reactions are exothermic reactions under normal environmental conditions and don't need external energy to maintain. Figure 9.3 depicts the heat evolution of the mineralization process of steel slag (Nakano et al., 2005). The reaction was conducted in a columnar cell and the CO_2

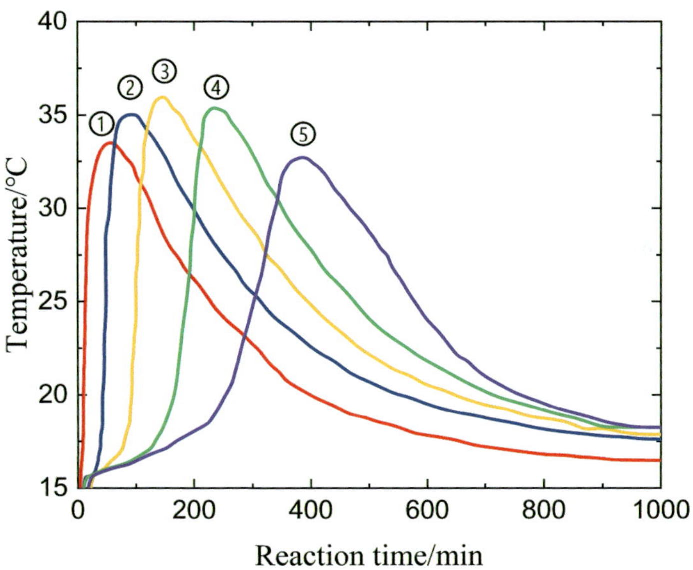

FIGURE 9.3 Changes in temperature of steel slag reacting with 100% CO_2 gas concentration (①②③④⑤ represent five thermocouples set from bottom to upper position of samples) (Nakano et al., 2005).

gas was supplied from the bottom to upper position. When the supply of CO_2 was sufficient, there was a rapid temperature rise and the peak temperature was relatively high.

As for any chemical reaction, the Gibbs free energy change determines whether it can spontaneously proceed or not. Figure 9.2(b) demonstrates that the mineralization reaction Gibbs free energy changes of C_3S, C_2S, CaO and $Ca(OH)_2$ are less than zero ($\Delta G < 0$) when the temperature is below 800 K (527°C) and that of $Ca_3Mg(SiO_4)_2$, MgO, and $Mg(OH)_2$ are less than zero when the temperature is below 600 K (327°C). It means the mineralization of above mineral phases spontaneously occurs at a relatively low temperature. When the temperature rises to threshold values, the mineralization reactions become non-spontaneous at standard atmospheric pressure. Zhang et al. (2021) used flue-gas (20% CO_2) mineralization to activate ladle slag at ambient (23°C) and elevated (55°C) temperatures. The results showed that higher CO_2 uptake content (12.1 wt.%) and compressive strength were achieved after 24 hours of mineralization compared with reference samples. Mineralogical analysis indicated that the major reactants of steel slag during mineralization are larnite, bredigite, calcio-olivine and cuspidine ($Ca_4Si_2O_7F_2$), which is consistent with the mineralization thermodynamics of steel slag.

On the contrary, the reaction enthalpy changes and Gibbs free energy changes of mineralization of FeO are always above zero ($\Delta G > 0$) during the temperature range of 273 K (0°C) to 1273 K (1000°C). It suggests that the mineralization of FeO can't proceed under normal conditions. This is also verified by Berryman et al. (2015). They found that srebrodolskite ($Ca_2Fe_2O_5$) and wustite (FeO) were preserved within the cores of reacted steel slag grains after mineralization in flow-through reactors. Moreover, the concentration of iron in the effluent was only 0.03 mmol/kg throughout the mineralization process. Chen et al. (2021) conducted mineralization experiment on synthesized RO phases containing MgO, FeO, and MnO. MgO in the RO phase can be mineralized into $MgCO_3$. However, when the content of FeO is too high, the RO phase becomes inert and cannot be mineralized.

The chemical equilibrium constant ($\log K$) is a mathematical way of expressing the ratio of product concentrations to reactant concentrations when the reaction reaches a steady state. It represents the maximum reaction degree of a reversible reaction at a certain temperature. Figure 9.4 shows the chemical equilibrium constants of mineralization reaction for different mineral phases in steel slag. It can be found that when the temperature is below 400 K, the order of different chemical equilibrium constants is $C_3S > Ca_3Mg(SiO_4)_2 > CaO > C_2S > Ca(OH)_2 > MgO > Mg_2SiO_4 > Mg(OH)_2 > FeO$. This indicates that the mineralization of mineral phases containing Ca and Mg can reach a relatively high degree. The chemical equilibrium constants of mineralization reaction decrease with increasing temperature, suggesting that high temperature hinders mineralization proceeding and reduces the reaction degree. Among the above mineral phases, the chemical equilibrium constants of CaO and $Ca(OH)_2$ maintain a relatively high level when temperature changes, so the mineralization reactions of CaO and $Ca(OH)_2$ are less sensitive to temperature. The chemical equilibrium constant of FeO is quite low within the temperature range of 273 K (0°C) to 1273 K (1000°C). This indicates the mineralization of FeO and

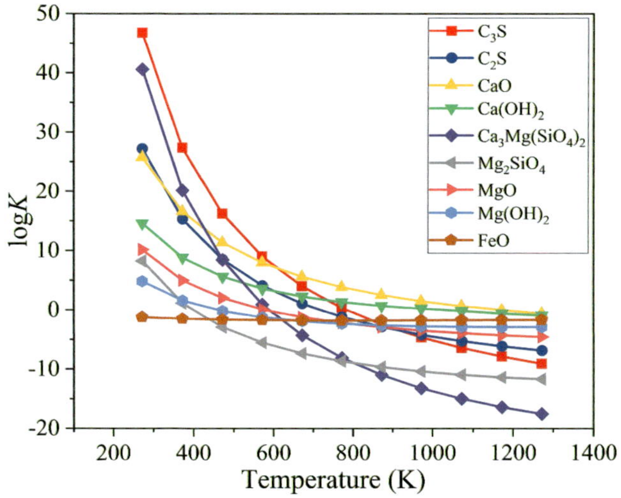

FIGURE 9.4 Mineralization equilibrium constants of different mineral phases in steel slag.

mineral phases containing Fe is hard to proceed. Fang et al. (2015) compared the CO_2 sequestration capacity of different minerals in steel slag. The results showed that $Ca(OH)_2$ had the highest CO_2 sequestration capacity, followed by C_2S, cuspidine, akermanite ($Ca_4Si_2O_7F_2$), and merwinite. On the other hand, srebrodolskite, brownmillerite, and calcium aluminate ($CaAl_2O_4$) performed poorly in mineralization activity. The mineral composition of steel slag significantly affects the potential and efficiency of CO_2 sequestration. Luo et al. (2021) reported that $Ca(OH)_2$, larnite, srebrodolskite, and calcium aluminate are the main mineralization reactive components in steel slag, while quartz, magnetite, and magnesium iron oxide are almost inert. In general, when the content of mineralization components in steel slag is high, the mineralization rate increases, and the mineralization degree is high.

9.4 MINERALIZATION DEGREE OF STEEL SLAG

The maximum theoretical CO_2 uptake capacity of steel slag can be predicted based on the chemical composition of the raw materials, through Steinour's formula, as shown in Equation (9.2) (Steinour, 1959) or Huntzinger's formula in Equation 2.3 (Huntzinger et al., 2009). It is assumed that all the CaO (except consisting of $CaCO_3$ and $CaSO_4$) was converted into $CaCO_3$, MgO was converted into $MgCO_3$, and Na_2O and K_2O were converted into Na_2CO_3 and K_2CO_3 after mineralization (Jang et al., 2016).

$$\max CO_2 \text{ uptake } (\%) = 0.785(CaO - 0.7SO_3) + 1.09\, Na_2O + 0.93\, K_2O \quad (9.2)$$

$$\max CO_2 \text{ uptake } (\%) = 0.785\,(CaO - 0.56\,CaCO_3 - 0.7\,SO_3) + 1.091\,MgO + 0.71\,Na_2O + 0.468\,(K_2O - 0.632\,KCl) \quad (9.3)$$

However, the actual CO_2 uptake depends on different mineralization conditions because of the non-homogeneous mineral phases and chemical compositions of steel slag (Huijgen and Ruijg, 2006). In general, mass change method is used to calculate the CO_2 uptake and there is a good linear relationship between CO_2 uptake and compressive strength.

It is a simple and effective method to evaluate CO_2 uptake or the degree of mineralization by monitoring the mass change of specimens during the mineralization process. No special technique is needed to prepare the specimens. Also, this non-destructive testing method makes it possible to monitor the development of the whole mineralization process over time, and this method has been used in many research studies (Hou et al., 2021; Shi et al., 2012c; Shi and Wu, 2008). The sample is normally placed in an airtight CO_2 mineralization chamber. The mass of the sample is measured before and after the CO_2 mineralization. The evaporated water from the exothermic mineralization reaction is collected using absorbent paper and added to the total mass gain as it is initially derived from the moisture content in the sample. The mass gain can then be calculated using the following equation (Shi et al., 2012a):

$$M_{CO_2}(\%) = \frac{(M_2 + M_W) - M_1}{M_C} \times 100 \qquad \cdot(9.4)$$

where M_{CO_2} is the mass percentage of CO_2 sequestered by steel slag, %. M_1 is the mass of the sample before accelerated mineralization, g. M_2 is the mass of the sample after accelerated mineralization, g. M_w is the mass of measured moisture evaporation during accelerated mineralization, g. M_C is the mass of steel slag in the sample.

Another mass change method is the mass loss method, which measures the CO_2 uptake by comparing the mass loss of mineralized specimens under different temperatures. This method measures the mass change of samples subjected to a calcination temperature range from room temperature to 1000°C with a steady heating rate. The content of $CaCO_3$ calculated from the decrease of CH was lower than that obtained from the TG results, indicating that mineralization of other phases such as calcium silicates, $Mg(OH)_2$, and C-S-H gel exists. Based on the TG data, the mass loss can be divided into three sections: (i) evaporation of adsorbed water and dehydration of C-S-H gel in the temperature range of 30–250°C (Y. Wang et al., 2021); (ii) dehydration of $Mg(OH)_2$ and $Ca(OH)_2$ in the temperature range of 300–470°C (Nielsen et al., 2020); and (iii) decomposition of carbonates in the temperature range of 500–680°C, 680–780°C, and 780–900°C, which correspond to $MgCO_3$ and amorphous $CaCO_3$, vaterite and aragonite, and well-crystallized calcite, respectively (Demir et al., 2003; Thiery et al., 2007; Tu et al., 2016). The equation to calculate the CO_2 uptake is expressed as in Equation (9.5).

$$CO_2\ uptake(\%) = \frac{M_{500°C} - M_{1000°C} - M_0}{M_1} \times 100\% \qquad (9.5)$$

where, M_{520} is the mass of pastes at 500°C,
M_{1000} is the mass of steel slag samples at 1000°C, and
M_0 is the initial CO_2 content in the non-mineralized specimen.

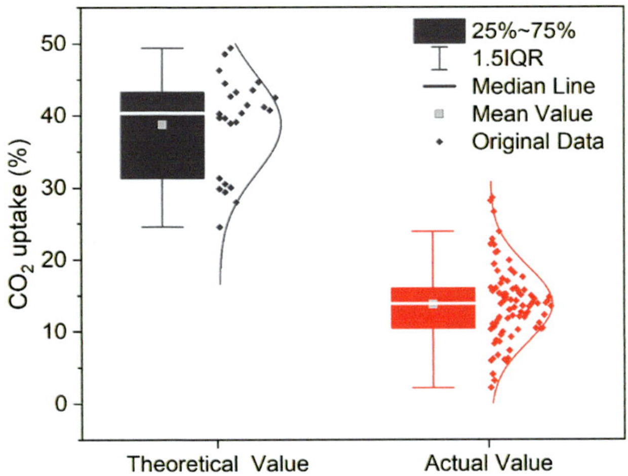

FIGURE 9.5 Theoretical value and actual value of CO_2 uptake of steel slag. (Ghouleh et al., 2015; Kim et al., 2024; Zhang et al., 2025; Zhang, Xiaozhe et al., 2024).

Figure 9.5 depicts the theoretical value and actual value of CO_2 uptake of steel slag. It is obvious that the theoretical value can be divided into two parts. The part ranging from 38% to 50% represents LFs. The high content of CaO and low content of FeO result in a high theoretical CO_2 uptake. The other part ranging from 25% to 33% represents BOFs and EAFs. Large amounts of FeO make the theoretical value lower than that of LFs. The actual CO_2 uptake distribution curve obeys a normal distribution. The median value is near 15% and far lower than the theoretical CO_2 uptake. This indicates significant amounts of minerals contained in steel slag haven't been mineralized. The results of XRD spectra reveal that the relative intensities of major CO_2 reactive phases in steel slag such as calcio-olivine, larnite, bredigite, and cuspidine all decreased after mineralization, suggesting these minerals were partially consumed (Zhang et al., 2022). This is because a dense calcium carbonate layer formed on the surface of steel slag grain, so the leaching of Ca and the migration of CO_2 were hindered and eventually caused the reaction to stop (Yi et al., 2021). Various mineralization conditions cause the differences in the actual values. The factors influencing the mineralization degree of steel slag will be discussed in the next section.

There is a good linear relationship between CO_2 uptake and compressive strength, suggesting that mineralization is the major contributor to the mechanical strength of steel slag specimens, which resulted from the rapid generation of C-S-H gels and $CaCO_3$. Figure 9.5 shows the relationship between compressive strength and CO_2 uptake of steel slag at normal and elevated mineralization temperature. When temperature rises from 20°C to 60°C, the slope of the fitted curve decreases from 0.23314 to 0.14907, which means the compressive strength at 60°C is lower than that at 20 when the CO_2 uptake values are the same. This is because high mineralization temperature promotes the formation of metas and polycrystalline forms of $CaCO_3$ (Fernandez Bertos et al., 2004). In addition, $CaCO_3$ mainly precipitated at

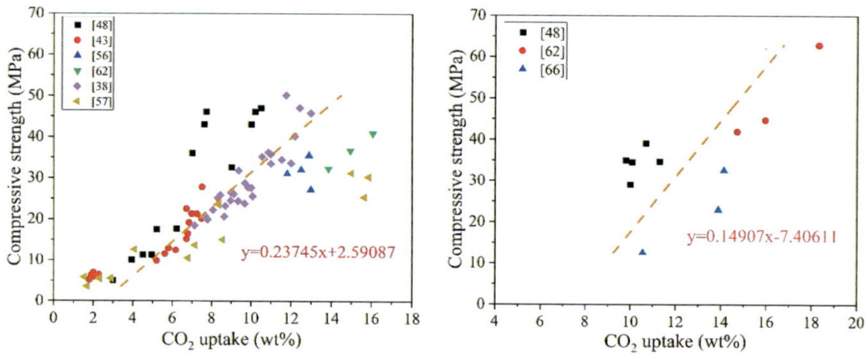

FIGURE 9.6 Relationship between compressive strength and CO_2 uptake of steel slag at different mineralization temperatures (20°C on the left, 60°C on the right).

the fluid-gas interfaces when the mineralization temperature is high, which leads to a microporous matrix and lower compressive strength for an equivalent CO_2 uptake (Nielsen et al., 2020).

9.5 MINERALIZATION TECHNIQUES

The mineralization of steel slag can be divided into direct and indirect mineralization. Direct mineralization is a single mineralization reaction step under a dry, moist, or aqueous environment without the aid of extraction agents and is convenient to be conducted (Bodor et al., 2013). However, high temperature, high gas pressure, and fine particle size are often employed to obtain a higher mineralization degree or speed up the mineralization process, which makes direct mineralization too expensive for large-scale industrial application as a CO_2 sequestration technology (Huijgen et al., 2007). On the contrary, the indirect mineralization method includes two successive steps: the extraction of Ca or Mg from steel slag by using various extraction agents and the reaction between alkaline elements and CO_2 to form carbonates (Liu et al., 2021). The indirect method has gained much more attention because the major limitation of direct mineralization for achieving a high reaction rate and CO_2 uptake, namely the extraction of alkaline elements from steel slag particles, can be more easily overcome using acid or salt extraction agents (Ragipani et al., 2021). Therefore, indirect method avoids the extreme environment (high temperature, high pressure, intensive crushing/milling) adopted by the direct method, and its mineralization efficiency is much higher than that of the direct method.

9.5.1 Direct mineralization

Direct mineralization is a straightforward method for steel slag mineralization, achieved by reacting steel slag directly with CO_2 in either the aqueous or gaseous phase. Unlike other methods, it does not require pre-treatment to extract active components (such as Ca^{2+} or Mg^{2+}) using chemical solvents. This approach offers several advantages, including simplicity, ease of operation, and minimal use of chemical extractants. However, it is limited by slow reaction kinetics, low conversion efficiency, and the need for an extended mineralization duration (Myers et al., 2019). In this process, a portion of the gaseous CO_2 dissolves in water, forming carbonic acid, which further ionizes into H^+, HCO_3^-, and CO_3^{2-}. These ions promote the dissolution of minerals and facilitate their mineralization. Consequently, the dissolution and diffusion of CO_2 and Ca^{2+} within the ore phase are critical factors influencing the mineralization rate. To enhance this process, direct mineralization typically employs high temperatures and pressures. Additionally, the mineralization efficiency of steel slag and the performance of the resulting cementitious materials are influenced by various factors, including particle size, raw material ratio, molding pressure, relative humidity, CO_2 concentration and pressure, as well as mineralization temperature and duration (Polettini et al., 2016b; Ukwattage et al., 2017).

9.5.1.1 Mineralization temperature

The mineralization reaction is highly influenced by temperature, which affects various processes differently, including the leaching of alkali metal ions (Ca^{2+} and Mg^{2+}), CO_2 migration through pores, CO_2 dissolution in water, carbonate precipitation, reaction rate, and mineralization equilibrium. The relationship between relative mineralization degree and temperature varies across mineralization stages.

Early Stage (0–0.5 h): As shown in Figure 9.7, there is a linear correlation between relative mineralization degree and temperature. At this stage, the reaction is far from equilibrium and is primarily controlled by reaction kinetics. Higher temperatures

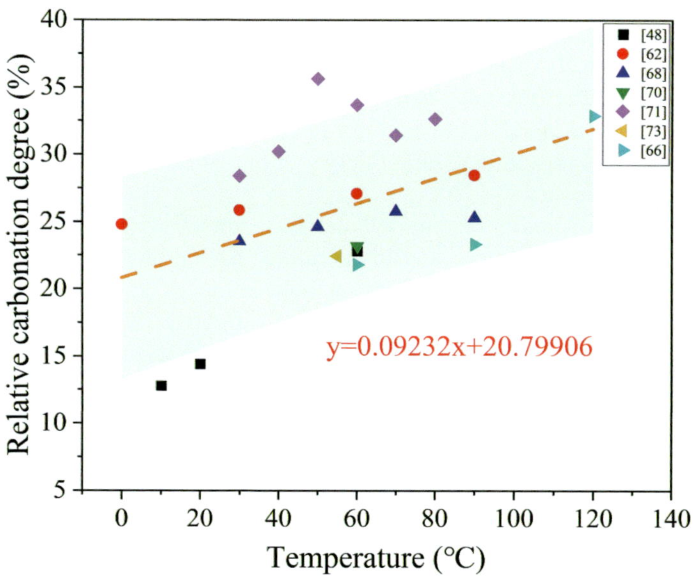

FIGURE 9.7 Relationship between relative mineralization degree and temperature at early age (0—0.5 h).

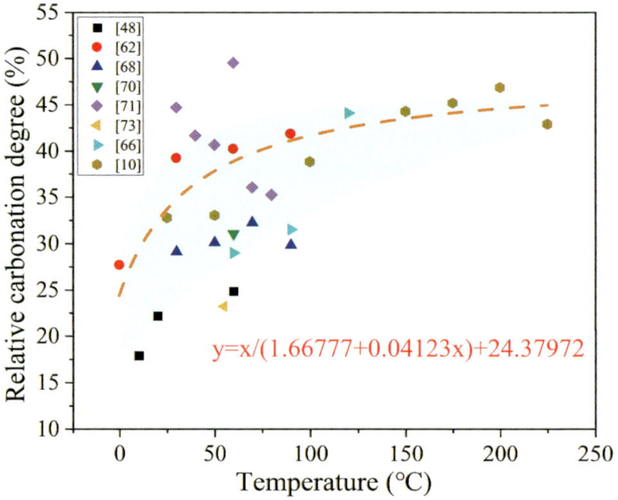

FIGURE 9.8 Relationship between relative mineralization degree and temperature at middle age (0.5–3 h).

accelerate the leaching of alkali metal ions, CO_2 migration, and carbonate precipitation rates (Drouet et al., 2019; Huijgen, 2005). Additionally, carbonate precipitation depletes dissolved inorganic carbon, amplifying the CO_2 concentration gradient between gaseous and liquid phases, which enhances CO_2 uptake (Back et al., 2011; Lekakh et al., 2008).

Middle Stage (0.5–3 h): In Figure 9.8, the relationship remains nearly linear at temperatures below 50°C. However, CO_2's retrograde solubility reduces its dissolution at higher temperatures, increasing the pH value of the liquid phase (Zhong et al., 2021) and lowering mineralization equilibrium (Nielsen et al., 2020). As a result, the temperature's effect on mineralization degree diminishes, and the curve flattens as temperature rises further.

Late Stage (3–24 hours): Figure 9.9 shows a downward parabolic relationship between mineralization degree and temperature. An optimal temperature (60°C) maximizes mineralization. Below this threshold, mineralization is kinetics-driven, and higher temperatures improve mineralization. Beyond 60°C, mineralization is equilibrium-controlled, where increased temperature reduces mineralization efficiency. Additionally, high temperatures may cause rapid $CaCO_3$ formation, creating a dense surface layer that hinders further mineralization (Liu et al., 2016). Luo et al. (2021) reported that excessive temperatures led to water evaporation, limiting calcium ion leaching and CO_2 dissolution, ultimately reducing CO_2 uptake in the later stages of mineralization.

Some studies indicated that elevated temperature produced a finer pore structure than ambient temperature (Hao, 2016; S. Zhang et al., 2021; X. Zhong et al., 2021). Zhong et al. (2021) found that raising mineralization temperature from 30°C to 70°C, the porosity of steel slag block decreased from 39.6% to 26.6%. Zhang et al. (2021)

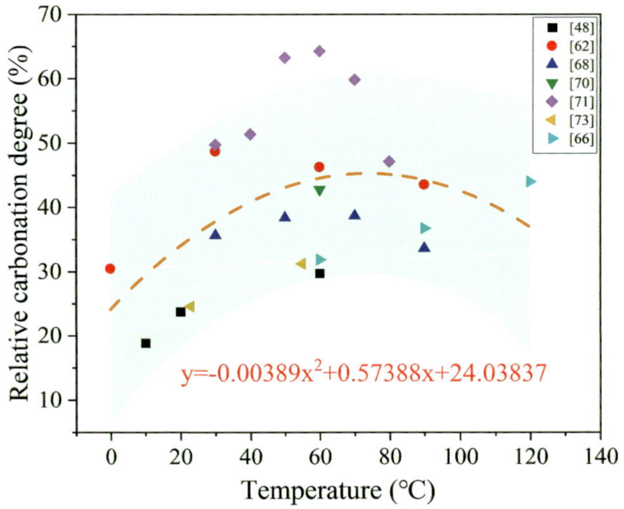

FIGURE 9.9 Relationship between relative mineralization degree and temperature at later age (3–24 h).

demonstrated that although the cumulative pore volume was comparable between elevated temperature and ambient temperature mineralized samples, the pore size distribution was refined. The intensity of the critical diameter the volume of capillary pores (<100 nm) increased at elevated temperature. Generally speaking, this pore structure refinement can lead to enhanced compressive strength, decreased permeability, and consequently, improved durability (Kumar and Bhattacharjee, 2003; Li et al., 2022). Nielsen et al. (Nielsen et al., 2020) studied the effect of temperature on carbonate precipitation; the results revealed that elevated temperature led to low dissolution of CO_2 and high dissolution of Ca^{2+}. As a result, the saturation degree in pore fluid became higher. It is reported that $CaCO_3$ nucleated spontaneously in the absence of growth substrates only at high saturation indexes (> 1) (Lioliou et al., 2007; Stockmann et al., 2014). So elevated temperature induced $CaCO_3$ to precipitate at the fluid-gas interfaces and reduced the critical diameter from 100 to 10 μm. Although there are very few large pores at elevated temperature, the matrix is still microporous which resulted in lower compressive strength.

Temperature also affects the crystallinity and morphology of mineralization products, especially for $CaCO_3$. Some research revealed that elevated temperature promotes the formation of aragonite or calcite with larger size after mineralization. Zhang et al. (2022) found strong calcite peaks in all mineralized samples and aragonite peaks only in elevated temperature mineralized samples by XRD analysis. The TGA results indicated that the principal decarbonization peaks of carbonates in elevated temperature mineralized samples were displaced to a higher temperature, which indicates a higher crystallinity of carbonates. However, the existence of aragonite cannot be detected under TG, because this polymorph will transform to calcite without mass change during the pyrolysis process (Scrivener et al., 2016). The SEM

FIGURE 9.10 SEM images of steel slag samples mineralized for 3 hours at different temperatures (Luo et al., 2021).

results manifested the presence of calcite with side lengths of rhombohedral crystals reaching 1–4 μm and aragonite with needlelike shape in elevated temperature mineralized samples, as shown in Figure 9.10. The growth in particle size of $CaCO_3$ leads to restrained stress at the microscopic level and finally enhances the strength of macro pastes (D. Wang et al., 2019). Drouet et al. (2019) demonstrated that the two metas Figure 9.11 forms of $CaCO_3$ (aragonite and vaterite) were always observed, and the decrease of relative humidity caused by elevated temperature inhibited their polymorphic transformation into calcite. However, Luo et al. (2021) found that the change of curing temperature did not alter the phase assemblage of mineralization products. However, elevated temperature promoted the formation and enhanced the particle size of mineralization products, resulting in a dense microstructure favoring the mechanical properties. More research is needed on how the mineralization temperature affects the formation process and crystal types of mineralization products of steel slag.

9.5.1.2 Concentration and pressure of CO_2

The concentration and pressure of CO_2 directly affect the transportation and solubility of CO_2. Zhang et al. (2024) studied the effects of CO_2 pressure on the performance and carbon sequestration efficiency of steel slag. Their findings showed that increasing mineralization pressure accelerated the mineralization reaction, with optimal efficiency achieved at 0.3 MPa, resulting in a carbon sequestration efficiency of up to 16.65%. However, excessively high pressures, such as 0.4 MPa, caused overly dense specimen surfaces, impeding internal mineralization and reducing specimen

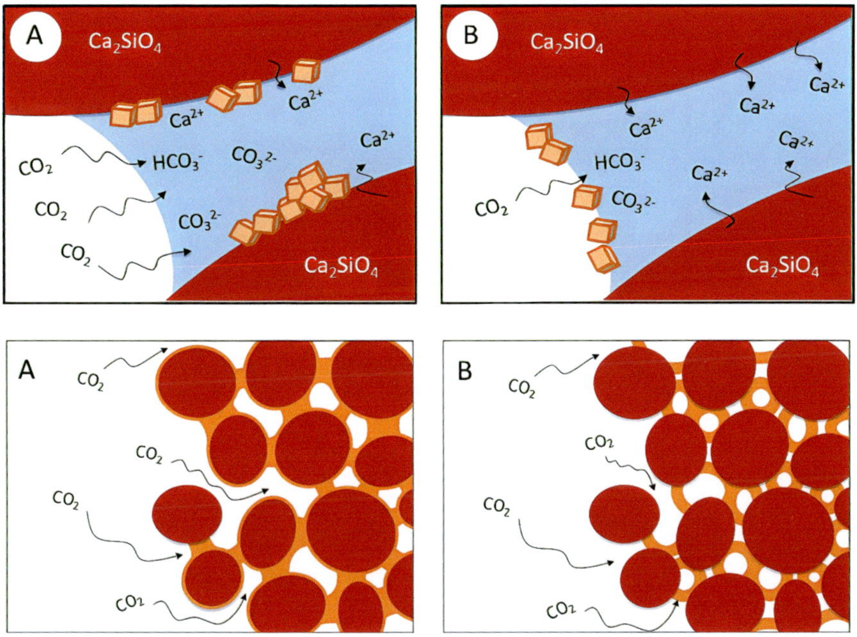

FIGURE 9.11 Schematic diagram of the mineralization process at different temperatures and CO_2 pressure (Nielsen et al., 2020).

strength. Mercury intrusion porosimetry (MIP) and scanning electron microscopy (SEM) analyses revealed that mineralization products primarily fill pores with diameters around 1 µm, significantly increasing the fractal dimensions of seepage and transition pores. These structural changes underline the balance required between mineralization pressure and pore accessibility for achieving optimal performance and efficiency. Humbert et al. (2019) found that the compressive strength of CO_2 mineralized steel slag blocks increased by 75% when the CO_2 partial pressure increased from 0.05 MPa to 0.15 MPa. However, when the CO_2 partial pressure increased from 0.15 MPa to 0.25 MPa, there was nearly no improvement in compressive strength. Zhong et al. (2021) reported a similar relationship between CO_2 pressure and compressive strength of CO_2 mineralized steel slag blocks. When CO_2 pressure increased from 0.05 MPa to 0.55 MPa, the compressive strength and Ca conversion of steel slag blocks increased by 45% and 28%, respectively. These results indicate that there is an optimal mineralization pressure, although the optimum pressure value varies from one investigator to another. The result behind this is that elevated CO_2 pressure accelerated the migration and dissolution of CO2, and more calcite was formed from the surface to the core of the samples, resulting in a dense structure of the blocks. However, the CO_2 quickly reached supersaturation in the pore solution at excessive pressure, which resulted in fast nucleation and growth of calcite. As a consequence, the pores on the surface were blocked, making CO_2 difficult to transport into the core of samples, which hindered further mineralization of steel slag. Figure 9.12 depicts

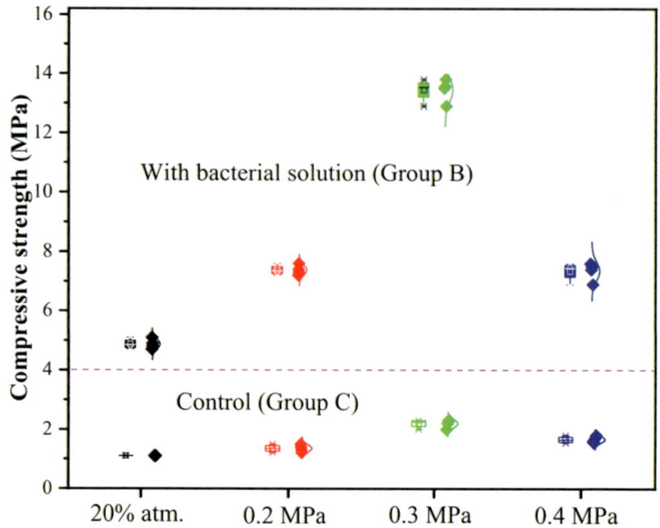

FIGURE 9.12 Impact of CO_2 pressure on weight increase and compressive strength of steel slag blocks (Hu, 2015).

that the microstructure of the core area is looser and there are more interconnected pores than in the surface area.

The CO_2 migration depends not only on porosity but also on pore size distribution. Moreover, the influence of pore size distribution on CO_2 migration also relates to CO_2 pressure. At low CO_2 pressure, mesopores and micropores dominate the CO_2 migration process, and high CO_2 pressure exhibits a positive effect. However, at high CO_2 pressure, ultra-micropore plays an important role in CO_2 migration instead, which weakens the relative contribution of CO_2 pressure to CO_2 migration (Zhu et al., 2020). The mineralization products, especially carbonate/bicarbonate equilibrium, are also affected by CO_2 pressure. Both thermodynamic calculations and experimental observations revealed that the content of bicarbonates increases with elevated CO_2 pressure. As a result, the pH value of the pore solution significantly decreases, which further decreases the solubility of CO_2 in water (Bernal et al., 2012).

Cui et al. (2015) studied the effects of CO_2 concentration on mineralization. The results showed that when samples were exposed to low CO_2 concentration (2–20%), the mineralization depth significantly increased with the increase of CO_2 concentration under the same mineralization periods. On the contrary, the mineralization depth almost had no change when the CO_2 concentrations were high (50–100%). This is because the microstructure of the surface became dense due to the formation of calcite under high CO_2 concentration, which reduced the migration rate of CO_2.

9.5.1.3 Particle size of steel slag

The particle size of steel slag affects the leaching of alkali metal ions (Ca^{2+} and Mg^{2+}) and the migration of CO_2 during the mineralization process. Tu et al. (Maoxia et al.,

2015) investigated the effect of particle size on mineralization efficiency of steel slag and found that the CO_2 uptake ratio of steel slag increased with the decrease of particle size. When the median diameter of steel slag was 17.1 μm, the CO_2 uptake ratio could reach 27.9%, which is greatly higher than 5% for the median diameter of 85.4 μm and 2.6% for the median diameter of 204.4 μm, respectively. Kim et al. (Kim and Azimi, 2021) also reported that the decrease of steel slag particle size led to a higher CO_2 uptake capacity. These results can be explained by a shrinking core model. Fine particles have a larger specific surface area, which is beneficial to the leaching of Ca^{2+} and the migration of CO_2 (Polettini et al., 2016a). Lekakh et al. (2008) studied the kinetics of aqueous mineralization of steel slag, and the fraction of Ca-leaching or carbonate precipitation was calculated to determine the rate-limiting step based on the shrinking core model. The results revealed that Ca-leaching is linear to particle size in the beginning (0.30 conversion), supporting a chemical reaction-controlled model (CaO dissolution). Afterwards, Ca-leaching is limited by the diffusion of Ca^{2+} through the porous surface layer of the particle, which exhibits a dependence on the square root of reaction time.

However, as mineralization progresses, the reaction products, especially $CaCO_3$, formed a layer on the outer shell of steel slag particles and hindered the succeeding mineralization process, as shown in Figure 9.13. Further research has shown that the thickness of this $CaCO_3$ layer is kept constant at 5.2 ± 1.6 μm regardless of the steel slag particle size. However, the mineralization layer of samples with smaller particle size is less dense, indicating that finer steel slag needs more time to be totally mineralized (Kim and Azimi, 2021).

Some research claimed that there is an optimal steel slag particle size according to the CO_2 uptake and compressive strength of samples. Liu et al. (2020) studied the diffusion process, leaching behavior, pore structure, and strength development of mineralized steel slag mortar with different particle size of steel slag. The results depicted that the range of optimal steel slag particle size is 21.75–24.13 μm, which led to a higher compressive strength (31.21 MPa) and CO_2 uptake (15.9%) compared to the larger or smaller steel slag sizes. On the one hand, small steel slag particles lead to a more reactive area for both mineralization and hydration; as a result, the CO_2 uptake and compressive strength increase (Kashef-Haghighi et al., 2015). On

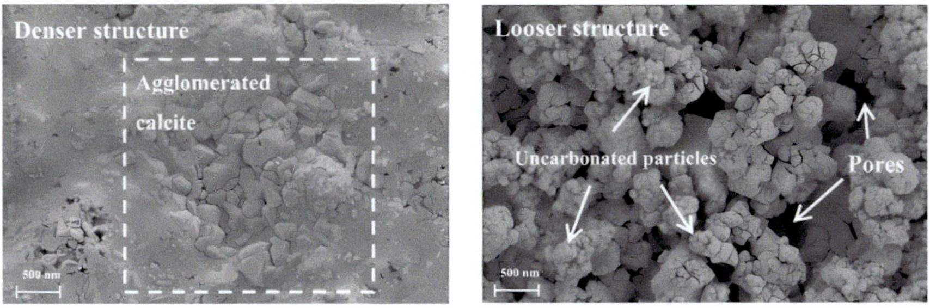

FIGURE 9.13 SEM images of the sample at CO_2 pressure of 1 MPa (left for surface and right for core) (Zhong et al., 2021).

the other hand, fine particles result in a more homogeneous, denser microstructure, which slows down the CO_2 migration and decreases the CO_2 uptake (Zhang et al., 2011).

The decrease of particle size is normally obtained by mechanical activation, which is environmentally clean because no chemicals are added during the pretreatment process (Li and Hitch, 2018). Mechanical activation can not only change the particle size but also alter the crystal form and pore structure of steel slag. Li et al. (022) found that mechanical activation didn't enhance the mineralization conversion of steel slag after 24 hours. However, the hydration rate and compressive strength were improved due to the increase in the amorphization degree. Furthermore, the hydration products such as C-S-H and C-A-S-H could catalyze the mineralization process.

9.5.1.4 Water-to-solid ratio (W/S)

Mineralization of steel slag is gas-liquid-solid reaction process, during which water is an essential substance for the dissolution of CO_2 and the leaching of alkaline ions. However, excess water content may block the pore structure and hinder the migration of CO_2. Hence, it is important to find out the influence mechanism of water and determine an optimal W/S ratio.

Wei et al. (2022) found that the compressive strength and the carbon capture of steel slag samples decrease when W/S decreases. When W/S reaches a maximum of 1:5, the compressive strength is 31.97 MPa, and the captured carbon content reaches 3.38%. Wang et al. (2021) investigated the influence of moist content on the bio-mineralized steel slag bricks. The results demonstrated that the residual water-cement ratio is a key factor for the preparation of mineralized steel slag bricks. The best value is 0.20 when the compressive strength reaches a maximum (33 MPa). However, the residual water-cement ratio cannot influence the types of mineralization or hydration products. Ukwattage et al. (2017) found that the CO_2 uptake of steel slag increased with liquid to solid (L/S) ratio (from 0.25 to 0.3). The formation of a calcium-depleted silicate layer and $CaCO_3$ on the surface of the steel slag particles hinders the diffusion of Ca^{2+} from the inside of the particles and the migration of CO_2 and CO_3^{2-} at low L/S, which is not conducive to subsequent mineralization. However, when there is too much water, it is easy to block the pores inside the steel slag and prevent the CO_2 gas from diffusing into the solids (Salman et al., 2014). The mineralization degree can be effectively increased by forming an optimal W/S and adopting a pre-curing method before the mineralization treatment (Shi et al., 2012b). Figure 9.14 depicts the influence mechanism of pre-curing on the mineralization process, as described in Chapter 2.

9.5.1.5 Mineralization period

The mineralization degree and strength of steel slag specimens increase with the mineralization period of time. The mineralization degree and compressive strength of steel slag/carbide slag blocks exhibit a three-stage progression: rapid initial increase, slower growth, and plateau (Liu et al., 2022). This trend closely aligns with compressive strength (Figure 9.14(b)), showing a strong linear correlation ($R^2 > 0.97$). The

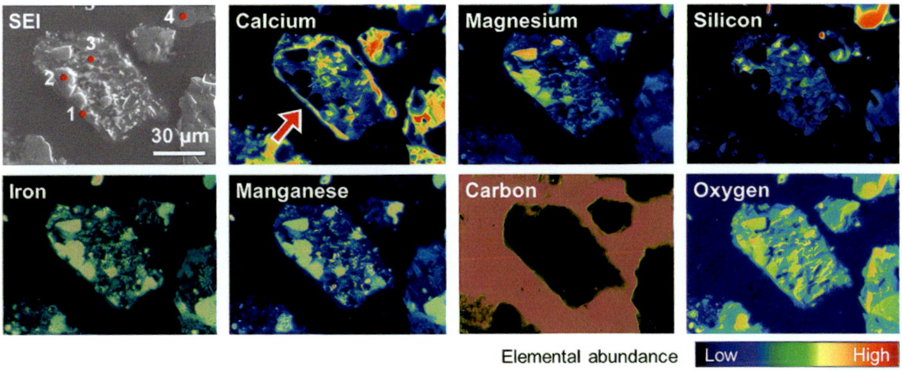

FIGURE 9.14 Secondary electron micrograph of mineralized steel slag with medium particle size (Kim and Azimi, 2021).

increase in compressive strength primarily results from the accumulation of mineralization products, which rises linearly with mineralization degree under identical conditions. However, discrepancies between the fitting equations here suggest compressive strength depends on more than mineralization product quantity. Intrinsic material properties, such as mechanical strength, initial pore structure, hydration state, as well as the packing density and bonding interactions between mineralization products and raw materials, also play critical roles (Zhang et al., 2025).

9.5.1.6 Additives

The mineralization rate of the steel slag system could also be affected by adding other raw material components, so the properties of mineralization products could change. Wang et al. (2019) added gypsum into steel slag pastes with the gypsum to steel slag ratio ranging from 1:16 to 1:2. The results showed that gypsum promoted an increase in both the compressive strength and the CO_2 uptake of steel slag. This is because gypsum was beneficial in promoting the rapid hydration of steel slag to form ettringite, which then reacted with CO_2 to produce monocarbonate. Ma et al. (2021) used some oxides as additives, such as Fe_2O_3, Fe_3O_4, SiO_2, MgO, and Al_2O_3, and investigated the influences of these oxides on the mineralization rate and efficiency of steel slag. The results showed that the addition of Fe_2O_3, Fe_3O_4, SiO_2, MgO, and Al_2O_3 could increase the mineralization rate by 5.5%, 13.2%, 13.2%, 6.9%, and 8.5%, respectively. The mineralization efficiency of steel slag with Fe_2O_3, Fe_3O_4, MgO, and Al_2O_3 increased by 3.6%, 4.3%, 10.1%, and 1.2%, respectively, and decreased by 2.4% with SiO_2. One reason is that the mass transfer was enhanced by nanoparticles attributed to the coalescence inhibition of CO_2 bubbles as well as micro-convection induced by Brownian motion. Another reason for improved mineralization rate and efficiency is that these oxides provided inert surfaces for $CaCO_3$ growth, which helped to avoid in-situ aggregation and reduced the coating effects of $CaCO_3$. Mo et al. (2016) found that the addition of 20% OPC as a substitute for steel slag would significantly increase the compressive strength of the steel slag slurry, mainly because the addition of OPC provides a rich calcium source for the mineralization of steel slag.

9.5.2 Indirect mineralization

The indirect mineralization comprises three successive reactions, i.e., the extraction of Ca or Mg from the feedstock by using various extraction agents, the dissolution of CO_2, and the precipitation of carbonates. It has attracted much attention because of its mild mineralization conditions (avoiding the high temperature and high gas pressure in the direct mineralization), higher mineralization efficiency, and purer byproducts. The most common extraction agents that could accelerate the mineralization process are strong acids (hydrochloric acid (HCl), nitric acid (HNO_3), sulfuric acid (H_2SO_4) (Jo et al., 2017; Tian et al., 2015; Zhao et al., 2015)), acetic acid (CH_3COOH) (Tian et al., 2016), and ammonium salt (Lee et al., 2016; Sun et al., 2011).

9.5.2.1 Strong acids

Compared with a weak acid, a strong acid can more effectively leach metal elements from steel slag, which is beneficial to mineralization. Under the condition of strong acid, the extraction rate of calcium and magnesium can reach 97% and 46%, respectively (Lee et al., 2021). Take HCl as an example, the processes of extracting calcium and magnesium from Ca/Mg-silicates and Ca-Mg-silicates are expressed by Equation (9.6) in acidic environment (Hemmati et al., 2014; Sanna et al., 2014).

$$(\mathrm{Ca,Mg})_x\,\mathrm{Si}_y\mathrm{O}_z\mathrm{H}_{2(z-x-2y)}(s)+2x\mathrm{HCL}(aq)\rightarrow x(\mathrm{Ca,Mg})\mathrm{Cl}_2(aq)+y\mathrm{SiO}_2(s)+(z-2y)\mathrm{H}_2\mathrm{O}(l) \tag{9.6}$$

After that, the Ca^{2+} or Mg^{2+} in the solution forms a $Ca(OH)_2$ precipitate or subsequently a $CaCO_3$ precipitate, which requires a high pH value (around 10) (Doucet, 2010). Equation (9.7) indicates its mineralization stage. This method requires extra chemicals to adjust the pH, which certainly raises the economic cost. To reduce the amount of extra chemicals, the solution can be heated to precipitate $Ca(OH)_2$ and $Mg(OH)_2$ and recover the acid as Equation (9.8) (Yadav and Mehra, 2021). Finally, the resulting $Ca(OH)_2$ and $Mg(OH)_2$ are dissolved in water, where they react with CO_2 to form $CaCO_3$ and $MgCO_3$ as shown in Equation (9.9) (Saran et al., 2018; Yadav and Mehra, 2021). In addition, Hoyong et al. used a sodium chloride (NaCl) electrolysis system with low energy requirements; it provided both acid and alkali solutions for indirect mineralization (Jo et al., 2017).

$$(\mathrm{Ca,Mg})\mathrm{Cl}_2(aq)+\mathrm{CO}_2(g)+2\mathrm{NaOH}(l)\rightarrow(\mathrm{Ca,Mg})\mathrm{CO}_3(s)+\mathrm{H}_2\mathrm{O}(l)+2\mathrm{NaCl}(l) \tag{9.7}$$

$$(Ca,Mg)Cl_2(aq)+2H_2O(l)\rightarrow(Ca,Mg)(OH)_2(s)+2HCl(aq) \tag{9.8}$$

$$(Ca,Mg)(OH)_2(s)+CO_2(g)\rightarrow(Ca,Mg)CO_3(s)+H_2O(l) \tag{9.9}$$

9.5.2.2 Acetic acid

The use of acetic acid as a solvent to accelerate mineralization was first proposed by Kakizawa et al. (Kakizawa et al., 2001). The extraction of metal elements from steelmaking slag is closely related to the concentration of acetic acid. Eloneva et al. (Eloneva et al., 2008) used acetic acid to extract calcium from steel slag. The results showed that high concentration acetic acid could extract various elements from steel slag; however, low concentration acetic acid could only selectively extract calcium. Bao et al. (2010) reported that only calcium and magnesium could be extracted under weak acetic acid conditions. Teir et al. (2007) used 33.3 wt.% acetic acid to leach different steel slags including BOFs and EAFs. More than 80% Ca could be extracted at 55°C within 2 hours. Higher temperature increased the volatilization of acetic acid, hindering the extraction. Maroto-Valer et al. (Sanna and Marotovaler, 2013) investigated the leaching of steel slag with recyclable NH_4HSO_4 (1.3 mol/L). The Mg and Fe were leached from the slag with extraction efficiencies of 85% and 90% at 90°C, while Ca was transferred to $CaSO_4$. The leachate was neutralized with ammonium water and generated $(NH4)_2SO_4$, which was used to regenerate NH_4HSO_4 by thermal decomposition. $Mg(OH)_2$ and $CaSO_4$ were then mineralized with $(NH_4)_2CO_3$ to realize the storage of CO_2. Bao et al. (2010) used the mixed solution of tributyl phosphate and acetic acid to leach steel slag. Around 75% of Ca and 35% Mg were extracted selectively. The purpose of adding tributyl phosphate was to absorb part of acetic acid generated during the mineralization of Ca into the organic phase by solvent extraction, thus enhancing the mineralization capacity.

9.5.2.3 Ammonium salt

The dissolution behavior of Ca/Mg-bearing mineral phases of iron and steel making slags by using four types of solution (HCl, CH_3COOH, NH_4Cl, and CH_3COONH_4 solutions) was studied by Zhao et al. (Zhao et al., 2019). The results showed that the order of mineral phases solubility in aqueous solution is as follows: (CaO and Ca_2SiO_4) > ($Ca_3MgSi_2O_8$, $Ca_2MgSi_2O_7$, and MgO) > $Ca_2Al_2SiO_7$ > $MgCr_2O_4$. Eloneva et al. (Eloneva et al., 2012) studied the indirect mineralization of steel slag, where various ammonium salts (NH_4Cl, NH_4NO_3, CH_3COONH_4) were utilized to leach calcium selectively, followed by mineralization by bubbling CO_2. It was calculated that the mineralization capacity reached approximately 213 kg/ton slag and 10 ~ 16 € was the cost for storing 1 ton CO_2. Kodama et al. (2008) used 1 mol/L NH_4Cl solution to leach converter slag at 80°C, and around 60% of calcium was extracted. The pH value of the leachate was adjusted to basic by ammonium water and then CO_2 gas was bubbled to obtain mineralized products and recyclable NH_4Cl solution. The total energy consumption in this process was roughly estimated to be 300 kWh/ton CO_2. Yong et al. (2011) used the same method but lower NH_4Cl concentration and leaching temperature. A maximum CO_2 mineral storage capacity reached 211 kg/ton slag. Pilot-scale experiment of this process was conducted by Said et al. (2016), but Ca extraction efficiency was declined obviously to 39% after 10 times recycle of the NH_4Cl solution. This was due to the evaporation of ammonia from the solution at high temperature. By decreasing the leaching temperature to 45°C, the extraction efficiency of Ca reached 80%. The NH_4Cl leaching process can also be improved by using mechanical agitation and sonication (Mo et al., 2019; Said et al., 2015). It was found that sonication can significantly intensify the leaching of calcium from 65% to 96% due to the effective breakage and removal of the Ca-depleted layer by sonication and at the same time renewing the surface area of the particle. Microwave irradiation can also enhance the leaching and mineralization process (Tong et al., 2019).

9.6 CO_2 MINERALIZED STEEL SLAG-BASED PRODUCTS

The CO_2 mineralization can solve most encountered barriers for steel slag utilization, such as low cementitious activity, heavy metal leaching, and volume instability. For instance, with the employment of accelerated mineralization, the expansive compounds like f-CaO/MgO in fresh slag can be eliminated, enabling the potential use of mineralized slag as building materials. Various types of applications for the mineralization of steel slag, such as aggregates, artificial reefs, abiotic catalysts, supplementary cementitious materials, and high value-added chemicals, can be considered. Several successful experiences in the applications of mineralization steel slag are illustrated as follows.

9.6.1 Supplementary cementitious materials with high reactivity

The use of steel slag as a supplementary cementitious material (SCM) of cement-based materials is an effective way to enhance clean production. But the poor volume stability caused by dead burned lime and periclase in steel slag is a crucial factor restricting its utilization. Some research has shown that mineralization could convert dead burned lime and periclase into carbonate. The mechanical properties of cement blended with mineralized steel slag as SCMs showed higher strength, especially early strength, than unmineralized slag. The hydration activity indexes of mineralized BOFS at 3 days and 28 days increased by 97% and 16% compared with non-mineralized BOFS, respectively (Liang et al., 2012).

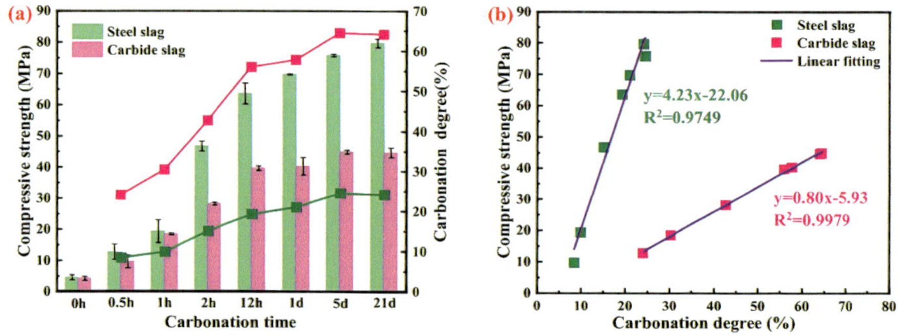

FIGURE 9.15 (a) Compressive strength and mineralization degree, (b) linear fitting relationship between mineralization degree and compressive strength at different mineralization time (Zhang et al., 2025).

Mineralized steel slag could accelerate the hydration of cement because $CaCO_3$ provides a favorable surface for nucleation and growth of hydration products (Zajac et al., 2020). This may also be due to the reaction between $CaCO_3$ and C_3A during cement hydration to form aluminum carbonate hydrates. The aluminum carbonate ($C_3A{\bullet}CaCO_3{\bullet}xH_2O$) has a larger volume and may be harder than the sulfoaluminate. In addition, the content of f-CaO and $Ca(OH)_2$ in the mineralized slag is reduced, which significantly improves the durability of mixed cement and concrete.

Particle size distribution plays a very important role in the development of compressive strength when steel slag is used as SCMs. Bodor et al. (2013) found that mortars prepared with the coarsest (<1.6 mm) or finest (<0.08 mm) fresh and mineralized BOFS showed lower compressive strength than the reference OPC mortar. Zhang et al. (2011) proposed a particle size mixed cement using a discontinuous particle size distribution (namely, fine and coarse SCMs) based on the compact packing theory. The advanced gap-graded method can increase the initial bulk density of the mixed cement slurry, thereby improving the development of the early and late strength of the cement, while also having a shorter set time, higher early strength, and better durability (Pan et al., 2016). Ghouleh et al. (2015) investigated curing scenarios combining mineralization and hydration, exploring the possibility of maximizing strength gain of the compacts, which was first cured by mineralization for a duration of 2 hours, followed by subsequent hydration for some days. The final strength of steel slag paste could reach 109 MPa, the highest ever reported for steel slag binder.

9.6.2 Aggregates

Steel slag could be used as aggregates after mineralization, which is an economic and environmentally friendly application of steel slag. Muhmood et al. (2009) suggested that the 28-day strengths of mortars with mineralized LFS sand were almost the same with those with natural river sand, which indicated that the LFS can successfully replace fine aggregates after mineralization to prepare mortar in precast products, such as hollow core slabs, paving stones, and masonry units. The formed $CaCO_3$ crystals in mineralized steel slag showed lower f-CaO content and pore spaces filled up with precipitation, thereby exhibiting good soundness and enhanced compactness (Wu et al., 2009). Results showed that concretes produced with total replacement of conventional aggregates by steel slag aggregates presented better performance compared to conventional concretes in accelerated mineralization tests. Some research used ethylenediamine tetra-acetic acid (EDTA) to enhance the mineralization efficiency and properties of artificial steel slag aggregates; results indicated that the addition of EDTA enhanced the mineralization of steel slag aggregates, yielding increases in CO_2 uptake, quantity of formed $CaCO_3$ and bulk density of aggregates, and reduction in total porosity. As a result, the addition of EDTA improved the mechanical properties of mineralized steel slag aggregates, and in particular, reduced the crushing value. Owing to the enhancement of EDTA on the property of mineralized steel slag aggregates, the compressive strengths of concretes cast with the mineralized steel slag aggregates reached 39.9–48.0 MPa at 28d, which were 5.2–13.3 MPa higher than that of the concrete cast with the mineralized steel slag aggregate without EDTA (Yang et al., 2021).

Figure 9.16 illustrates the Vickers microhardness distribution at the interfacial transition zone (ITZ) of concrete specimens with different aggregates. The ITZ in this study measured approximately 50 μm and exhibited significantly lower microhardness compared to the aggregate and cement paste. The ITZ hardness in the Control sample was higher than that in the Normal sample, attributed to the high water absorption of SSA. Proximity to the aggregate reduced the water-to-cement (w/c) ratio, resulting in higher ITZ hardness and fewer defects. The ITZ hardness of the E0.1-50%-24h sample was further enhanced by dry mineralization, likely due to densification from carbonate formation on the SSA surface. This aligns with

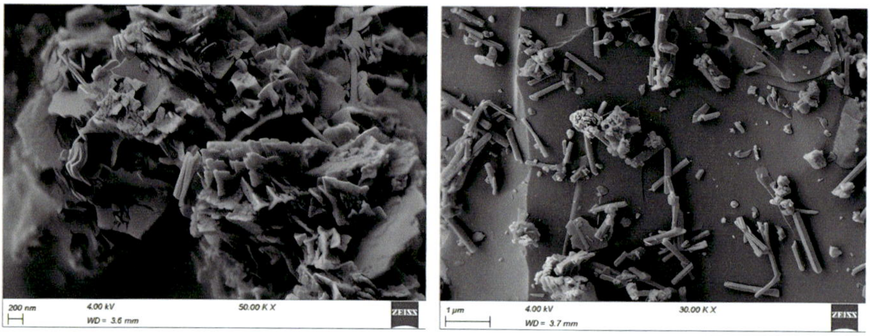

FIGURE 9.16 SEM morphologies of cement pastes blended with BOFS and mineralized BOFS (Li and Unluer, 2023b).

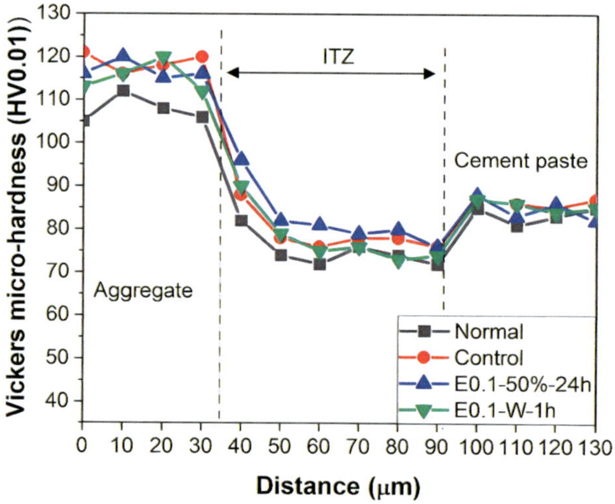

FIGURE 9.17 Vickers microhardness distribution at ITZ of specimens with different aggregate (Li and Unluer, 2023a).

previous studies showing a 5–30 μm mineralized layer forming on mineralized SSA surfaces, which increases SSA hardness and reduces the w/c ratio near SSA, thereby improving ITZ hardness. Conversely, the ITZ in the E0.1-W-1h sample did not show improvement, potentially due to the formation of brucite. As brucite shares structural and property similarities with portlandite, its presence can weaken the ITZ (Nežerka et al., 2019).

It can be concluded that accelerated mineralization can be employed to manufacture steel slag-based building materials with high early compressive strength and eligible soundness, along with the simultaneous benefit of CO_2 capture.

9.6.3 Construction blocks/bricks

Wu et al. (2009) explored accelerated mineralization technology to treat steel slag-slag material to produce a high-strength mineralized brick. However, its preparation requires high-pressure molding and curing by high-pressure high-purity CO_2, which greatly limits the energy-saving and emission-reduction effect of the technology and its popularization and application. Cao et al. (Cao and Yang, 2013) investigated steel slag with additional slaked lime material as raw material, adopting a conventional pouring molding method mineralized by industrial exhaust gas (with CO_2 concentration of 20%). The strengths of the mineralized steel slag-slaked lime bricks were increased by three to four times, water absorption was significantly reduced, and stability problems were resolved. Steel slag bricks were also prepared by both alkali-activated and accelerated mineralization. The steel slag brick showed a compressive strength of 35.0 MPa and uptake of CO_2 with 10.19% due to the synergistic effect between alkali activation and accelerated mineralization (Zhang et al., 2011). Furthermore, a test block with gypsum/steel slag can reach 32 MPa at one day of mineralization. However, the 28-day strength of CO_2 mineralized samples with gypsum was weaker than their 14-day strength because of the decomposition of C-S-H gels and monocarbonate. Therefore, short-term mineralization is suitable for this system (Zhang et al., 2019).

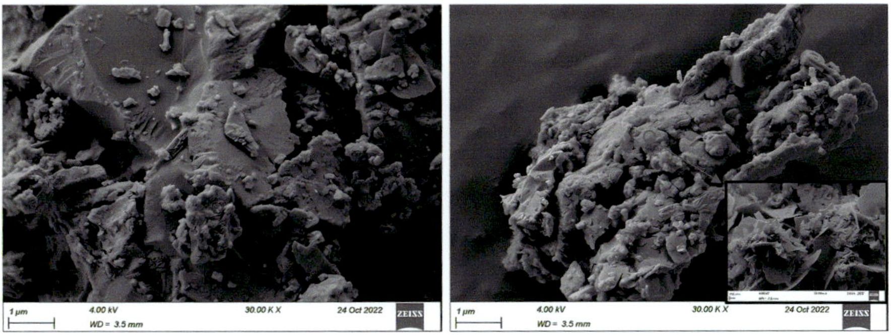

FIGURE 9.18 SEM images of steel slag aggregate and mineralized steel slag aggregate (Li and Unluer, 2023a).

TABLE 9.2
Costs of different composites of AAC and SS-AAC products

Materials	Cost calculation statistics (CNY/t) (G. Chen et al., 2021)	Mass required (kg/m³)	
		Control AAC-2 (Gu et al., 2024)	This study (Sun et al., 2022)
Steel slag	35	0	244.8
Quartz sand	147.2	432.9	214.2
Cement	350	92.3	91.8
Lime	280	74.8	49.2
Gypsum	30	15.7	12.6
Al powder	7500	0.8	1.08
Superplasticizer	7700	0	0.36
Water	2.5	270	240
Total CNY/ m³		**124.12**	**97.86**

Source: (G. Chen et al., 2021; Gu et al., 2024; Sun et al., 2022).

The raw material costs of steel slag-autoclaved aerated concrete (SS-AAC) are 26.26 CNY/m³ lower than conventional AAC, as shown in Table 9.2. While mineralization techniques involve initial energy and monetary investments, they are one-time expenditures. Modified steaming plants can feasibly transition to mineralization plants with minor adjustments, such as gas resource and sensor updates, leveraging existing systems. Furthermore, CO_2 for mineralization can be sourced from power plant flue gases, enhancing sustainability. Despite requiring further research, mineralization technology shows significant potential for CO_2 utilization and sequestration. Recycling steel slag in SS-AAC reduces reliance on natural raw materials and minimizes solid waste landfilling. Combining mineralization and autoclaving offers a more sustainable approach, promoting a greener AAC industry by integrating CO_2 sequestration, waste recycling, and cost efficiency into the production process.

9.7 SUMMARY

Steel slag has complex components, high hardness, low cementitious reactivity, and poor volume stability, thus limiting its scale application in building materials. Accelerated mineralization can effectively reduce the content of f-CaO so that the soundness problem of steel slag can be solved. In addition, this technique can capture partial CO_2 and mitigate the greenhouse effect. However, the technology of preparing steel slag products via mineralization is still in the laboratory stage. Some specific conclusions and suggestions for future research are as follows:

(1) Most minerals in steel slag can spontaneously react with CO_2. Thermodynamically, the variability of composition results in a complicated mineralization mechanism and diversity of mineralization products.

(2) The mineralization process consists of the diffusion and dissolution of CO_2, the leaching of alkaline ions, and the precipitation of carbonates. Each process may be predominant according to different conditions.
(3) Many factors such as temperature, CO_2 concentration and pressure, steel slag particle size, water content, and extraction agent affect the mineralization rate, degree, and products of steel slag. Although plenty of work has been done to study their influence mechanism on mineralization, there are still many knowledge gaps about the synergistic effect between different factors.
(4) There is still a long way to extend and promote the industrial application of accelerated mineralized steel slag-based products. New technologies and equipment need to be developed to manufacture products with better performance and lower costs.

REFERENCES

Back, M., Bauer, M., Stanjek, H., Peiffer, S., 2011. Sequestration of CO_2 after reaction with alkaline earth metal oxides CaO and MgO. *Applied Geochemistry* 26(7), 1097–1107.

Bao, W., Li, H., Zhang, Y., 2010. Selective leaching of steelmaking slag for indirect CO_2 mineral sequestration. *Industrial & Engineering Chemistry Research* 49(5), 2055–2063.

Bernal, S.A., Provis, J.L., Brice, D.G., Kilcullen, A., Duxson, P., van Deventer, J.S.J., 2012. Accelerated mineralization testing of alkali-activated binders significantly underestimates service life: The role of pore solution chemistry. *Cement and Concrete Research* 42(10), 1317–1326.

Berryman, E.J., Williams-Jones, A.E., Migdisov, A.A., 2015. Steel slag mineralization in a flow-through reactor system: the role of fluid-flux. *Journal of Environmental Sciences* (China) 27, 266–275.

Bodor, M., Santos, R., Gerven, T., Vlad, M., 2013. Recent developments and perspectives on the treatment of industrial wastes by mineral mineralization — a review. *Open Engineering* 3(4), 566–584.

Cao, W., Yang, Q., 2013. Characteristics of carbonated steel slag-slaked lime brick. *Journal of the Chinese Ceramic Society* 41(10), 1395–1400.

Chang, J., Wu, H., 2010. Study on mineralization mechanism of steel slag. *Journal of the Chinese Ceramic Society* 16(9), 5.

Chen, G., Li, F., Geng, J., Jing, P., Si, Z., 2021. Identification, generation of autoclaved aerated concrete pore structure and simulation of its influence on thermal conductivity. *Construction and Building Materials* 294.

Chen, Z., Li, R., Zheng, X., Liu, J., 2021. Carbon sequestration of steel slag and mineralization for activating RO phase. *Cement and Concrete Research* 139, 106271.

Cui, H., Tang, W., Liu, W., Dong, Z., Xing, F., 2015. Experimental study on effects of CO_2 concentrations on concrete mineralization and diffusion mechanisms. *Construction and Building Materials* 93, 522–527.

Demir, F., Dönmez, B., Okur, H., Sevim, F., 2003. Calcination kinetic of magnesite from thermogravimetric data. *Chemical Engineering Research and Design* 81(6), 618–622.

Dhoble, Y.N., Ahmed, S., 2018. Review on the innovative uses of steel slag for waste minimization. *Journal of Material Cycles & Waste Management* 20(3), 1–10.

Doucet, F.J., 2010. Effective CO_2-specific sequestration capacity of steel slags and variability in their leaching behaviour in view of industrial mineral mineralization. *Minerals Engineering* 23(3), 262–269.

Drouet, E., Poyet, S., Le Bescop, P., Torrenti, J.-M., Bourbon, X., 2019. mineralization of hardened cement pastes: Influence of temperature. *Cement and Concrete Research* 115, 445–459.

Eloneva, S., Said, A., Fogelholm, C.J., Zevenhoven, R., 2012. Preliminary assessment of a method utilizing carbon dioxide and steelmaking slags to produce precipitated calcium carbonate. *Applied Energy* 90(1), 329–334.

Eloneva, S., Teir, S., Salminen, J., Fogelholm, C.J., Zevenhoven, R., 2008. Steel converter slag as a raw material for precipitation of pure calcium carbonate. *Industrial & Engineering Chemistry Research* 47(18), 7104–7111.

Fan, C., Wei, R., Cheng, T., Sun, R., Zhang, H., Long, H., 2024. The positive contributions of steel slag in reducing carbon dioxide emissions in the steel industry: Waste heat recovery, carbon sequestration, and resource utilization. *Chemical Engineering Journal* 498.

Fernandez Bertos, M., Simons, S.J., Hills, C.D., Carey, P.J., 2004. A review of accelerated mineralization technology in the treatment of cement-based materials and sequestration of CO_2. *Journal of Hazardous Materials* 112(3), 193–205.

Fu, S., Kwon, E.E., Lee, J., 2024. Upcycling steel slag into construction materials. *Construction and Building Materials* 444.

Ghouleh, Z., Guthrie, R.I.L., Shao, Y., 2015. High-strength KOBM steel slag binder activated by mineralization. *Construction and Building Materials* 99, 175–183.

Ghouleh, Z., Guthrie, R.I.L., Shao, Y., 2017. Production of carbonate aggregates using steel slag and carbon dioxide for carbon-negative concrete. *Journal of CO_2 Utilization* 18, 125–138.

Gu, X., Wang, S., Liu, J., Zhu, Z., Wang, H., Ge, X., Hu, Z., Xu, X., Nehdi, M.L., 2024. Improving physical properties, microstructure and actual carbon sequestration of steel slag based autoclaved aerated concrete by accelerated mineralization. *Journal of Building Engineering* 94.

Hao, X.J., Lu, B., Jiang, W. L., 2016. Study on carbonization process of steel slag under pressure steam conditions. *Materials Review B: Research Papers* 30(4), 106–110.

He, Y., Jia, S., Yi, H., Tang, X., Yu, Q., Gao, F., Kang, D., Zhao, S., 2024. Utilization of steel slag in air pollution and greenhouse gas emission reduction-application, mechanism and challenge: A review. *Journal of Environmental Chemical Engineering* 12(5).

Hemmati, A., Shayegan, J., Sharratt, P., Yeo, T.Y., Bu, J., 2014. Solid products characterization in a multi-step mineralization process. *Chemical Engineering Journal* 252, 210–219.

Hou, G., Yan, Z., Sun, J., Naguib, H.M., Lu, B., Zhang, Z., 2021. Microstructure and mechanical properties of CO_2-cured steel slag brick in pilot-scale. *Construction and Building Materials* 271, 121581.

Huijgen, W., Ruijg, G.J., 2006. Energy consumption and net CO_2 sequestration of aqueous mineral mineralization. *Industrial & Engineering Chemistry Research* 45(26), 9184–9194.

Huijgen, W.J.J., Comans, R.N.J., Witkamp, G.-J., 2007. Cost evaluation of CO_2 sequestration by aqueous mineral mineralization. *Energy Conversion and Management* 48(7), 1923–1935.

Huijgen, W.J.J., Witkamp, G. J., Comans, R. N. J., 2005. Mineral CO_2 sequestration by steel slag mineralization. *Environmental Science & Technology* 39(24), 9676–9682.

Humbert, P.S., Castro-Gomes, J.P., Savastano, H., 2019. Clinker-free CO_2 cured steel slag based binder: Optimal conditions and potential applications. *Construction and Building Materials* 210, 413–421.

Huntzinger, D.N., Gierke, J.S., Kawatra, S.K., Eisele, T.C., Sutter, L.L., 2009. Carbon dioxide sequestration in cement kiln dust through mineral mineralization. *Environmental Science & Technology* 43(6), 1986–1992.

Jang, J.G., Kim, G.M., Kim, H.J., Lee, H.K., 2016. Review on recent advances in CO_2 utilization and sequestration technologies in cement-based materials. *Construction and Building Materials* 127, 762–773.

Jeon, M.K., Shin, J.M., Park, J.J., Park, G.I., 2012. Simulation of Cs behavior during the high temperature voloxidation process using the HSC chemistry code. *Journal of Nuclear Materials* 430(1–3), 37–43.

Jiajie Li, Wen Ni, Xue Wang, Sitao Zhu, Xinlei Wei, Fuxing Jiang, Hui Zeng, Hitch, M., 2022. Mechanical activation of medium basicity steel slag under dry condition for mineralization curing. *Journal of Building Engineering*(50), 104123.

Jo, H., Lee, M.-G., Park, J., Jung, K.-D., 2017. Preparation of high-purity nano-$CaCO_3$ from steel slag. *Energy* 120, 884–894.

Jose-Luis, P., Abadias, A., Valero, A., Valero, A., Reuter, M., 2019. The energy needed to concentrate minerals from common rocks: The case of copper ore. *Energy* 181, 494–503.

Kakizawa, M., Yamasaki, A., Yanagisawa, Y., 2001. A new CO_2 disposal process via artificial weathering of calcium silicate accelerated by acetic acid. *Energy* 26(4), 341–354.

Kashef-Haghighi, S., Shao, Y., Ghoshal, S., 2015. Mathematical modeling of CO_2 uptake by concrete during accelerated mineralization curing. *Cement and Concrete Research* 67, 1–10.

Kim, J., Azimi, G., 2021. The CO_2 sequestration by supercritical mineralization of electric arc furnace slag. *Journal of CO2 Utilization* 52.

Kim, S., Kim, J., Jeon, D., Yang, J., Moon, J., 2024. Enhanced mechanical property of steel slag through glycine-assisted hydration and mineralization curing. *Cement and Concrete Composites* 149.

Kodama, S., Nishimoto, T., Yamamoto, N., Yogo, K., Yamada, K., 2008. Development of a new pH-swing CO_2 mineralization process with a recyclable reaction solution. *Energy* 33(5), p.776–784.

Kriskova, L., Pontikes, Y., Cizer, Ö., Mertens, G., Veulemans, W., Geysen, D., Jones, P.T., Vandewalle, L., Van Balen, K., Blanpain, B., 2012. Effect of mechanical activation on the hydraulic properties of stainless steel slags. *Cement and Concrete Research* 42(6), 778–788.

Kumar, R., Bhattacharjee, B., 2003. Porosity, pore size distribution and in situ strength of concrete. *Cement and Concrete Research* 33(1), 155–164.

Lee, S., Kim, J.-W., Chae, S., Bang, J.-H., Lee, S.-W., 2016. CO_2 sequestration technology through mineral mineralization: An extraction and mineralization of blast slag. *Journal of CO2 Utilization* 16, 336–345.

Lee, Y.H., Eom, H., Lee, S.M., Kim, S.S., 2021. Effects of pH and metal composition on selective extraction of calcium from steel slag for $Ca(OH)_2$ production. *RSC Advances* 11(14), 8306–8313.

Lekakh, S.N., Rawlins, C.H., Robertson, D.G.C., Richards, V.L., Peaslee, K.D., 2008. Kinetics of aqueous leaching and carbonization of steelmaking slag. *Metallurgical and Materials Transactions B* 39(1), 125–134.

Li, J., Hitch, M., 2018. Mechanical activation of magnesium silicates for mineral mineralization, a review. *Minerals Engineering* 128, 69–83.

Li, J., Jin, Q., Zhang, W., Li, C., Monteiro, P.J.M., 2022. Microstructure and durability performance of sustainable cementitious composites containing high-volume regenerative biosilica. *Resources, Conservation and Recycling* 178.

Li, N., Unluer, C., 2023a. A comparative study of ethylenediamine tetraacetic acid induced gas-solid and liquid-solid accelerated mineralization for enhancement of steel slag aggregates. *Construction and Building Materials* 400.

Li, N., Unluer, C., 2023b. Development of high-volume steel slag as cementitious material by ethylenediamine tetraacetic acid induced accelerated mineralization. *Powder Technology* 428.

Li, W., Cao, M., Wang, D., Chang, J., 2023a. Increase in volume stability of RO phases in steel slag by combined treatment of alkali and dry mineralization. *Construction and Building Materials* 396.

Li, W., Cao, M., Wang, D., Zhao, J., Chang, J., 2023b. Improving the hydration activity and volume stability of the RO phases in steel slag by combining alkali and wet mineralization treatments. *Cement and Concrete Research* 172.

Li, Y., Liu, F., Yu, F., Du, T., 2024. A review of the application of steel slag in concrete. *Structures* 63.

Liang, X., Ye, Z., Chang, J., 2012. Early hydration activity of composite with carbonated steel slag. *Journal of the Chinese Ceramic Society* 40(2), 8.

Lioliou, M.G., Paraskeva, C.A., Koutsoukos, P.G., Payatakes, A.C., 2007. Heterogeneous nucleation and growth of calcium carbonate on calcite and quartz. *Journal of Colloid and Interface Science* 308(2), 421–428.

Liu, G., Schollbach, K., van der Laan, S., Tang, P., Florea, M.V.A., Brouwers, H.J.H., 2020. Recycling and utilization of high volume converter steel slag into CO_2 activated mortars – The role of slag particle size. *Resources, Conservation and Recycling* 160, 104883.

Liu, P., Mo, L., Zhong, J., Tang, M., 2024. In-situ investigation on the mineralization behaviors of various mineral phases in steel slag: The role of RO phase. *Cement and Concrete Composites* 149.

Liu, Q., Liu, J., Qi, L., 2016. Effects of temperature and mineralization curing on the mechanical properties of steel slag-cement binding materials. *Construction and Building Materials* 124, 999–1006.

Liu, W., Teng, L., Rohani, S., Qin, Z., Zhao, B., Xu, C.C., Ren, S., Liu, Q., Liang, B., 2021. CO_2 mineral mineralization using industrial solid wastes: A review of recent developments. *Chemical Engineering Journal* 416, 129093.

Liu, X., Feng, P., Cai, Y., Yu, X., Yu, C., Ran, Q., 2022. mineralization behavior of calcium silicate hydrate (C-S-H): Its potential for CO_2 capture. *Chemical Engineering Journal* 431.

Luo, Z., Wang, Y., Yang, G., Ye, J., Zhang, W., Liu, Z., Mu, Y., 2021. Effect of curing temperature on mineralization behavior of steel slag compacts. *Construction and Building Materials* 291, 123369.

Ma, Z., Liao, H., Wang, L., Cheng, F., 2021. Effects of iron/silicon/magnesium/aluminum on CaO mineralization of CO_2 in steel slag-based building materials during mineralization curing. *Construction and Building Materials* 298, 123889.

Maoxia, T., Ze, L., Xiaofang, L., Hongxin, Z., Lina, W., Junling, Z., Desheng, C., Wenwan, S., Tao, Q., 2015. Carbon dioxide sequestration using water quenched steel slag. *Chinese Journal of Environmental Engineering* 9(9), 4514–4518.

Martins, A.C.P., Franco de Carvalho, J.M., Costa, L.C.B., Andrade, H.D., de Melo, T.V., Ribeiro, J.C.L., Pedroti, L.G., Peixoto, R.A.F., 2021. Steel slags in cement-based composites: An ultimate review on characterization, applications and performance. *Construction and Building Materials* 291, 123265.

Mayes, William, Matthew, Riley, Alex, L., Gomes, Helena, I., Brabham, 2018. Atmospheric CO_2 sequestration in iron and steel slag: Consett, County Durham, United Kingdom. *Environmental Science & Technology.*

Mo, A., Mj, A., Pt, B., As, A., 2019. Experimental study on the extraction of calcium, magnesium, vanadium and silicon from steelmaking slags for improved mineral mineralization of CO_2. *Journal of CO2 Utilization* 31, 1–7.

Mo, L., Yang, S., Huang, B., Xu, L., Deng, M., 2020. Preparation, microstructure and property of carbonated artificial steel slag aggregate used in concrete. *Cement and Concrete Composites* 113, 103715.

Mo, L., Zhang, F., Deng, M., 2016. Mechanical performance and microstructure of the calcium carbonate binders produced by carbonating steel slag paste under CO_2 curing. *Cement and Concrete Research* 88, 217–226.

Muhmood, L., Vitta, S., Venkateswaran, D., 2009. Cementitious and pozzolanic behavior of electric arc furnace steel slags. *Cement and Concrete Research* 39(2), 102–109.

Myers, C.A., Nakagaki, T., Akutsu, K., 2019. Quantification of the CO_2 mineralization potential of ironmaking and steelmaking slags under direct gas-solid reactions in flue gas. *International Journal of Greenhouse Gas Control* 87, 100–111.

Nakano, Y., Ioku, K., Fujimori, H., Goto, S., Takahashi, T., 2005. Heat liberation of mineralization reaction of steel-making slag. *Journal- Ceramic Society Japan* 112(5), S1381–1383.

Nežerka, V., Bílý, P., Hrbek, V., Fládr, J., 2019. Impact of silica fume, fly ash, and metakaolin on the thickness and strength of the ITZ in concrete. *Cement and Concrete Composites* 103, 252–262.

Nielsen, P., Boone, M.A., Horckmans, L., Snellings, R., Quaghebeur, M., 2020. Accelerated mineralization of steel slag monoliths at low C_{O2} pressure – microstructure and strength development. *Journal of CO2 Utilization* 36, 124–134.

Pan, S.-Y., Adhikari, R., Chen, Y.-H., Li, P., Chiang, P.-C., 2016. Integrated and innovative steel slag utilization for iron reclamation, green material production and CO_2 fixation via accelerated mineralization. *Journal of Cleaner Production* 137, 617–631.

Polettini, A., Pomi, R., Stramazzo, A., 2016a. Carbon sequestration through accelerated mineralization of BOF slag: Influence of particle size characteristics. *Chemical Engineering Journal* 298, 26–35.

Polettini, A., Pomi, R., Stramazzo, A., 2016b. CO_2 sequestration through aqueous accelerated mineralization of BOF slag: A factorial study of parameters effects. *Journal of Environmental Management* 167, 185–195.

Ragipani, R., Bhattacharya, S., Suresh, A.K., 2021. A review on steel slag valorisation via mineral mineralization. *Reaction Chemistry & Engineering* 6(7), 1152–1178.

Roslan, N.H., Ismail, M., Abdul-Majid, Z., Ghoreishiamiri, S., Muhammad, B., 2016. Performance of steel slag and steel sludge in concrete. *Construction and Building Materials* 104, 16–24.

Said, A., Järvinen, M., Laukkanen, T., 2016. Pilot-scale experimental work on carbon dioxide sequestration using steelmaking slag. *Applied Energy* 177, 602–611.

Said, A., Mattila, O., Eloneva, S., Järvinen, M., 2015. Enhancement of calcium dissolution from steel slag by ultrasound. *Chemical Engineering & Processing Process Intensification* 89, 1–8.

Salman, M., Cizer, Ö., Pontikes, Y., Santos, R.M., Snellings, R., Vandewalle, L., Blanpain, B., Van Balen, K., 2014. Effect of accelerated mineralization on AOD stainless steel slag for its valorisation as a CO_2-sequestering construction material. *Chemical Engineering Journal* 246, 39–52.

Sanna, A., Marotovaler, M.M., 2013. Dissolution of steel slag and recycled concrete aggregate in ammonium bisulphate for CO_2 mineral mineralization. *Fuel Processing Technology* 113, 8.

Sanna, A., Uibu, M., Caramanna, G., Kuusik, R., Maroto-Valer, M.M., 2014. A review of mineral mineralization technologies to sequester CO_2. *Chemical Society Reviews* 43(23), 8049–8080.

Saran, R.K., Arora, V., Yadav, S., 2018. CO_2 Sequestration by mineral mineralization: A review. *Global NEST Journal* 20(3), 497–503.

Scrivener, K., Snellings, R., Lothenbach, B., 2016. *A practical guide to microstructural analysis of cementitious materials*. CRC Press.

Setién, J., Hernández, D., González, J.J., 2009. Characterization of ladle furnace basic slag for use as a construction material. *Construction and Building Materials* 23(5), 1788–1794.

Shen, Z., Xiao, Y., Hou, K., Chen, Y., Meng, F., 2024. Investigating the impact of porous steel slag aggregates on the low-temperature crack resistance of asphalt concrete. *Construction and Building Materials* 453.

Shi, C., 2002. Characteristics and cementitious properties of ladle slag fines. *Cement & Concrete Research* 32, 459–462.

Shi, C., 2004. Steel slag—its production, processing, characteristics, and cementitious properties. *Journal of Materials in Civil Engineering* 16(3), 230–236.

Shi, C., He, F., Wu, Y., 2012a. Effect of pre-conditioning on CO_2 curing of lightweight concrete blocks mixtures. *Construction and Building Materials* 26(1), 257–267.

Shi, C., Meng, L., He, P., Ou, Z., 2012b. Factors affecting kinetics of CO_2 curing of concrete. *Journal of Sustainable Cement-Based Materials* 1(1–2), 24–33.

Shi, C., Wang, D., He, F., Liu, M., 2012c. Weathering properties of CO_2-cured concrete blocks. *Resources, Conservation and Recycling* 65, 11–17.

Shi, C., Wu, Y., 2008. Studies on some factors affecting CO_2 curing of lightweight concrete products. *Resources, Conservation and Recycling* 52(8–9), 1087–1092.

Steinour, H.H., 1959. Some effects of carbon dioxide on mortars and concrete-discussion. *Journal of the American Concrete Institute* 30(2), 905–907.

Stockmann, G.J., Wolff-Boenisch, D., Bovet, N., 2014. The role of silicate surfaces on calcite precipitation kinetics. *Geochimica et Cosmochimica Acta* 135, 231–250.

Sun, X., Liu, J., Zhao, Y., Zhao, J., Li, Z., Sun, Y., Qiu, J., Zheng, P., 2022. Mechanical activation of steel slag to prepare supplementary cementitious materials: A comparative research based on the particle size distribution, hydration, toxicity assessment and carbon dioxide emission. *Journal of Building Engineering* 60.

Sun, Y., Yao, M.-S., Zhang, J.-P., Yang, G., 2011. Indirect CO_2 mineral sequestration by steelmaking slag with NH_4Cl as leaching solution. *Chemical Engineering Journal* 173(2), 437–445.

Tanaka, H., 2014. Resources trend and use of directly reduced iron in steelmaking process. *Research and Development Kobe Steel Engineering Reports* 64(1), 2–7.

Teir, S., Eloneva, S., Fogelholm, C.J., Zevenhoven, R., 2007. Dissolution of steelmaking slags in acetic acid for precipitated calcium carbonate production. *Energy* 32(4), 528–539.

Thiery, M., Villain, G., Dangla, P., Platret, G., 2007. Investigation of the mineralization front shape on cementitious materials: Effects of the chemical kinetics. *Cement and Concrete Research* 37(7), 1047–1058.

Tian, S., Jiang, J., Hosseini, D., Kierzkowska, A.M., Imtiaz, Q., Broda, M., Muller, C.R., 2015. Development of a steel-slag-based, iron-functionalized sorbent for an autothermal carbon dioxide capture process. *ChemSusChem* 8(22), 3839–3846.

Tian, S., Li, K., Jiang, J., Chen, X., Yan, F., 2016. CO_2 abatement from the iron and steel industry using a combined Ca–Fe chemical loop. *Applied Energy* 170, 345–352.

Tong, Z., Ma, G., Zhou, D., Yang, G., Peng, C., 2019. The indirect mineral mineralization of electric arc furnace slag under microwave irradiation. *Scientific Reports* 9(1), 1–7.

Tossavainen, M., Engstrom, F., Yang, Q., Menad, N., Lidstrom Larsson, M., Bjorkman, B., 2007. Characteristics of steel slag under different cooling conditions. *Waste Manag* 27(10), 1335–1344.

Tu, Z., Guo, M.-Z., Poon, C.S., Shi, C., 2016. Effects of limestone powder on $CaCO_3$ precipitation in CO_2 cured cement pastes. *Cement and Concrete Composites* 72, 9–16.

Ukwattage, N.L., Ranjith, P.G., Li, X., 2017. Steel-making slag for mineral sequestration of carbon dioxide by accelerated mineralization. *Measurement* 97, 15–22.

Wang, D., Fang, Y., Zhang, Y., Chang, J., 2019. Changes in mineral composition, growth of calcite crystal, and promotion of physico-chemical properties induced by mineralization of β-C_2S. *Journal of CO2 Utilization* 34, 149–162.

Wang, H., Gu, X., Xu, X., Liu, J., Zhu, Z., Wang, S., 2024. The effects of iron tailings on steel slag-based cementitious systems: Strength, hydration products, and volume stability. *Journal of Building Engineering* 91.

Wang, Q., Wang, D., Zhuang, S., 2017. The soundness of steel slag with different free CaO and MgO contents. *Construction and Building Materials* 151, 138–146.

Wang, R., Jin, P., Dong, H., Liu, Y., Ding, Z., Zhang, W., 2021. Effect of moist content on the bio-carbonated steel slag bricks. *Construction and Building Materials* 269, 121313.

Wang, X., Ni, W., Li, J., Zhang, S., Hitch, M., Pascual, R., 2019. mineralization of steel slag and gypsum for building materials and associated reaction mechanisms. *Cement and Concrete Research* 125, 105893.

Wang, Y., Lu, B., Hu, X., Liu, J., Zhang, Z., Pan, X., Xie, Z., Chang, J., Zhang, T., Nehdi, M.L., Shi, C., 2021. Effect of CO_2 surface treatment on penetrability and microstructure of cement-fly ash–slag ternary concrete. *Cement and Concrete Composites* 123, 104194.

Wei, X., Ni, W., Zhang, S., Wang, X., Li, J., Du, H., 2022. Influence of the key factors on the performance of steel slag-desulphurisation gypsum-based hydration-mineralization materials. *Journal of Building Engineering* 45.

Worldsteel, 2025. *December 2024 crude steel production and 2024 global crude steel production totals*. https://worldsteel.org/media/press-releases/2025/december-2024-crude-steel-production-and-2024-global-totals/. Accessed 24 January 2025.

Wu, H.Z., Chang, J., Pan, Z.Z., Cheng, X., 2009. Carbonate steelmaking slag to manufacture building materials. *Advanced Materials Research* 79–82, 1943–1946.

Yadav, S., Mehra, A., 2021. A review on ex situ mineral mineralization. *Environmental Science and Pollution Research International* 28(10), 12202–12231.

Yan-Feng, F., Ting-Ting, Z., Jun, C., 2015. Accelerated mineralization of different minerals of steel slag. *Journal of Chinese Electron Microscopy Society* 34(6), 464–469.

Yang, C., Xie, J., Wu, S., Amirkhanian, S., Zhou, X., Kong, D., Ye, Q., Hu, J., 2020. Revelation and characterization of selective absorption behavior of bitumen to basic oxygen furnace slag. *Construction and Building Materials* 253.

Yang, S., Mo, L., Deng, M., 2021. Effects of ethylenediamine tetra-acetic acid (EDTA) on the accelerated mineralization and properties of artificial steel slag aggregates. *Cement and Concrete Composites* 118, 103948.

Yang, S., Tang, Z., Gao, S., Mo, L., Lu, D., 2024. Elevating mineralization efficiency and CO_2 capture capacity of steel slag by sodium tripolyphosphate (STP) The role of $CaCO_3$ morphology modification. *Cement and Concrete Composites* 153, 105716.

Ye, Y., Wu, S., Li, C., Kong, D., Shu, B., 2019. Morphological discrepancy of various basic oxygen furnace steel slags and road performance of corresponding asphalt mixtures. *Materials* (Basel) 12(14).

Yi, H., Xu, G., Cheng, H., Wang, J., Wan, Y., Chen, H., 2012. An overview of utilization of steel slag. *Procedia Environmental Sciences* 16, 791–801.

Yi, Y.-R., Lin, Y., Du, Y.-C., Bai, S.-q., Ma, Z.-l., Chen, Y.-g., 2021. Accelerated mineralization of ladle furnace slag and characterization of its mineral phase. *Construction and Building Materials* 276, 122235.

Yildirim, I.Z., Prezzi, M., 2011. Chemical, mineralogical, and morphological properties of steel slag. *Advances in Civil Engineering* 2011, 1–13.

Yong, S., Yao, M.S., Zhang, J.P., Yang, G., 2011. Indirect CO_2 mineral sequestration by steelmaking slag with NH_4Cl as leaching solution. The *Chemical Engineering Journal* 173(2), 437–445.

Zajac, M., Skocek, J., Durdzinski, P., Bullerjahn, F., Skibsted, J., Ben Haha, M., 2020. Effect of carbonated cement paste on composite cement hydration and performance. *Cement and Concrete Research* 134, 106090.

Zhang, Q., Feng, P., Shen, X., Cai, Y., Zhen, H., Liu, Z., 2025. Comparative analysis of mineralization strengthening mechanisms in full solid waste materials: Steel slag vs. carbide slag. *Cement and Concrete Composites* 157.

Zhang, S., Ghouleh, Z., Liu, J., Shao, Y., 2021. Converting ladle slag into high-strength cementing material by flue gas mineralization at different temperatures. *Resources, Conservation and Recycling* 174, 105819.

Zhang, S., Ghouleh, Z., Mucci, A., Bahn, O., Provençal, R., Shao, Y., 2022. Production of cleaner high-strength cementing material using steel slag under elevated-temperature mineralization. *Journal of Cleaner Production* 342.

Zhang, T., Yu, Q., Wei, J., Zhang, P., Chen, P., 2011. A gap-graded particle size distribution for blended cements: Analytical approach and experimental validation. *Powder Technology* 214(2), 259–268.

Zhang, X., Qian, C., Yi, H., Ma, Z., 2021. Study on mineralization reactivity of silicates in steel slag accelerated by Bacillus mucilaginosus. *Construction and Building Materials* 292, 123433.

Zhang, X., Su, Y., Liu, S., 2024. The effect of mineralization pressure on microbial assisted steel slag-based carbon sequestration material. *Journal of Building Engineering* 86.

Zhang, X., Zhao, J., Liu, Y., Li, J., 2024. Use of steel slag as mineralization material: A review of mineralization methods and evaluation, environmental factors and carbon conversion process. *Journal of CO2 Utilization* 88.

Zhang, X., Zhao, S., Liu, Z., Wang, F., 2019. Utilization of steel slag in ultra-high performance concrete with enhanced eco-friendliness. *Construction and Building Materials* 214, 28–36.

Zhao, Q., Li, J., You, K., Liu, C., 2019. Recovery of calcium and magnesium bearing phases from iron– and steelmaking slag for CO_2 sequestration. *Process Safety and Environmental Protection* 135(85).

Zhao, Q., Liu C.-j., Jiang, M.-f., Saxén, H., Zevenhoven, R., 2015. Preparation of magnesium hydroxide from serpentinite by sulfuric acid leaching for CO_2 mineral mineralization. *Minerals Engineering* 79, 116–124.

Zhao, W., Wen, W., Li, H., Hu, J., 2024. Research on the performance of asphalt mixture with acid-treated steel slag based on microscopic properties. *Construction and Building Materials* 455.

Zhong, X., Li, L., Jiang, Y., Ling, T.-C., 2021. Elucidating the dominant and interaction effects of temperature, CO_2 pressure and mineralization time in carbonating steel slag blocks. *Construction and Building Materials* 302, 124158.

Zhu, X., Tsang, D.C.W., Wang, L., Su, Z., Hou, D., Li, L., Shang, J., 2020. Machine learning exploration of the critical factors for CO_2 adsorption capacity on porous carbon materials at different pressures. *Journal of Cleaner Production* 273.

10 CO_2 mineralization in magnesium-based materials

10.1 INTRODUCTION

Magnesium-based materials have gained significant interest due to their high CO_2 capturing capability and enhanced early strength development. CO_2 mineralized magnesium-based materials represent a transformative and potential low-carbon alternative to conventional Portland cement, offering a viable pathway to mitigate the carbon footprint of the cement and concrete industry. Notably, magnesium-based materials exhibit comparable or even superior CO_2 mineralization potential compared to their calcium-based counterparts. Among various natural minerals, magnesium silicate-based minerals have become the first choice for CO_2 mineralization due to their excess abundance in nature, high magnesium content, and wide global distribution. It is estimated that the resources of magnesium silicate-based minerals are capable of storing all anthropogenic CO_2 emissions in the next few decades or even 1,000 years (Scott et al., 2021). Apart from this, magnesium silicate-based minerals can act as an excellent alternative for sustainable binder production in regions where limestone and clay minerals are scarce. For instance, MgO-based cement has been put into industrial practice for about 160 years. These materials show an excellent CO_2 absorption potential (1.09 metric tons per metric ton of MgO) (Winnefeld et al., 2022). Especially, reactive magnesia cement exhibits a CO_2 uptake potential of up to 92.8 wt% (Wang et al., 2020), which is significantly higher than that of ordinary Portland cement (46–58 wt%) (Fang et al., 2017).

The mechanism and product formation of CO_2 mineralization in magnesium-based materials show a complex behavior. These materials tend to form a series of anhydrous and hydrated magnesium carbonates during CO_2 mineralization. The formed carbonates are well-developed crystals and show different microscopic morphologies, thus contributing to a distinctive manner in the densification of the matrix. CO_2 mineralization is capable of effectively enhancing the mechanical properties and durability of cementitious materials. In comparison with calcium-based carbonate binders, magnesium-based carbonate binders have been reported to exhibit higher compressive strength (De Silva et al., 2009), stronger binding ability (D. Wang et al., 2019b), and superior resistance to acid and sulfate attack (Liska and Al-Tabbaa, 2012a). However, the amount of CO_2 captured and stored by CO_2 mineralization of magnesium-based materials in practice is much lower than the theoretical value due

DOI: 10.1201/9781003627531-10

to various factors (Shen et al., 2016), like CO_2 concentration and pressure, relative humidity, temperature, mineralization time, incorporated additives, and seeds.

10.2 MAGNESIUM SOURCES WITH CO_2 MINERALIZATION POTENTIAL

The magnesium sources with CO_2 mineralization potential can be mainly divided into two categories: (1) magnesium silicate-based minerals (such as Olivine, Serpentine); and (2) magnesia-based binders (such as the Mg oxychloride cements (MOC), Mg phosphate cements (MPC), Mg oxysulfate cements (MOS), and reactive magnesia cements (RMC)).

10.2.1 Magnesium silicate-based minerals

Magnesium silicate-based minerals (xMgO·ySiO$_2$·zH$_2$O) are the most studied natural minerals for CO_2 mineralization due to their natural abundance, wide geographic distribution with global reserves of over 30 trillion tons and, more importantly, higher CO_2 mineralization capacity than calcium silicate-based minerals (Lackner, 2003). The main subgroups of magnesium silicates are olivine $(Mg, Fe)_2SiO_4$ and serpentine $Mg_3Si_2O_5(OH)_4$, accounting for around two-thirds of all magnesium silicate-based minerals (Lackner, 2003). Forsterite (Mg_2SiO_4), being the magnesium end member of the olivine family, has the highest Mg content and thus exhibits the highest CO_2 mineralization potential.

Olivine has an orthorhombic crystal structure and is mainly found in columnar or lamellar crystal form. It is a nesosilicate that consists of isolated silicate tetrahedrons with no shared oxygen bonds and linked with their divalent cations occupying the octahedral sites between the tetrahedral sheets (Kerisit et al., 2013). Olivine has a relatively high hardness (6.5–7) and density (3.2–3.4 g/cm^3). Serpentine is classified as a phyllosilicate in which sheet-like layers are formed due to shared oxygen bonds across silica tetrahedra and held together by van der Waals forces and hydrogen bonds. Serpentine tends to form various arrangements and has three polymorphs, including chrysotile, lizardite, and antigorite, whose structures are cyclic, cylindrical, planar, and wavy, respectively. Serpentine has a relatively low hardness of 2.5–4 and a density of 2.2–2.9 g/cm^3. Olivine possesses higher reactivity and CO_2 uptake potential than serpentine (63 wt% vs. 48 wt%).

The thermodynamic parameters of forsterite and serpentine, as well as their CO_2 mineralization reaction, are listed in Table 10.1. As demonstrated in Table 10.1, the total free energies of the CO_2 mineralization reaction for forsterite and serpentine are both negative, which indicates that the CO_2 mineralization reaction can take place spontaneously and give off energy under ambient conditions. However, the CO_2 mineralization reaction of magnesium silicate-based minerals is mainly controlled by kinetics with quite a slow rate and reaction degree (only 1–4%) in ambient environments. The realization of reasonable reaction rates and CO_2 mineralization degree relies on high temperatures and high CO_2 partial pressures resulting in energy consumption. The practical industrial application of magnesium silicate-based minerals

TABLE 10.1
Gibbs energy of CO_2 mineralization for forsterite and serpentine calculated from tabulated thermodynamic data at 25°C and 1 bar

Phase	Formula	ΔG_f^0 (kJ·mol⁻¹)	Reaction	ΔG_r^0 (kJ·mol⁻¹)
forsterite	Mg_2SiO_4	–2053.6	$Mg_2SiO_4 + 2CO_2 \rightarrow 2MgCO_3 + SiO_2$	–66.80
serpentine	$Mg_3Si_2O_5(OH)_4$	–4032.4	$Mg_3Si_2O_5(OH)_4 + 3CO_2 \rightarrow 3MgCO_3 + 2SiO_2 + 2H_2O$	–39.20

Source: Renforth et al., 2011; Xu et al., 2001.

for CO_2 sequestration faces challenges related to economic viability and environmental sustainability.

10.2.2 Magnesia-based binders

The most desirable materials for CO_2 mineralization reactions are oxides and hydroxides. MgO-based binders have been identified as a potential alternative to PC and have actually been used in industrial practice. Magnesium occurs in nature mainly in solid minerals as carbonates, silicates, and as chlorides in brines or seawater. Magnesium oxide is usually obtained by two methods: (1) calcination of magnesia-based minerals (dry-route) and (2) extraction from brine or seawater (wet-route). The reactivity of MgO is determined by physical and chemical properties such as specific surface area, particle size, agglomeration ratio (i.e., the ratio of the primary particle size to the crystallite size), crystal structure and morphology, density, pore structure, and morphology. Magnesium oxide tends to exhibit different properties depending on the source and manufacturing process. Aramendia et al. (Aramendia et al., 1996) have reported that magnesium oxide from the calcination of brucite exhibited a higher specific surface area and alkalinity compared to that from the calcination of hydromagnesite at the same calcination temperature. Magnesite is prevalently exploited to produce MgO on an industrial scale by calcination, known as the dry route. The calcination of magnesite is carried out in two different steps: (1) At lower calcination temperatures, magnesite decomposes into magnesium oxide and CO_2, and the escape of CO_2 gas results in the generation of MgO with a porous structure and therefore a high specific surface area reactivity. The complete decomposition temperature of magnesite is around 500 °C, and the difference in complete decomposition temperature is mainly affected by the impurity content, crystal structure, and microstructure of magnesite. (2) With the further increase of the calcination temperature above 900°C, magnesium oxide will begin to sinter, which leads to an increase in the size of magnesium oxide grains and a decrease in the specific surface area and reactivity. Birchal et al. (2000) has indicated that the effect of calcination

TABLE 10.2
Different grades, properties, and applications of MgO

Grade	Calcination temperature (°C)	Reactivity	SSA (m^2/g)	Crystallinity	Applications
Light-burned (reactive, caustic calcined)	700–1000 (600–1300)[a]	Highest	>20	Lowest	RMC, MOS, and MOC
Hard-burned	1000–1500	Lower	1–20	Higher	MOC and MOS; MEA in concrete
Dead-burned (periclase)	1500–2000 (1600–2000)[a]	Even lower	<1	Even higher	Refractory, MPC, MOC, and MOS
Fused [a]	>2800[a]	Lowest	0.01–0.1	Highest	Refractory

[a] proposed by the European Commission.

temperature on the specific surface area of magnesium oxide was most significant. Therefore, MgO is typically classified into four categories according to calcination temperature via the dry process route. Various grades of MgO exhibit different properties and applications in construction industries (see Table 10.2).

At present, the major magnesium-based binders in construction materials include magnesium oxysulfate (MOS) cement, magnesium phosphate cement (MPC), magnesium oxychloride cement (MOC), and reactive magnesium cement (RMC).

The reactivity of MPC is based on an acid-base reaction between dead-burnt MgO and a soluble acid phosphate (e.g., ammonium and potassium phosphate) to form magnesium phosphate salts (such as $NH_4MgPO_4 \cdot 6H_2O$ or $MgKPO_4 \cdot 6H_2O$) to gain cementitious properties.

MOC was first reported by Sorel in a 1866 patent. The reactivity of MOC is based on the aqueous reaction between MgO and $MgCl_2$ to form basic magnesium chloride salts which have binding capacity expressed as $x\text{Mg(OH)}_2 \cdot y\text{MgCl}_2 \cdot z\text{H}_2\text{O}$ or $x\text{MgO} \cdot y\text{MgCl}_2 \cdot (z + x)\text{H}_2\text{O}$, including two $\text{Mg(OH)}_2 \cdot \text{MgCl}_2 \cdot 4\text{H}_2\text{O}$ (phase 2), three $\text{Mg(OH)}_2 \cdot \text{MgCl}_2 \cdot 8\text{H}_2\text{O}$ (phase 3), five $\text{Mg(OH)}_2 \cdot \text{MgCl}_2 \cdot 8\text{H}_2\text{O}$ (phase 5), and nine $\text{Mg(OH)}_2 \cdot \text{MgCl}_2 \cdot 5\text{H}_2\text{O}$ (phase 9).

MOS is similar in concept to MOC and is based on the aqueous reaction between MgO and $MgSO_4$. The cementitious phases in MOS are $x\text{Mg(OH)}_2 \cdot y\text{MgSO}_4 \cdot z\text{H}_2\text{O}$ phases including $3\text{Mg(OH)}_2 \cdot \text{MgSO}_4 \cdot 8\text{H}_2\text{O}$ (3·1·8), five $\text{Mg(OH)}_2 \cdot \text{MgSO}_4 \cdot 3\text{H}_2\text{O}$ (5·1·3), $\text{Mg(OH)}_2 \cdot 2\text{MgSO}_4 \cdot 3\text{H}_2\text{O}$ (1·2·3), and $\text{Mg(OH)}_2 \cdot \text{MgSO}_4 \cdot 5\text{H}_2\text{O}$ (1·1·5).

MPC has the lowest carbon sequestration potential due to the low reactivity of dead-burned MgO and magnesium phosphate salts. Though MOS and MOC can use high-reactivity reactive magnesia as raw materials, magnesia in MOS and MOC mainly undergoes hydration instead of CO_2 mineralization reactions.

Further, xMg(OH)$_2$·yMgSO$_4$·zH$_2$O phases are reported to be almost unable to be CO_2 mineralized while xMg(OH)$_2$·yMgCl$_2$·zH$_2$O has limited CO_2 mineralization capacity.

John Harrison first proposed and obtained a patent for "Reactive magnesium oxide cements" in 2001. The hydration rate of reactive magnesium oxide is equal to Portland cement, which enables reactive magnesium oxide to be used as a cementitious material together with PC or even separately (Vandeperre et al., 2008). However, the low hydration degree (40–80%) due to the low solubility of magnesia and the porous structure of the hydrated phase, brucite, sets limits on the strength of RMC. For example, it has been reported that the compressive strength of reactive magnesium oxide cement can reach only half of that of OPC under sealed conditions (Bhagath Singh et al., 2020). On the contrary, RMC exhibits great CO_2 sequestration capacity (1.09 metric tons per metric ton of reactive MgO) (Winnefeld et al., 2022). Further, RMC itself and its hydration product $Mg(OH)_2$ both have high CO_2 mineralization reactivity. The CO_2 mineralization of RMC leads to the formation of magnesium carbonates, which create continuous networks, fill in the pores, densify the matrix microstructure, and thereby facilitate strength development. The compressive strength of CO_2 mineralized RMC (CRMC) can reach around 20–85 MPa, which is comparable to that of conventional (Grünhäuser Soares and Castro-Gomes, 2021). Thus, CRMC has aroused great interest as one of the most promising low-carbon substitutes for PC.

The decomposition temperature to produce reactive MgO from magnesite is quite lower compared to the production of Portland cement from limestone (700–1000 vs. 1450°C). However, the reported total CO_2 emissions for the production of MgO from magnesite were higher than that of the PC (1360–1600 kg CO_2-eq vs. 811–977 kg CO_2-eq) (Ruan and Unluer, 2016; Zhang et al., 2021). Nevertheless, considering the CO_2 offsetting capacity of 430 kg CO_2/t RMC, the net CO_2 emissions are reduced to 930–1170 kg CO_2-eq.

The net CO_2 emissions of CRMC can be decreased if reactive magnesia is obtained by the wet route, i.e., from magnesium-rich brines and seawater. The wet process consists of two steps: Precipitation of magnesium ions from seawater/brine by raising the pH to an appropriate value by the addition of alkali, i.e., lime (CaO), dolomite lime (CaO-MgO), sodium hydroxide, and ammonium hydroxide. (2) Generation of magnesium oxide by calcination of attained precipitates. Thus, no direct CO_2 emissions are involved in the wet route. Various magnesium-rich brines and seawater containing high concentrations of Mg^{2+} constitute another significant source of magnesium. Seawater contains an average Mg^{2+} concentration of 1.29–1.35 g/L, thus around 125 gallons of seawater can produce 1 kg of MgO (Shand, 2006). Natural brines present in arid and semiarid environments are more salt-concentrated solutions than seawater. The salt concentration of reject brines, from oil and gas production, mineral extraction, cooling towers in power stations, desalination plants, and other chemical industries, may vary from 1000–3000000 ppm. An estimated amount of around 14% of the total worldwide magnesia supply is extracted from seawater and brines through the wet route (John Wiley and Sons, 2006).

10.3 MECHANISM, PRODUCTS, AND MICROSTRUCTURES OF CO_2 MINERALIZATION IN MAGNESIUM-BASED MATERIALS

10.3.1 CO_2 MINERALIZATION MECHANISM

CO_2 mineralization can be divided into two categories: direct CO_2 mineralization and indirect CO_2 mineralization (also known as the pH swing process), which are differentiated by the process configuration. Direct CO_2 mineralization involves a one-step process, while indirect CO_2 mineralization employs a multi-stage process.

Direct CO_2 mineralization can be further classified into dry CO_2 mineralization and aqueous CO_2 mineralization, depending on the phases present in the system. When the moisture level is typically below 0.2, CO_2 mineralization can occur as a gas-solid reaction, where gaseous CO_2 reacts directly with solid magnesium-based minerals to form magnesium carbonates. This reaction mechanism is the simplest, as shown below for the CO_2 mineralization of magnesia and forsterite olivine in reactions (10.1) and (10.2). However, under natural reaction conditions, the gas-solid CO_2 mineralization reaction is limited by poor reaction kinetics and diffusion, resulting in slow reaction rates and CO_2 mineralization efficiency. Dry CO_2 mineralization reactions rely on high pressure and high temperature for reasonable reaction rates, which is energy-intensive. It has been reported that around 10% MgO is only converted at 300°C and 20 bar after 400 minutes (Fagerlund et al., 2012).

$$\mathrm{MgO}(s)+\mathrm{CO_2}(g)\rightarrow \mathrm{MgCO_3}(s) \tag{10.1}$$

$$\mathrm{Mg_2SiO_4}(s)+2\mathrm{CO_2}(g)\rightarrow 2\mathrm{MgCO_3}(s)+\mathrm{SiO_2}(s) \tag{10.2}$$

$$\mathrm{Mg_3Si_2O_5(OH)_{4(s)}}+\mathrm{CO_{2(g)}}\rightarrow \mathrm{MgCO_{3(aq)}}+\mathrm{SiO_{2(aq)}}+\mathrm{H_2O} \tag{10.3}$$

The rates of CO_2 mineralization reactions involving liquid phases significantly increase but are accompanied by complex processes and mechanisms. In contrast to the one-step gas-solid CO_2 mineralization process, aqueous CO_2 mineralization involves three steps. The detailed steps are as follows: (Step 1) gaseous CO_2(g) dissolves in water to form CO_2(aq), which then reacts with OH^- and H_2O molecules to dissociate into H_2CO_3(aq), HCO_3^-(aq), CO_3^{2-}(aq), and H_3O^+(aq) corresponding to reaction (10.4); (Step 2) Magnesium-bearing minerals dissolve with the leaching of Mg^{2+} into the solution; (Step 3) Mg^{2+} reacts with CO_3^{2-} to form magnesium carbonate and hydrated magnesium carbonates (HCMs) corresponding to reactions (10.5)–(10.10).

$$CO_{2(g)}=CO_{2(aq)}+H_2O_{(l)}\rightarrow HCO_{3(aq)}+H^+{}_{(aq)}\rightarrow CO^{2-}{}_{3(aq)}+2H^+ \tag{10.4}$$

$$Mg^{2+}+CO_3{}^{2-}\rightarrow \mathrm{MgCO_3}\quad (Magnesite) \tag{10.5}$$

$$Mg^{2+} + CO_3^{2-} + 3H_2O \rightarrow MgCO_3 \cdot 3H_2O \quad (\textit{Nesquehonite}) \tag{10.6}$$

$$Mg^{2+} + CO_3^{2-} + 5H_2O \rightarrow MgCO_3 \cdot 5H_2O \quad (\textit{Landsfordite}) \tag{10.7}$$

$$Mg^{2+} + CO_3^{2-} + 3H_2O + 2OH^- \rightarrow Mg_2CO_3(OH)_2 \cdot 3H_2O \quad (\textit{Artinite}) \tag{10.8}$$

$$5Mg^{2+} + 4CO_3^{2-} + 4H_2O + 2OH^- \rightarrow Mg_5(CO_3)_4(OH)_2 \cdot 4H_2O \quad (\textit{Hydromagnesite}) \tag{10.9}$$

$$5Mg^{2+} + 4CO_3^{2-} + 5H_2O + 2OH^- \rightarrow Mg_5(CO_3)_4(OH)_2 \cdot 5H_2O \quad (\textit{Dypingite}) \tag{10.10}$$

In aqueous CO_2 mineralization reaction systems, the overall reaction kinetics can be controlled by these processes, especially at the boundary layers. The solubility of CO_2 in aqueous solutions is a critical factor in the initial step of aqueous CO_2 mineralization, especially when the process occurs under extreme conditions, including high temperatures and pressures. The dissolution of CO_2 is influenced by gas pressure, temperature, and alkalinity. The following step, the leaching of Mg^{2+} from the minerals, is usually the rate-determining step of the overall aqueous CO_2 mineralization. The solubility of MgO is about five orders of magnitude lower than that of CaO ($K_{SP} = -4.92$), and the MgO dissolution rate is reported to be two orders of magnitude slower than the dissolution rate of CaO, which strongly affects the velocity and potential for magnesium oxide (Back et al., 2011).

The dissolution of magnesium silicate-based minerals includes two stages. The first stage is a rapid dissolution controlled by surface reaction and usually lasts for about 5 min. The second phase is controlled by diffusion of the silica-rich product layer (F. Wang et al., 2019b). The leaching of Mg^{2+} from magnesium silicate-based minerals is accompanied by the formation of silicic acid due to the preferential release of Mg^{2+} from minerals during this stage, which is facilitated by the rapid breakdown of ionic octahedral Mg-O bonds relative to covalent Si-O bonds (Oelkers et al., 2018). The formed silicic acid is unstable and tends to decompose into aqueous or solid amorphous silica by-products before diffusing into aqueous solution, which will then gather on the surface of the mineral to form a Si-rich product layer (F. Wang et al., 2019a), as shown in Figure 10.1. The Si-rich product layer prevents the inner mineral from contacting the solution and hinders further dissolution of the mineral (Kelemen et al., 2019). The thickness of the Si-rich product layer has a critical value. When the Si-rich product layer grows to the critical thickness, stresses between the product layer and minerals occur and cause exfoliation of the silica layer (Jarvis et al., 2009), as shown in Figure 10.2. Si-rich product layer and dense carbonate product layer are significant factors influencing the CO_2 mineralization kinetics of magnesium silicate-based minerals.

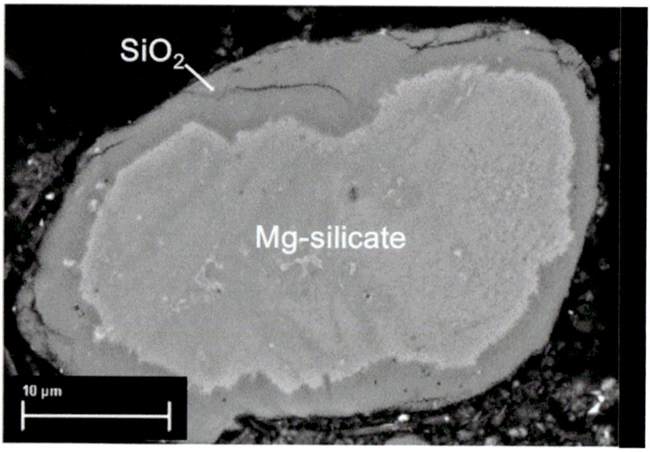

FIGURE 10.1 Cross-section SEM image of silica-rich layer formed on the surface of olivine particle (Rashid et al., 2020).

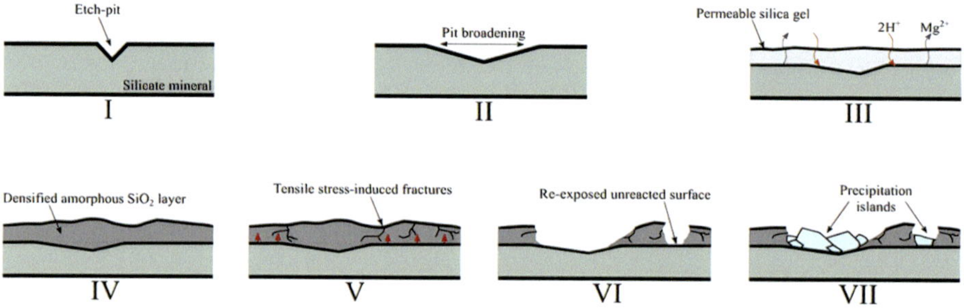

FIGURE 10.2 The formation and evolution of Si-rich product layer on surfaces of magnesium silicate minerals (Saleh et al., 2025).

Indirect CO_2 mineralization route may solve the challenges faced using direct aqueous CO_2 mineralization. Solution pH and temperature are primary controls on the rate at which dissolution reactions progress. While low pH is preferred to maximize the leaching of Mg^{2+}, on the other hand, it impedes the dissolution of CO_2 into the aqueous solution during the subsequent precipitation step, resulting in a low CO_2 mineralization conversion rate. Indirect CO_2 mineralization typically involves the extraction of Mg^{2+} from minerals by using solvents like acids, followed by the reaction of the extracted components with CO_2 in either the gaseous or aqueous phase. The involvement of acids changes the dissolution mechanism of olivine and effectively accelerates the leaching of Mg^{2+}, as shown in Figure 10.3. The separate leaching and precipitation processes occurring in distinct reactors permit the removal of impurities, enabling indirect CO_2 mineralization generally to yield higher CO_2 mineralization efficiency compared to direct CO_2 mineralization. Apart from this, the by-products like SiO_2 derived from multistep CO_2 mineralization can provide additional value. Further, indirect CO_2 mineralization allows more moderate processing

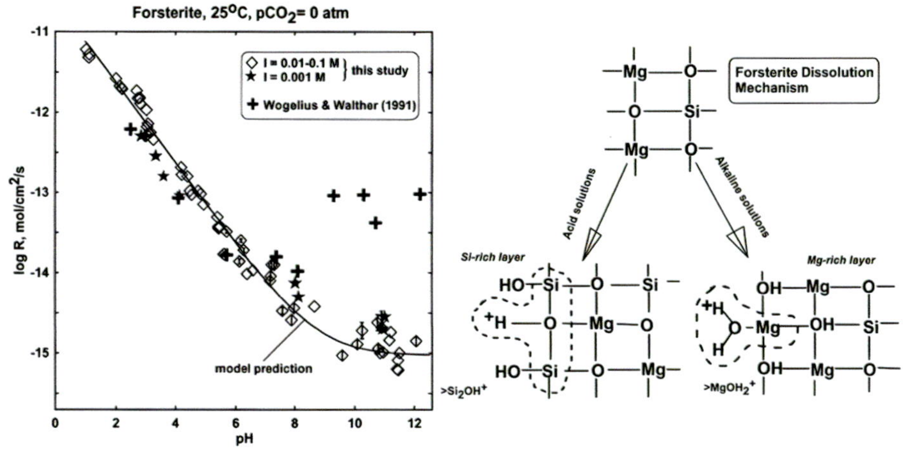

FIGURE 10.3 Dissolution mechanism and rate of forsterite in acidic and basic conditions (Pokrovsky and Schott, 2000).

condition instead of high temperature and high pressure as compared to direct CO_2 mineralization. However, the need for abundant chemicals leads to the expensive cost of indirect CO_2 mineralization.

10.3.2 CO_2 MINERALIZATION PRODUCTS

The CO_2 mineralization products of magnesium-based materials include a series of anhydrous and hydrated magnesium carbonate products (HMCs) which can be represented by a general formula $xMgCO_3 \cdot yMg(OH)_2 \cdot zH_2O$. Table 10.3 shows

TABLE 10.3
Different HMCs formed in the MgO-CO_2-H_2O system

Group	Moles of MgO	Moles of H_2O	Moles of CO_2	Compound	Chemical formula
I	1	1	—	Brucite	$Mg(OH)_2$
		—	1	Magnesite	$MgCO_3$
		2	1	Barringtonite	$MgCO_3{\cdot}2H_2O$
		3	1	Nesquehonite	$MgCO_3{\cdot}3H_2O$
		5	1	Lansfordite	$MgCO_3{\cdot}5H_2O$
II	2	1.5	1	Pokrovskite	$Mg_2(CO_3)(OH)_2{\cdot}0.5H_2O$
		4	1	Artinite	$Mg_2(CO_3)(OH)_2{\cdot}3H_2O$
III	5	5	4	Hydromagnesite	$Mg_5(CO_3)_4(OH)_2{\cdot}4H_2O$
		6	4	Dypingite	$Mg_5(CO_3)_4(OH)_2{\cdot}5H_2O$
		6–7	4	Giorgiosite	$Mg_5(CO_3)_4(OH)_2{\cdot}5\text{-}6H_2O$
IV	7	25	5	Shelkovite	$Mg_7(CO_3)_5(OH)_4{\cdot}24H_2O$

TABLE 10.4
Main phases and their stand thermodynamical parameters in the MgO-CO_2-H_2O system

Phase	ΔG_f^0 (kg mol^{-1})	Crystal habit
Magnesite	−1029.3	Massive trigonal
Nesquehonite	−1724.0	Elongated, orthorhombic
Lansfordite	−2197.8	Platy crystal, pseudo–orthorhombic
Artinite	−2568.6	–
Hydromagnesite	−5856.8	Globular aggregates, monoclinic
Dypingite	−6081.7	Globular aggregates, Monoclinic

Source: Harrison et al., 2019; Hummel et al., 2002; Johnson et al., 1992.

the different HMCs formed in the mineralized magnesium cement system and their corresponding chemical composition. The types of magnesium carbonates formed in the MgO-CO_2-H_2O system depend mainly on temperature and CO_2 concentration and are affected by several factors. Table 10.4 lists the common magnesium carbonates and their thermodynamic parameters and crystal habit in the MgO-CO_2-H_2O system. All the hydrated magnesium carbonates are thermodynamically metastable, except magnesite. In accordance with Ostwald's rule of stages, hydrated magnesium carbonates precipitate much more easily than magnesite but have a thermodynamic tendency to convert to magnesite in the MgO-CO_2-H_2O system.

Magnesite is almost impossible to form at ambient temperature and pressure due to the reason that its formation is kinetically limited (Hänchen et al., 2008; Hopkinson et al., 2012; Königsberger et al., 1999). There are two main reasons: (1) the high hydrated complexion of Mg^{2+} due to the high charge density and a small ionic radius (0.65Å),Mg^{2+} can bind tightly with six H_2O molecules to form [Mg^{2+}–$6H_2O$] complex, (2) the high dehydration energy of [Mg^{2+}–$6H_2O$] ($\Delta_{hyd}G^0 = -1903.72$kj.mol^{-1}) (Giammar et al., 2005; Xu et al., 2013). Direct nucleation and growth of magnesite from solution has been reported to occur at high partial CO_2 pressure (99–148 atm) and high temperature (80–120 °C) (Giammar et al., 2005). The indirect pathway to magnesite formation via intermediate phases is quite slow, especially below 100°C.

Hydrated magnesium carbonates have a complex interconversion relationship. Lansfordite has been reported to form at relatively low temperatures, around <10°C. As the temperature rises, the transformation from lansfordite to nesquehonite would take place, which is spontaneous and irreversible. Artinite can merely exist at temperatures lower than 15°C and low CO_2 pressure. At temperatures higher than 15°C and lower than 52°C, precipitations in the system are mostly in the form of nesquehonite, dypingite, and hydromagnesite. Nesquehonite is more readily formed at 25°C and high CO_2 partial pressure. Further, nesquehonite has been reported to be the dominant magnesium carbonate forming in ambient conditions, while hydromagnesite and dypingite tend to form at relatively higher temperatures. Starting from 52°C, the transition to hydromagnesite occurs, which can be directly from nesquehonite, or via a transitory mineral dypingite.

$$\textit{Lansfordite} \xrightarrow{T\uparrow} \textit{nesquehonite} \xrightarrow{T\uparrow} \textit{dypingite} \xrightarrow{T\uparrow} \text{hydromagnesite} \xrightarrow{T\uparrow} \text{magnesite.} \quad (10.11)$$

10.3.3 Microstructures of CO_2 mineralization products

The effect of CO_2 mineralization on the microstructure of magnesium-based materials is mainly discussed in two aspects: (1) change of pore structure; CO_2 mineralization is an expansive reaction that can reduce the overall pore volume. The solid volume increases by 1.8–3.1 times due to the conversion of MgO into various types of hydrated magnesium carbonate, (2) Microstructural evolution; magnesium carbonates have microstructures with varying morphologies, as shown in Figure 10.4. Magnesite is typically hexagonal prisms (Figure 10.4(a)). Nesquehonite is prismatic

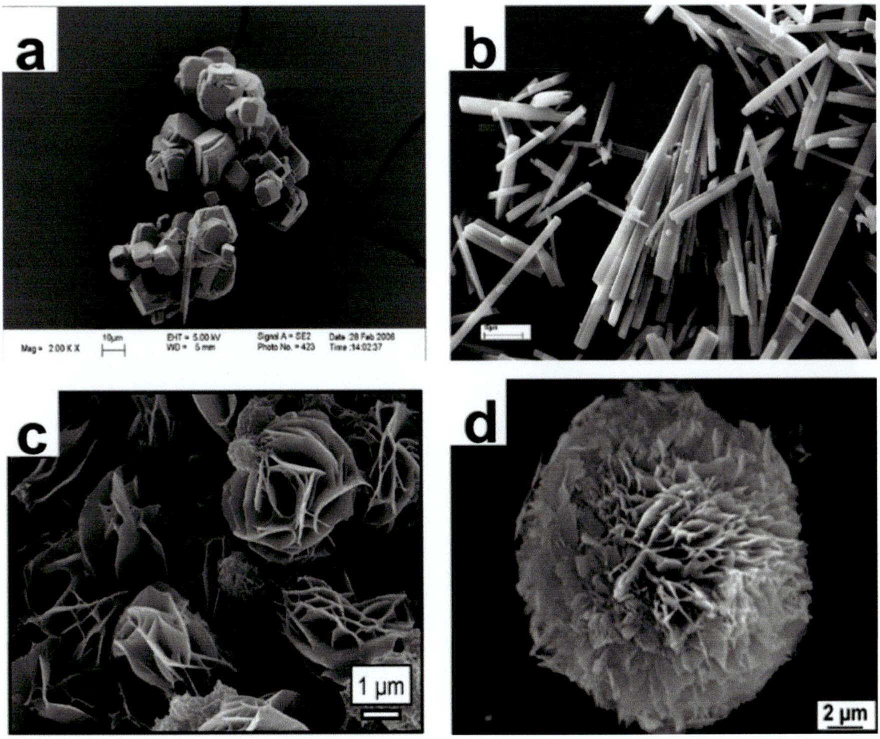

FIGURE 10.4 SEM images of different CO_2 mineralization products from the magnesia system (a) magnesite (Hänchen et al., 2008), (b) nesquehonite (Guermech et al., 2022), and (c) hydromagnesite (Teir et al., 2009), (d) dypingite (Power et al., 2007).

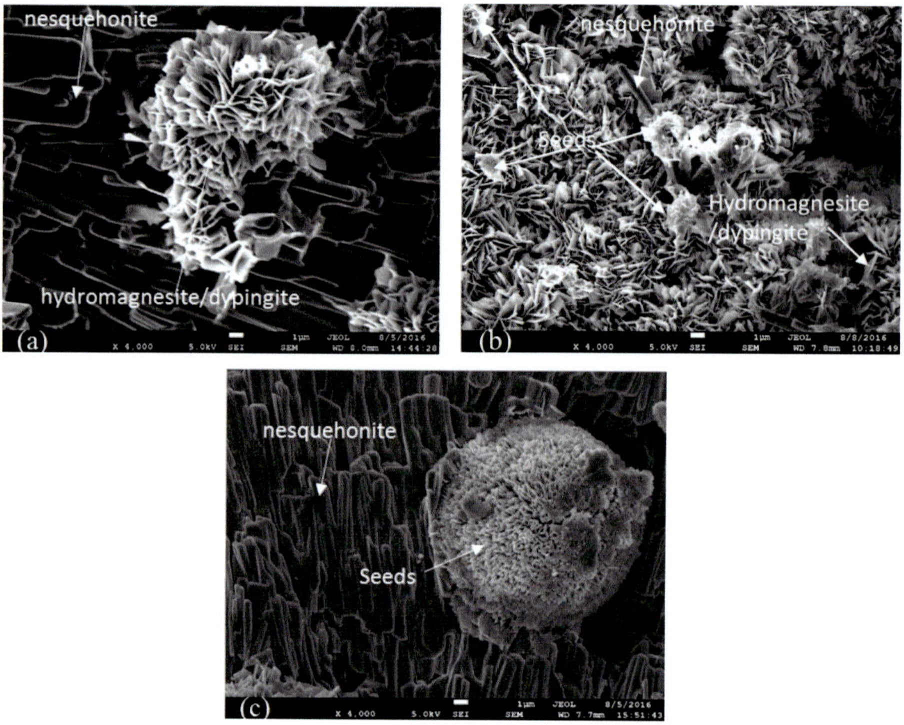

FIGURE 10.5 cementitious matrix containing Nesquehonite, Hydromagnesite, and Dypingite crystals (Dung and Unluer, 2018).

and forms star-like clusters (Figure 10.4(b)). Dypingite and hydromagnesite are composed of an agglomerate of very thin platelets, exhibiting rounded rosette-like clusters (Figure 10.4(c) and Figure 10.4(d)).

Inter-connected and well-developed crystals of magnesium carbonates form a compact and interlocked network structure (see Figure 10.5) which controls the overall performance of cementitious materials. Several studies (De Silva et al., 2006; Mo and Panesar, 2012; Panesar and Mo, 2013) indicated that the morphology and amount of magnesium carbonates controlled the overall performance of cementitious materials. Fibrous, acicular or otherwise elongated nature can increase the interlocking effects occurring between the crystals. Thus, the elongated morphology of nesquehonite crystals with their 3D structures enables the connection of the various phases and reduces the initially available pore space, thereby leading to higher strengths than disk-like hydromagnesite/dypingite. Although the hardness of nesquehonite is close to that of brucite (~2.5), the nesquehonite system is more compact due to its monolithic nature (Unluer and Al-Tabbaa, 2015).

10.4 MECHANICAL PROPERTIES AND DURABILITY OF CO_2 MINERALIZED MAGNESIUM-BASED MATERIALS

The CO_2 mineralization can remarkably increase the mechanical properties of magnesium-based materials. Though the hydration of reactive magnesia is a volume-expansive process where 1 cm^3 can transform into 2.18 cm^3 brucite after complete hydration, the compressive strength of reactive magnesia cement under sealed conditions is quite low. This is attributed to the reason that the hydration product, brucite, is a porous and low-strength phase. After CO_2 mineralization, the matrix of RMC becomes denser and its compressive strength is equal to or even exceeds Portland cement.

The CO_2 mineralized magnesium-based materials show excellent corrosion resistance to hydrochloric acid and magnesium sulfate. Liska and Al-Tabbaa (2012a) found that reactive MgO under ambient CO_2 mineralization conditions showed high resistance to sulfate attack with little change in strength. Even reactive MgO under accelerated CO_2 mineralization still showed good resistance to sulfate corrosion, maintaining 50% strength after 12 months of exposure to sulfate.

Most magnesium carbonate-based phases are stable under chloride attack (Liska and Al-Tabbaa, 2012b). Figure 10.6 shows SEM images of CRMC samples before and after 12 weeks of exposure to NaCl through wet and dry cycles (D. Wang et al., 2019a). It can be seen that RMC samples retain a significant portion of the dense network composed of rosette-like hydromagnesite and needle-like nesquehonite after wet and dry cycles. The addition of an appropriate amount of fly ash can improve

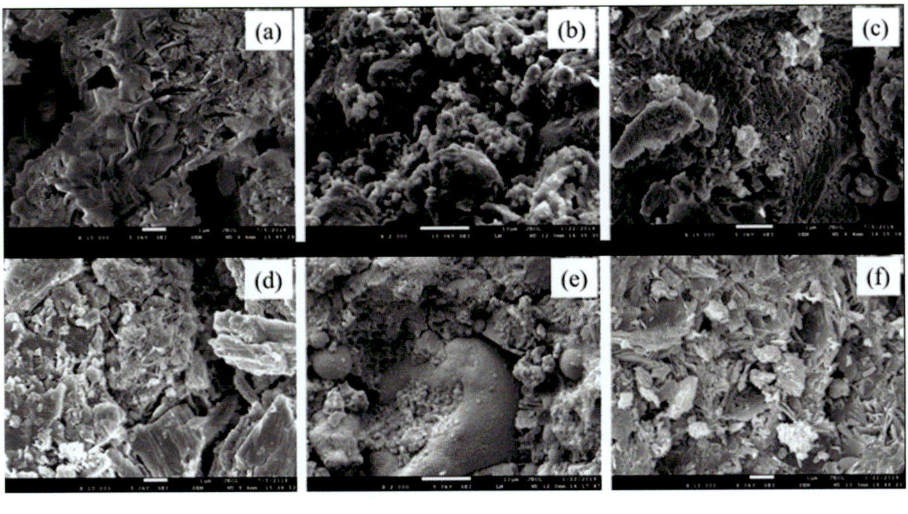

FIGURE 10.6 SEM images before exposure to cyclic wetting and drying (a) 40%RMC, (b) 20%RMC +20%FA, and (c) 20%RMC+20%GGBS, and after exposure (d) 40%RMC, (e) 20%RMC +20%FA, and (f) 20%RMC+20%GGBS (D. Wang et al., 2019a).

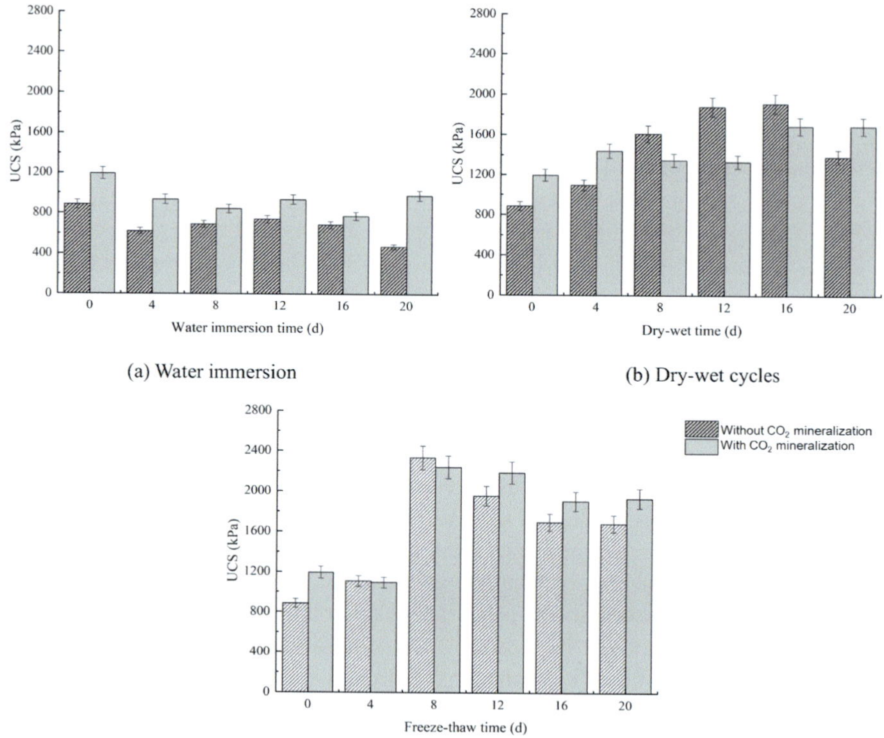

(c) Freeze-thaw cycles

FIGURE 10.7 Evolution of the strength of RMC with and without CO_2 mineralization to resist (a) water immersion, (b) dry-wet cycles, (c) freeze-thaw cycles (D. Wang et al., 2019a).

the chloride ion resistance of RMC, as it does not alter the dense network formed by HMCs and reduces the content of permeable pores. Although a decrease in compactness was found in samples mixed with slag, a large amount of needle-like nesquehonite with excellent chloride binding capacity could still be observed.

Wang et al. (2019a) studied the mechanical properties and microstructure of mineralized MgO-fly ash solidified sludge and found that the coupling skeleton constructing-filling-cementing effects make the mineralized sample a higher strength, better water stability, and stronger resistance to dry-wet cycles and freeze-thaw cycles, as shown in Figure 10.7. The phase transition from dypingite and nesquehonite to hydromagnesite happened during the continuous water immersion and dry-wet cycle experiments. Although no obvious phase transition was detected during the freeze-thaw cycle, the content of macropores with a radius greater than 10 microns increased significantly.

10.5 FACTORS INFLUENCING CO_2 MINERALIZATION OF MAGNESIUM-BASED MATERIALS

10.5.1 CO_2 CONCENTRATION AND PRESSURE

CO_2 concentration can affect the CO_2 mineralization rate and degree by directly influencing the diffusion of CO_2. The available literature suggests that high CO_2 concentrations are favorable for the CO_2 mineralization process (Jang et al., 2016). The study conducted by Unluer and Al-Tabbaa (2014) suggested that accelerated CO_2 mineralization with a relatively low CO_2 concentration (5%) could significantly increase the strength of the RMC samples compared to samples under ambient atmosphere. However, as the CO_2 concentration further increased from 5% to 20%, the RMC samples did not show a significant increase in strength. Dung et al. (2021) reported that the effect of CO_2 concentration on strength was more significant in the first 14 days, with the strength of the 20% CO_2 samples at 7 days being comparable to that of the 10% CO_2 samples at 14 days. However, the enhanced effect diminished as the time was further extended to 28 days. The findings of these studies indicate that beyond a specific duration, CO_2 concentration no longer serves as a limiting factor for the CO_2 mineralization efficiency of reactive MgO. A plausible explanation for this behavior lies in the initially elevated CO_2 concentration that substantially accelerates the mineralization rate in the early phase, while consequently the swift formation of a product layer ultimately would restrict further CO_2 mineralization of MgO (Dung et al., 2021). Wang and his co-workers (2020) have proved that the compressive strength of reactive MgO cement samples increased as CO_2 concentration increased. However, the increasing rate decreased significantly when the CO_2 concentration was higher than 10%. This suggests that a 10% CO_2 concentration is relatively most effective. Figure 10.8 summarizes the effect of CO_2 concentration on the compressive strength of RMC samples in related studies.

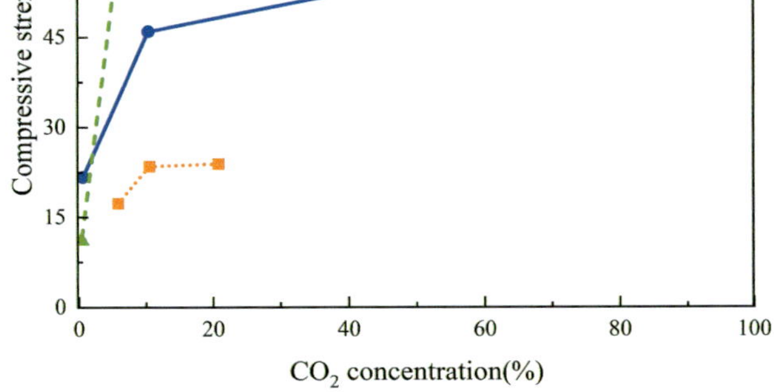

FIGURE 10.8 The effect of CO_2 concentration on compressive strength of RMC.

The utilization of pressurized CO_2 can effectively enhance the CO_2 mineralization reaction. Experimental results have shown that the use of relatively high CO_2 pressures could more effectively achieve comparable mechanical performance in a shorter time (Mo et al., 2016). However, a further increase in CO_2 pressure resulted in a modest increase in strength. Hay and Celik (2020) investigated the use of supercritical CO_2 ($scCO_2$, 35 ± 2 °C, 9.5 MPa) for CO_2 mineralization of RMC. They found that $scCO_2$ could significantly increase the compressive strength, and RMC samples reached the mature strength level (46 MPa) within a few hours, which would otherwise be achieved after 3 days under 20% CO_2, 80% RH, 30°C. In addition, the CO_2 sequestration ratio increased threefold. It can be concluded that moderately elevated pressure could improve the CO_2 mineralization efficiency, while increasing the pressure further may not be effective. $scCO_2$ mineralization could lead to significantly rapid strength gain in just a few hours.

The study conducted by Gerdeman et al. (2007) has suggested that the CO_2 mineralization degree of magnesium silicate-based minerals increased with CO_2 partial pressure. The conversion ratio of serpentine reached a maximum when the CO_2 partial pressure increased to about 150 atm, while olivine requires a high partial pressure of 250 atm. Considering that increasing CO_2 partial pressure above 150 atm for CO_2 mineralization in olivine and serpentine is considered impractical, it has been suggested that the optimal CO_2 partial pressure for olivine is 150 atm and for serpentine is 115 atm, where the minerals have reached 80% of the maximum conversion ratio in practice. The increase in the CO_2 mineralization yield with respect to the increase in pressure is less at pressures greater than the determined optimal CO_2 partial pressure.

10.5.2 CO_2 MINERALIZATION TIME

CO_2 mineralization time is another key factor in ensuring full CO_2 mineralization of activated magnesium oxide cement. Determining the optimum CO_2 mineralization time can reduce the time and costs associated with CO_2 mineralization and oxidation. It was found that although CO_2 concentration has a significant effect on the efficiency of CO_2 mineralization and oxidation, after a certain period of time, CO_2 concentration is no longer the limiting factor for the efficiency of magnesium oxide CO_2 mineralization (Bhagath Singh et al., 2020). The study by Wang et al. (2020) indicated that under low CO_2 concentration (e.g., no more than 20%), a longer CO_2 mineralization time (e.g., at least 7 days) was required to achieve a higher degree of CO_2 mineralization. The experimental results of Dung et al. (2019) showed that the compressive strength of RMC samples after 3, 7, 14, and 28 days of CO_2 mineralization reached 40, 51, 58, and 60 MPa, respectively. But the CO_2 sequestration rate remained constant at 11% from 14 to 28 days. However, Hay and Celik (2020) found that the compressive strength might decrease as the CO_2 mineralization time increases. CO_2 mineralization can lead to volume expansion of the material which may cause internal stresses and cracking.

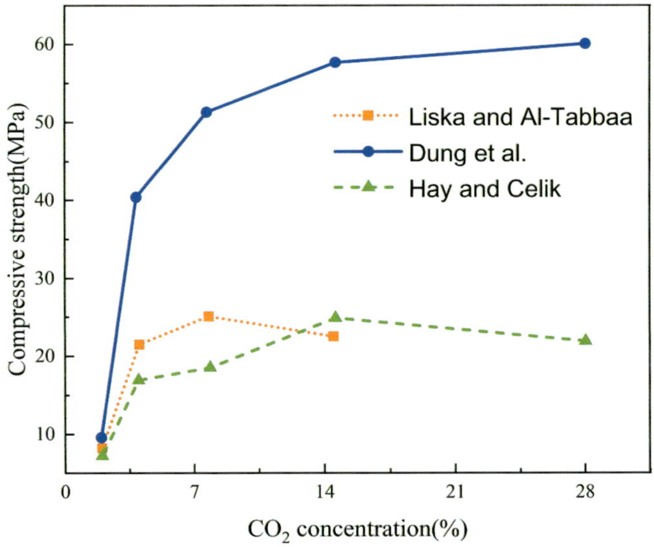

FIGURE 10.9 The effect of CO_2 mineralization time on compressive strength of RMC.

In summary, the appropriate extension of CO_2 mineralization time can promote the efficiency of RMC and improve the compressive strength of the material. However, excessive CO_2 mineralization may lead to early cracking of the material, which is detrimental to the development of strength, and increases the cost, which should be avoided as much as possible in the actual project. Current research (Liska and Al-Tabbaa, 2009) suggests that CO_2 mineralization time should be limited to 7–14 days. Figure 10.9 summarizes the effect of CO_2 mineralization time on the compressive strength of RMC samples in related studies.

10.5.3 Relative Humidity and Temperature

CO_2 mineralization of magnesium-based cementitious materials to produce hydrated magnesium carbonates requires the direct participation of water. Dissolution and ionic diffusion of CO_2 and magnesium-based minerals also rely on water as a medium. However, excess water can saturate the capillary pores and inhibit CO_2 diffusion because CO_2 diffuses faster in dry environments (the diffusion coefficient of CO_2 in air is ~1390 × 10–8 m^2/s, whereas the diffusion rate in water is four to five orders of magnitude slower than in air). This means that too high and too low relative humidity are detrimental to the CO_2 mineralization of magnesium-based materials. Determining the optimum relative humidity is essential to achieve efficient CO_2 mineralization. Unluer and Al-Tabbaa (2014) found the optimum relative humidity for CO_2 mineralization of reactive magnesium oxide is 78%, and wet/dry cycling at a certain cycling frequency may be more favorable for the CO_2 mineralization degree to increase. However, the study conducted by Wang et al. (2020) showed that 98%

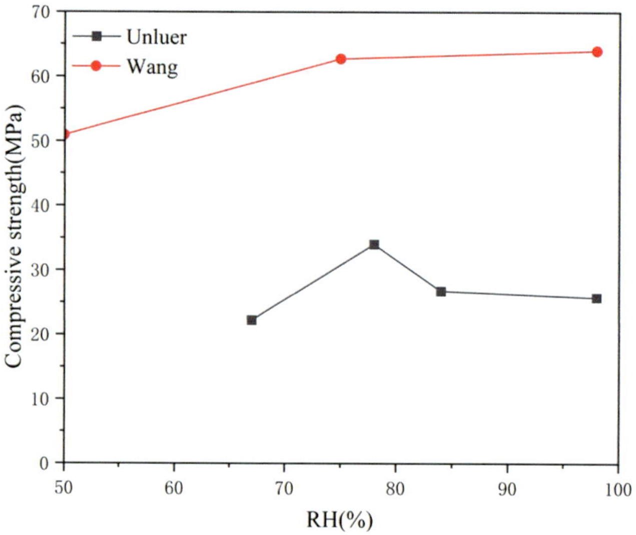

FIGURE 10.10 The effect of RH on compressive strength of RMC.

relative humidity was more favorable. The difference in the results of the two studies may be ascribed to the fact that the water-binder ratio of the former study (0.5–0.7) is higher than the latter. Figure 10.10 summarizes the effect of relative humidity on the compressive strength of RMC samples in related studies.

Elevated temperature can accelerate the rate of dissolution, diffusion, and chemical reaction and promote nucleation and growth of products. However, elevated temperature can also reduce the dissolution of CO_2 in the solution and further decrease the availability of bicarbonate ions and carbonate ions. Thus, there is a critical temperature for the utmost enhancement of CO_2 mineralization. When the temperature increased from 10 °C to 40 °C, the CO_2 mineralization rate of reactive MgO increased by almost 60% (Dheilly and Sebaibi, 1998). However, the increase in temperature leads to a decrease in the solubility of CO_2 in water (Grünhäuser Soares and Castro-Gomes, 2021). As mentioned in the former chapter, the formation and transformation of CO_2 mineralization products are controlled by temperature, while the morphology and amount of magnesium carbonates control the overall performance of cementitious materials. Pu and Unluer (Pu and Unluer, 2020) determined that a pretreatment temperature of 50°C–60°C was the most effective temperature for enhancing the hydration of MgO, which increases the rate of the reaction and the degree of hydration, improves the dissolution of the MgO and the CO_2 mineralization precipitation of products, and reduces porosity. This resulted in an increase of 25–35% in the compressive strength at 28 days (50–54 MPa) as compared to the control sample (40 MPa). On the other hand, magnesium silicate-based minerals demonstrate different optimal temperatures for CO_2 mineralization. Olivine reaches an optimal reaction temperature at approximately 185°C while serpentine is optimally CO_2 mineralized at 150°C (Gerdemann et al., 2007; Kelemen and Matter, 2008).

10.5.4 Additives

Additives used for CO_2 mineralization of reactive MgO include hydrochloric acid, magnesium chloride, and magnesium acetate. The presence of hydrochloric acid can increase the concentration of H^+, which promotes the dissolution of magnesium oxide and the precipitation of $Mg(OH)^+_2$ particles by the following steps (Matabola et al., 2010; Raschman and Fedoročková, 2004):(1) diffusion of H^+ ions through the liquid-solid interface at the surface of the solid particles via the liquid film; (2) surface chemical reaction between MgO and H^+; (3) diffusion of liquid products from the interface into the bulk solution via the liquid film; and (4) precipitation of $Mg(OH)_2$ away from the original particles. Magnesium acetate is considered to be the most effective additive. Acetate acts as a bidentate ligand with strong chelating properties and can promote the dissolution of magnesium oxide by forming complexes with magnesium ions as shown in reaction (10.12). The magnesium acetate ions migrate and dissociate in solution enabling the CO_2 mineralization products to precipitate in the bulk solution away from the magnesium oxide.

$$MgO_{(s)} + CH_3COO^-{}_{(aq)} + H_2O_{(l)} \rightarrow CH_3COOMg^+{}_{(aq)} + 2OH^-{}_{(aq)} \quad (10.12)$$

Kamala et al. (Kamala Ilango et al., 2024) found that magnesium acetate could be adsorbed on the [001] surface of magnesium hydroxide to weaken the interlayer force, impede the crystal growth, and change the morphology of magnesium hydroxide, which is manifested by the increase of diameter and decrease of thickness. The specific mechanism is shown in Figure 10.11. Co-precipitation of ligands of acetic acid with magnesium hydroxide can also result in a decrease in the degree of crystallinity of magnesium hydroxide.

The study of Du et al. (Du and Unluer, 2024) indicated that the structure of magnesium hydroxide strongly affected the potential of its CO_2 mineralization reaction. Nguyen et al. (Nguyen et al., 2022) further investigated the effect of magnesium acetate on the CO_2 mineralization process of magnesium hydroxide. They found that a 1 M magnesium acetate solution resulted in the complete CO_2 mineralization of magnesium hydroxide after one month, whereas the CO_2 mineralization degree of a control sample mineralized in pure water only reached about 50%. The

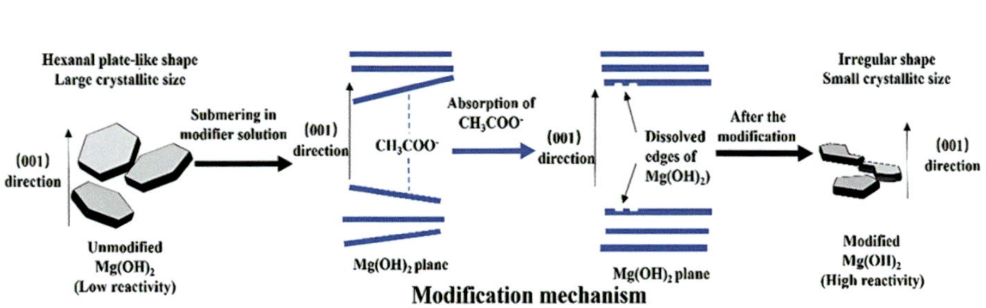

FIGURE 10.11 Mechanism of influence of Mg-acetate on the CO2 mineralization of $Mg(OH)_2$ (Du and Unluer, 2024).

complex can play the function of nucleation sites for the growth of HMCs, facilitating the HMCs to precipitate away from the surface of magnesium hydroxide due to competition between acetate and carbonate for magnesium ions. Thus, HMCs formed in the solution can also serve as nucleation sites. It is noteworthy that the presence of magnesium acetate promotes the conversion of nesquehonite into giorgiosite, $Mg_5(CO_3)_4(OH)_2$-5-$6H_2O$ instead of hydromagnesite and dypingite as shown in Figure 10.12. However, giorgiosite is a relatively unknown HMC, and its relevant properties need to be further determined.

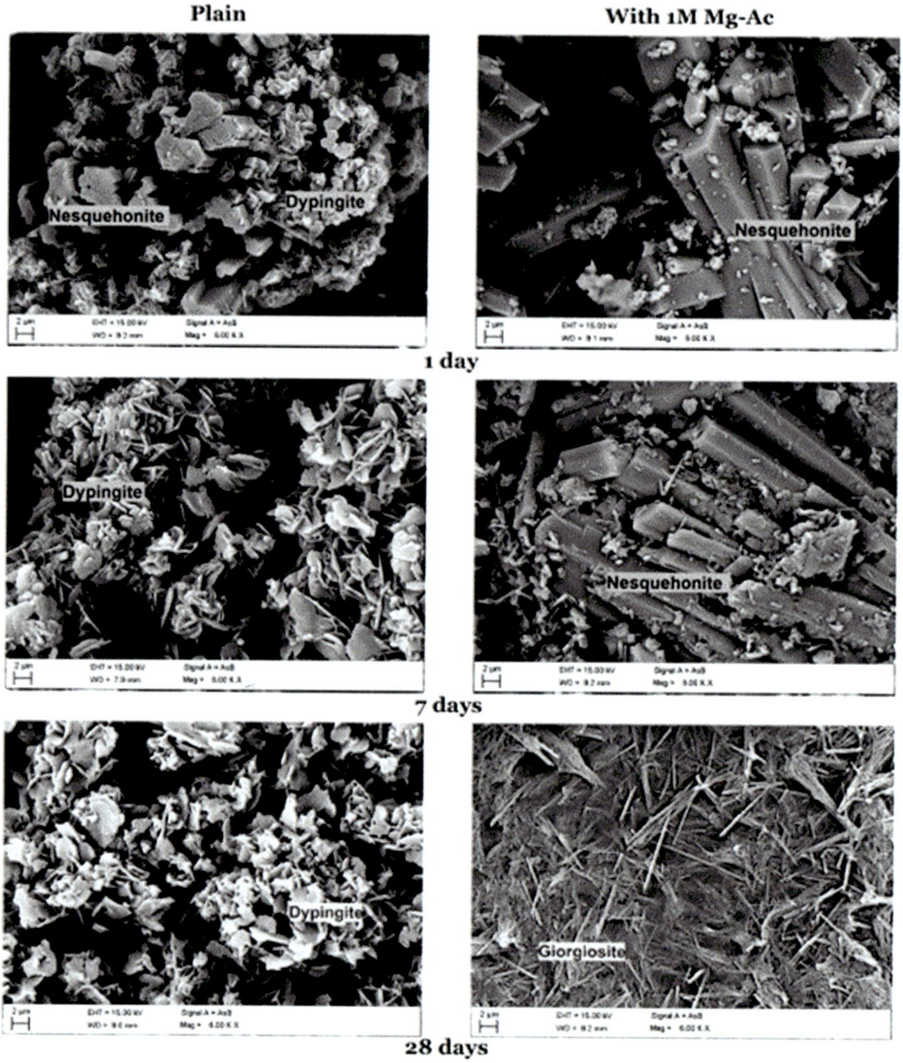

FIGURE 10.12 Morphology of the precipitates of CO_2 mineralization products without and with 1 M Mg-Ac 1 (Nguyen et al., 2022).

Sodium bicarbonate ($NaHCO_3$) and sodium chloride (NaCl) are reported to effectively accelerate CO_2 mineralization of olivine and serpentine. By the addition of $NaHCO_3$, the time required for mineralization of 80% of olivine was reduced from 24 hours to 6 hours (Oconnor et al., 2005). $NaHCO_3$ has been proposed to act as a CO_2 carrier accelerating its dissolution in water. The addition of NaCl further improved conversion at constant $NaHCO_3$ concentration, temperature, and pressure. The addition of sodium ions and chloride also provides a role by increasing the ionic strength of the aqueous solution, modifying the surface charge of silicate particles, aiding in ion exchange at the solid/liquid interface, inhibiting the development of extensive Si-rich passivating layers, promoting the dissolution of magnesium and allowing CO_2 mineralization with fewer kinetic constraints (F. Wang et al., 2019b). The most studied and reported concentrations of these additives are 1 M NaCl and 0.64 M $NaHCO_3$.

10.5.5 Nucleating seeds

Nucleation seeds play an important role in enhancing the CO_2 mineralization efficiency of minerals. Figure 10.13 reveals the mechanism of nucleating agents on the promotion of hydration and CO_2 mineralization of magnesium oxide. In the absence of nucleating seeds, the formation of HMCs on the surface resulted in the encapsulation of unreacted MgO particles, thus preventing them from participating in the CO_2 mineralization reaction. Nucleating seeds can alleviate this problem by forming CO_2 mineralization products on the surface of the seed rather than magnesium oxide and increasing the opportunity for unreacted magnesium oxide particles to contact with the bulk solution. Nucleation within the pores away from the seed then removes Mg^{2+} and OH^- ions from solution, creating a local concentration gradient which accelerates

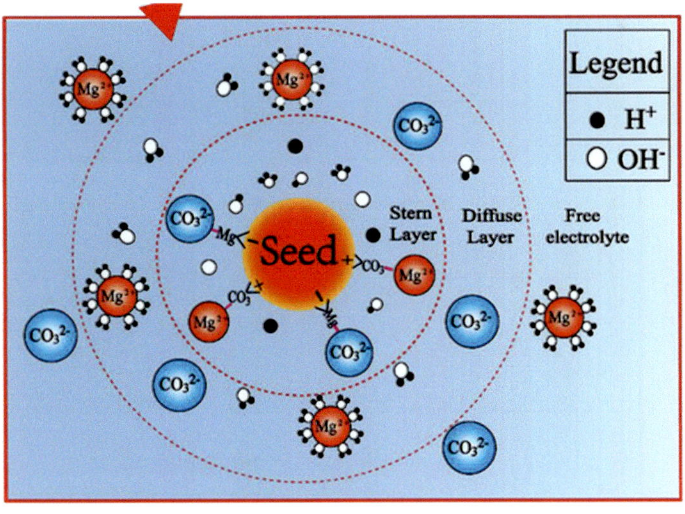

FIGURE 10.13 Mechanism of nucleating seeds on the promotion of CO_2 mineralization of magnesium oxide (Song et al., 2023).

the dissolution of MgO. The thickness of the red arrows reflects the difference in the local concentration gradient in the seeded and unseeded groups. The formation of hydration products dispersed in the pores can further enhance the CO_2 mineralization process. The nucleation effect of the seeds mainly depends on the surface area of the seeds, whereas the subsequent growth of the phase depends on ion adsorption interactions.

Unluer and Al-Tabbaa (Unluer and Al-Tabbaa, 2013) found that 20 wt% hydromagnesite was effective in increasing the strength of the samples, while 50 wt% hydromagnesite was rather unfavorable for the strength development of the samples. This difference in strength development was attributed to the formation of different CO_2 mineralization products. The main products observed in the 20 wt% samples were hydromagnesite, while in the 50 wt% samples the products were mainly nesquehonite and dypingite. Dung and Unluer (2017) found that 1 wt% hydromagnesite increased the CO_2 mineralization degree of reactive magnesium oxide by 96% and the compressive strength by 33%.

German et al. (2023) studied the microstructure of hydromagnesite-reactive MgO mixtures. They have reported that a new phase, hydrous carbonate-containing brucite (HCB), has been formed. They have suggested further investigation of the new phase. In another study, the authors (Unluer and Al-Tabbaa, 2013) demonstrated that using dypingite as a magnesium oxide substitute could significantly reduce the porosity of RMC and increase the compressive strength by about 20%. The CO_2 mineralization degree of the specimens was almost 100%. In comparison with the control specimens, the main CO_2 mineralization products of the specimens doped with dypingite are dypingite/hydromagnesite rather than nesquehonite and had smaller crystal sizes and different morphology.

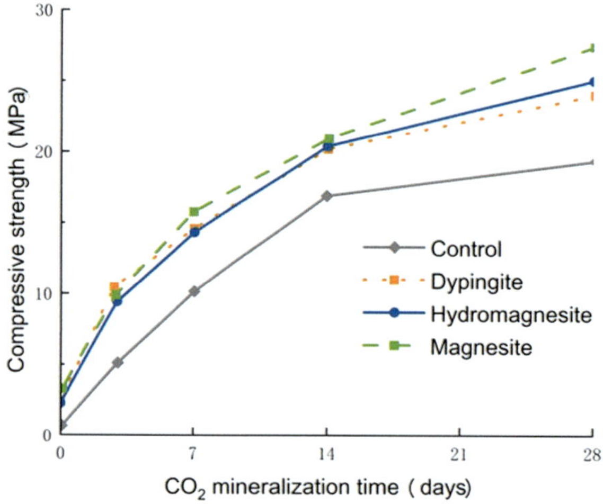

FIGURE 10.14 The influence of different nucleating seeds on the CO_2 mineralization of RMC (Song et al., 2023).

Song et al. (2023) investigated the influence of different nucleating seeds on the CO_2 mineralization of reactive magnesium oxide. They found that magnesite exhibited a superior promotion effect on the CO_2 mineralization degree and strength of RMC than hydromagnesite and dypingite, as shown in Figure 10.14. The RMC with 2 wt% magnesite can attain a 28-day strength of 27.4 MPa, which was 41% greater than that of pure RMC (19.4 MPa). This may be attributed to the reason that magnesite has a smaller particle size and molar mass, providing more nucleation sites.

10.6 SUMMARY

This chapter reviewed the potential sources, mechanisms, products, properties, and factors of CO_2 mineralized magnesium-based materials, with the goal of coalescing existing knowledge and providing improved understanding of the known capabilities and limitations. The conclusions are delineated as follows:

1. Magnesium silicate-based minerals are abundant in resources, possessing great CO_2 mineralization potential. However, they require high temperature and pressure. Magnesia-based binders are relatively limited in quantity but generally more reactive. RMC is the most promising low-carbon substitute for PC if using a greener production routine.
2. The products of CO_2 mineralized magnesium-based materials are a series of well-developed crystals among which nesquehonite, dypingite, and hydromagnesite are the predominant phases. These crystals exhibit different micromorphology and interconnect to form a network structure.
3. CO_2 mineralization can remarkably increase the mechanical properties and durability of magnesium-based materials.
4. Many factors including CO_2 concentration and pressure, curing time, RH, and temperature will affect the properties development of magnesium-based materials. Moreover, the influencing factors are linked to one another. Determining the optimal factors is essential. Using extra chemicals like additives and nucleating seeds can effectively promote CO_2 mineralization reactions.

REFERENCES

Aramendía, M.A., Borau, V., Jiménez, C., Marinas, J.M., Porras, A., Urbano, F.J., 1996. Synthesis and characterization of various MgO and related systems. *Journal of Materials Chemistry.* 6, 1943–1949.

Back, M., Bauer, M., Stanjek, H., Peiffer, S., 2011. Sequestration of CO2 after reaction with alkaline earth metal oxides CaO and MgO. *Applied Geochemistry* 26, 1097–1107.

Bhagath Singh, G.V.P., Sonat, C., Yang, E.H., Unluer, C., 2020. Performance of MgO and MgO–SiO2 systems containing seeds under different curing conditions. *Cement and Concrete Composites* 108, 103543.

Birchal, V.S.S., Rocha, S.D.F., Ciminelli, V.S.T., 2000. The effect of magnesite calcination conditions on magnesia hydration. *Minerals Engineering* 13, 1629–1633.

De Silva, P., Bucea, L., Moorehead, D.R., Sirivivatnanon, V., 2006. Carbonate binders: Reaction kinetics, strength and microstructure. *Cement and Concrete Composites* 28, 613–620.
De Silva, P., Bucea, L., Sirivivatnanon, V., 2009. Chemical, microstructural and strength development of calcium and magnesium carbonate binders. *Cement and Concrete Research* 39, 460–465.
Dheilly, R.-M., Sebaibi, Y., 1998. Importance de la présence de magnésie dans le stockage de la chaux: Carbonatation de l'oxyde et de l'hydroxyde de magnésium. *Canadian Journal of Chemistry* 76(8), 1188–1196.
Du, Z., Unluer, C., 2024. Modification of magnesium hydroxide for improved performance in CO2 sequestration. *Cement and Concrete Research* 177, 107418.
Dung, N.T., Hay, R., Lesimple, A., Celik, K., Unluer, C., 2021. Influence of CO concentration on the performance of MgO cement mixes2. *Cement and Concrete Composites* 115, 103826.
Dung, N.T., Lesimple, A., Hay, R., Celik, K., Unluer, C., 2019. Formation of carbonate phases and their effect on the performance of reactive MgO cement formulations. *Cement and Concrete Research* 125, 105894.
Dung, N.T., Unluer, C., 2017. Influence of nucleation seeding on the performance of carbonated MgO formulations. *Cement and Concrete Composites* 83, 1–9.
Dung, N.T., Unluer, C., 2018. Development of MgO concrete with enhanced hydration and carbonation mechanisms.*Cement and Concrete Research* 103, 160–169.
Fagerlund, J., Highfield, J., Zevenhoven, R., 2012. Kinetics studies on wet and dry gas–solid carbonation of MgO and mg(OH)2 for CO2 sequestration. *RSC Advances* 2, 10380–10393.
Fang, X., Xuan, D., Poon, C.S., 2017. Empirical modelling of CO2 uptake by recycled concrete aggregates under accelerated carbonation conditions. *Material Structure* 50, 200.
Gerdemann, S.J., O'Connor, W.K., Dahlin, D.C., Penner, L.R., Rush, H., 2007. Ex situ aqueous mineral carbonation. *Environmental Science & Technology* 41, 2587–2593.
German, A., Winnefeld, F., Lura, P., Rentsch, D., Lothenbach, B., 2023. Hydrous carbonate-containing brucite (HCB) in MgO/hydromagnesite blends. *Cement and Concrete Research* 173, 107304.
Giammar, D.E., Bruant, R.G., Peters, C.A., 2005. Forsterite dissolution and magnesite precipitation at conditions relevant for deep saline aquifer storage and sequestration of carbon dioxide. *Chemical Geology, Geochemical Aspects of CO2 Sequestering* 217, 257–276.
Grünhäuser Soares, E., Castro-Gomes, J., 2021. Carbonation curing influencing factors of Carbonated Reactive Magnesia Cements (CRMC) – A review. *Journal of Cleaner Production* 305, 127210.
Guermech, S., Mocellin, J., Tran, L.-H., Mercier, G., Pasquier, L.-C., 2022. A study of hydromagnesite and nesquehonite precipitation in indirect aqueous carbonation of thermally-activated serpentine in a batch mode. *Journal of Crystal Growth* 584, 126540.
Hänchen, M., Prigiobbe, V., Baciocchi, R., Mazzotti, M., 2008. Precipitation in the Mg-carbonate system—effects of temperature and CO2 pressure. *Chemical Engineering Science* 63, 1012–1028.
Harrison, A.L., Mavromatis, V., Oelkers, E.H., Bénézeth, P., 2019. Solubility of the hydrated mg-carbonates nesquehonite and dypingite from 5 to 35 °C: Implications for CO2 storage and the relative stability of mg-carbonates. *Chemical Geology* 504, 123–135.
Hay, R., Celik, K., 2020. Accelerated carbonation of reactive magnesium oxide cement (RMC)-based composite with supercritical carbon dioxide (scCO2). *Journal of Cleaner Production* 248, 119282.
Hopkinson, L., Kristova, P., Rutt, K., Cressey, G., 2012. Mineral carbonation of ultramafic tailings: A review of reaction mechanisms and kinetics, industry case studies, and modelling. *Geochimica et Cosmochimica Acta* 76, 1–13.

Hummel, W., Berner, U., Curti, E., Pearson, F.J., Thoenen, T., 2002. Nagra/PSI chemical thermodynamic data base 01/01. *Radiochimica Acta* 90, 805–813. 5

Jang, J.G., Kim, G.M., Kim, H.J., Lee, H.K., 2016. Review on recent advances in CO_2 utilization and sequestration technologies in cement-based materials. *Construction Building Materials* 127, 762–773.

Jarvis, K., Carpenter, R.W., Windman, T., Kim, Y., Nunez, R., Alawneh, F., 2009. Reaction mechanisms for enhancing mineral sequestration of CO_2. *Environmental Science & Technology* 43, 6314–6319.

Johnson, J.W., Oelkers, E.H., Helgeson, H.C., 1992. SUPCRT92: A software package for calculating the standard molal thermodynamic properties of minerals, gases, aqueous species, and reactions from 1 to 5000 bar and 0 to 1000°C. *Computers & Geosciences* 18, 899–947.

Kamala Ilango, N., Nguyen, H., German, A., Winnefeld, F., Kinnunen, P., 2024. Role of magnesium acetate in hydration and carbonation of magnesium oxide-based cements. *Cement and Concrete Research* 175, 107357.

Kelemen, P., Benson, S.M., Pilorgé, H., Psarras, P., Wilcox, J., 2019. An overview of the status and challenges of CO2 storage in minerals and geological formations. *Frontiers in Climate* 1, 9.

Kelemen, P.B., Matter, J., 2008. In situ carbonation of peridotite for CO_2 storage. *Proceedings of the National Academy of Sciences of the United States of America* 105, 17295–17300.

Kerisit, S., Bylaska, E.J., Felmy, A.R., 2013. Water and carbon dioxide adsorption at olivine surfaces. *Chemical Geology* 359, 81–89.

Königsberger, E., Königsberger, L.-C., Gamsjäger, H., 1999. Low-temperature thermodynamic model for the system Na2CO3–MgCO3–CaCO3–H2O. *Geochimica et Cosmochimica Acta* 63, 3105–3119.

Lackner, K.S., 2003. A guide to CO_2 sequestration. *Science* 300, 1677–1678.

Liska, M., Al-Tabbaa, A., 2009. Ultra-green construction: Reactive magnesia masonry products. *Proceedings of Institution of Civil Engineers* 162, 185–196.

Liska, M., Al-Tabbaa, A., 2012a. Performance of magnesia cements in porous blocks in acid and magnesium environments. *Advances in Cement Research* 24, 221–232.

Liska, M., Al-Tabbaa, A., 2012b. Performance of magnesia cements in porous blocks in acid and magnesium environments. *Advances in Cement Research* 24, 221–232.

Matabola, K.P., van der Merwe, E.M., Strydom, C.A., Labuschagne, F.J.W., 2010. The influence of hydrating agents on the hydration of industrial magnesium oxide. *Journal of Chemical Technology & Biotechnology* 85, 1569–1574.

Mo, L., Panesar, D.K., 2012. Effects of accelerated carbonation on the microstructure of Portland cement pastes containing reactive MgO. *Cement and Concrete Research* 42, 769–777.

Mo, L., Zhang, F., Deng, M., Panesar, D.K., 2016. Effectiveness of using CO_2 pressure to enhance the carbonation of portland cement-fly ash-MgO mortars. *Cement and Concrete Composites* 70, 78–85.

Nguyen, H., Santos, H., Sreenivasan, H., Kunther, W., Carvelli, V., Illikainen, M., Kinnunen, P., 2022. On the carbonation of brucite: Effects of Mg-acetate on the precipitation of hydrated magnesium carbonates in aqueous environment. *Cement and Concrete Research* 153, 106696.

Oconnor, W., Dahlin, D.C., Rush, G.E., Gerdemann, S.J., Penner, L.R., Nilsen, D.N., 2005. Aqueous mineral carbonation: Mineral availability, pretreatment, reaction parametrics, and process studies. DOE/ARC-TR-04-002.

Oelkers, E.H., Declercq, J., Saldi, G.D., Gislason, S.R., Schott, J., 2018. Olivine dissolution rates: A critical review. *Chemical Geology* 500, 1–19.

Panesar, D.K., Mo, L., 2013. Properties of binary and ternary reactive MgO mortar blends subjected to CO_2 curing. *Cement and Concrete Composites* 38, 40–49.
Pokrovsky, O.S., Schott, J., 2000. Kinetics and mechanism of forsterite dissolution at 25°C and pH from 1 to 12. *Geochimica et Cosmochimica Acta* 64, 3313–3325.
Power, I.M., Wilson, S., Thom, J.M., Dipple, G.M., Southam, G., 2007. Biologically induced mineralization of dypingite by cyanobacteria from an alkaline wetland near Atlin, British Columbia, Canada. *Geochemical Transactions* 8, 13.
Pu, L., Unluer, C., 2020. Effect of preconditioning on carbonated reactive MgO-based concrete samples. *Journal of Materials in Civil Engineering* 32, 04020314.
Raschman, P., Fedoročková, A., 2004. Study of inhibiting effect of acid concentration on the dissolution rate of magnesium oxide during the leaching of dead-burned magnesite. *Hydrometallurgy* 71, 403–412.
Rashid, M.I., Benhelal, E., Farhang, F., Oliver, T.K., Stockenhuber, M., Kennedy, E.M., 2020. Application of a concurrent grinding technique for two-stage aqueous mineral carbonation. *Journal of CO_2 Utilization* 42, 101347.
Renforth, P., Washbourne, C.-L., Taylder, J., Manning, D.A.C., 2011. Silicate production and availability for mineral carbonation. *Environmental Science & Technology.* 45, 2035–2041.
Ruan, S., Unluer, C., 2016. Comparative life cycle assessment of reactive MgO and portland cement production. *Journal of Cleaner Production* 137, 258–273.
Saleh, M.A., Ryan, M.P., Trusler, J.P.M., Krevor, S., 2025. The interfacial processes controlling carbon dioxide mineralisation in magnesium and calcium silicates. *Fuel* 380, 132969.
Scott, A., Oze, C., Shah, V., Yang, N., Shanks, B., Cheeseman, C., Marshall, A., Watson, M., 2021. Transformation of abundant magnesium silicate minerals for enhanced CO2 sequestration. *Communications Earth Environment* 2, 1–6.
Shand, M.A., 2006. *The chemistry and technology of magnesia*, 1st ed. Wiley.
Shen, W., Cao, L., Li, Q., Wen, Z., Wang, J., Liu, Y., Dong, R., Tan, Y., Chen, R., 2016. Is magnesia cement low carbon? Life cycle carbon footprint comparing with Portland cement. *Journal of Cleaner Production* 131, 20–27.
Song, Y., Qian, X., Yan, D., Unluer, C., Peng, Y., Kong, D., Hu, C., Wang, S., Ruan, S., 2023. Understanding the role of seeds in reactive magnesia cement (RMC) formulations. *Journal of the American Ceramic Society* 106, 3812–3831.
Synthetic Magnesia, 2006. *The chemistry and technology of magnesia.* John Wiley & Sons, Ltd, pp. 39–61.
Teir, S., Eloneva, S., Fogelholm, C.-J., Zevenhoven, R., 2009. Fixation of carbon dioxide by producing hydromagnesite from serpentinite. *Applied Energy, IGEC III* 86, 214–218.
Unluer, C., Al-Tabbaa, A., 2013. Impact of hydrated magnesium carbonate additives on the carbonation of reactive MgO cements. *Cement and Concrete Research* 54, 87–97.
Unluer, C., Al-Tabbaa, A., 2015. The role of brucite, ground granulated blastfurnace slag, and magnesium silicates in the carbonation and performance of MgO cements. *Construction Building Materials* 94, 629–643.
Unluer, C., Al-Tabbaa, A., 2014. Enhancing the carbonation of MgO cement porous blocks through improved curing conditions. *Cement and Concrete Research* 59, 55–65.
Vandeperre, L.J., Liska, M., Al-Tabbaa, A., 2008. Microstructures of reactive magnesia cement blends. *Cement and Concrete Composites* 30, 706–714. 2
Wang, D., Xiao, J., He, F., Zhou, Y., 2019a. Durability evolution and associated micro-mechanisms of carbonated reactive MgO-fly ash solidified sludge from east lake, China. *Construction Building Materials* 208, 1–12.
Wang, D., Zhu, J., He, F., 2019b. CO_2 carbonation-induced improvement in strength and microstructure of reactive MgO-CaO-fly ash-solidified soils. *Construction Building Materials* 229, 116914.

Wang, F., Dreisinger, D., Jarvis, M., Hitchins, T., 2019a. Kinetics and mechanism of mineral carbonation of olivine for CO_2 sequestration. *Minerals Engineering* 131, 185–197.
Wang, F., Dreisinger, D., Jarvis, M., Hitchins, T., Dyson, D., 2019b. Quantifying kinetics of mineralization of carbon dioxide by olivine under moderate conditions. *Chemical Engineering Journal* 360, 452–463.
Wang, L., Chen, L., Provis, J.L., Tsang, D.C.W., Poon, C.S., 2020. Accelerated carbonation of reactive MgO and Portland cement blends under flowing CO2 gas. *Cement and Concrete Composites* 106, 103489.
Winnefeld, F., Leemann, A., German, A., Lothenbach, B., 2022. CO2 storage in cement and concrete by mineral carbonation. *Current Opinion in Green and Sustainable Chemistry* 38, 100672.
Xu, J., Yan, C., Zhang, F., Konishi, H., Xu, H., Teng, H.H., 2013. Testing the cation-hydration effect on the crystallization of Ca–Mg–CO3 systems. *Proceedings of the National Academy of Sciences* 110, 17750–17755.
Xu, T., Apps, J.A., Pruess, K., 2001. *Analysis of mineral trapping for CO_2 disposal in deep aquifers* (No. LBNL-46992). Lawrence Berkeley National Lab. (LBNL).
Zhang, R., Arrigoni, A., Panesar, D.K., 2021. Could reactive MgO cement be a green solution? The effect of CO_2 mineralization and manufacturing route on the potential global warming impact. *Cement and Concrete Composites* 124, 104263.

11 Durability of CO_2 mineralized materials and products

11.1 INTRODUCTION

As acknowledged, carbonation is one of the primary durability issues of reinforced concrete. Gaseous CO_2 in the atmosphere reacts with cement hydration products such as $Ca(OH)_2$ and C-S-H, reducing the alkalinity of cement matrix. This destroys the passive oxide film on the surface of rebars embedded in concrete, initiating corrosion of rebars and thus shortening the service life of reinforced concrete. However, CO_2 mineralization is generally applied to unhardened cement-based materials and products. CO_2 mainly reacts with C_3S and C_2S to form carbonates, which do not induce shrinkage and cracking. Besides, the lasting time commonly ranges from several hours to several days (Kim et al., 2022; Liang et al., 2022; Zhang et al., 2017). Some unreacted cement clinkers remain in the cement matrix after CO_2 mineralization. These cement clinkers can rehydrate once they are exposed to moisture. This subsequent hydration of calcium silicates recovers the alkalinity of the cement matrix, preventing the corrosion of steel bars (Zhang and Shao, 2016a). Thus, differing from weathering carbonation of mature concrete, CO_2 mineralization does not induce corrosion of rebars.

The durability of cementitious materials is mainly governed by their porosity and pore structure, as well as the phase assemblages. The molar volume of mineralization products of silicates (C_3S and C_2S) is greater than that of hydration products $Ca(OH)_2$ (Mu et al., 2018), so that CO_2 mineralized cement-based materials exhibited lower porosity and finer pore structure in comparison to conventional moist cured cement-based materials (Qin et al., 2019a). This decreases the ingress of water and external aggressive ions such as sulfate ions, chloride ions, etc. In addition, CO_2 mineralization alters the phase assemblages of cementitious materials. Conventional moist cured cementitious materials are mainly composed of hydration products such as C-S-H, $Ca(OH)_2$, and aluminate hydrates. In addition to hydration products, a considerable amount of $CaCO_3$ is present in CO_2 mineralized cementitious materials. Furthermore, there is less $Ca(OH)_2$ and C-S-H with a lower Ca–Si ratio in CO_2 mineralized cementitious materials than in conventional moist cured cementitious materials (Qin et al., 2019a). $Ca(OH)_2$ can react with sulfate to form expansive products gypsum and ettringite, inducing expansive stress in the cement matrix, and thus leading to cracking and deterioration. Also, it can promote the dissolution of silica from alkali-reactive aggregate, causing alkali silica reaction (ASR). C-S-H can be attacked by magnesium ions to form magnesium silicate hydrates (M-S-H), losing

 DOI: 10.1201/9781003627531-11

cementing capacity. Therefore, CO_2 mineralized cementitious materials are less attackable when they are exposed to aggressive ions. As a result, CO_2 mineralized materials show superior long-term performance, such as sulfate attack resistance (Rostami et al., 2011), ASR resistance (Luo et al., 2024), and chloride permeability resistance (Zhang et al., 2017) in comparison to conventional moist cured cement-based materials.

In this chapter, the long-term durability performance of CO_2 mineralized materials and products, including chloride transport and binding, sulfate attack, leaching, weathering, carbonation, and aging, ASR, and reinforcement corrosion is discussed in detail.

11.2 CHLORIDE TRANSPORT AND BINDING

11.2.1 Chloride transport

Chloride-induced reinforcement corrosion is one of the main durability issues for reinforced concrete exposed to chloride-bearing solutions such as seawater and de-icing salt. Chloride transport is the deciding factor governing the corrosion resistance of reinforced concrete structures. The faster the chloride transport in the cement matrix, the faster the corrosion of steel bars in reinforced concrete. Thus, understanding the transportation of chloride in CO_2 mineralized cementitious materials and products is critical. Intensive studies revealed that CO_2 mineralization decreased the chloride permeability of cement mortar and concrete according to the results of NT Build 443 and 492 (Zhang and Shao, 2016b; Qin et al., 2019a; Song et al., 2022). Figure 11.1 compares the rapid chloride migration coefficients (D_{RCM}) and apparent diffusion coefficients (D_{app}) of CO_2 mineralized and conventional water-cured cement mortars. CO_2 mineralized mortars show smaller D_{RCM} and D_{app} compared to conventional moist cured mortars, both at 28 and 90 days, regardless of the w/c ratio. At the curing age of 28 days, the value of D_{RCM} of conventional water cured cement mortar with w/c ratios of 0.3, 0.4, and 0.5 is 6.69, 9.88, and 17.28 × 10^{-12}

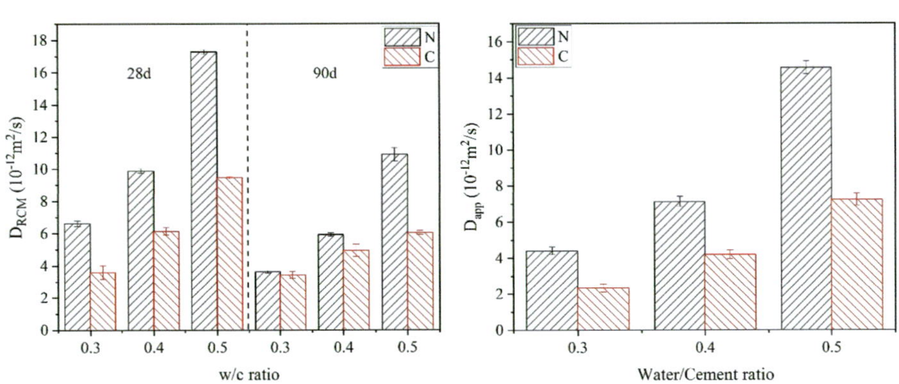

FIGURE 11.1 D_{RCM} and D_{app} of cement mortars (Song et al., 2024).

m^2/s, while it reduces by 45.8%, 37.9%, and 45.1% respectively when they were subjected to 3 days of CO_2 mineralization. This is opposite to the effect of weathering carbonation on the chloride permeability of mature concrete, which usually causes an increase in chloride permeability (Liu et al., 2014). In the course of weathering carbonation, the decalcification of C-S-H increases the porosity and even leads to cracking (Justnes et al., 2020). This affords more transportation channels for the water and aggressive ions such as chloride ions. However, CO_2 mineralization is applied to unhardened mortar and concrete. The gaseous CO_2 mainly reacts with calcium silicates (i.e., C_3S and C_2S), which enhances the volume of solid phases and thus densifies the microstructure of mortars. Besides, the continuing hydration of unreacted cement clinkers during subsequent moist curing further densifies the microstructure of the cement matrix (He et al., 2016). This leads to the porosity of CO_2 mineralized mortars being lower than that of conventional moist cured ones, as shown in Figure 11.2. Consequently, the chloride permeability of cement mortar is lowered by CO_2 mineralization.

Similar to the conventional moist cured mortars, the chloride permeability of CO_2 mineralized mortar reduces as the curing ages are extended. Nevertheless, the gap of D_{RCM} between CO_2 mineralized and conventional moist cured mortars decreases as the curing age increases. For example, the D_{RCM} of CO_2 mineralized mortar with a w/c ratio of 0.4 is only 17% lower than that of conventional moist cured counterparts at 90 days, while this gap is 37.9% at 28 days. This is due to the conventional moist cured cement matrix showing a greater further hydration degree than the CO_2 mineralized cement matrix between 28 and 90 days, leading to greater microstructure improvement. However, Pan et al. (Pan et al., 2017, 2019) found that the CO_2 mineralized mortar and conventional moist cured mortar showed similar reduction in D_{RCM} as the curing ages were extended from 28 days to 90 days. The probable cause is that the further hydration of the cement matrix is less affected due to a much smaller carbonation depth and shorter CO_2 mineralization duration (3 hours and 6 hours) in their study.

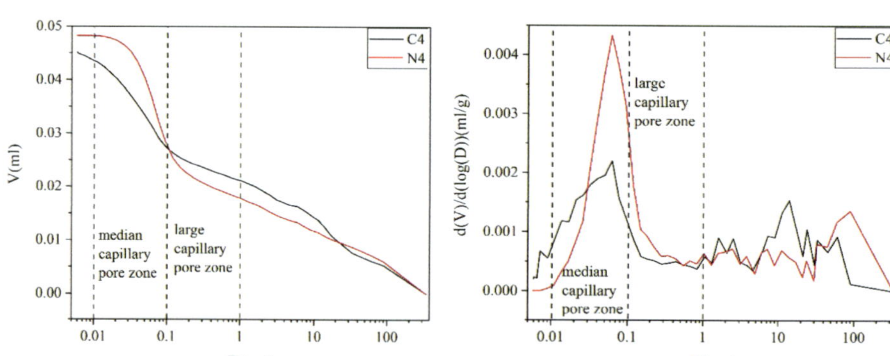

FIGURE 11.2 Pore size distribution of CO_2 mineralized and conventional moist cured mortars with w/c ratio of 0.4 (Song et al., 2024).

Supplementary cementitious materials (SCMs) such as fly ash and GGBS have been commonly used to prepare mortar and concrete due to lower cost, environmental friendliness, and improved performance of mortar and concrete. It has been an important component of modern concrete. Incorporating a proper level of fly ash and GGBS generally decreases the chloride permeability of traditional moist cured cement mortar and concrete. This can be attributed to the "filling" and pozzolanic effect of them, which densifies the microstructure and refines the pore structure of the cement matrix. However, the incorporation of fly ash and GGBS increases the chloride ion transport rate in CO_2 mineralized mortar and concrete, as shown in Figures 11.3 and 11.4. In fact, CO_2 mineralization significantly eliminates $Ca(OH)_2$,

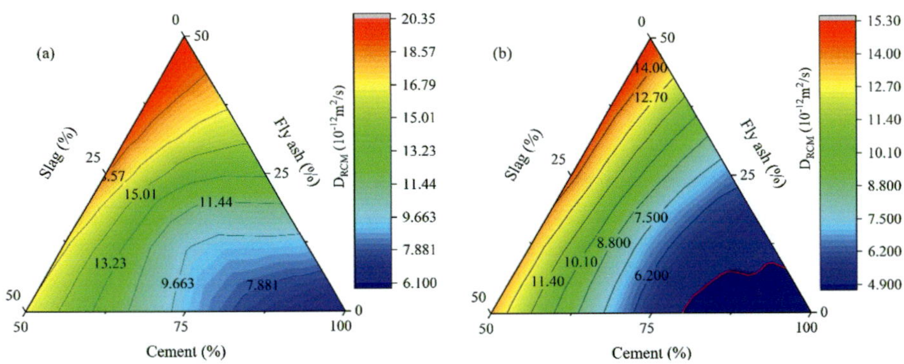

FIGURE 11.3 D_{RCM} contour maps of CO_2 mineralized Portland cement-fly ash-GGBS mortar specimens at 28 days (a) and 90 days (b) (Song et al., 2022).

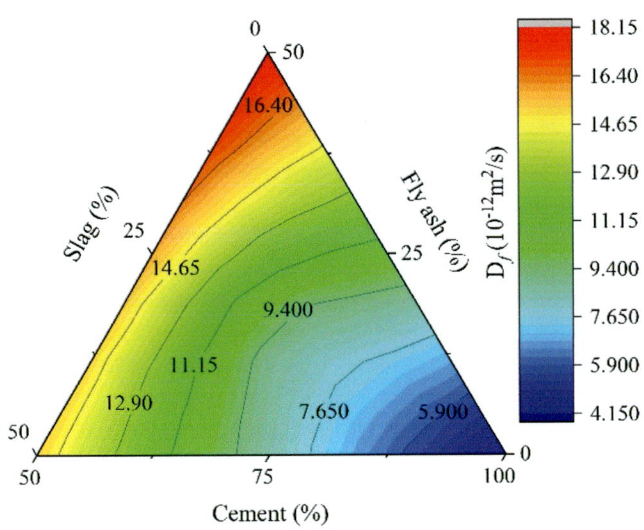

FIGURE 11.4 Chloride diffusion coefficient contours of CO_2 mineralized Portland cement-fly ash-GGBS mortar specimens at 28 days (Song et al., 2022).

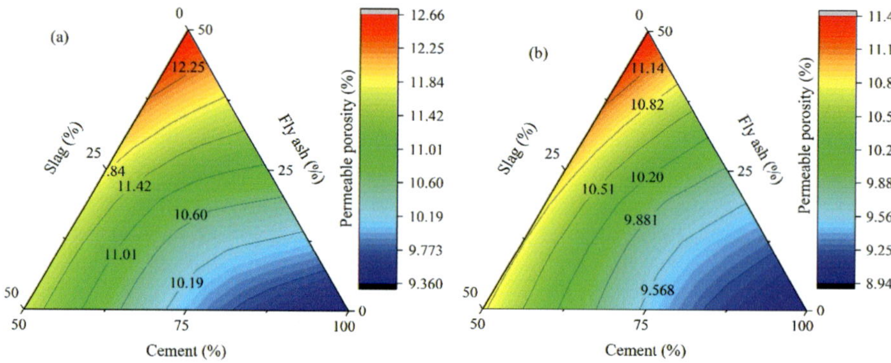

FIGURE 11.5 Contour maps of volume of permeable pores of CO_2 mineralized Portland cement-fly ash-GGBS mortars with curing ages of 28 days (a) and 90 days (b) (Song et al., 2022).

limiting the pozzolanic reaction of fly ash and GGBS. Thus, the incorporation of fly ash and GGBS increases the pore volume in CO_2 mineralized mortar and concrete, as shown in Figure 11.5, and thus increases the chloride permeability of mortar and concrete. Moreover, the chloride permeability of CO_2 mineralized mortars increases with the increase of fly ash and GGBS. The D_{RCM} of mortars containing 50% SCMs, including fly ash and GGBS, is 2.7–3.3 times as large as that of plain Portland cement mortars. Therefore, high substitution rates of SCMs are not suggested in CO_2 mineralized cement mortars regarding the chloride permeability resistance. Furthermore, the D_{RCM} of fly ash blended mortar is smaller than that of GGBS blended mortar when the addition level of them is identical, which suggests that fly ash exerts more obvious effect on the chloride permeability of CO_2 mineralized mortar compared to GGBS. Therefore, the value of D_{RCM} decreases with the increase of fly ash when the total SCMs are constant. Zhang and Shao (2016b) also found that the addition of 20% fly ash increased the chloride permeability of CO_2 mineralized mortar and concrete. The investigation carried out by Qin et al. (Qin and Gao, 2019a) revealed that the incorporation of waste autoclaved aerated concrete powder amplifies the chloride permeability of CO_2 mineralized mortar.

The chloride permeability of CO_2 mineralized mortar containing fly ash and GGBS shows a reduction as the curing age is extended from 28 days to 90 days. It is noted that the incorporation of SCMs exerts little influence on D_{RCM} of mortars when the fly ash content is smaller than around 10% and GGBS content is smaller than around 20% (the lower right area of the red curve). Although early CO_2 mineralization consumes almost all of $Ca(OH)_2$, it is reproduced during the subsequent lime water curing process. Thus, when the replacement level of SCMs is small, it can react with $Ca(OH)_2$ to produce secondary hydration products depositing in the pores, which compacts the microstructure of mortars. However, the pozzolanic reaction degree of SCMs in the early CO_2 mineralized mortars is much smaller than that in the moist cured mortars, which induces less microstructure improvement for the former. As a result, the D_{RCM} of the mortars containing SCMs is slightly lower than that of plain Portland cement mortars. On the other hand, the "filling" effect also contributes to this observation.

11.2.2 Chloride binding

Chloride binding is defined as the interaction of chloride ions and the cement matrix. As the external chloride ions penetrate cement-based materials, a portion of the chloride ions may either chemically react with aluminate hydrates to form Friedel's salt (Fs) (Paul et al., 2015) or Kuzel's salt (Balonis et al., 2010; Wang et al., 2019), or be physically adsorbed onto the surface of C–S–H (Chang et al., 2019). The free chloride ions dissolved in the pore solution (Shi et al., 2017) are generally movable and considered the main reason leading to steel bar corrosion (Suryavanshi et al., 1998). As the free chloride concentration in the pore solution around the surface of steel bars reaches a critical value, which is defined as the threshold chloride concentration, the corrosion of steel bars can be initiated. The chloride binding declines the quantity of free chloride ions in the pore solution, which thus lowers the transport speed of Cl^- and corrosion risk of rebars in the concrete. Thus, chloride binding of the cement matrix also plays a critical role in the corrosion resistance of concrete.

CO_2 mineralization altered the chloride binding behavior and decreased the chloride binding capacity of cementitious materials due to changes in phase assemblage. Figures 11.6 and 11.7 compare the chloride binding behavior of CO_2 mineralized and traditional moist cured cement paste with the curing ages of 28 days and 90 days, respectively. The fitting coefficients are presented in Table 11.1. Both Langmuir and Freundlich isotherms reasonably describe the chloride binding behavior of CO_2 mineralized and conventional moist cured cement paste. Furthermore, Freundlich is more suitable to describe the chloride binding behavior of conventional moist cured cement paste, while the Langmuir isotherm more reasonably characterizes the chloride binding behavior of CO_2 mineralized cement paste. Moreover, the chloride

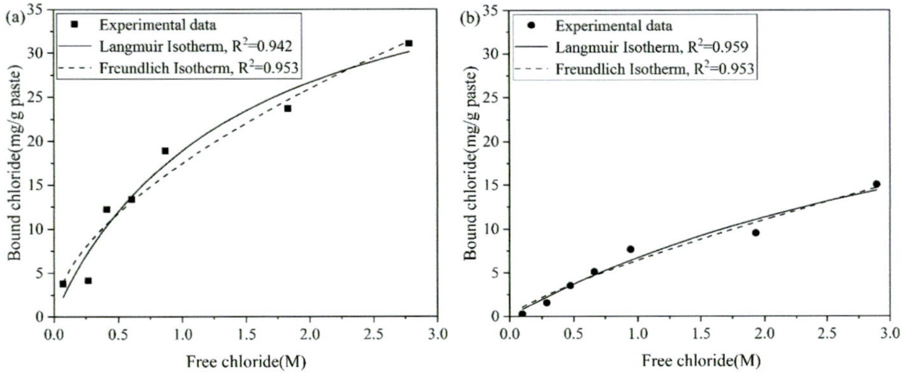

FIGURE 11.6 Fitted isotherms for unmineralized (a) and mineralized (b) pure Portland cement pastes at 28 days (Song et al., 2022).

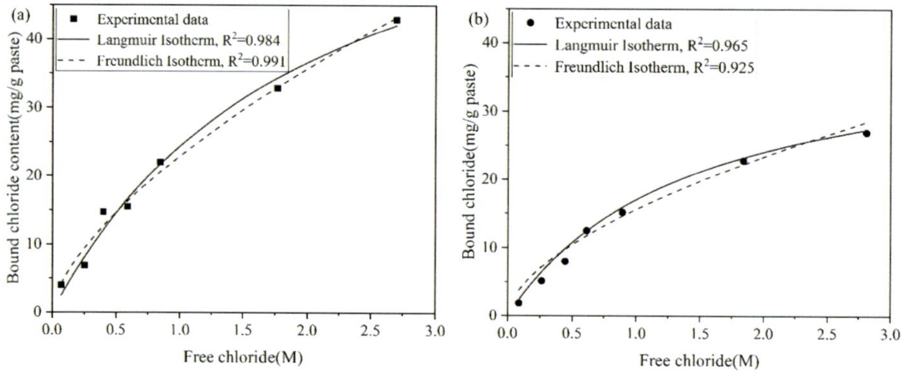

FIGURE 11.7 Fitted isotherms for unmineralized (a) and mineralized (b) pure Portland cement pastes at 90 days (Song et al., 2022).

TABLE 11.1
Fitted binding constants for Langmuir and Freundlich isotherms of cement pastes

		Langmuir			Freundlich		
		α	β	R^2	α	β	R^2
Unmin-	28 d	32.4	0.72	0.953	17.5	0.57	0.942
	90 d	36.3	0.49	0.984	22.9	0.64	0.991
Min-	28 d	11.5	0.16	0.959	9.4	0.85	0.953
	90 d	29.2	0.71	0.965	15.6	0.58	0.925

Source: Song et al., 2022.

binding of CO_2 mineralized cement paste is much lower than that of conventional moist cured cement paste.

The bound chloride within cement paste can be classified into chemically bound as Friedel's salt or Kuzel's salt, and physically absorbed on the surface of C-S-H gel. Figure 11.8 illustrates the XRD patterns of cement paste exposure to chloride ion solution with different concentrations. Obvious diffraction peaks corresponding to Friedel's salt were detected in the conventional moist cured cement paste after exposure to chloride ion solution, while only minor Friedel's salt diffraction peaks were detected in the CO_2 mineralized cement paste after exposure to 1 M NaCl and $CaCl_2$ solution. In other words, the chemical chloride binding of CO_2 mineralized cement paste was negligible, and the chloride binding capacity primarily depends on the physical chloride binding associated with C-S-H gel. The chloride binding isotherm of C-S-H gel can be fitted with the Langmuir isotherm, while the chloride binding associated with the formation of Friedel's salt shows a better fit to the Freundlich

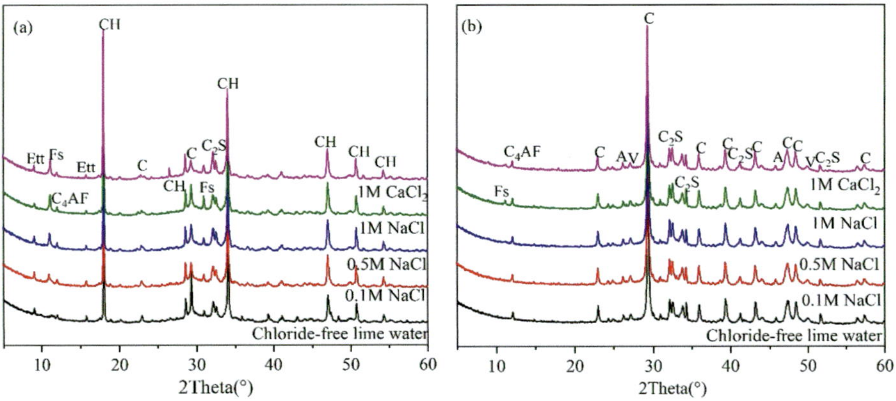

FIGURE 11.8 XRD patterns of pure Portland cement pastes (at the curing age of 28) without (a) and with (b) early CO_2 curing post-exposure to chloride solutions (Ett: ettringite; CH: calcium hydroxide; C: calcite; A: aragonite; V: vaterite; Fs: Friedel's salt) (Song et al., 2022).

isotherm. Therefore, chloride binding of CO_2 mineralized cement paste is a better fit to the Langmuir isotherm, while the continuous water cured cement paste is more suitable for the Freundlich isotherm.

Since six water molecules in the main layer of Friedel's salt structure release between 230°C and 410°C (Grishchenko et al., 2013), the quantity of chloride ions in Friedel's salt can be accessed via DTG curves as in Equation (11.8):

$$X_{c,b}\left(\text{mg}/\text{g paste}\right) = \frac{2\times 35.453\left(m_{H,\text{total}} - m_{H,\text{ref}}\right)}{6\times 18} \tag{11.8}$$

where: $m_{H,ref}$ (mg/g paste) and $m_{H,total}$ (mg/g paste) are the samples' mass loss between 230°C and 410°C after being exposed to chloride-free lime water and chloride bearing solution, respectively.

Both chemical and physical chloride binding capacity of cement paste are dramatically decreased by CO_2 mineralization, as shown in Figure 11.9. Upon exposure to 1M NaCl solution, the chemically and physically bound chloride contents of conventional water cured reference cement paste with curing age of 28 days are 4.21 and 14.71 mg/g paste, respectively. However, when CO_2 mineralization is applied, they reduce to 1.03 and 6.62 mg/g paste, which are about 76% and 55% lower than that of the reference cement paste, respectively. The amount of chloride absorbed by C-S-H gel is more than that chemically bound to form Friedel's salt, in particular when CO_2 mineralization is applied. For CO_2 mineralized cement paste, the chemically bound chloride ions are about 10% of the total bound chloride ions after being immersed in 1M NaCl solution and can be ignored when the chloride concentration of NaCl solution is smaller than 0.5M. Thus, the chloride binding of CO_2 mineralized cement paste mainly depends on the physical chloride absorption of C-S-H.

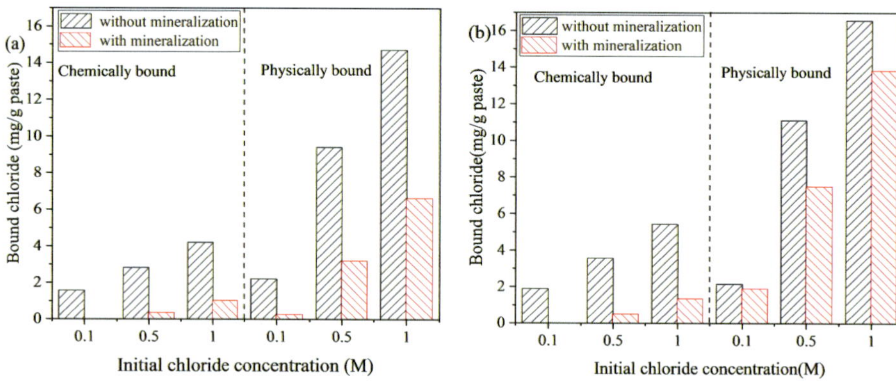

FIGURE 11.9 Physically and chemically bound chloride ions of plain Portland cement pastes with and without early CO_2 mineralization at curing ages of 28 days (a) and 90 days (b) (Song et al., 2022).

TABLE 11.2

Chloride desorption results of cement pastes with and without CO_2 mineralization

	Initial Cl^- concentration (M)	Water added (mL)	Initial C_b (mg/g paste)	Remaining $C_{b,r}$ (mg/g paste)	Remaining bound rate (%)
Umin-	1	80	22.00	13.33	60.6
CO_2Min-	1	80	15.17	7.54	49.7

Source: Song et al., 2022.

A portion of the bound chloride ions is reversible. Some bound chloride ions will desorb when the pore solution is diluted with chloride-free water. Table 11.2 compares the chloride desorption of CO_2 mineralized and conventional moist cured cement paste. The quantity of bound chloride ions within CO_2 mineralized and conventional moist cured cement pastes was reduced from 15.17 and 22.00 mg/g paste to 7.54 and 13.33 mg/g paste, representing 50.3% and 39.4% desorption, respectively. The chemically bound chloride ions are more stable than physically absorbed chloride ions. Thomas et al. (2012) pointed out that a considerable amount of chemically bound chloride ions was irreversible even when the chloride ion concentration approached zero. Thus, the chloride desorption of CO_2 mineralized cement pastes with negligible chemical chloride binding was higher than that of conventional moist cured cement pastes.

To enhance the chloride binding capacity of traditional moist cured cement matrix, the addition of fly ash and GGBS is often applied. Fly ash and GGBS are silica and alumina-rich admixtures with high pozzolanic activity. The increased reactive

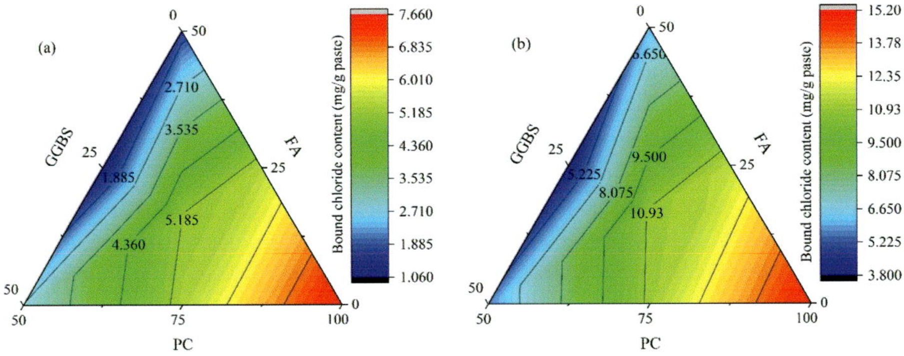

FIGURE 11.10 Bound chloride contour maps of early CO_2-cured Portland cement-fly ash-GGBS pastes after exposure to 1 M NaCl solution at the curing ages of (a) 28 days and (b) 90 days (Song et al., 2022).

SiO_2 and Al_2O_3 can react with $Ca(OH)_2$ to form additional hydrates such as C–S–H gel and AFm phases that are conducive to chloride binding, thereby improving the chloride binding capacity of the cement matrix. On the contrary, the addition of fly ash and GGBS decreases the chloride binding of CO_2 mineralized cement paste, as shown in Figure 11.10. All of the fly ash and GGBS blended cement pastes show lower bound chloride contents than that of neat Portland cement paste, and the higher the replacement level, the lower the bound chloride content. When the replacement level of fly ash increases from 0 to 25% and 50%, the bound chloride ions in cement paste with a curing age of 28 days and exposure to 1 M NaCl solution decrease from 7.66 to 5.45 and 1.63 mg/g paste, respectively. The bound chloride contents of fly ash blends and GGBS blends are similar, as the replacement levels of them are identical. In fact, CO_2 mineralization remarkably reduced $Ca(OH)_2$, which is essential for the pozzolanic reaction of fly ash and GGBS. As a result, the pozzolanic reactions of fly ash and GGBS are hindered in CO_2 mineralized cement paste. In other words, the active Al_2O_3 and SiO_2 resources from fly ash and GGBS cannot contribute to chloride binding. The chloride binding of fly ash and/or GGBS blended cement paste mainly depends on the Portland cement. Thus, the higher the incorporation of fly ash and/or GGBS, the lower the chloride binding of CO_2 mineralized cement paste.

The subsequent water curing period after CO_2 mineralization strongly impacts the chloride binding of cement paste. Specifically, the chloride binding of cement paste increases markedly with the extension of the water curing period after CO_2 mineralization. For the plain cement paste without SCMs soaking in 1 M NaCl solution, the chemically and physically bound chloride contents increase by 0.32 and 7.20 mg/g paste as the water curing period extends from 27 days to 89 days. The cement pastes containing SCMs show no chemical chloride binding, both after 27 days and 89 days of water curing. Thus, it can be concluded that the extension of the water curing period mainly enhances the physical chloride adsorption ability of CO_2 mineralized cement paste associated with C–S–H. In fact, the formation of Friedel's salt is restrained due to the lack of portlandite and the presence of carbonates in the cement. Nevertheless, further hydration of unreacted cement clinkers increases the quantity of C–S–H, which in turn increases the physically bound chloride content. Besides, the extension of the subsequent water curing period induces greater chloride binding improvement for neat Portland cement paste compared to SCMs-blended cement pastes. As a result, the gaps of bound chloride between SCMs-blended and neat Portland cement pastes are enlarged. This is due to the further hydration of SCMs-blended cement pastes being smaller than that of neat cement paste. Hence, in order to enhance the chloride binding of CO_2 mineralized cementitious materials exposed to seawater or de-icing salt, the water curing period after CO_2 mineralization should be long enough.

11.3 CARBONATION AND AGING

11.3.1 Weathering Carbonation

Hydration products such as $Ca(OH)_2$ and C-S-H in the mature cement matrix can react with gaseous CO_2 in the atmosphere, leading to a reduction in alkalinity. This destroys the passive film on the surface of steel bars, thereby initiating corrosion. Besides, it may also cause shrinkage, inducing cracking, which facilitates the diffusion of aggressive ions. Thus, the weathering carbonation of the mature cement matrix is a type of detrimental reaction. The weathering carbonation resistance of the cement matrix mainly depends on the gas permeability and attackable phases. Carbonation resistance of the cement matrix is often assessed with carbonation depth and coefficient. The weathering carbonation rate of the mature cement matrix is slow, requiring decades to progress to the rebars; thereby, an accelerated carbonation test with high CO_2 concentration is mostly recommended in the laboratory as specified in many specifications.

Zhang et al. (2020) preliminarily compared the weathering carbonation performance of conventional moist cured and CO_2 mineralized concrete using an accelerating carbonation test. Their results suggested that the carbonation resistance of CO_2 mineralized concrete was linked to the w/c ratio, and it was not more vulnerable to weathering carbonation than conventional moist cured concrete. When the w/c ratio is 0.4, the weathering carbonation depth of CO_2 mineralized concrete and conventional moist cured concrete is similar. Nevertheless, when the w/c ratio is 0.65, the weathering carbonation depth of CO_2 mineralized concrete is greater than that of traditional moist cured concrete in the first several weeks due to a higher initial carbonation depth. However, the carbonation coefficients of concrete reduce by 44% and 59% if 2 hours and 12 hours CO_2 mineralization is adopted, respectively.

The authors systematically studied the weathering carbonation of CO_2 mineralized Portland cement-fly ash-GGBS ternary mortars with different initial mineralization depths and subjected to different subsequent moist curing times. Figures 11.11 and 11.12 illustrate the weathering carbonation depth evolution of cement mortars with different initial mineralization depths and subsequent water curing time. The weathering carbonation depth of CO_2 mineralized mortar is greater than that of conventional moist cured counterparts, regardless of the initial mineralization depth

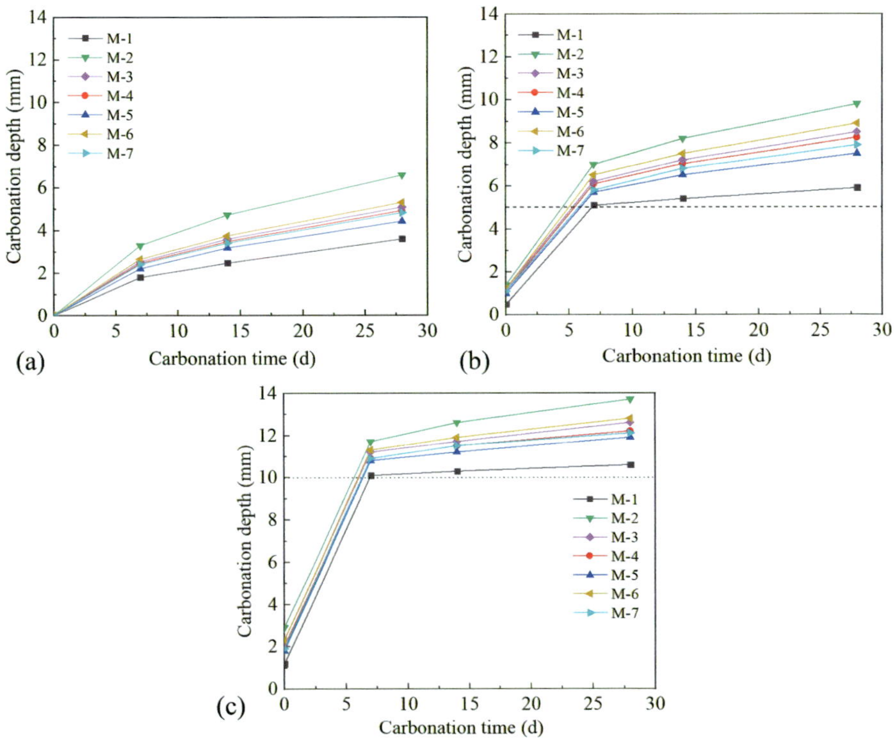

FIGURE 11.11 Weathering carbonation depth of mortars with initial CO_2 mineralization depth of 0 (a), 5 (b) and 10 mm (c) and moist cured to 28 days.

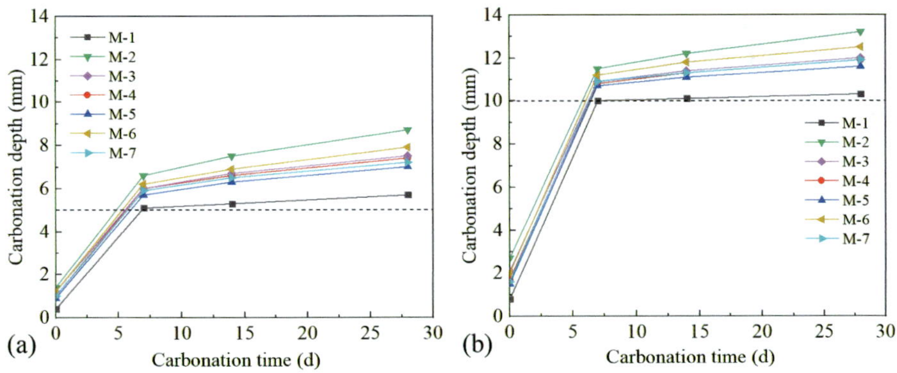

FIGURE 11.12 Weathering carbonation depth of mortars with initial CO_2 mineralization depth of 5 (a) and 10 mm (b) and moist cured to 90 days.

and subsequent water curing time. This is consistent with the results reported by Zhang et al. (2020). The weathering carbonation depth progresses fast in the zone that is CO_2 mineralized before the carbonation test. The weathering carbonation depths of all of the samples exceed the initial mineralized zone after around 7 days of accelerating carbonation. However, the progress rate of carbonation depth slows down sharply after the carbonation depth exceeds the initial mineralized zone, and it is slower than that in conventional moist cured mortars. Furthermore, the greater the initial mineralized depth, the slower the carbonation depth progress. In fact, the carbonatable phases $Ca(OH)_2$ and C-S-H are dramatically eliminated in the initial CO_2 mineralized zone. This significantly promotes the carbonation depth progress in this zone. In the meantime, CO_2 mineralization lowers the gas permeability of the cement matrix. Besides, the quantity of carbonatable phases in the unmineralized zone is comparable to that of conventional moist cured mortars. Therefore, the carbonation depth progress rate is slowed down in this zone. Similar to moist cured mortars, the incorporation of fly ash and/or GGBS increases the carbonation depth of mortars. This can be attributed to the fact that the addition of fly ash and/or GGBS reduces $Ca(OH)_2$ in the cement matrix.

Compared to the CO_2 mineralized mortars with 28 days of subsequent water curing, the carbonation depth of the mortars with 90 days of water curing progresses more slowly. Some unreacted cement clinkers remain in the cement matrix, and they will further hydrate with the extension of curing time. The resultant hydration products deposit in the pores, refining the pore structure and decreasing the porosity, which in turn lowers the gas permeability. In addition, this further hydration offers more carbonatable phases $Ca(OH)_2$ and C-S-H. As a result, the carbonation depth progress is slowed down.

Based on Fick's second law of diffusion, the carbonation depth of concrete can be described as: $x=kt^{0.5}$, x (mm) is the carbonation depth, k (mm/day$^{0.5}$) is the carbonation coefficient, and t (days) is the duration of carbonation. Carbonation coefficient encompasses the quality of concrete and environmental conditions. Thus, it is commonly used to assess the carbonation resistance of concrete and to predict carbonation depth. According to the carbonation depth presented in Figures 11.11 and 11.12, the carbonation coefficients of mortars are obtained, as shown in Figures 11.13 and

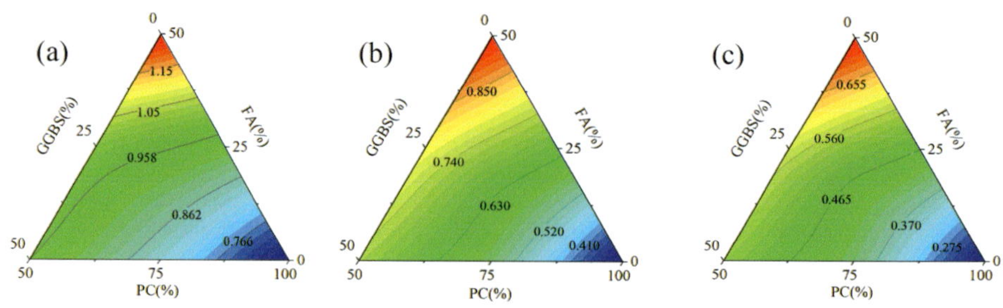

FIGURE 11.13 Weathering carbonation coefficients of mortars with initial CO_2 mineralization depth of 0 (a), 5 (b), and 10 mm (c) and moist cured to 28 days.

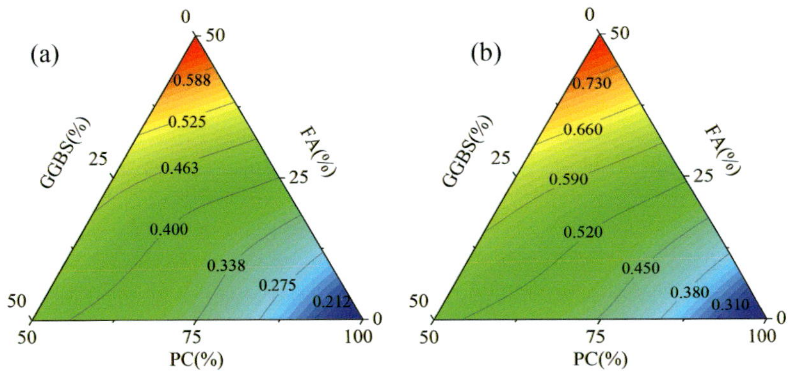

FIGURE 11.14 Weathering carbonation coefficients of mortars with initial CO_2 mineralization depth of 0 (a), 5 (b), and 10 mm (c) and moist cured to 90 days.

11.14. The carbonation coefficients of CO_2 mineralized mortar are smaller than those of conventional moist cured mortars, and they decrease with increasing initial CO_2 mineralized depth, although the carbonation depth progresses faster in CO_2 mineralized mortar. For the mortars with 28 days of curing age, the carbonation coefficients of plain Portland cement mortar with the initial mineralized depth of 0, 5, and 10 mm are 0.678, 0.304, and 0.243, respectively. The carbonation coefficients of mortars with the initial mineralized depth of 5 and 10 mm are 45% and 64% lower than that of conventional moist cured mortars, respectively. According to the carbonation coefficients, it can be predicted that the carbonation depth of conventional moist cured mortars exceeds the carbonation depth of mortars with an initial mineralized depth of 5 mm after 20 weeks. Thus, CO_2 mineralization will not increase the carbonation-induced corrosion risk of rebars. Moreover, the carbonation coefficients further decrease as the subsequent water curing duration is prolonged. For example, the carbonation coefficient of plain Portland cement mortar with an initial mineralized depth of 5 mm decreases from 0.304 to 0.243 as the water curing is extended from 28 to 90 days, representing a 20% decrement. Overall, CO_2 mineralization does not exert a negative effect on the weathering carbonation resistance of cement mortar and concrete.

Similar to the conventional moist cured mortars, the carbonation coefficients of CO_2 mineralized mortar increase with increasing content of fly ash and/or GGBS. For example, the carbonation coefficients of Portland mortars with 10 mm of initial mineralized depth and cured to 90 days in water are 0.157, while they increase to 0.645 as 50% fly ash is incorporated, representing a 310% growth. Besides, the carbonation coefficients of fly ash blends are greater than those of GGBS blends. As a result, the carbonation coefficients of mortars increase with the increasing content of fly ash as the cement content is fixed. This implies that fly ash exerts more obvious effect on the carbonation resistance of mortars. This may be caused by the lower calcium content in fly ash than in GGBS.

11.3.2 Aging resistance

CO_2 mineralized fibers reinforced cementitious composites exhibit dramatically superior aging performance in comparison to the moist cured reference (Wang et al., 2017; Soroushian et al., 2012; Almeida et al., 2013; Tolêdo Filho et al., 2003). In the aging process, alkali attack and mineralization through the migration of hydration products to lumens and spaces weaken the fibers, leading to a reduction in strength and toughness. However, CO_2 mineralization consumes $Ca(OH)_2$ and lowers the alkalinity of the pore solution, mitigating the corrosion risk of fibers. Besides, CO_2 mineralization densifies the cement matrix and improves the bonding between the cement matrix and fibers, which drops the permeability of liquids and gases from the external environment into the cement matrix. As a result, CO_2 mineralization delays the degradation of cement composites containing fibers. Figure 11.15 presents the statistical results of cellulose fibers-reinforced cement composites obtained via a three-point flexural test. Accelerating aging in warm water (60±2°C) declines the flexural strength and toughness as well as initial stiffness of cellulose fibers-reinforced cement composites, regardless of the curing regimes applied. Nevertheless, CO_2 mineralized specimens exhibit less reduction in the flexural strength and toughness, as well as a similar reduction in stiffness, in comparison to the moist cured specimens.

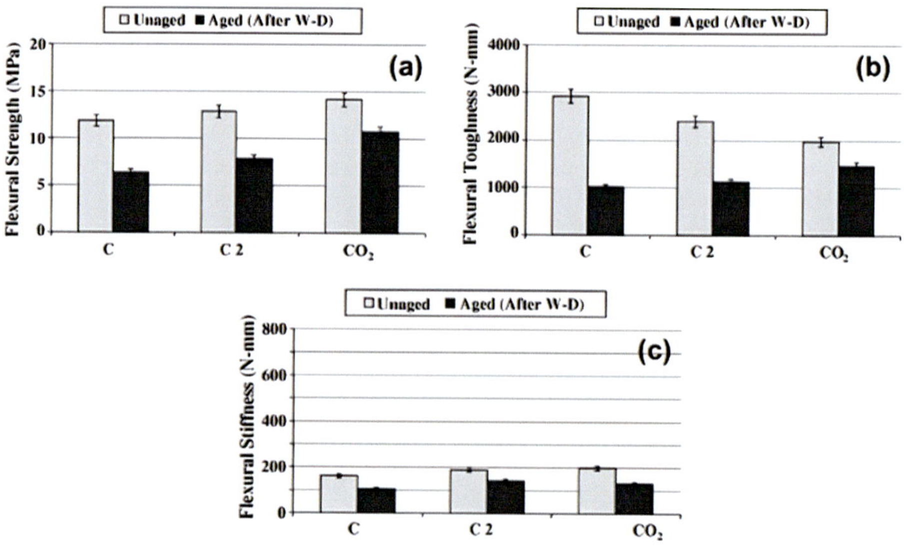

FIGURE 11.15 Flexural performance evolution of cellulose fibers reinforced cement composites: (a) flexural strength, (b) flexural toughness, and (c) initial stiffness (Soroushian et al., 2012).

11.4 REINFORCEMENT CORROSION

Reinforcement corrosion reduces the cross-sectional area and carrying capacity of rebars. Besides, the formation of rust increases the volume of rebars, which leads to expansive stress around the rebars. The concrete cover will crack and even spall when the expansive stress exceeds the tensile strength of the cement matrix, accelerating the ingress of moisture and aggressive ions such as chloride ions and sulfate ions. Thus, reinforcement corrosion is one of the severe durability issues of reinforced concrete, which significantly reduces the lifetime of reinforced concrete structures. CO_2 mineralization alters the alkalinity, phases assemblage, and diffusivity of the cement matrix, so that the corrosion behavior of rebars in CO_2 mineralized concrete may differ from that in conventional moist cured concrete.

Zhang and Shao (2016a) assessed the feasibility of using CO_2 mineralization to produce precast concrete beams. The carbonation depth due to CO_2 mineralization is around 10 mm that is smaller than the cover thickness of rebars in their study. After 27 days of subsequent hydration, the pH value around the rebars in CO_2 mineralized concrete and conventional moist cured concrete is similar, and the weathering carbonation of the cement matrix with and without CO_2 mineralization is also similar. However, CO_2 mineralization reduces the chloride permeability of the cement matrix. Thus, they believed that CO_2 mineralization will not increase the weathering carbonation and chloride ions-induced corrosion vulnerability of rebars. Therefore, they deduced that CO_2 mineralization can be applied to produce precast reinforced concrete.

Xian et al. (2022) examined the corrosion of rebars embedded in CO_2 mineralized concrete. The electrochemical measurements revealed that the rebars embedded in CO_2 mineralized concrete exhibited a lower corrosion rate than in conventional moist cured concrete under the ponding corrosion test, as shown in Figure 11.16(a). The rebars embedded in conventional moist cured concrete are completely corroded, while little rust is observed on the surface of rebars embedded in CO_2 mineralized concrete after 33 months of ponding, as shown in Figure 11.16(b). Moreover,

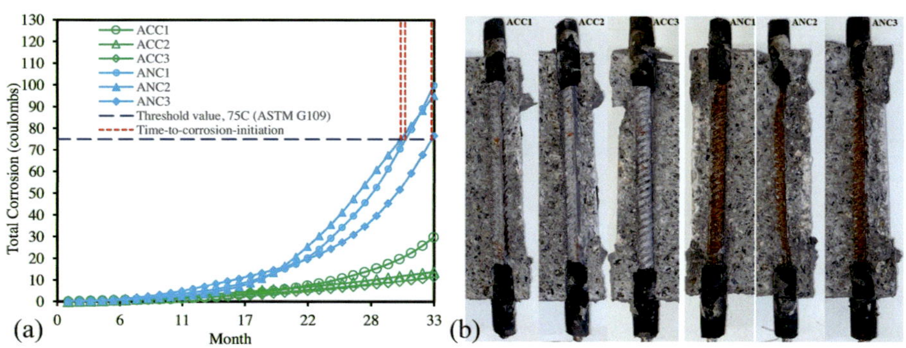

FIGURE 11.16 Total corrosion (a) and optical photo of rebars under natural ponding test (ACC: CO_2 mineralized concrete; ANC: non-mineralized concrete) (Xian et al., 2022).

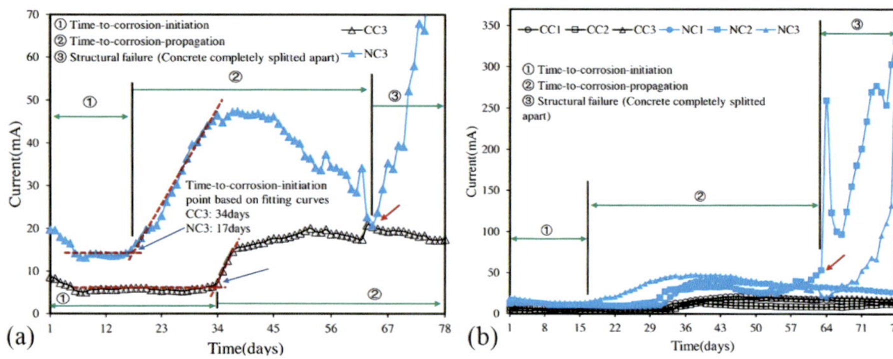

FIGURE 11.17 Typical current curves (a) and summary of current curves (b) of CO_2 mineralized concrete and moist cured concrete subjected to 78 days accelerated corrosion test with sodium chloride solution (Xian et al., 2022).

the IC-accelerated test indicated that CO_2 mineralization delayed the initiation of rebars' corrosion and lowered the corrosion rate in propagation phases, as shown in Figure 11.17. The corrosion was initiated after 17 days of the accelerating test in conventional moist cured concrete, while it was initiated after 34 days of accelerated test in CO_2 mineralized concrete, representing a 100% increase. The corrosion current increased more quickly in conventional moist cured concrete than in CO_2 mineralized concrete after corrosion was initiated. As acknowledged, only does the corrosion initiate after the accumulated chloride on the surface of the steel bars reaches the threshold. The chloride ions diffuse much slower in CO_2 mineralized concrete than that in conventional moist cured concrete due to the densified microstructure and refined pore structure. Consequently, a longer time is needed to initiate corrosion, and the corrosion rate is slower in the former than in the latter.

The authors further systematically studied the corrosion of rebars embedded in intensively mineralized mortar. The passive behaviors and de-passive behavior of rebars were studied in a simulated pore solution. Although CO_2 mineralization decreases the pH value of the cement matrix, the rebars in CO_2 mineralized mortar and concrete can be passivated over a longer time. The threshold chloride concentration was dramatically decreased by CO_2 mineralization. Figure 11.18 compares the results determined by electrical measurements of rebars embedded in CO_2 mineralized mortars (C3, C4, C5) and conventional moist cured mortars (N3, N4, N5). The results show that the corrosion of rebars is initiated at a similar time in the mortars subjected to different curing regimes. In other words, CO_2 mineralization exerts a negligible influence on the corrosion initiation. This can be ascribed to the combination of the following factors: (1) CO_2 mineralization reduces the chloride ions' diffusivity in the cement matrix, which increases their resistance to chloride-induced rebar corrosion; (2) on the contrary, the reduced threshold chloride concentration accelerates rebars' corrosion. Figure 11.19 presents the mass loss of severely

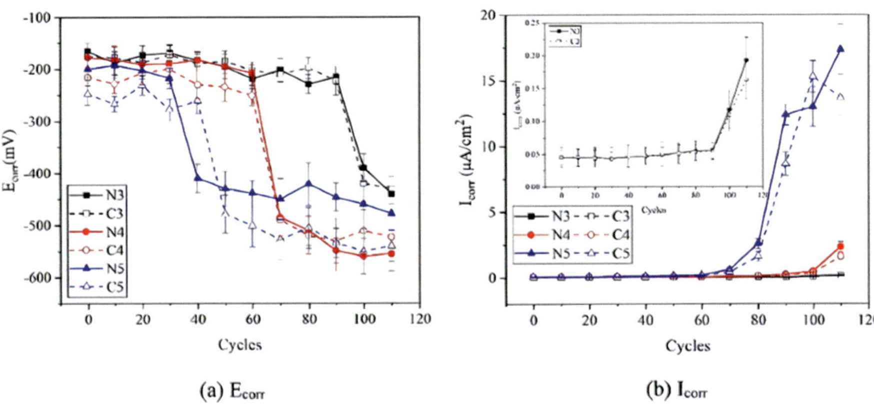

FIGURE 11.18 E_{corr} (a) and I_{corr} (b) of rebars in mortar exposure to sodium chloride solution (Song et al., 2024).

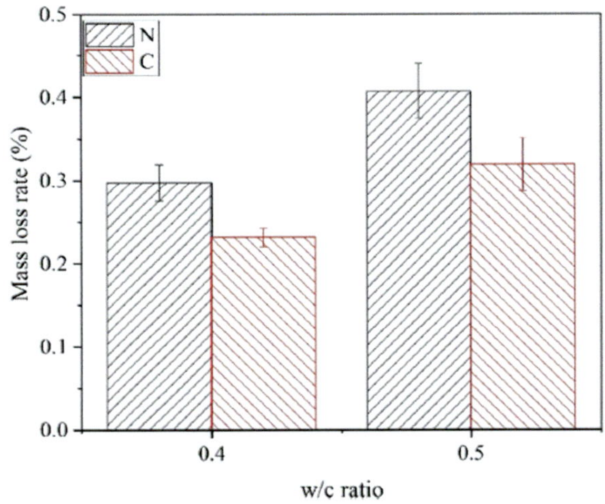

FIGURE 11.19 Mass loss of steel bars in conventional moist cured and CO_2 mineralized cement mortar exposure to sodium chloride (Song et al., 2024).

corroded steel bars embedded in mortars. The mass loss of rebar in CO_2 mineralized cement mortar is around 20% lower than that in conventional moist cured cement mortar. Thus, it can be concluded that CO_2 mineralization lowers the corrosion rate of rebars in the propagation phases and thus overall corrosion of rebars due to much lower chloride diffusivity. In summary, either CO_2 mineralization improved the chloride-induced corrosion resistance of cement mortar and concrete.

11.5 SULFATE ATTACK

11.5.1 Expansion sulfate attack

The sulfate ions sourced from the external environment can react with $Ca(OH)_2$ and mono-sulfate (SO_4-AFm) to form gypsum and ettringite with a higher molar volume than their precursor. This causes expansion and induces expansive stress within the cement matrix. As a result, the cement matrix cracks and its surface will scale and deteriorate, finally. The sulfate resistance of cement-based materials is mainly governed by the content of attackable phases and the transport rate of ions in the cement matrix.

In the course of CO_2 mineralization, hydration products $Ca(OH)_2$ and C-S-H are partially mineralized. Although the subsequent hydration of unreacted cement clinkers forms additional hydrates, there is less $Ca(OH)_2$ in CO_2 mineralized cement-based materials than in conventional moist cured ones. In other words, CO_2 mineralization reduces the sulfate attackable phases in the cement matrix. Besides, the CO_2 mineralization products refine the pore structure and decrease the porosity of the cement matrix. This slows down the transport of sulfate ions and water in the cement matrix. Thus, CO_2 mineralization decreases the formation of expansive products gypsum and ettringite and in turn improves the sulfate attack resistance of cement-based materials, as indicated by several investigations (Rostami et al., 2011; Su et al., 2020; Qin et al., 2021). Figure 11.20 compares the expansion of mortar and concrete samples with and without CO_2 mineralization. All of the reported results demonstrate that the expansion of CO_2 mineralized mortar and concrete is much lower than that of their conventional, moist cured reference. XRD patterns, thermogravimetry, and SEM-EDS results identified that fewer erosion products,

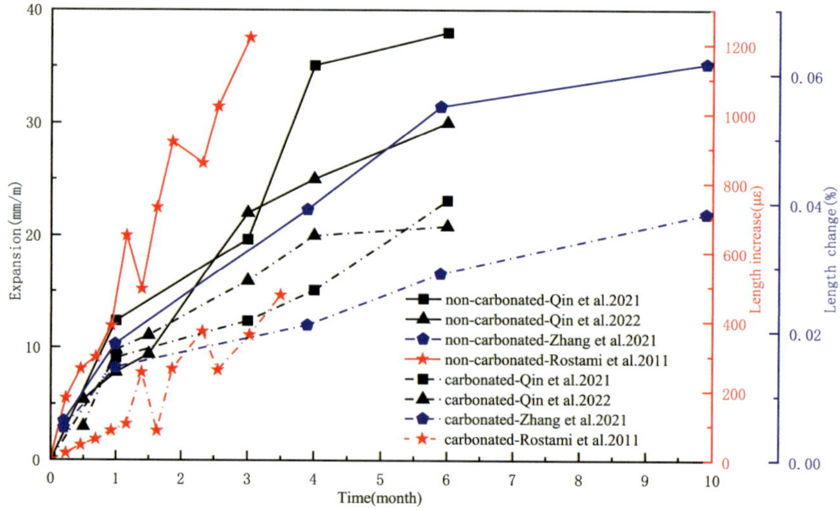

FIGURE 11.20 Expansion of concrete exposed to sulfate attack (Liang et al., 2022).

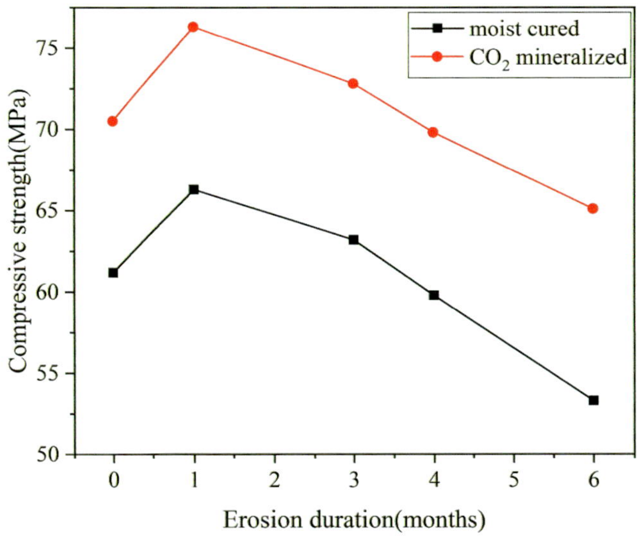

FIGURE 11.21 Compressive strength evolution of cement paste exposed to $MgSO_4$ (Qin et al., 2021).

including gypsum and ettringite, formed in CO_2 mineralized cement matrix than in conventional moist cured cement matrix. The expansion of the cement matrix will lead to cracking, scaling, and strength loss (Qin et al., 2021). Figure 11.21 shows the compressive strength evolution of cement paste exposure to $MgSO_4$ solution. Both the conventional moist cured and CO_2 mineralized cement paste samples show an increase first and then a decrease in compressive strength. The compressive strength reduction of conventional moist cured and CO_2 mineralized cement paste samples is 13% and 7.6% for 6 months exposure to $MgSO_4$ solution, respectively. Furthermore, the compressive strength of CO_2 mineralized cement paste subjected to 6 months of sulfate attack is higher than the initial compressive strength of conventional moist cured cement paste. This also implies that CO_2 mineralized cement paste exhibits superior sulfate attack resistance than conventional moist cured cement paste. Figure 11.22 illustrates the visual appearance of cement paste samples after 6 months

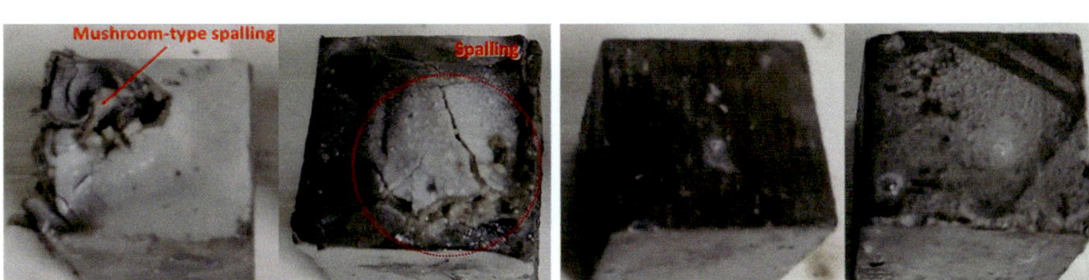

FIGURE 11.22 Optical photos of paste after 6 months exposure to $MgSO_4$ (NC-normal curing, CC-CO_2 curing) (Qin et al., 2021).

of exposure to sulfate attack. The conventional moist cured cement paste shows serious degradation with severe cracking at the edges and the occurrence of mushroom-type spalling after 6 months of sulfate attack. This damage is attributed to both the formation of sulfate-containing chemical products and the decomposition of C-S-H. Nevertheless, no visible cracks are observed on the surface of CO_2 mineralized cement paste, and the sample is intact after 6 months of sulfate attack. The visual inspection further identifies that the deterioration of CO_2 mineralized cement paste is much less severe than that of conventional moist cured cement paste.

The sulfate attack resistance of CO_2 mineralized mortar and concrete containing SCMs is also investigated. CO_2 mineralization improves the sulfate attack resistance of SCMs blended cement paste or mortar, either at normal or low temperature (Su et al., 2020; Qin et al., 2021). CO_2 mineralization reduces the expansion, compressive strength loss, and damage of coal gangue blended paste exposed to sulfate attack (Qin et al., 2021). However, the coal gangue incorporated cement paste subjected to CO_2 mineralization showed more severe deterioration compared to the CO_2 mineralized paste without coal gangue, and it increased with increasing incorporation level of coal gangue. This is similar to that of conventional moist cured paste. The reason is that the incorporation of coal gangue increases the porosity of cement paste, facilitating the transport of SO_4^{2-} and formation of erosion products. Upon 120 days of exposure, the CO_2 mineralized steel slag blended mortar showed better sulfate attack resistance in terms of corrosion coefficients compared to moist cured Portland cement mortar, as reported by Chang et al. (2020).

11.5.2 Thaumasite sulfate attack

In addition to the expansive sulfate attack, there is another type of sulfate attack forming thaumasite ($Ca_3[Si(OH)_6]\cdot(CO_3)_2(SO_4)_2\cdot 24H_2O$). Thaumasite sulfate attack (TSA) is often found in concrete containing $CaCO_3$ and served at low temperature. The $CaCO_3$ in the cement composites can be presented in the form of aggregate (e.g., limestone, dolomite, etc.), filler, as well as nuberalization/carbonation products of the cement matrix. There are two scenarios of thaumasite formation: direct route and indirect (woodfordite) route. The direct route of thaumasite formation is the reaction between C–S–H, gypsum, calcite, and excess water. The woodfordite route was named after the partial solid solution, from which ettringite and thaumasite can arise. Both the reactions of the direct and indirect routes are very slow. The formation of thaumasite results in a white mushy paste without bonding strength instead of expansion. Thus, the characteristic of TSA is less notable than expansive sulfate attack, and the timescale of deterioration of the former is much longer than that of the latter. The essential conditions of thaumasite formation include (Zatloukalová et al., 2024): (1) presence of carbonates and sulfates sourced from external environment or internal components; (2) cold environment, temperature below 15°C, ideally 0–5°C; (3) abundance of moisture or water; (4) pH value of sulfate bearing solutions above 10.5. To mitigate the TSA deterioration, the following strategies are often applied: (1) limiting the use of limestone aggregates and fillers; (2) modifying the phase assemblage of cementitious materials such as reducing content of ettringite, Portlandite and Ca–Si ratio of C-S-H gel; (3) modifying the pore structure to lower the transport of sulfate ions in cement matrix.

CO_2 mineralization introduces abundant $CaCO_3$ into cement-based materials. This favors the formation of thaumasite. On the contrary, CO_2 mineralization consumes Portlandite, lowers the permeability and pH value of the cement matrix as well as the Ca–Si ratio of C-S-H, which limits the formation of thaumasite. By far, the investigations implied that CO_2 mineralization improved the TSA resistance of cementitious materials (Zhang et al., 2021; Qin et al., 2022; Azar, 2013). Zhang et al. (2021) studied the effect of CO_2 mineralization on the formation of thaumasite and performance deterioration of Portland limestone cement (PLC) subjected to low-temperature sulfate attack. Thaumasite was detected in severely damaged conventional moist cured paste samples upon low-temperature exposure. However, it was not detected in CO_2 mineralized cement paste. Moreover, the compressive strength reduction of CO_2 mineralized paste is lower than conventional moist cured paste. It also implies that CO_2 mineralization enhances the resistance to low-temperature sulfate attack of cement-based materials. Qin et al. (2022) investigated the effect of CO_2 mineralization on the sulfate attack performance of Portland cement paste containing coal gangue. Thaumasite was not detected both in the paste with and without coal gangue upon 6 months of low temperature sulfate exposure, although $CaCO_3$ was introduced through CO_2 mineralization. In addition, the compressive strength reduction of CO_2 mineralized cement paste is lower than that of moist cured reference cement paste under the sulfate attack, irrespective of the incorporation level of coal gangue. Both Zhang et al. (2021) and Qin et al. (2022) believed that the enhanced resistance to low temperature sulfate attack was attributed to the reduction in permeability, consumption of Portlandite, and decrease in pH value of pore solution.

11.6 FREEZE-THAW RESISTANCE

The water in the pores of cement composites can freeze once the temperature is lower than the freezing temperature. The molar volume of ice is greater than that of water. Thus, the formation of ice crystals will lead to expansion, which in turn creates surplus pressure in the cement matrix. The accumulation of pressure will result in cracking and scaling as well as performance degradation. Freeze-thaw damage is one of the durability issues of concrete used in cold regions such as the Northeast and Northwest of China. To mitigate the freeze-thaw damage of concrete, many strategies have been developed, such as introducing extra space using air bubbles, reducing porosity and refining pore structures, as well as reducing water absorption.

As stated above, CO_2 mineralization reduces the porosity and refines the pore structure of cement composites. This decreases the penetration of water and the saturation degree of pores in the cement matrix. Consequently, the expansion stress due to the formation of ice crystals is reduced when cement composites are subjected to freezing. In addition, CO_2 mineralization strengthens the cement matrix, which can

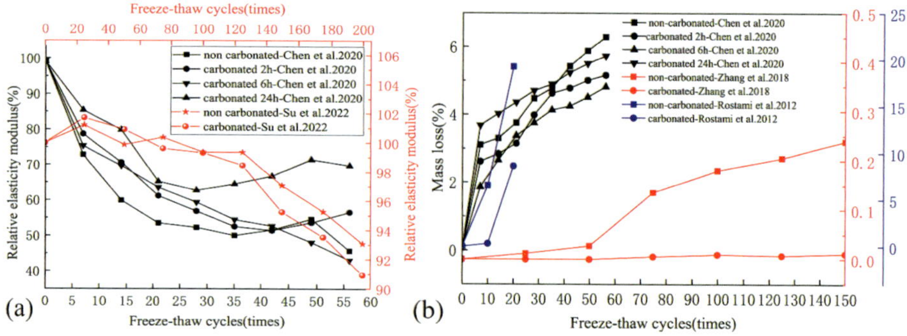

FIGURE 11.23 Relative dynamic elastic modulus (a) and the mass loss (b) of concrete subjected to freeze-thaw cycling (Liang et al., 2022).

bear greater stress. This lowers the cracking and degradation of the cement matrix. Thus, CO_2 mineralized cement composites are less susceptible to freeze-thaw damage (Rostami et al., 2011, 2012; Su et al., 2020; Zhang and Shao, 2018; Monkman and Shao, 2010). This was supported by higher relative dynamic elastic and less mass loss of CO_2 mineralized concrete than that of conventional moist cured ones, as presented in Figure 11.23. Farnam et al. (Farnam et al., 2016) investigated the freeze-thaw performance of CO_2 mineralized cement paste in NaCl and $MgCl_2$ deicing salts. Conventional moist cured cement paste exhibited more severe degradation compared to CO_2 mineralized cement paste. In addition to ice crystals, low-temperature differential scanning calorimetry identified that destructive products calcium/magnesium oxychloride also formed in conventional moist cured cement paste when freeze-thawing in $MgCl_2$ solution, while no destructive products were identified in CO_2 mineralized cement paste. This also contributes to the enhanced freeze-thaw performance of CO_2 mineralized paste in deicing salts. Su et al. (2020) studied the effect of CO_2 mineralization on the freeze-thaw performance of cement mortars in $MgSO_4$ solution. They found that the relative dynamic elastic moduli of CO_2 mineralized mortar were higher in comparison to those of conventional moist cured mortar samples.

The addition of fly ash increases the freeze-thaw damage of CO_2 mineralized mortar and concrete (Zhang and Shao, 2018), which is different from conventional moist cured samples. This can be attributed to the fact that incorporating fly ash increases the porosity and coarsens the pore structure of CO_2 mineralized cement composites. After 200 cycles of freeze-thaw, the compressive strength of Portland cement concrete blended with 0%, 10%, and 30% fly ash decreased by around 15%, 19%, and 23%, respectively (Su et al., 2020). However, the final strength of CO_2 mineralized concrete containing 30% fly ash was comparable to that of moist cured neat Portland cement concrete. Thus, CO_2 mineralization is an effective strategy to mitigate the decreased freeze-thaw resistance of moist cured concrete due to the incorporation of SCMs such as fly ash.

11.7 LEACHING

The leaching of cement-based materials in the immersion environment is the outward diffusion of calcium ions under the action of a concentration gradient. In a soft water environment, the concentration of calcium ions in the pore fluid is much higher than that in outside environments, and the concentration gradient promotes the continuous precipitation of internal calcium to the outside environments, further causing the successive decalcification of hydration products such as $Ca(OH)_2$, C-S-H gel, and ettringite. This induces a growth in porosity, coarsening of pore structure, loss in cementing capacity and strength of the cement matrix as well as loss in integrity ultimately. Thus, the leaching properties of cementitious materials play an important role in the long-term durability of concrete structures as they are exposed to water or underground water, such as the dam concrete. The leaching performance of cementitious materials is mainly dependent on the solubility and amount of the soluble mineral phases as well as pore structure of the cement matrix that controls the outward diffusion of calcium ions.

Previous investigations reported that appropriate CO_2 mineralization decreased the leaching of calcium ions, as shown in Figure 11.24. Watanabe et al. (2006) predicted that the leaching depth of the CO_2 mineralized Portland cement concrete was around 60 mm after 10,000 years of exposure to water, which was only a quarter of the leaching depth of moist cured concrete. This can be mainly attributed to two factors. Firstly, a part of the domain leachable phases, $Ca(OH)_2$ and C-S-H in hydrated cementitious materials, was transformed into insoluble $CaCO_3$. Secondly, proper CO_2 mineralization decreases the porosity and refines the pore size distribution of

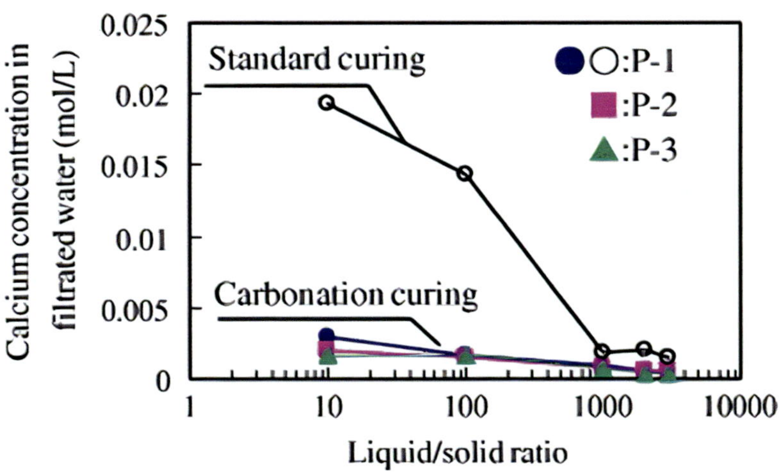

FIGURE 11.24 Calcium concentration in filtered water after immersion test (P-1: PLC with water–cement ratio of 0.4, P-2: PLC with water–cement ratio of 0.58, P-3: PLC with water–cement ratio of 0.89) (Watanabe et al. 2006).

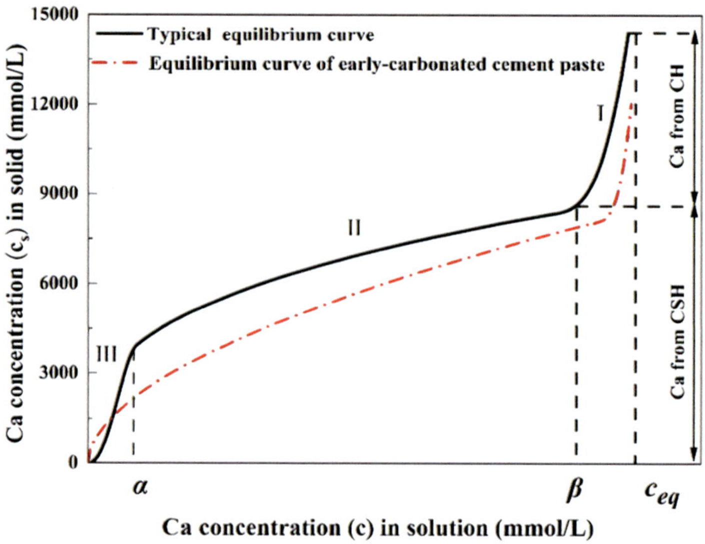

FIGURE 11.25 The solid–liquid equilibrium relationships of cement paste (2023).

cement paste, which prevents the outward diffusion of Ca^{2+}. The CO_2 mineralized CSA cement also showed improved anti-leaching performance compared to the conventional moist cured ones, and the variation of reaction products in the former is smaller upon leaching (Seo et al., 2024). Nevertheless, CO_2 mineralization also lowers the Ca–Si ratio of C-S-H, which favors the decalcification tendency. As such, the leaching behavior of calcium ions is changed, as shown in Figure 11.25. Fewer Ca^{2+} ions leach from CO_2 mineralized cement paste in stage I due to the elimination of $Ca(OH)_2$, while Ca^{2+} ions leach much more quickly from C-S-H in CO_2 mineralized cement paste in stage II compared to conventional moist cured cement paste. As a result, excessive CO_2 mineralization may be detrimental to the anti-leaching performance of cementitious materials, which has been proven by Chen et al. (2023). They found that calcium ions leaching decreased with the increase of CO_2 mineralization time, and the cement paste exhibited the most excellent anti-leaching performance as the CO_2 mineralization time was 8 hours based on the loss in Ca–Si ratio and compressive strength, and evolution of pore structure as well as porosity. However, the calcium leaching increased in the paste as the CO_2 mineralization time further extended after 8 hours. They believed that the growth in porosity and critical pore diameter due to excessive CO_2 mineralization also contributed to the increased calcium ions leaching. Das et al. (2024) also reported that more significant calcium leaching occurred in CO_2 mineralized cement paste than conventional moist cured cement paste, as it was exposed to $MgSO_4$ solution when 28 days of CO_2 mineralization was employed to cement paste. Chen et al. (2023) proposed that 12–15% of CO_2 uptake was effective in improving the leaching performance of cementitious materials.

Apart from Ca^{2+}, heavy metals in the cement composites containing industrial by-products will also leach when the cement composites are exposed to water, which pollutes the water and soil. The previous studies demonstrated that the leaching of heavy metals could be mitigated via CO_2 mineralization (Chang et al., 2013; Boone et al., 2014; Yao et al., 2024). This can be attributed to the following factors: (1) fewer channels are available for heavy metals' transportation outward due to decreased porosity and refined pore structure; (2) the formation of carbonated rim surrounding cement or by-product grains enhances the immobility of heavy metals. Therefore, CO_2 mineralization is an effective route to alleviate the environmental pollution risk of industrial by-products used as cementitious materials regarding the leaching of heavy metals.

11.8 ALKALI-SILICA REACTION

Alkali-silica reaction (ASR), defined as the interaction between the poorly crystallized (or amorphous) silica minerals in aggregates and the alkali species in cement, is one of the most severe and complicated deteriorations of concrete, commonly known as a "concrete virus" (Luo et al., 2022). The formation and swelling of ASR products induce deleterious pressure inside the concrete matrix, leading to volume expansion and cracks, which serve as channels for external aggressive ions and moisture to trigger multiple deteriorations and result in irreversible damage to concrete structures. The ASR expansion primarily depends on the formation of ASR products and the swelling properties of these products. Many efforts have been made to reduce the ASR expansion in mortar and concrete, such as limiting the alkali in cement and incorporating SCMs.

Shoji et al. (2015) believed that the reduced pH value due to CO_2 mineralization suppressed ASR expansion of mortar bars as it reduced the solubility of reactive silica from aggregate. Figure 11.26 summarizes the ASR expansion of mortars at 14 days. As CO_2 mineralization was applied, the ASR expansion of mortar bars decreased

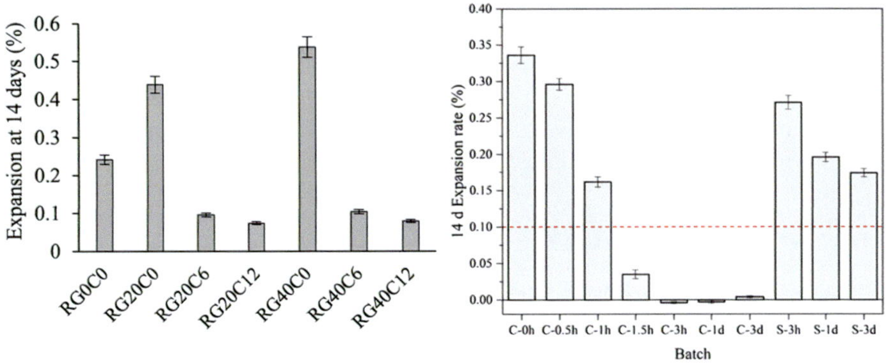

FIGURE 11.26 Mortar bars' expansion at 14 days (Liu et al., 2022; Wang et al., 2022).

by 50–98% depending on the CO_2 mineralization regimes and materials. When the CO_2 mineralization duration is longer than 3 hours, the expansion of mortar bars cast with alkali-active aggregate can be reduced to less than the permissible threshold of 0.1% (specified in ASTM C1260) at 14 days. Liu et al. (2022) discussed the ASR expansion suppression mechanism of cement-based materials with CO_2 mineralization. They believed the following factors are responsible for the reduction in ASR expansion: (1) CO_2 mineralization considerably consumes $Ca(OH)_2$ and drops the pH value of pore solution, which leads to a reduction in reactive silica dissolution; (2) the reduced Ca–Si ratio due to removal of interlayer Ca^{2+} in the course of CO_2 mineralization enhanced the absorption of alkali metal ions from pore solution; (3) densified microstructure prevents the transportation of moisture and alkali metals. Wang et al. (2022) noted that CO_2 mineralization also altered the distribution of ASR gels, as shown in Figure 11.27, and the mechanism was described in Figure 11.28. ASR gel formed at the glass-cement interface in CO_2 mineralized mortars, while it formed at microcracks within the glass aggregate in conventional moist cured mortars. This was also identified as an important reason for the decreased expansion. The swelling of the surface ASR gel applied inward pressure on the glass aggregate, reducing the aggregate cracking progression, which also played an important role in reducing the ASR expansion. Luo et al. (2024) compared the amount of ASR products and the swelling capacity in CO_2 mineralized mortar and conventional moist cured mortars. Roman spectroscopy identified that CO_2 mineralization reduced the formation of ASR products in the mortar bars. The Ca–Si ratio of ASR products is higher, but the [Na + K]/Si ratio of it is lower in CO_2 mineralized mortars compared to that in conventional moist cured mortars. The increased Ca–Si ratio and decreased [Na + K]/Si ratio reduce the water absorption and swelling potential of ASR products (Saha and Sarker, 2018). The reduced expansion and enhanced strength of the cement matrix reduced and narrowed the cracks on the surface of cement mortars. Therefore, CO_2 mineralization significantly mitigates the ASR deterioration.

Almost all of the reported results indicated that the higher the CO_2 mineralization degree, the higher the ASR expansion mitigating efficacy. Luo et al. (2024) found that the reduction in ASR expansion nearly linearly increased with the increase of CO_2 uptake. Extending the CO_2 mineralization time and increasing the CO_2 concentration are effective in enhancing ASR expansion, mitigating efficacy. As CO_2 mineralization suppressed the formation and swelling of ASR products, the amount of alkali reactive aggregate has little effect on the ASR expansion mortar bars, which differs from the conventional moist cured ones.

11.9 VOLUME STABILITY

The porous C-S-H gel with a huge specific surface can absorb and release water freely when the moisture in the atmosphere ascends and descends. This leads to an increase or decrease in the volume of solid phases in cementitious materials. Thus, drying shrinkage is a common phenomenon in hydrated cementitious materials exposed to the environment with low moisture. However, Shi et al. (2012) found that drying shrinkage of Portland cement concrete can be alleviated with CO_2 mineralization.

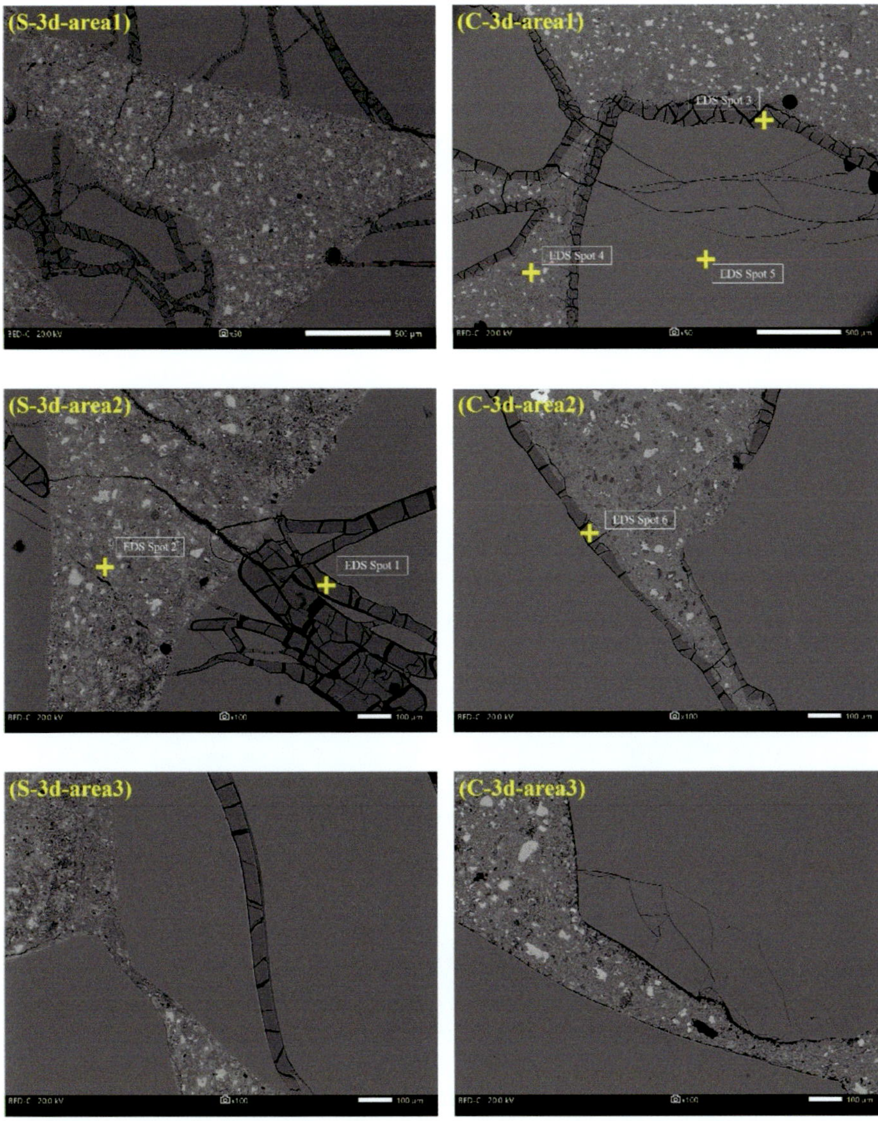

FIGURE 11.27 SEM-BSE images of mortar bars after immersion in 80°C alkali solution for 28 days (Wang et al., 2022).

Qin et al. (Qin and Gao, 2019b) found that CO_2 mineralization increased the initial drying shrinkage of mortars due to intensive evaporation of water in the course of CO_2 mineralization. However, the final drying shrinkage of CO_2 mineralized mortar at 56 days was decreased by 2.3–17.8% for specimens containing 0–50% coal gangue when compared with that of the control specimen. This can be attributed to two factors. First, a considerable amount of C-S-H gel is converted into $CaCO_3$

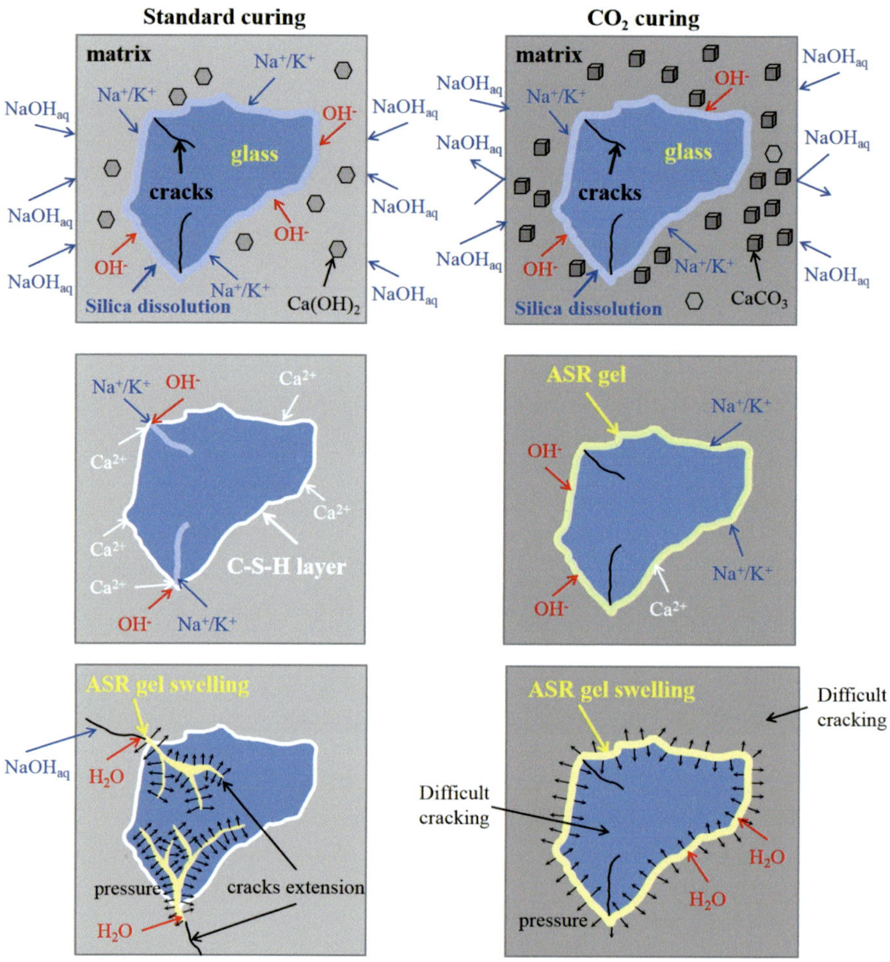

FIGURE 11.28 Schematic diagram of the ASR mechanism in standard cured and CO_2 cured specimens (Wang et al., 2022).

crystals with excellent dimensional stability. Second, the decreased porosity and permeability of CO_2 mineralized cement matrix restrict the reversible volume changes due to variation of relative humidity.

The reaction of free CaO and MgO is much slower than calcium silicate clinkers, and generally the reactions are initiated in the hardened cement matrix. The molar volumes of the hydration products of them (i.e., $Ca(OH)_2$ and $Mg(OH)_2$) are much higher than those of their precursors, leading to volume instability issues in concrete. This issue often occurs in the concrete containing calcium and/or magnesia-rich by-products such as steel slag and magnesia slag. However, the mineralization of free CaO and MgO is much faster than their hydration, and they can react with CO_2 instantly, consuming it in a short time. This avoids cracking in the hardened cement

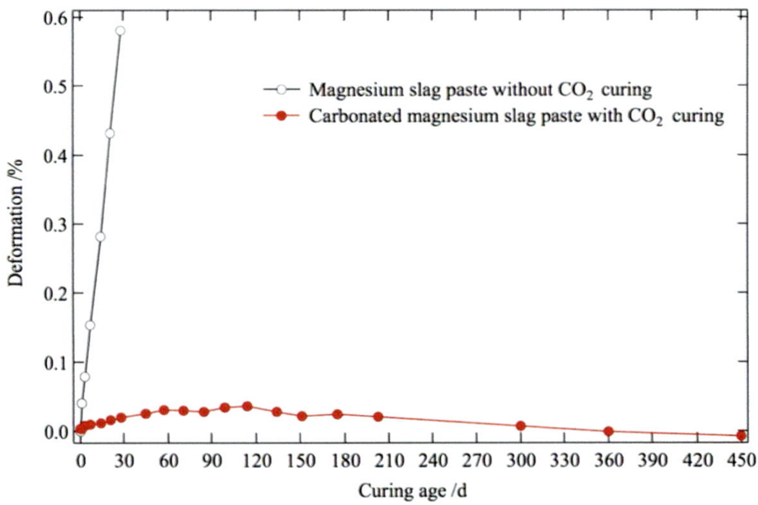

FIGURE 11.29 Deformation of the mineralized or non-mineralized magnesium GGBS pastes exposed to accelerated curing in water under 60°C (Mo et al., 2019).

matrix. Many studies have proved that CO_2 mineralization was an effective strategy to mitigate instability issues in concrete containing steel slag and magnesia slag. Mo et al. investigated industrial by-product magnesia slag with high CaO and MgO content, and successfully prepared carbonate binders with a remarkable strength of up to 119.5 MPa using an accelerated CO_2 mineralization method (Mo et al., 2019). Additionally, it was observed that magnesia slag specimens treated with CO_2 mineralization exhibited volume stability and no expansion after curing in water at 60°C (Figure 11.29). As can be seen from Figure 11.29, the expansion rate of magnesium paste without mineralization reaches 0.04% when it is cured for 1 day, and 0.58% after 28 days. However, after being treated with CO_2 mineralization, the expansion of the magnesium paste is smaller than 0.04% even after 450 days of curing in hot water. Similarly, Zhang et al. (2023) developed a new continuous CO_2 mineralization-hydration curing method that improves the volume stability problem caused by the late hydration reaction of MgO. Xie et al. (2024) used magnesium slag as a raw material and accelerated the CO_2 mineralization process to produce a new type of CO_2-solidified fiber cement board. The results show that CO_2 mineralization can not only strengthen the bonding between fiber and matrix but also improve the volume stability of magnesium slag.

Besides, CO_2 mineralization is a simple yet effective way to address the dimensional stability of calcium aluminate cement (Park et al., 2017). The hydrates CAH_{10} and C_2AH_8 are metastable phases; they will convert into stable hydrates C_3AH_6 and AH_3, which causes a dramatic reduction in the volume of hardened paste and in turn results in a progressive increase in porosity and loss of structural integrity. However, if CO_2 mineralization is applied, the metastable hydrates can be rapidly carbonated and transformed into stable hydrates, accompanied by no increase in porosity.

Therefore, the dimensional instability issue of calcium aluminate cement products is mitigated.

11.10 SUMMARY

In the course of CO_2 mineralization, the primary products are $CaCO_3$ and silica-rich amorphous phases instead of hydration products. This alters the phase assemblage and modifies the pore structure of cementitious materials, as well as decreases the alkalinity of the pore solution. Mineralization products show different reactivity to aggressive ions such as SO_4^{2-} and Cl^- ions, and decrease the permeability of the cement matrix. Thus, CO_2 mineralized cementitious materials show different long-term durability from traditional moist cured cementitious materials. Specifically, CO_2 mineralization decreases the chloride binding capacity of the cement matrix but improves their resistance to chloride-induced rebar corrosion due to decreased chloride diffusivity. Proper CO_2 mineralization also mitigates the carbonation-induced rebar corrosion because of decreased weathering carbonation coefficients. Besides, CO_2 mineralized cementitious materials exhibit superior sulfate resistance, freeze-thaw resistance, ASR resistance, anti-leaching performance, aging performance, and dimensional stability compared to traditional moist cured cementitious materials due to the elimination of $Ca(OH)_2$ and refined pore structure.

REFERENCES

Almeida, A.E.F.S., Tonoli, G.H.D., Santos, S.F., Jr, H.S., 2013. Improved durability of vegeTable 11. fiber reinforced cement composite subject to accelerated carbonation at early age. *Cement and Concrete Composites* 42, 49–58.

Azar, A., 2013. *Improved resistance to thaumasite formation in cement pastes by early age carbonation*. McGill University.

Balonis, M., Lothenbach, B., Saout, G.L., Glasser, F.P., 2010. Impact of chloride on the mineralogy of hydrated Portland cement systems. *Cement and Concrete Research* 40(7), 1009–1022.

Boone, M., Nielsen, P., De Kock, T., Boone, M., Mieke, Q., Veerle, C., 2014. Monitoring of Stainless-Steel GGBS Carbonation Using X-ray Computed Microtomography. *Environmental Science & Technology* 48, 674–80.

Chang, E.E., Chen, T.L., Pan, S.Y., Chen, Y.H., Chiang, P.H., 2013. Kinetic modeling on CO_2 capture using basic oxygen furnace GGBS coupled with cold-rolling wastewater in a rotating packed bed. *Journal of Hazardous Materials* 260, 937–946.

Chang, H., Feng, P., Lyu, K., Liu, J., 2019. A novel method for assessing C-S-H chloride adsorption in cement pastes. *Construction and Building Materials* 225, 324–331.

Chang, J., Gu, Y., Ansari, W.S., Mechanism of blended steel GGBS mortar with CO_2 curing exposed to sulfate attack. *Construction and Building Materials* 251(2020), 118880.

Chen, T., Li, L., Gao, X., Guo, M., Qin, L., 2023. New insights into the role of early accelerated carbonation on the calcium. *Cement and Concrete Composites* 140, 105103.

Das, K.K., Sharma, R., Wu, X., Lee, J.H., Kang, J.G., 2024. Effects of carbonation curing on chemical attack in cement paste. *Case Studies in Construction Materials* 20, e03251.

Farnam, Y., Villani, C., Washington, T., Spence, M., Jain, J., Weiss, W.J., 2016. Performance of carbonated calcium silicate based cement pastes and mortars exposed to NaCl and $MgCl_2$ deicing salt. *Construction and Building Materials* 111, 63–71.

Grishchenko, R.O., Emelina, A.L., Makarov, P.Y., 2013. Thermodynamic properties and thermal behavior of Friedel's salt. *Thermochimica Acta* 570, 74–79.
He, P., Shi, C., Tu, Z., Poon, C.S., Zhang, J., 2016. Effect of further water curing on compressive strength and microstructure of CO_2-cured concrete. *Cement and Concrete Composites* 72, 80–88.
Justnes, H., Skocek, J., Østnor, T.A., Engelsen, C.J., Skløsvold, O., 2020. Microstructural changes of hydrated cement blended with fly ash upon carbonation. *Cement and Concrete Research* 137, 106192.
Kim, H., Siddique, S., Jang, J.G., 2022. Effect of carbonation curing on the mechanical properties of belite-rich cement mortar exposed to elevated temperatures. *Journal of Building Engineering* 52, 104453.
Liang, C., Li, B., Guo, M., Hou, S., Wang, S., Gao, Y., Wang, X., 2022. Effects of early-age carbonation curing on the properties of cement-based materials: A review. *Journal of Building Engineering* 52, 104453.
Liu, R., Jiang, L., Xu, J., Xiong, C., Song, Z., 2014. Influence of carbonation on chloride-induced reinforcement corrosion in simulated concrete pore solutions. *Construction and Building Materials* 56, 16–20.
Liu, Z., Shi, C., Shi, Q., Tan, X., Meng, W., 2022. Recycling waste glass aggregate in concrete Mitigation of alkali-silica (ASR) by carbonation curing. *Journal of Cleaner Production* 370, 133545.
Luo, D., Sinha, A., Adhikari, M., Wei, J., 2022. Mitigating alkali-silica reaction through metakaolin-based internal conditioning: New insights into property evolution and mitigation mechanism. *Cement and Concrete Research* 159, 106888.
Luo, D., Sinha, A., Mulcahy, R., Wei, J., 2024. Influence of enforced carbonation on alkali-silica reaction: Performance and multi-scale mechanisms. *Cement and Concrete Composites* 153, 105721.
Monkman, S., Shao, Y., 2010. Carbonation curing of GGBS-cement concrete for binding CO_2 and improving performance. *Journal of Materials in Civil Engineering* 22(4), 296–304.
Mo, L., Hao, Y., Liu, Y., Wang, F., Deng, M., 2019. Preparation of calcium carbonate binders via CO_2 activation of magnesium GGBS. *Cement and Concrete Research* 121, 81–90.
Mu, Y., Liu, Z., Wang, F., Huang, X., 2018. Carbonation characteristics of γ-dicalcium silicate for low-carbon building material. *Construction and Building Materials* 177, 322–331.
Pan, X. Shi, C., Farzadnia, N., Hu, X., Zheng, J., 2019. Properties and microstructure of CO_2 surface treated cement mortars with subsequent lime-saturated water curing. *Cement and Concrete Composites* 99, 89–99.
Pan, X. Shi, C., Hu, X., Ou, Z., 2017. Effects of CO_2 surface treatment on strength and permeability of one-day-aged cement mortar. *Construction and Building Materials* 154, 1087–1095.
Park, S.M., Jang, J.G., Son, H.M., Lee, H.K., 2017. STable 11. conversion of metasTable 11. hydrates in calcium aluminate cement by early carbonation curing. *Journal of CO2 Utilization* 21, 51–60.
Paul, G., Boccaleri, E., Buzzi, L., Canonico, F., Gastald, D., 2015. Friedel's salt formation in sulfoaluminate cements: A combined XRD and 27 Al MAS NMR study. *Cement and Concrete Research* 67, 93–102.
Qin, L., Gao, X., 2019a. Recycling of waste autoclaved aerated concrete powder in Portland cement by accelerated carbonation. *Waste Management* 89, 254–264.
Qin, L., Gao, X., 2019b. Properties of coal gangue-Portland cement mixture with carbonation. *Fuel* 245, 1–12.
Qin, L., Gao, X., Chen, T., 2019a. Influence of mineral admixtures on carbonation curing of cement paste. *Construction and Building Materials* 212, 653–662.

Qin, L., Gao, X., Su, A., Li, Q., 2021. Effect of carbonation curing on sulfate resistance of cement-coal gangue paste. *Journal of Cleaner Production* 278, 123897.

Qin, L., Mao, X., Gao, X., Zhang, P., Chen, T., Li, Q., Cui, Y., 2022. Performance degradation of CO_2 cured cement-coal gangue pastes with low-temperature sulfate solution immersion. *Case Studies in Construction Materials* 17, e01199.

Rostami, V., Shao, Y., Boyd, A.J., 2011. Durability of concrete pipes subjected to combined steam and carbonation curing. *Construction and Building Materials* 25(8), 3345–3355.

Rostami, V., Shao, Y., Boyd, A.J., 2012. Carbonation Curing versus Steam Curing for Precast Concrete Production. *Journal of Materials in Civil Engineering* 24, 1221–1229.

Saha, A.K., Sarker, P.K., 2018. Potential alkali silica reaction expansion mitigation of ferronickel GGBS aggregate by fly ash. *Structure Concrete* 19(5), 1376–1386.

Seo, J., Bae, J.H., Kim, N., Jang, D., Lee, H.K., 2024. Investigation on leaching behaviors of carbonation-cured CSA cements using an electrically accelerated leaching test. *Construction and Building Materials* 411, 134278.

Shi, C., Wang, D., He, F., Liu, M., 2012. Weathering properties of CO_2-cured concrete blocks. *Resources, Conservation and Recycling* 65, 11–17.

Shi, Z., Geiker, M.R., Lothenbach, B., Weerd, K.D., Garzon, S.F., Enemark-Rasmussen, K., Skibsted, J., 2017. Friedel's salt profiles from thermogravimetric analysis and thermodynamic modelling of Portland cement-based mortars exposed to sodium chloride solution. *Cement and Concrete Composites* 78, 73–83.

Shoji, M., Higuchi, T., Morioka, M., Yokozeki, K., 2015. Inhibitory effect of alkali-silica reaction by the carbonation reaction. *Cement Science Concrete Technology* 69(1), 504–510.

Song, B., Hu. X., Pang, S.D., Du, H., Shi, C., 2024. Effect of early CO_2 curing on the chloride induced steel bars corrosion in cement mortars. *Journal of Sustainable Cement-Based Materials* 13(7), 1063–1078.

Song, B., Hu, X., Liu, S., Shi, C., 2022. Chloride binding of early CO_2-cured Portland cement-fly ash-GGBS ternary pastes. *Cement and Concrete Composites* 134, 104793.

Song, B., Liu, S., Hu, X., Ouyang, K., Li, G., Shi, C., 2022. Compressive strength, water and chloride transport properties of early CO_2-cured Portland cement-fly ash-GGBS ternary mortars. *Cement and Concrete Composites* 134, 104786.

Soroushian, P., Won, J., Hassan, M., 2012. Durability characteristics of CO_2-cured cellulose fiber reinforced cement composites. *Construction and Building Materials* 34, 44–53.

Suryavanshi, A.K., Scantlebury, J.D., Lyon, S.B., 1998. Corrosion of reinforcement steel embedded in high water-cement ratio concrete contaminated with chloride. *Cement and Concrete Composites* 20(4), 263–281.

Su, A., Chen, T., Gao, X., Li, Q., Qin, L., 2020. Effect of carbonation curing on durability of cement mortar incorporating carbonated fly ash subjected to Freeze-Thaw and sulphate attack. *Construction and Building Materials* 251, 118880.

Thomas, M.D.A., Hooton, R.D., Scott, A., Zibara, H., 2012. The effect of supplementary cementitious materials on chloride binding in hardened cement paste. *Cement and Concrete Research* 42(1), 1–7.

Tolêdo Filho, R.D., Ghavami, K., England, G.L., Scrivener, K., 2003. Development of vegeTable 11. fiber-mortar composites of improved durability. *Cement and Concrete Composites* 25(2), 185–196.

Wang, L., Chen, S.S., Tsang, D.C.W., Poon, C.S., Dai, J.G., 2017. CO_2 curing and fiber reinforcement for green recycling of contaminated wood into high-performance cement-bonded particleboards. *Journal of CO2 Utilization* 18, 107–116.

Wang, Y., Mo, K.H., Du, H., Ling, T.C., 2022. Effects of CO_2 curing treatment on alkali-silica reaction of mortars containing glass aggregate. *Construction and Building Materials* 323, 126637.

Wang, Y., Shui, Z., Gao, X., Huang, Y., Yu, R., Ling, G., 2019. Chloride binding behaviors of metakaolin-lime hydrated blends: Influence of gypsum and atmospheric carbonation. *Construction and Building Materials* 201, 380–390.

Watanabe, K., Yokozeki, K., Ashizawa, R., Sakata, N., Morioka, M., Sakai, S., Daimon, M., 2006. High durability cementitious material with mineral admixtures and carbonation curing. *Waste Management* 26(7), 752–757.

Xian, X., Zhang, D., Lin, H., Shao, Y., 2022. Ambient pressure carbonation curing of reinforced concrete for CO_2 utilization and corrosion resistance. *Journal of CO2 Utilization* 56, 101861.

Xie, D., Zhang, Z., Liu, Z., Wang, F., Hu, S., Fu, J., 2024. Utilization of magnesium GGBS to prepare CO_2 solidified fiber cement board. *Construction and Building Materials* 411, 134345.

Yao, X., Song, B., Huan, Q., Hu, Y., Song, M., 2024. Ambient pressure carbonation curing of cold-bonded fly ash lightweight aggregate using coal-fired flue gas for properties enhancement and CO_2 sequestration. *Journal of Environmental Chemical Engineering* 12, 114208.

Zatloukalová, J., Zatloukal, J., Kolář, K., Konvalinka, P., 2024. Thaumasite formation in cementitious composites used in underground repositories-conditions and impacts on composite's properties: A review. *Progress in Nuclear Energy* 176, 105356.

Zhang, C., Liu, S., Tang, P., Guan, X., Shi, C., 2023. Enhancing the hardening properties and microstructure of magnesium GGBS blocks by carbonation-hydration sequential curing. *Journal of Building Engineering* 76, 107414.

Zhang, D., Ghouleh, Z., Shao, Y., 2017. Review on carbonation curing of cement-based materials. *Journal of CO2 Utilization* 21, 119–131.

Zhang, D., Liu, T., Shao, Y., 2020. Weathering carbonation behavior of concrete subject to early-age carbonation curing. *Journal of Materials in Civil Engineering* 32(4), 10.1061.

Zhang, D., Shao, Y., 2016a. Early age carbonation curing for precast reinforced concretes. *Construction and Building Materials* 113, 134–143.

Zhang, D., Shao, Y., 2016b. Effect of early carbonation curing on chloride penetration and weathering carbonation in concrete. *Construction and Building Materials* 123, 516–526.

Zhang, D., Shao, Y., 2018. Surface scaling of CO_2-cured concrete exposed to freeze-thaw cycles. *Journal of CO2 Utilization* 217, 137–144.

Zhang, S., Ghouleh, Z., Azar, A., Shao, Y., 2021. Improving concrete resistance to low temperature sulfate. *Materials and Structures* 54(1), 10.1617.

12 Methods for determination of CO_2 mineralization degree

12.1 INTRODUCTION

CO_2 can be captured and stored in cement-based materials due to CO_2 mineralization reactions. The major reactive phases in cement-based materials for CO_2 mineralization are calcium species (i.e., C_3S, C_2S, C-S-H, $Ca(OH)_2$, and AFt), which can react with CO_2 and form thermodynamically stable calcium carbonate. Both the hydration products and the unhydrated cement clinker minerals can thus be mineralized in an environment containing CO_2. CO_2 sequestration and the properties of CO_2 mineralization products are highly correlated with the CO_2 mineralization degree. A higher CO_2 mineralization degree represents more CO_2 captured and stored, which is important for the cement industry's low carbon and ecological sustainability. In addition, a higher CO_2 mineralization degree allows more mineralization products, implying a denser microstructure, higher cementitious activity, and better mechanical performance. However, currently, there is no uniformity in the evaluation and calculation methods for CO_2 mineralization degree. This chapter outlines the main measurement techniques that are currently applied to determine the CO_2 mineralization degree of cement-based materials, including the mineralized area method and the CO_2 uptake calculation method.

The CO_2 mineralization degree of a cement-based material can be expressed as the ratio of the amount of actual CO_2 mineralized cement-based material to the amount of the theoretical maximum cement-based material that can be CO_2 mineralized, which is defined as follows:

$$CO_2\ \text{mineralization degree} = \frac{\begin{array}{c}\text{Amount of actual } CO_2 \text{ mineralized}\\ \text{cement-based material}\end{array}}{\begin{array}{c}\text{Amount of the theoretical maximum}\\ \text{cement-based material that}\\ \text{can be } CO_2 \text{ mineralized}\end{array}} \tag{12.1}$$

In this study, the methodologies for evaluating the CO_2 mineralization of cement-based materials used in existing studies were divided into the mineralized area method and the CO_2 uptake calculation method.

DOI: 10.1201/9781003627531-12

12.2 MINERALIZED AREA METHOD

12.2.1 Principle

The mineralized area method converts the amount of actual mineralized cement-based material and the amount of the theoretical maximum cement-based material that can be mineralized in Equation (12.1) into the actual mineralized cross-area and the theoretical maximum cross-area that can be mineralized in the cross-section of the cement-based material, respectively (Eq. 12.2). Existing studies have taken the theoretical maximum area that can be mineralized as the entire cross-sectional area of the cement paste material, i.e., the whole cement-based material was considered to be mineralized. In addition, the particles of the cement-based material in the actual cross-section zone were considered completely mineralized when using this method. Obviously, these are too idealistic, but the need to determine only the actual volume of CO_2 mineralized cementitious material makes the test easy (Zhang et al., 2017).

$$CO_2 \text{ mineralization degree} = \frac{\begin{array}{c}\text{Actual cross-section area of actual}\\ \text{mineralized cement-based material}\end{array}}{\begin{array}{c}\text{Maximum theoretical cross-section area of}\\ \text{cement-based material that can be mineralized}\end{array}} \tag{12.2}$$

The mineralization area method is suitable for determining the mineralization of block cement-based materials, such as cement paste, mortar, concrete, and recycled aggregates, since the method requires measurement of the cross-sectional area of the specimen. It cannot be used for powdered materials such as clinker powder and RCF. In addition, because it is usually necessary to crush and split the specimen, this method is unsuitable for specimens that need to be preserved intact. Regarding measurement accuracy, two aspects lead to a significant error in the method. One is the idealized assumptions mentioned above. The second is the inaccuracy of the current evaluation technique for determining the cement-based material's actual mineralized cross-sectional area. Therefore, the method is usually considered a semi-quantitative method (Bui et al., 2023; Zhang et al., 2017).

12.2.2 Determination of actual mineralized cross-section area

To determine the actual cross-sectional area of the CO_2 mineralized cement-based material, it usually splits the cement-based material specimen so that its cross-sectional area is exposed. Figure 12.1 shows a typical sampling process of the CO_2 mineralized concrete specimens. The CO_2 mineralization reaction is an outside-in reaction. The actual cross-sectional area of the CO_2 mineralized cement-based material can be obtained by subtracting the unmineralized cross-sectional area from the CO_2 mineralization depth or total cross-sectional area. Many codes (2004, 2007a, 2007b; China, 2009) specify using an indicator of 70% ethanol mixed with 1% phenolphthalein to determine the mineralized cross-sectional area. The pH discoloration

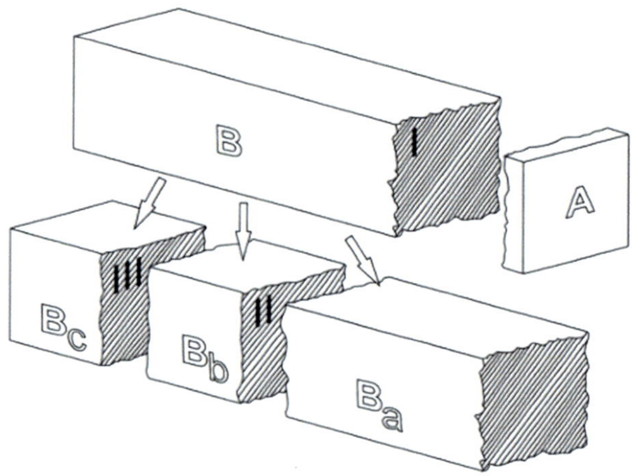

FIGURE 12.1 Schematic diagram of the sample preparation and splitting for the determination of CO_2 mineralized cross-section area (Vogler et al., 2020).

of the phenolphthalein indicator ranges from 8.2 to 10.0, while the pH ranges of fully mineralized, partially mineralized, and non-mineralized zones of cement-based materials are 8.5 to 9.0, 9.0 to 12.5, 12.5 to 13.0, respectively (Mi et al., 2020, 2021). Therefore, the phenolphthalein indicator covered the completely mineralized area and part of the partially mineralized area of the cement-based material (Figure 12.2).

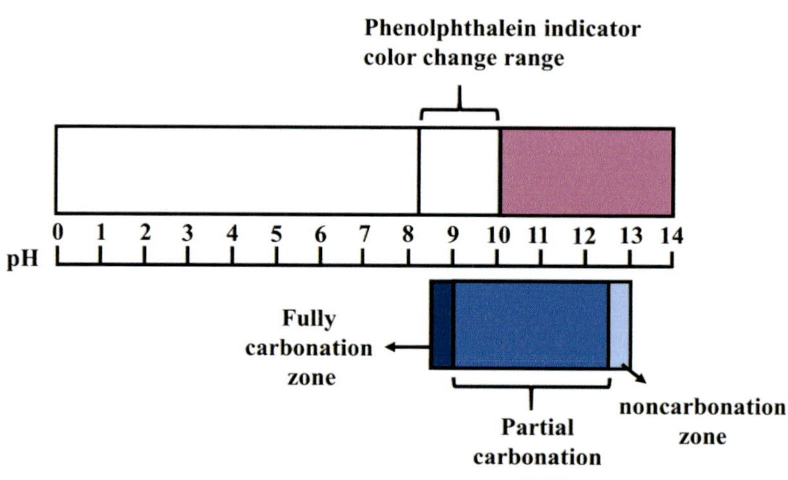

FIGURE 12.2 pH range of color change of phenolphthalein indicator, pH range of mineralized zones, and non-mineralized zone of cement-based materials.

Many researchers (Cheng Zhang, 2022; Giulietti et al., 2021; Mi et al., 2021) criticized the inaccuracy of the phenolphthalein indicator for the determination of CO_2 mineralization depth, which was mainly related to the inability of the phenolphthalein color change range to cover part of the mineralization zone of cement-based materials (Figure 12.2). To address this problem, Zhang et al. (Cheng Zhang, 2022) determined the CO_2 mineralization depth by the pH measurement combined with layer-by-layer grinding method. The layer in which the mineralized concrete with the same pH as the unmineralized counterpart concrete was located was determined as the junction between the mineralized and non-mineralized zones. The results indicated that the CO_2 mineralization depth obtained by this method was about 1.2–1.3 times higher than that obtained by the phenolphthalein indicator method (Figure 12.3). Similarly, researchers have used FTIR (Kachko et al., 2016) and NMR (Li et al., 2020) to determine the boundaries of the mineralized zone, and the CO_2 mineralization depth determined using these two methods is about 1.5 mm deeper than that determined with phenolphthalein indicator. Another advantage of the NMR method is that it is non-destructive. Another criticism of phenolphthalein

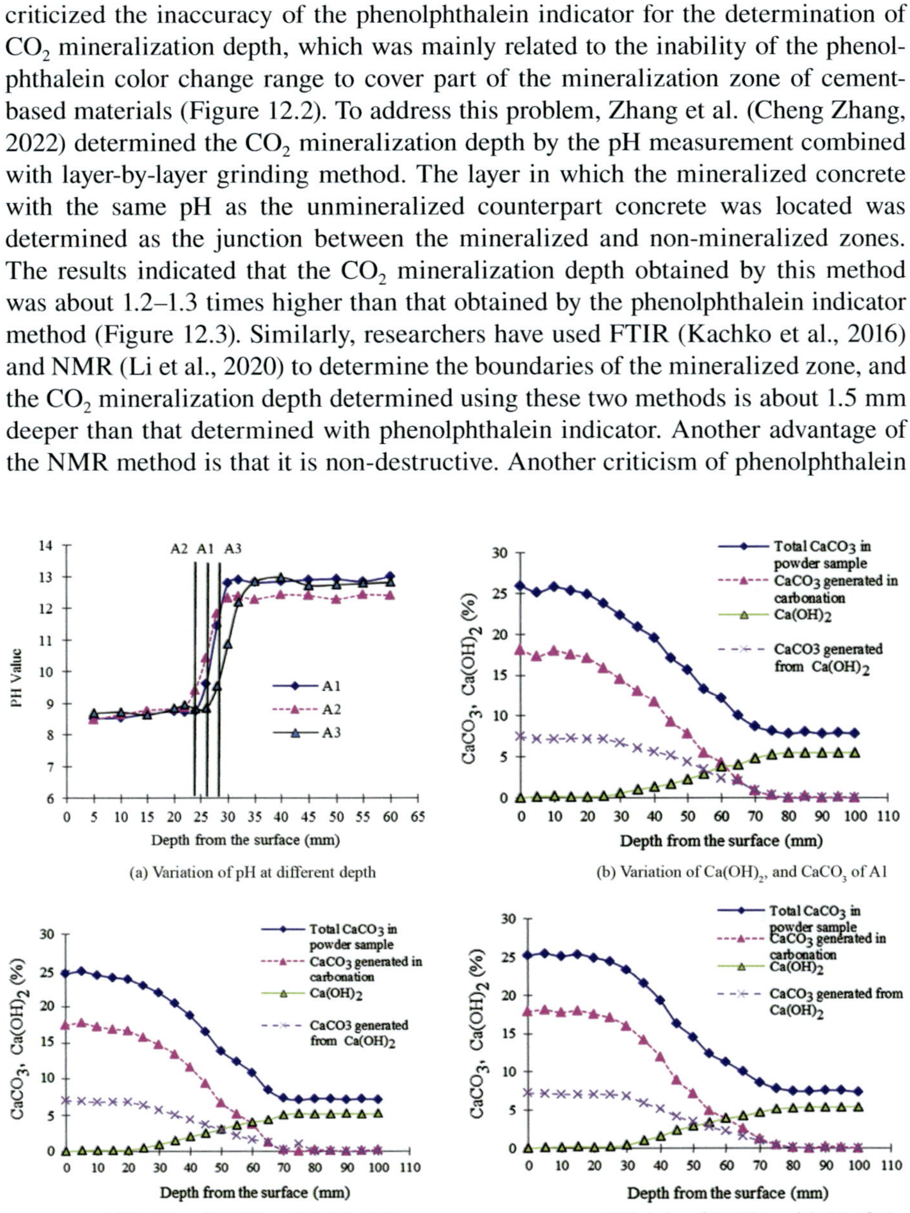

FIGURE 12.3 pH value profiles, $Ca(OH)_2$, and $CaCO_3$ relative contents of CO_2 mineralized concrete at different depths (Ji et al., 2014). (a) Variation of pH at different depths; (b) Variation of $Ca(OH)_2$, and $CaCO_3$ of A1; (c) Variation of $Ca(OH)_2$, and $CaCO_3$ of A2; (d) Variation of $Ca(OH)_2$, and $CaCO_3$ of A3.

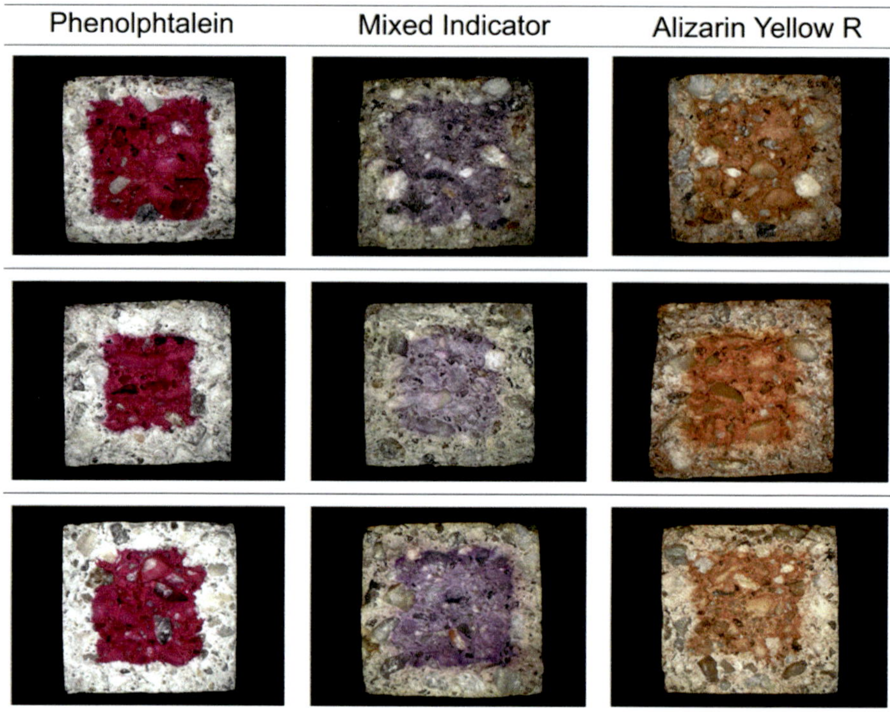

FIGURE 12.4 CO_2 mineralization concrete samples at 28 days and subsequently sprayed with different pH-indicator solutions (Vogler et al., 2020).

indicators for determining the area of mineralization is that phenolphthalein causes genetic defects, including cancer (Chinchón-Payá et al., 2016; Cui et al., 2022; Vogler et al., 2020). Therefore, Chinchón-Payá et al. (2016), Vogler et al. (2020), and Cui et al. (2022) evaluated a variety of possible indicators, including curcumin-based, Alizarin Yellow R, Thymol Blue, and anthocyanin extracted via the petals. The results indicated that these novel indicators tested comparable to phenolphthalein for the range of CO_2 mineralization zones (Figure 12.4) (Chinchón-Payá et al., 2016; Vogler et al., 2020) or more accurately (Cui et al., 2022).

On the other hand, using the CO_2 mineralization depth to determine the CO_2 mineralization degree has a significant error because the CO_2 mineralized area is often not regular in shape (Cheng Zhang, 2022; Mi et al., 2021). To increase the accuracy of measuring the CO_2 mineralized area, several studies (Choi et al., 2017; Ghantous et al., 2017; Mi and Pan, 2022; Segura et al., 2010; Sharma et al., 2023) used the total cross-sectional area of cement-based materials minus the unmineralized cross-sectional area determined by image recognition techniques to determine the CO_2 mineralized cross-sectional area. These technologies include dedicated vision systems and convolutional neural network strategies. Figure 12.5 shows a typical use case of this method (Sharma et al., 2023).

FIGURE 12.5 Segmentation of CO_2 mineralized area of cement paste samples via phenolphthalein spray test at (a) 1 day, (b) 3 days, (c) 14 days, and (d) 28 days (Sharma et al., 2023).

12.3 CO_2 UPTAKE CALCULATION METHOD

12.3.1 Principle

The CO_2 uptake calculation method converts the amount of actual CO_2 mineralized cement-based material and the amount of the maximum theoretical cement-based material that can be mineralized in Equation (12.1) into the actual CO_2 uptake of cement-based material and the maximum theoretical CO_2 uptake of cement-based material, respectively (Eq. 12.3). Maximum CO_2 uptake of cement-based material is calculated by theoretical equations. The actual CO_2 uptake of cement-based material is obtained by test.

$$CO_2 \text{ mineralization degree} = \frac{\text{Actual } CO_2 \text{ uptake of cement-based material}}{\text{Maximum theoretical } CO_2 \text{ uptake of cement-based material}} \tag{12.3}$$

12.3.2 Determination of maximum CO_2 uptake

The European Committee for Standardization, in its 2016 publication *CO_2 Uptake by Mineralization-Guidance on Calculation* ((CEN), 2016), gives a method for calculating the maximum CO_2 uptake of clinker as follows:

$$CO_{2,\text{max.uptake}} = \left(CaO_{\text{reactive}}/100\right) \times \text{binder} \times \left(M_{CO_2}/M_{CaO}\right) \tag{12.4}$$

where $CO_{2,\text{max.uptake}}$ is the maximum amount of CO_2 that can be absorbed; CaO_{reactive} is the CaO portion in binder; binder is the amount of cement; M_{CO2} is the molar mass of CO_2 (44.0 g/mol); M_{CaO} is the molar mass of CaO (56.1 g/mol).

Many researchers (Sanjuán et al., 2020; Sevelsted and Skibsted, 2015; Zhou and Glasser, 2000) have questioned the accuracy of this formula from the European Committee for Standardization because it ignores the possible CO_2 mineralization of cement components other than CaO. A classical formula for CO_2 uptake considering CaO, SO_3, MgO, Na_2O, and K_2O proposed by Steinour (Steinour, 1959) in 1959 is currently the most common formula used by researchers to determine the maximum CO_2 uptake of cement-based materials:

$$\begin{aligned} CO_2\text{,max.uptake} = & \left(M_{CO_2}/M_{CaO}\right)\times\left(CaO - 0.7\times\left(SO_3\right)_{CaSO_4}\right) \\ & +\left(M_{CO_2}/M_{MgO}\right)\times MgO+\left(M_{CO_2}/M_{Na_2O}\right) \\ & \times Na_2O+\left(M_{CO_2}/M_{K_2O}\right)\times K_2O \end{aligned} \quad (12.5)$$

where M_{MgO} is the molar mass of MgO (40.3 g/mol); M_{Na_2O} is the molar mass of Na_2O (62.0 g/mol); M_{K_2O} is the molar mass of K_2O (94.2 g/mol)

Bringing in the molar mass of each oxide gives the following equation:

$$CO_2\text{,max.uptake}=0.785\left(CaO\text{-}0.7SO_3\right)+1.091MgO+1.420Na_2O+0.935K_2O \quad (12.6)$$

However, in a slight departure from Steinour (1959), Sanjuánit et al. (2020) considered that only 50% of the C-S-H gel might be mineralized. Therefore, in their study (Sanjuán et al., 2020), a modification was made to the European Committee for Standardization formula. Based on the cement parameters in this study, the calculation equation is as follows:

$$CO_2\text{,max.uptake} = CCaO\times\left(CaO_{reactive}/100\right)\times binder\times\left(M_{CO_2}/M_{CaO}\right) \quad (12.7)$$

where C_{CaO} is the portion of CaO in C-S-H that can be mineralized; it is 0.76 in the study by Sanjuánit et al. (2020).

In addition, Huntzinger et al. (2009) suggested that very old recycled aggregates and recycled cement paste powders may contain some calcium carbonate due to natural CO_2 mineralization, which can no longer absorb CO_2. Therefore, Huntzinger et al. (2009) modified Steinour's equation (Sanjuán et al., 2020) by removing CaO from $CaCO_3$. The calculation equation is as follows:

$$CO_2\text{,max.uptake}=0.785\left(CaO\text{-}0.56Ca\left(CO_3\right)\text{-}0.7SO_3\right)+1.091MgO+0.71Na_2O+0.468K_2O \quad (12.8)$$

However, the present study's authors believe that the natural CO_2 mineralization fraction in recycled cement-based materials should be included in the total mineralization degree. Overall, the determination of the fraction that can be CO_2 mineralized is currently controversial, and more research is needed in the future to harmonize the calculation equations.

12.3.3 Determination of Actual CO_2 Uptake

12.3.3.1 Mass change method

Testing the mass change of a specimen during CO_2 mineralization makes it feasible to measure the total sequestered CO_2. No special technique is needed to prepare the specimens. Also, this non-destructive testing method makes it possible to monitor the development of the whole CO_2 mineralization process over time (Villain et al., 2007).

(1) Mass gain of the specimen

The mass change method is a simple method to determine the actual CO_2 uptake of cement-based materials. This method was used in many previous studies (Huntzinger et al., 2009; Ramezanianpour et al., 2014; Shi and Wu, 2009; Zhang and Shao, 2016). The sample is normally placed in an airtight chamber. The mass of the sample is measured before and after the CO_2 mineralization. The evaporated water from the exothermic CO_2 mineralization reaction is collected using absorbent paper and added to the total mass gain as it is initially derived from the moisture content in the sample. In a study by Tu et al. (2016), the sample was embedded with a thermocouple to monitor the temperature profile during accelerated CO_2 mineralization, as shown in Figure 12.6. The mass gain can then be calculated using the following equation:

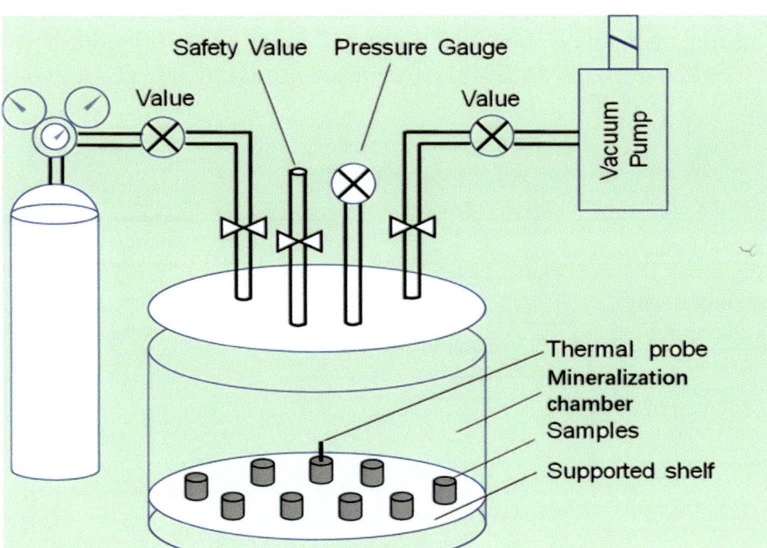

FIGURE 12.6 Schematic of the CO_2 mineralization equipment (He et al., 2016).

$$CO_2\ \text{uptake}(\%) = \frac{(M_1 + M_W) - M_2}{M_3} \times 100\% \tag{12.9}$$

where, M_1 is the mass of the sample after CO_2 mineralization; M_w is the mass of the evaporated water; M_2 is the mass of the sample before CO_2 mineralization; and M_3 is the cement content in the sample.

Shao et al. (Zhang and Shao, 2016) examined the actual maximum CO_2 uptake of a wet-mixed concrete using the mass gain method. The initial mineralization of the wet-mixed concrete was first conducted in the mold for 5 hours as it had a high flowability. The preconditioning was then performed by a fan drying process. The preconditioning was critical for the CO_2 mineralization process of concrete, as it can remove part of the free water and provide more voids for the penetration of CO_2. The maximum CO_2 uptake occurred when the water removal was approximately 40%. Further water removal reduced the CO_2 uptake as the mineralization process needs water as an essential reactant. For different concrete mixtures, the appropriate water removal would be different, while an optimal moisture content was needed for a high degree of CO_2 mineralization.

(2) Mass change of the whole test-up

When subjected to acceleration CO_2 mineralization, the cumulative mass change method of the whole test set-up with the sample can also be used to represent the rate of CO_2 sequestration of the sample over time (Figure 12.7) (Rostami et al., 2011). Similar to the mass gain method, this method is also a non-destructive test method. The sample is put in an airtight chamber, and the entire mineralization chamber is put on a digital balance, which is zeroed after vacuuming the chamber. At the end of the CO_2 mineralization process, CO_2 is released, and the final mass, M, is measured. Before testing, calibration should be conducted using a CO_2-insensitive styrofoam sample (El-Hassan and Shao, 2015) or nonreactive dummy sample (Rostami et al.,

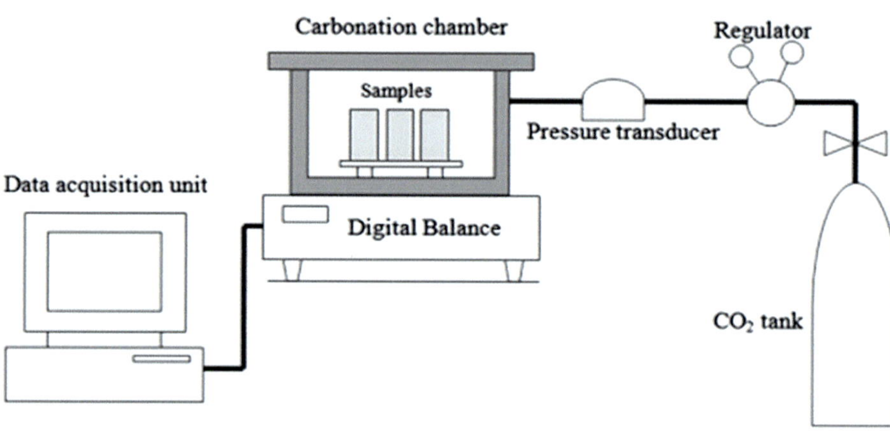

FIGURE 12.7 Schematic of CO_2 mineralization set-up using the cumulative mass change method (Rostami et al., 2011).

2011) with the same volume of the tested concrete sample to obtain the second residual mass, m. The CO_2 absorption is then calculated as follows:

$$CO_2\ \text{uptake}(\%) = \frac{M - m}{\text{Mass of cement}} \times 100\% \tag{12.10}$$

where, M is the mass of the total after the CO_2 mineralization; m is the second residual mass

A typical cumulative mass development curve for a concrete sample during a CO_2 mineralization process is presented in Figure 12.8 (Rostami et al., 2011). The CO_2 uptake showed a sharp rise in the first 20 minutes, and its rate decreased gradually until reaching a steady level of around 9% at 2 hours. This behavior should be attributed to the formed $CaCO_3$ covering the non-mineralized cement phases and preventing the further mineralization reaction. This trend is consistent with the temperature development profile (Shi et al., 2012a) and the reaction kinetics (Shi et al., 2012b) of the concrete sample subject to the CO_2 mineralization, which indicates the feasibility of using the cumulative mass development method to evaluate the CO_2 uptake process.

Normally, the mass gain of the specimen and cumulative mass development curve are measured simultaneously from the same process. Zhang et al. (2016) reported that the results of these two methods were consistent with a deviation of less than 5%. There are two advantages to using the cumulative mass development curve method over the mass change method. First, the cumulative mass development curve method can be used to monitor the reaction kinetics. Besides, the cumulative mass development curve method is more accurate than the mass change method. The CO_2 uptake obtained by the mass gain method was slightly lower than the cumulative mass

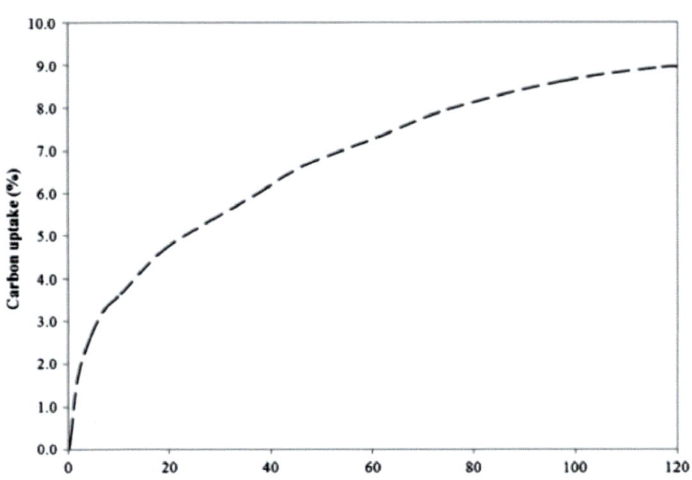

FIGURE 12.8 Cumulative mass development curve of a concrete product during 2 hours of CO_2 mineralization (Rostami et al., 2011).

development method because the water evaporation due to mineralization could not be fully collected when using Equation (12.8) as reported by Zhang et al. (Zhang and Shao, 2016). However, the loss of evaporated water was not considered in this method when releasing the CO_2 gas, which could decrease the calculated degree of CO_2 mineralization.

(3) Concentration variation of the inlet–outlet CO_2 gas

This method is used to determine actual CO_2 uptake based on the difference in the CO_2 concentration between the inlet and outlet gas (which can represent the change in CO_2 mass) as follows:

$$CO_2\ \text{uptake}(\%) = \frac{M_{\text{in}} - M_{\text{out}}}{M_c} \times 100\% \tag{12.11}$$

where, M_{in} is the mass of CO_2 measured by inlet; M_{out} is the mass of the CO_2 measured by outlet; and M_c is the cement content of the sample.

The amount of CO_2 that has reacted with cement was measured by the two mass flow meters once every minute (Kashef-Haghighi and Ghoshal, 2013). Figure 12.9 shows the rapid CO_2 uptake took place in the first 10 minutes, where the mass of the outlet CO_2 was almost zero. After 30 minutes, the concentration ratio of the outlet and the inlet CO_2 was approximately 93%, which indicates the rapid reaction at the beginning of the accelerated mineralization process has ended. In order to improve the accuracy, Phung et al. (2015) adopted two mass flow meters. One of them with a higher measurement range (0–50 ml/min) was used at the beginning of the CO_2 mineralization process; the other one with a lower measurement range (0–10 ml/min) was used when the reaction rate became slower.

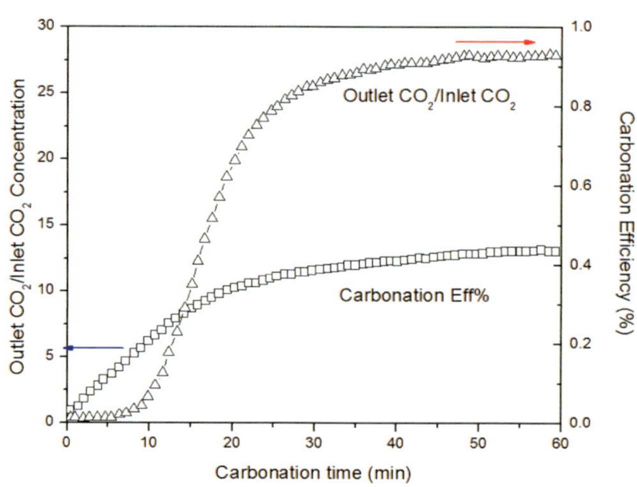

FIGURE 12.9 Ratio of outlet to inlet CO_2 concentration and mineralization efficiency in a vertical flow-through reactor (data from Kashef-Haghighi and Ghoshal, 2013).

The Chinese standard GB/T 45126—2025 *Determination of Carbon Dioxide Content of Steel Slag during Carbonation Fixation* (Administration, 2025) proposes a method for measuring CO_2 concentration variation during steel slag carbonation using an infrared carbon analyzer or a high-frequency infrared carbon-sulfur analyzer. In this method, the steel slag sample is combusted in an oxygen stream within a high-frequency induction furnace. The resulting CO_2 is carried by oxygen to the measurement chamber of the infrared analyzer, where it absorbs infrared energy at a specific wavelength. The absorption intensity is proportional to the CO_2 concentration, allowing the determination of the total carbon mass fraction based on the detected energy variation. For free carbon content measurement, the sample is first decomposed with hydrochloric acid, assisted by sodium fluoride (or hydrofluoric acid) for dissolution. The residue is then subjected to acid washing, water rinsing, and drying on acid-washed asbestos before being analyzed via infrared absorption. The fixed carbon mass fraction is derived from the difference between the total carbon and free carbon mass fractions, which is subsequently used to calculate the carbonation degree of the sample.

12.3.3.2 Ignition testing method

Ignition testing method is regarded as an easy way to assess the amount of precipitated $CaCO_3$ in the specimen based on its decomposition and the mass loss at high temperatures, and thus actual CO_2 absorbed by the cement-based materials can be estimated. Before igniting the sample, small pieces of the sample are first pre-treated in an acetone solution to stop the cement hydration. After that, the sample is dried in an oven at 65°C until the sample mass remains constant. Then the sample is ground into powder, heated from room temperature to 1000°C, and the mass loss between approximately 500°C and 850°C is regarded mainly as the decomposition of $CaCO_3$.

The advantage of this method is that the degree of CO_2 mineralization of the sample at different depths can be examined. Rostami et al. (2012) compared the degree of mineralization at the surface and in the core (2–4 mm depth) of a cement paste. They found the content of $CaCO_3$ at the surface was much higher than that at the core. Phung et al. (2015) also found that no calcite was detected beyond 9 mm inside the cement paste sample. Besides, it should be noted that the initial $CaCO_3$ content is included in the amount of the determined $CaCO_3$.

Several disadvantages are also associated with this method. First, it is difficult to accurately determine the amount of $CaCO_3$ because C-S-H gel decomposes from 110°C to 1000°C as well. The mass loss could include a part of non-evaporable waste from the C-S-H gel. Second, the heating rate also affects the mass loss value. Third, limestone or other initial calcium carbonate content may influence the results. In addition, no strict measures are taken to prevent air mineralization during sampling and preparation processes (Thiery et al., 2007; Villain et al., 2007). In El-Hassan's study, the CO_2 content obtained using a furnace method was 6% higher than that of the TGA method, which was due to the presence of 13–15% limestone in the original specimen (El-Hassan and Shao, 2015). So they used the average value of these methods to examine the CO_2 uptake and found the CO_2 uptake of the mineralized OPC sample exceeded that of the Portland limestone cement by up to 6 mass%, which was due to the addition of limestone.

(1) Ignition by Furnace method

In general, the sample subjected to CO_2 mineralization is dried in a muffle furnace at 350°C to a constant mass to remove free water and combined water. The sample is then heated to 520°C to remove the non-evaporable water in some hydration products and $Ca(OH)_2$. After that, the sample is heated to 1000°C and the weight difference between 520°C and 1000°C is equal to the weight of CO_2 released. The CO_2 sequestration can be estimated as follows (Tu et al., 2016):

$$CO_2 \text{ uptake}(\%) = \frac{M_{520°C} - M_{1000°C} - M_0}{M_{\text{cement}}} \times 100\% \tag{12.12}$$

where, $M_{520°C}$ is the mass of pastes at 520°C; $M_{1000°C}$ is the mass of pastes at 1000°C; M_0 is the initial CO_2 content in the non-mineralized specimen.

For the furnace method, the exact starting temperature of calcium carbonate decomposition is still a subject of debate. Some researchers stated that the decomposition of $CaCO_3$ starts at around 520°C (Morandeau et al., 2014; Rostami et al., 2012). But the mass loss between 520°C and 720°C could also be contributed partly by the water loss from the hydrates as well (Morandeau et al., 2014; Villain et al., 2007). Generally, the mass loss between 520°C and 1000°C can be roughly regarded as proportional to the quantity of $CaCO_3$ present in the sample (Tu et al., 2016).

(2) Thermogravimetric analysis (TGA)

Another method often used to determine the actual CO_2 uptake of cement-based material is thermogravimetric analysis (TGA), which most researchers favor because it shows more accurate values compared with the other methods, owing to the high-sensitivity detector (the thermal scale of this method has 0.001 μg accuracy) (Bui et al., 2020). Figure 12.10 shows the TG and DTG curves of a typical mineralized

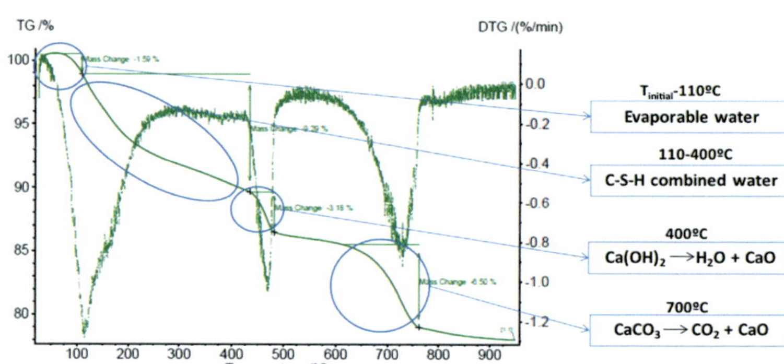

FIGURE 12.10 TG and DTG curve of a cement paste with the main decompositions depicted: evaporable water, C-S-H gel water, portlandite, and calcium carbonate (Andrade, 2020).

cement paste. The peaks between 500°C and 900°C, representing the thermal decomposition of CO_2 bound in $CaCO_3$ (Lu et al., 2018, 2022), were focused on, i.e., the actual CO_2 uptake of a cement-based material. Dividing it by the residual mass at 900°C is the actual CO_2 uptake by the cement-based material:

$$CO_2\text{uptake}(\%) = \frac{\left(M_{500°C} - M_{900°C}\right)}{M_{900°C}} \times 100 \tag{12.13}$$

where $M_{500°C}$ and $M_{900°C}$ are the residual mass at 500°C and 900°C.

Moreover, the TGA method can be used to investigate whether C-S-H has reacted with CO_2 by calculating the portlandite content between the mineralized and the reference sample. Phung et al. (2015) found that the C-S-H mineralization contributed 41.8%, 33.3%, and 4.6% to the calcite formation in the depth intervals 0–3 mm, 3-6mm, and 6–9 mm for the cement pastes designed with a water/binder ratio of 0.425.

12.3.3.3 Gamma densitometry method

The *gamma densitometry* is a non-destructive testing method used to measure the density of civil engineering materials, which is based on the absorption of gamma rays emitted by a radioactive source of Cesium Cs137 (Villain and Thiery, 2006; Villain et al., 2007). The basic principle of the test is shown in Figure 12.11. This method makes it possible to record the CO_2 mineralization process of cement-based materials by monitoring the evolution of their density according to depth and time. The increased density can be related to the actual CO_2 uptake of cement-based materials, assuming that no significant water evaporation occurs during the accelerated CO_2 mineralization process as the mineralization temperature is kept constant.

The density ρ (kg/m^3) is calculated using the following equation:

$$\rho = \frac{-1}{\mu \cdot l} ln\left(\frac{N}{N_0}\right) \tag{12.14}$$

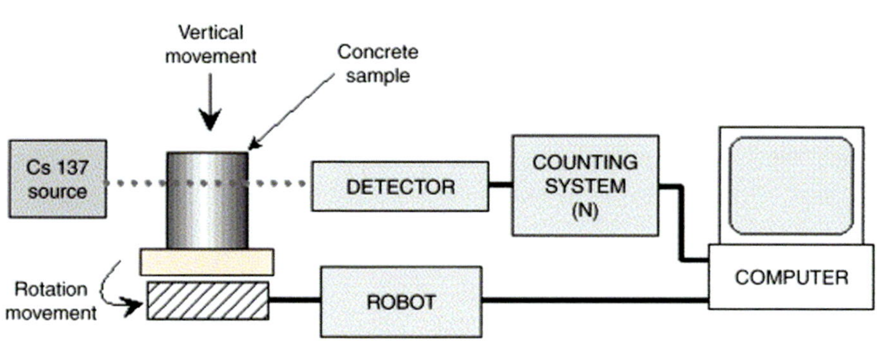

FIGURE 12.11 Gamma-ray measurement set-up (Villain and Thiery, 2006).

where, N_0 is the number of the incidental photons in the air; N is the number of incidental photons through concrete; l is the thickness across the sample; μ is the mass absorption coefficient of the material in accordance with the US National Bureau of Standards.

The density variation of the sample before and after the CO_2 mineralization corresponds to the amount of CO_2 sequestration. The CO_2 mineralization profile at different depths is calculated based on the density of the sample before ($\rho_0(t_0)$) and the density of the CO_2 mineralized sample at time t ($\rho(t)$) according to Equation (12.5).

$$\frac{\Delta\rho}{\rho} = \frac{\rho(t) - \rho_0(t_0)}{\rho_{0(t_0)}} \tag{12.15}$$

where, $\rho_0(t_0)$ is the density of the sample before CO_2 mineralization; $\rho(t)$ is the density of CO_2 mineralized sample at time t. The quantity of the penetrated CO_2 can also be expressed as a concentration C_{CO_2} in mol/L as follows:

$$C_{CO_2} = \frac{\rho(t) - \rho_0(t_0)}{M_{CO_2}} \tag{12.16}$$

where, M_{CO_2} is the molar concentration of the sequestrated CO_2.

Figure 12.12 shows the CO_2 uptake measured by the gamma densitometry method (Thiery et al., 2007). This method detects all forms of CO_2 in the CO_2 mineralized specimens: in gaseous phase, in aqueous solution as H_2CO_3, HCO_3^- and CO_3^{2-}, as absorbed CO_2 in the C-S-H gel, and as chemically bound in $CaCO_3$. It thus includes free CO_2 at equilibrium in the gaseous and aqueous phases. However, their concentrations are very small when the CO_2 pressure is higher than 5000 kPa (50% CO_2 to 50% air). Furthermore, the CO_2 uptake measured using the gamma densitometry method is consistent with that measured from the decomposition test of $CaCO_3$ by thermogravimetric analysis (TGA). Therefore, the CO_2 content in aqueous and gaseous phases can be regarded as negligible, and the CO_2 content profile obtained by using the gamma densitometry method can be used to investigate the accelerated CO_2 mineralization process of cement-based materials.

As a non-destructive test method, this method can monitor the CO_2 mineralization process at different mineralization times and different depths on the same specimen, as illustrated in Figure 12.13 (Villain et al., 2007). The density of the sample increased with mineralization duration, which indicates the gradual CO_2 capture and storage. While the surface density of the sample was higher than that of the core, indicating a lower degree of mineralization occurred in the center of the sample. By using this method, the effects of water/cement ratio and fly ash on the CO_2 uptake of cement were detected, as shown in Figure 12.14 (Morandeau et al., 2015). The density increase of CN30 was higher than that of CN at the same depth, which indicates that the incorporation of fly ash can promote the CO_2 uptake. The density decreased for CN60 after the 20-mm depth. This could be due to the release of the free water

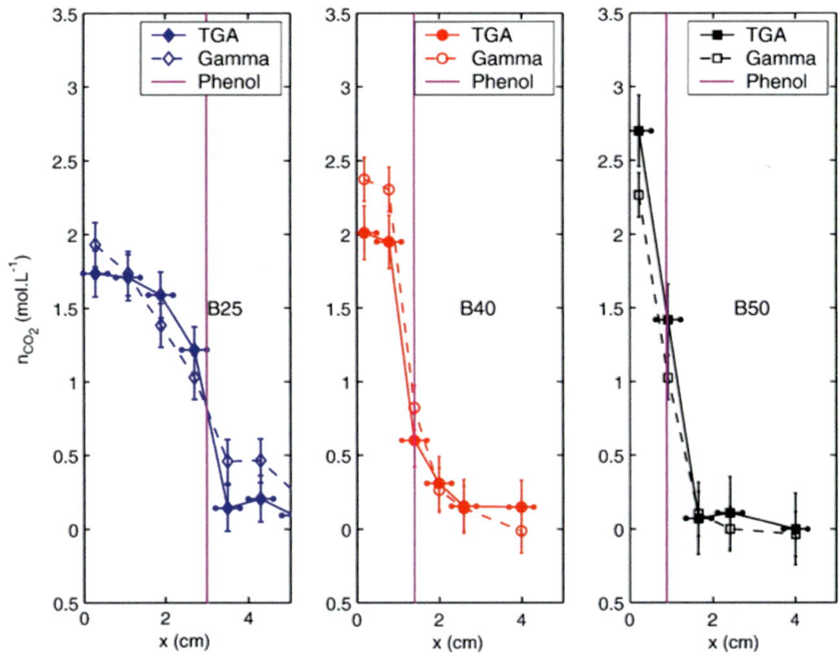

FIGURE 12.12 CO_2 content profiles of concrete samples after 14 days mineralization, determined by TGA and gamma densitometry method (Thiery et al., 2007).

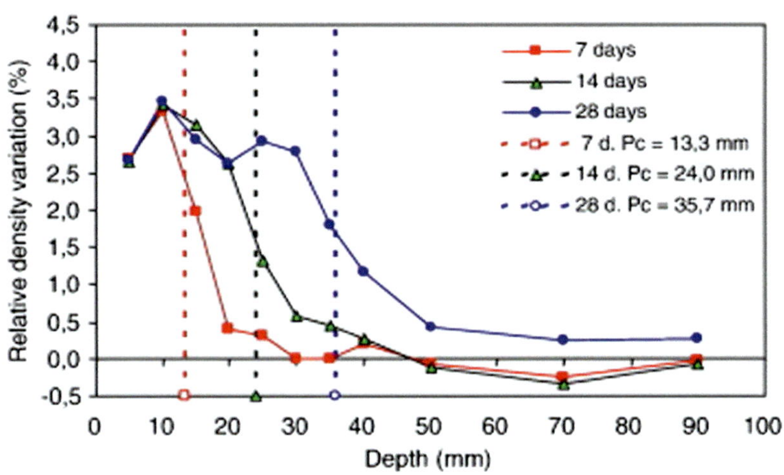

FIGURE 12.13 Mineralization profiles of samples subjected to different mineralization days (Villain et al., 2007).

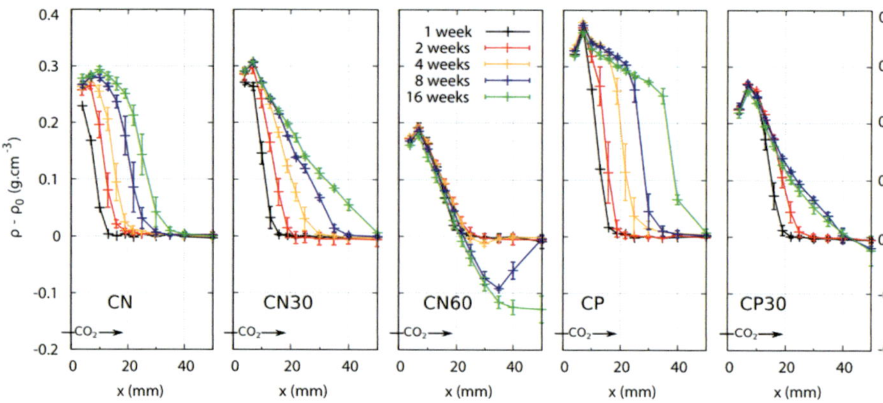

FIGURE 12.14 Density evolution after 1 week, 2 weeks, 4 weeks, 8 weeks, and 16 weeks of accelerated mineralization for the five cement pastes (CN means cement paste with w/c ratio of 0.45. CP means cement paste with w/c ratio of 0.6. CN30 means cement paste incorporating 30% fly ash) (Morandeau et al., 2015).

from the center of the sample to the surface as the surface is in equilibrium with the outer environment, as well as the exothermic reaction of the mineralization reactions. The density of CP was higher than that of CN at the same depth, indicating the benefit for mix designs with a higher w/c ratio for CO_2 mineralization. For this method, a reference non-mineralized sample is required. That is one of the reasons that this method would not accurately determine the CO_2 content in an aged concrete sample. However, this method can be used to test the accelerated mineralization process in the laboratory (Rostami et al., 2012).

12.3.3.4 Quantitative X-ray diffraction

Quantitative X-ray diffraction is an effective method to quantify the content of well-crystalline $CaCO_3$, and further determine the actual CO_2 uptake. In addition, this method can be used to distinguish the content of calcite, vaterite, and aragonite, as shown in Figure 12.15 (Tu et al., 2016). The sample preparation is the same as the ignition method. Tu et al. (2016) used the Bruker D8 AXS diffractometer equipped with a LynxEye detector and a Cu $K\alpha_{1,2}$ X-ray radiation source to quantify the content of calcium carbonate in Portland limestone cement. About 20% CaF_2 was used as an internal standard to calculate the amount of amorphous and poorly crystalline phases. The quantitative analysis was carried out using TOPAS 4.2, as shown in Figure 12.16. The amount of C_3S and C_2S significantly decreased, and a great amount of calcite was observed after 24 hours of CO_2 mineralization for all mixtures. The intensity of aragonite and vaterite was relatively low and not detected from the QXRD results. When the pure C-S-H gel was mineralized for 2 hours, calcite, aragonite, and vaterite were observed with content of 33.9%, 18.7%, and 17.1%, respectively (Chang and Fang, 2015).

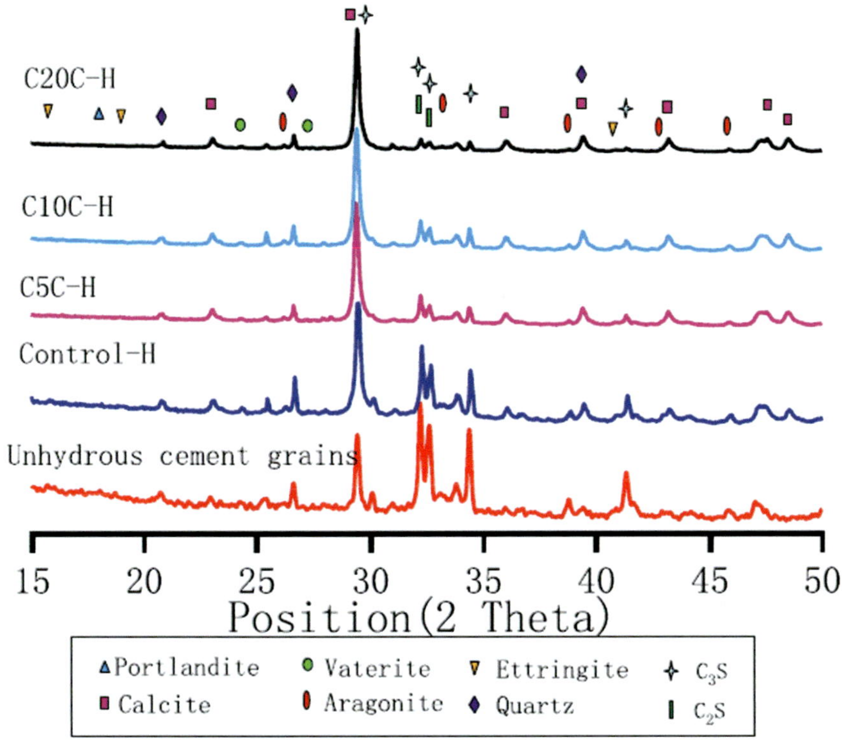

FIGURE 12.15 XRD patterns of mineralized cement powder and cement pastes after 24-hour 4-bar CO_2 mineralization (Tu et al., 2016).

12.4 APPLICATION SCOPE OF METHODS FOR THE DETERMINATION OF MINERALIZED CO_2 AND THEIR LIMITATIONS

The mineralization area method is suitable for determining the mineralization of block cement-based materials, such as cement paste, mortar, concrete, and recycled aggregates, since the method requires measurement of the cross-sectional area of the specimen. It cannot be used for powdered materials such as clinker powder and recycled cement paste powder. In addition, because it is usually necessary to crush and split the specimen, this method is unsuitable for specimens that need to be preserved intact. Regarding measurement accuracy, two aspects lead to a significant error in the method. One is the entire cross-sectional area of the cement-based material, considered to be the area that can be mineralized. The second is the inaccuracy of the current evaluation technique for determining the cement-based material's actual mineralized cross-sectional area.

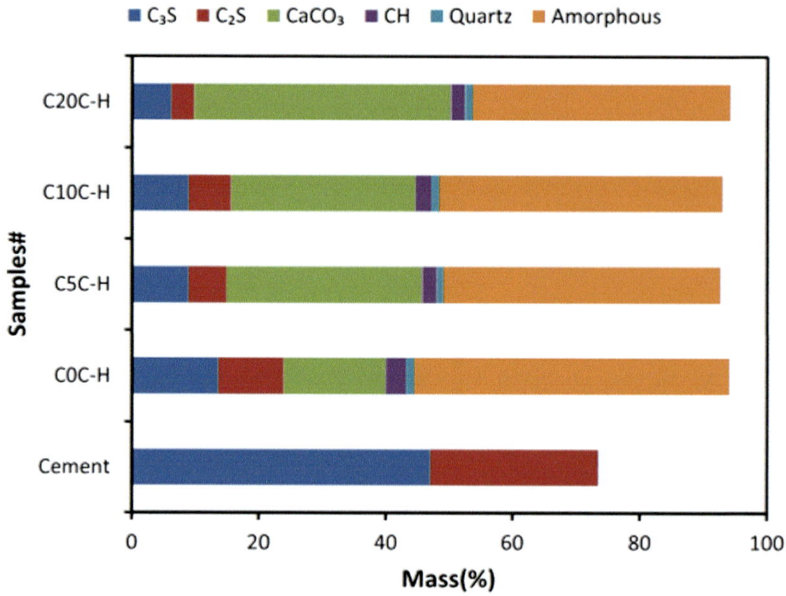

FIGURE 12.16 QXRD results of 24-hour 4-bar CO_2 mineralized cement pastes (the content of limestone in the cement mixture was 0%, 5%, 10%, 20% from the bottom) (Tu et al., 2016).

The CO_2 uptake calculation method can be used to determine the mineralization degree of all types of cement-based materials. However, the sampling location for cement paste, mortar, concrete, and mortar adhering to recycled aggregates is a crucial issue. Previous studies (Andrade, 2020; Rostami et al., 2012; Zhang and Shao, 2016) often first identified the mineralized area and then sampled the cement-based samples in the mineralized area. However, according to the discussion above, it can also lead to errors since the imprecision in the determination of the boundary of the mineralized area. The CO_2 uptake calculation method is very suitable for powdered cement-based materials, such as clinker powder and RCF, since it is not necessary to distinguish between the mineralized and non-mineralized zones. In general, because theoretical equations calculate the maximum CO_2 uptake, the actual CO_2 uptake is often obtained by high-accuracy testing methods. The CO_2 uptake calculation method is more accurate than the mineralization area method. It is suitable for calculating the mineralization degree of cement-based materials that require a certain level of accuracy.

The existing methods for evaluating the mineralization degree of cement-based materials can be divided into the mineralization area method and the CO_2 uptake calculation method. The mineralization area method applies to large bulk cement-based materials. The irregularity of the mineralization area and the low accuracy of the existing determination techniques are the reasons that result in errors in this method. The CO_2 uptake calculation method is more suitable for powdered cement-based materials. The determination of the maximum CO_2 uptake of cement-based materials is still controversial. The high accuracy and uniformity of methods for evaluating the mineralization degree of cement-based materials are urgent issues to be addressed.

12.5 SUMMARY

This chapter aims to present different measurement methods for assessing the degree of CO_2 mineralization (CO_2 uptake) by mineralized cement-based materials. The existing methods for evaluating the degree of CO_2 mineralization of cement-based materials can be divided into the CO_2 mineralization area method and the CO_2 uptake calculation method. The CO_2 mineralization area method applies to large bulk cement-based materials. The irregularity of the mineralization area and the low accuracy of the existing determination techniques resulted in the method being considered semi-quantitative. There is some controversy on the formula for calculating the theoretical maximum CO_2 uptake of cement-based materials when using the CO_2 uptake calculation method. Currently, four main methods, including mass change, gamma densitometry, ignition, and Q-XRD, are used to determine the actual CO_2 uptake of cement-based materials.

REFERENCES

2004. DIN EN 13295:2004–08, Products and systems for the protection and repair of concrete structures – test methods – determination of resistance to carbonation.

2007a. DIN EN 14630:2007–01, Products and systems for the protection and repair of concrete structures – test methods – determination of carbonation depth in hardened concrete by the phenolphthalein method.

2007b. E DIN EN 12390–10:2017–04, Testing hardened concrete – part 10: determination of the carbonation resistance of concrete at atmospheric levels of carbon dioxide.

Andrade, C., 2020. Evaluation of the degree of carbonation of concretes in three environments. *Construction and Building Materials* 230, 116804.

Bui, H., Delattre, F., Levacher, D., 2023. Experimental methods to evaluate the carbonation degree in concrete: State of the art review. Applied Sciences 13(2), Article 812.

Bui, H., Sebaibi, N., Boutouil, M., Levacher, D., 2020. Determination and review of physical and mechanical properties of raw and treated coconut fibers for their recycling in construction materials. Fibers 8(6), 37.

Chang, J., Fang, Y., 2015. Quantitative analysis of accelerated carbonation products of the synthetic calcium silicate hydrate (C–S–H) by QXRD and TG/MS. *Journal of Thermal Analysis and Calorimetry* 119(1), 57–62.

Cheng Zhang, L.W., Yan Yao, Xinyu Shi, 2022. Determination of concrete carbonation depth by testing the pH value of layer-by-layer grinding concrete samples. *Materials Reports* 36(7), 21030009.

Chinchón-Payá, S., Andrade, C., Chinchón, S., 2016. Indicator of carbonation front in concrete as substitute to phenolphthalein. *Cement and Concrete Research* 82, 87–91.

Choi, J.-I., Lee, Y., Kim, Y.Y., Lee, B.Y., 2017. Image-processing technique to detect carbonation regions of concrete sprayed with a phenolphthalein solution. *Construction and Building Materials* 154, 451–461.

Cui, D., Liu, W., Wang, J., Hu, J., Shan, D., Wan, Y., Wang, Q., Wang, J., 2022. Use of a novel pH indicator extracted from petals to investigate the carbonation behavior in cementitious materials. *Cement and Concrete Composites* 134, 104804.

El-Hassan, H., Shao, Y., 2015. Early carbonation curing of concrete masonry units with Portland limestone cement. *Cement and Concrete Composites* 62, 168–177.

European Committee for Standardization, 2016. *Sustainability of construction works—environmental product declarations—product category rules for concrete and concrete elements. Annex BB (Informative). CO2 uptake by carbonation—Guidance on calculation.*

Ghantous, R.M., Poyet, S., L'Hostis, V., Tran, N.-C., François, R., 2017. Effect of crack openings on carbonation-induced corrosion. *Cement and Concrete Research* 95, 257–269.

Giulietti, N., Chiariotti, P., Cosoli, G., Mobili, A., Pandarese, G., Tittarelli, F., Revel, G.M., 2021. Automated measurement system for detecting carbonation depth: Image-processing based technique applied to concrete sprayed with phenolphthalein. *Measurement* 175, 109142.

Huntzinger, D.N., Gierke, J.S., Kawatra, S.K., Eisele, T.C., Sutter, L.L., 2009. Carbon dioxide sequestration in cement kiln dust through mineral carbonation. *Environmental Science & Technology* 43(6), 1986–1992.

Ji, Y.-S., Wu, M., Ding, B., Liu, F., Gao, F., 2014. The experimental investigation of width of semi-carbonation zone in carbonated concrete. *Construction and Building Materials* 65, 67–75.

Kachko, A., van der Ham, L.V., Bardow, A., Vlugt, T.J.H., Goetheer, E.L.V., 2016. Comparison of Raman, NIR, and ATR FTIR spectroscopy as analytical tools for in-line monitoring of CO2 concentration in an amine gas treating process. *International Journal of Greenhouse Gas Control* 47, 17–24.

Kashef-Haghighi, S., Ghoshal, S., 2013. Physico–chemical processes limiting CO2 uptake in concrete during accelerated carbonation curing. *Industrial & Engineering Chemistry Research* 52(16), 5529–5537.

Li, Y., Liu, W., Xing, F., Wang, S., Tang, L., Lin, S., Dong, Z., 2020. Carbonation of the synthetic calcium silicate hydrate (C-S-H) under different concentrations of CO2: Chemical phases analysis and kinetics. *Journal of CO2 Utilization* 35, 303–313.

Lu, B., Drissi, S., Liu, J., Hu, X., Song, B., Shi, C., 2022. Effect of temperature on CO_2 curing, compressive strength and microstructure of cement paste. *Cement and Concrete Research* 157, 106827.

Lu, B., Shi, C., Zhang, J., Wang, J., 2018. Effects of carbonated hardened cement paste powder on hydration and microstructure of Portland cement. *Construction and Building Materials* 186, 699–708.

Mi, R., Pan, G., 2022. Inhomogeneities of carbonation depth distributions in recycled aggregate concretes: A visualisation and quantification study. *Construction and Building Materials* 330, 127300.

Mi, R., Pan, G., Li, Y., Kuang, T., 2021. Carbonation degree evaluation of recycled aggregate concrete using carbonation zone widths. *Journal of CO2 Utilization* 43, 101366.

Mi, R., Pan, G., Liew, K.M., 2020. Predicting carbonation service life of reinforced concrete beams reflecting distribution of carbonation zones. *Construction and Building Materials* 255, 119367.

Ministry of Housing and Urban–Rural Development of the People's Republic of China, 2009. *GB/T 50082–2009, Standard for test methods of long-term performance and durability of ordinary concrete.* China Architecture & Building Press.

Morandeau, A., Thiery, M., Dangla, P., 2014. Investigation of the carbonation mechanism of CH and CSH in terms of kinetics, microstructure changes and moisture properties. *Cement and Concrete Research* 56, 153–170.

Morandeau, A., Thiéry, M., Dangla, P., 2015. Impact of accelerated carbonation on OPC cement paste blended with fly ash. *Cement and Concrete Research* 67, 226–236.

National Standards Administration, 2025. *Determination of carbon dioxide content of steel slag during carbonation fixation.*

Phung, Q.T., Maes, N., Jacques, D., Bruneel, E., Van Driessche, I., Ye, G., De Schutter, G., 2015. Effect of limestone fillers on microstructure and permeability due to carbonation of cement pastes under controlled CO_2 pressure conditions. *Construction and Building Materials* 82, 376–390.

Ramezanianpour, A., Ghahari, S., Esmaeili, M., 2014. Effect of combined carbonation and chloride ion ingress by an accelerated test method on microscopic and mechanical properties of concrete. *Construction and Building Materials* 58, 138–146.

Rostami, V., Shao, Y., Boyd, A.J., 2011. Durability of concrete pipes subjected to combined steam and carbonation curing. *Construction and Building Materials* 25(8), 3345–3355.

Rostami, V., Shao, Y., Boyd, A.J., He, Z., 2012. Microstructure of cement paste subject to early carbonation curing. *Cement and Concrete Research* 42(1), 186–193.

Sanjuán, M.Á., Andrade, C., Mora, P., Zaragoza, A., 2020. Carbon dioxide uptake by cement-based materials: A Spanish case study. Applied Sciences 10(1), 339.

Segura, I., Molero, M., Aparicio, S., Moragues, A., 2010. Measurement of the degraded depth in cementitious materials by automatic digital image processing. *Measurement Science and Technology* 21(5), 055103.

Sevelsted, T.F., Skibsted, J., 2015. Carbonation of C–S–H and C–A–S–H samples studied by 13C, 27Al and 29Si MAS NMR spectroscopy. *Cement and Concrete Research* 71, 56–65.

Sharma, R., Kim, H., Lee, N.K., Park, J.-J., Jang, J.G., 2023. Microstructural characteristics and CO_2 uptake of calcium sulfoaluminate cement by carbonation curing at different water-to-cement ratios. *Cement and Concrete Research* 163, 107012.

Shi, C., He, F., Wu, Y., 2012a. Effect of pre-conditioning on CO_2 curing of lightweight concrete blocks mixtures. *Construction and Building Materials* 26(1), 257–267.

Shi, C., Liu, M., He, P., Ou, Z., 2012b. Factors affecting kinetics of CO_2 curing of concrete. *Journal of Sustainable Cement-Based Materials* 1(1–2), 24–33.

Shi, C., Wu, Y., 2009. CO_2 curing of concrete blocks. *Concrete International* 31(02), 39–43.

Steinour, H.H., 1959. Some effects of carbon dioxide on mortars and concrete-discussion. *Journal of the American Concrete Institute* 30(2), 905–907.

Thiery, M., Villain, G., Dangla, P., Platret, G., 2007. Investigation of the carbonation front shape on cementitious materials: effects of the chemical kinetics. *Cement and Concrete Research* 37(7), 1047–1058.

Tu, Z., Guo M.-z., Poon, C.S., Shi, C., 2016. Effects of limestone powder on $CaCO_3$ precipitation in CO_2 cured cement pastes. *Cement and Concrete Composites* 72, 9–16.

Villain, G., Thiery, M., 2006. Gammadensimetry: A method to determine drying and carbonation profiles in concrete. *NDT & E International* 39(4), 328–337.

Villain, G., Thiery, M., Platret, G., 2007. Measurement methods of carbonation profiles in concrete: Thermogravimetry, chemical analysis and gammadensimetry. *Cement and Concrete Research* 37(8), 1182–1192.

Vogler, N., Lindemann, M., Drabetzki, P., Kühne, H.-C., 2020. Alternative pH-indicators for determination of carbonation depth on cement-based concretes. *Cement and Concrete Composites* 109, 103565.

Zhang, D., Cai, X., Shao, Y., 2016. Carbonation curing of precast fly ash concrete. *Journal of Materials in Civil Engineering* 28(11), 04016127.

Zhang, D., Ghouleh, Z., Shao, Y., 2017. Review on carbonation curing of cement-based materials. *Journal of CO2 Utilization* 21, 119–131.

Zhang, D., Shao, Y., 2016. Early age carbonation curing for precast reinforced concretes. *Construction and Building Materials* 113, 134–143.

Zhou, Q., Glasser, F.J.A.I.C.R., 2000. Kinetics and mechanism of the carbonation of ettringite. *Advances in Cement Research* 12(3), 131–136.

13 Applications and standardizations of CO_2 mineralized products

13.1 INTRODUCTION

In the past two decades, CO_2 mineralization has attained a global surge of interest as a promising method to reduce the carbon footprint of the cement industry. A multitude of laboratory research has been emerging in this realm which acts as a theoretical underpinning of the feasibility of CO_2 mineralization in commercial application. The most significant step is to take actions to apply CO_2 mineralization technology in industrial production, turning theory into a realistic solution. Over the past decade, some pioneer startups, spin-offs, and large corporations have initiated the first step to develop proprietary carbonate binder products or processes by utilizing CO_2 mineralization technologies worldwide.

Standardizations were developed and continued to exist to regularize commercial production. Minimum specifications are set for materials, while standard test methods are proposed to determine compliance of materials and products with the specification limits. New materials and technologies call for the development of standardizations. The American Society of Testing Materials (ASTM) has developed specifications and tests for cements that are set by CO_2 mineralization. China is authorizing sectoral technical specifications for CO_2 mineralized products.

13.2 APPLICATIONS OF CO_2 MINERALIZED MATERIALS AND PRODUCTS

13.2.1 Applications of CO_2 mineralized Portland cement-based materials

13.2.1.1 CJS technology

Shi first proposed the theory of "moisture regulation-gas migration-CO_2 mineralization reaction" and developed the pre-conditioning technology, which significantly increased the mineralization rate and degree of cement-based materials and made the large-scale production of CO_2 mineralization concrete products possible (Shi et al., 2012; Shi and Wu, 2008). Then Hunan University and Huaxin Cement Co., Ltd. established the world's first commercial-scale CO_2 mineralization concrete blocks production line by using cement kiln exhaust gas in 2021 (Liu et al., 2022). The brick preparation and curing procedure has been divided into four stages: Mixing, Compacting, CO_2 mineralization, and Transportation, which are basically automated, as shown in Figure 13.1. The raw materials are mixed in the mixer and

DOI: 10.1201/9781003627531-13

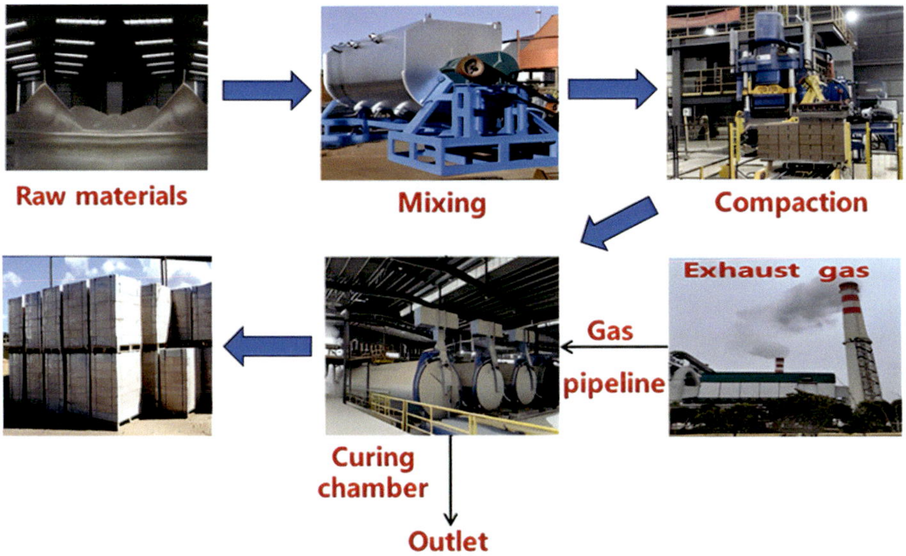

FIGURE 13.1 Production process of CO_2 mineralized concrete blocks using CJS technology (Liu et al., 2022).

transported to the brick making machine. The concrete bricks are compacted by the brick making machine. The demolded bricks were placed into a slide rail car and rested for 4–6 hours, allowing them to gain some strength before they were moved to the curing chamber. The gas generated in the cement plant with a volume concentration of CO_2 20–30% and a relative humidity larger than 95% was drawn into the curing chamber. The CO_2 mineralization degree of the carbonatable materials in concrete brick can reach 50%.

The CO_2 mineralized bricks possess qualified compressive strength and excellent durability, as shown in Table 13.1. The average compressive strength of bricks can exceed 15 MPa and reach up to 18.1 MPa, meeting the requirements of the

TABLE 13.1
Performance of the bricks after autoclaved curing and CO_2 mineralization

Performance		CO_2 mineralization	Autoclaved curing
Compressive strength	Average value	18.1	19.2
	standard value	16.6 MPa	16.8 MPa
Frost resistance	Compressive strength after freezing and thawing	15.5 Mpa	13.8 Mpa
	Mass loss	0.02%	0.10%
28 days drying shrinkage		203 μm/m	230 μm/m

Source: Liu et al., 2022.

relevant national standards. Moreover, the compressive strength even significantly increases when stockpiled under a proper environment. The CO_2 mineralized bricks by exhaust gas curing exhibit better performance in durability than those with autoclaved curing, with drying shrinkage reduced by 15% and frost resistance increased by 12% due to higher calcium carbonate content.

Since the establishment of the production line for CO_2-mineralized concrete bricks using cement kiln exhaust gas, 100 million bricks have been produced per year, which can utilize about 26,000 tons of CO_2 and save 3.79 million kW·h of electricity. The carbon tax in China's trading market is around 76 yuan/ton, with the highest carbon tax in a recent auction reaching 485 yuan/ton. Considering the price of Portland cement to be 280 yuan/ton, the cost to produce one concrete brick can be decreased from 0.21 yuan (steam cured bricks) to 0.08 yuan (carbon tax 76 yuan) and −0.09 yuan (carbon tax 485 yuan). However, because the carbon tax in Europe is almost 800 yuan/ton, the cost to produce a concrete brick is −0.23 yuan. At present, the CO_2 mineralized concrete bricks have been adopted for various applications, as shown in Figure 13.2. The wall made of CO_2 mineralized bricks exhibits less building settlement and fewer cracks; the pavement made of CO_2 mineralized bricks is wear-resistant and has a longer service life.

(a) Building wall (b) Retaining wall

(c) Pavement brick (d) Masonry fence

FIGURE 13.2 Applications of CO_2 mineralized concrete bricks of CJS technology.

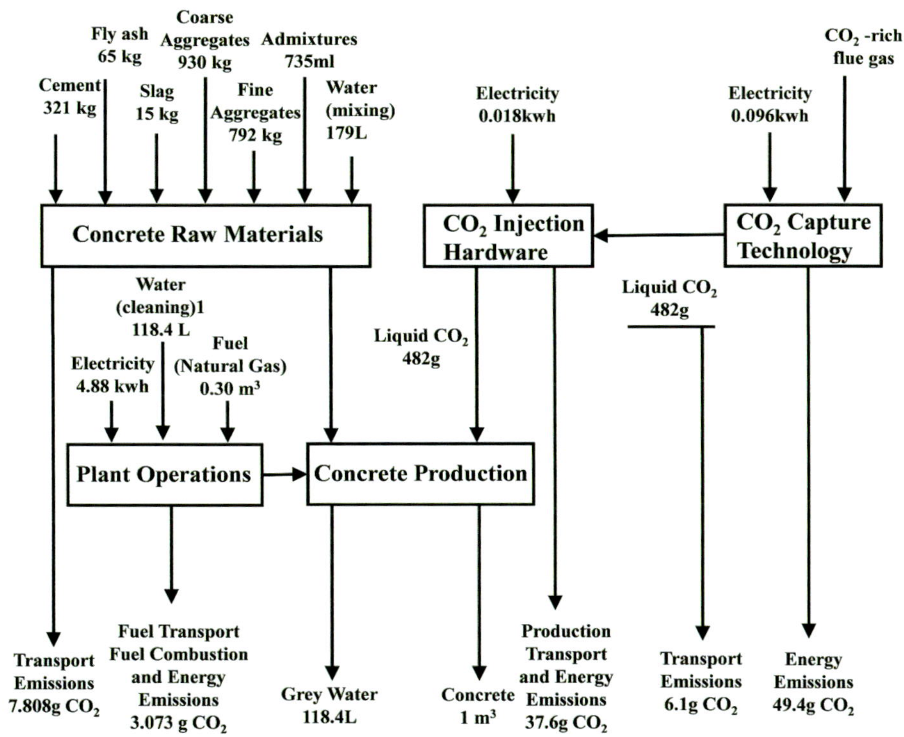

FIGURE 13.3 Process flow diagram for CarbonCure technology (Monkman and MacDonald, 2017).

13.2.1.2 CarbonCure technology

CarbonCure Technologies (Canada) developed a CO_2 utilization approach where liquid CO_2 transported from pressurized tanks is injected into fresh concrete during mixing. The process flow diagram for the utilization of CO_2 in the production of ready-mixed concrete by CarbonCure's technology is shown in Figure 13.3. The amount of CO_2 injected is usually limited to up to 1.5% by weight of cement. The Ca^{2+} from cement reacts with the CO_2 to form nanoscale $CaCO_3$, which is embedded in the concrete, thereby increasing the strength of concrete (Monkman and MacDonald, 2016). The cement content in the concrete mixture can also be reduced by 5% due to the strengthening effect of CO_2 (Monkman and MacDonald, 2017). It has been proved that the CO_2-injection curing process has a neutral to positive effect on concrete durability with suitable chloride penetration resistance, drying shrinkage, freeze-thaw, and de-icer salt scaling resistance performance assured through testing (Monkman et al., 2016).

CarbonCure's process ultimately reduces net CO_2 emissions from concrete manufacturing by 18 kg/m^3 when compared to that of PC concrete. CarbonCure's technology has been approved by several countries, such as Canada, the USA, Germany, and France. CarbonCure's technology has been adopted in construction cases, with 2.5 million truckloads (about 13.92 million tons) of concrete and 180,000 tons of

(a) (b)

FIGURE 13.4 Applications of CarbonCure technology (a) Office buildings employing this technology and (b) concrete pavement applications.

CO_2 saved (relative to the equivalent amount of PC concrete produced). A seven-story residential building in Argentina was completed in 2023 using CarbonCure concrete (483 m^3), enabling the sequestration of 5,600 kg of CO_2. Another example is a 12-story building in America completed in 2018. This structure involves the consumption of 367,000 m^3 of CarbonCure concrete, resulting in the sequestration of 680 tons of CO_2 (CarbonCure, 2025) (Figure 13.4).

13.2.1.3 Carboclave technology

Carboclave (Canada) offers a retrofit that strengthens prefabricated cement products by using CO_2 instead of steam or moisture, which can make use of existing autoclave curing equipment (Carboclave, 2025). The technology is suitable for a variety of dry-cast and precast concrete products. Blocks using CarboClave technology are reported to have 50% higher compressive strength (35.6 MPa vs 23.6 MPa at 1 day and 52.5 MPa vs 27.2 MPa at 28 days), higher splitting-tensile strength (2.04 MPa vs 1.83 MPa at 1 day), lower absorption (4.990% vs 5.924%), and little or no efflorescence and streaking compared with conventional autoclave curing (Hargest and Al-Ghouleh, 2020). It is estimated that one carboclave concrete block can sequester up to 250 g CO_2, save 25% cement, and reduce at least 50% embodied carbon footprint compared with conventional blocks. At present, 500 systems of carboclave technology are adopted by concrete plants around the world and can produce nearly 1 million truckloads (about 6 million m^3) with 66,000 tons of CO_2 saved annually (Figure 13.5).

13.2.1.4 CO_2-SUICOM technology

Kajima Corporation in Japan developed a low-carbon concrete technology named CO_2-SUICOM (CO_2 Storage under Infrastructure by Concrete Materials) with three features: (1) using CO_2 from waste flue gases produced by industrial sources such as steel plants and power plants for concrete curing where CO_2 concentration is approximately 15% to 20%, (2) utilizing low-carbon and high-mineralization activity γ-C_2S, and (3) incorporating industry by-products as supplementary cementitious

FIGURE 13.5 Equipment and products of Carboclave technology.

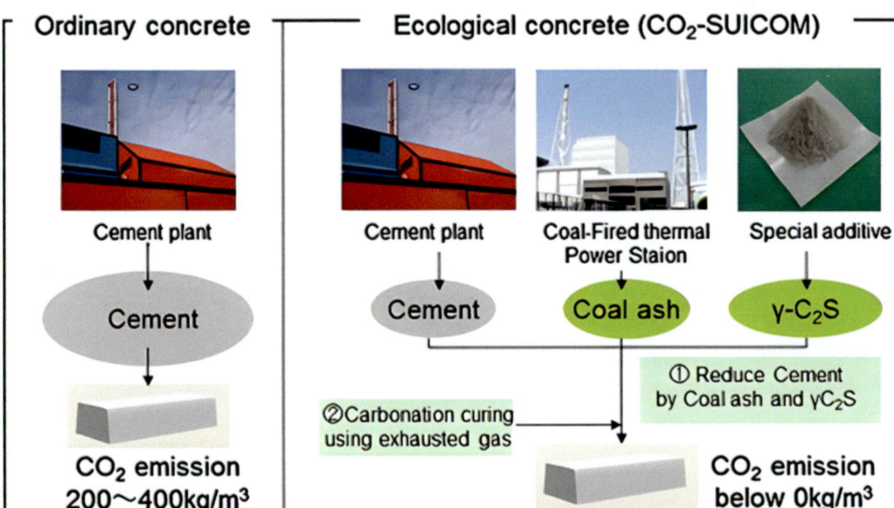

FIGURE 13.6 CO_2-SUICOM concept (Higuchi et al., 2014).

materials such as fly ash and GGBS to replace cement, as shown in Figure 13.6 (Higuchi et al., 2014; Yoshioka et al., 2013). A typical mixture for CO_2-SUICOM consists of 161 kg/m^3 of water, 101 kg/m^3 of cement, 82 kg/m^3 of GGBS, 53 kg/m^3 of fly ash, 32 kg/m^3 of γ-C_2S, and 844 kg/m^3 and 965 kg/m^3 fine and coarse aggregates, respectively. The compressive strength of this concrete could reach 21.6 MPa, comparable to that of PC-based concrete products. CO_2-SUICOM concrete can reduce 197 kg/m^3 CO_2 due to cement substitutes and fix 109 kg/m^3 CO_2. CO_2-SUICOM makes carbon-negative concrete possible with −18 kg/m³ of CO_2 net emission. The main products are currently precast concrete components such as factory-made road-blocks and paving blocks.

This newly developed concrete has already been used in construction work. In February 2011, CO_2-SUICOM was used for boundary, foundation, and paving blocks at the information facilities of the Fukuyama Photovoltaic Power Station of the Chugoku Electric Power Co., Inc., as shown in Figure 13.7. In January 2012, building elements using CO_2-SUICOM were used in the terrace ceiling of a condominium named NAKANO Central Park Residence, as shown in Figure 13.8. By the application of CO_2-SUICOM, 100–200 kg/m³ of CO_2 was absorbed in the hardening process of concrete.

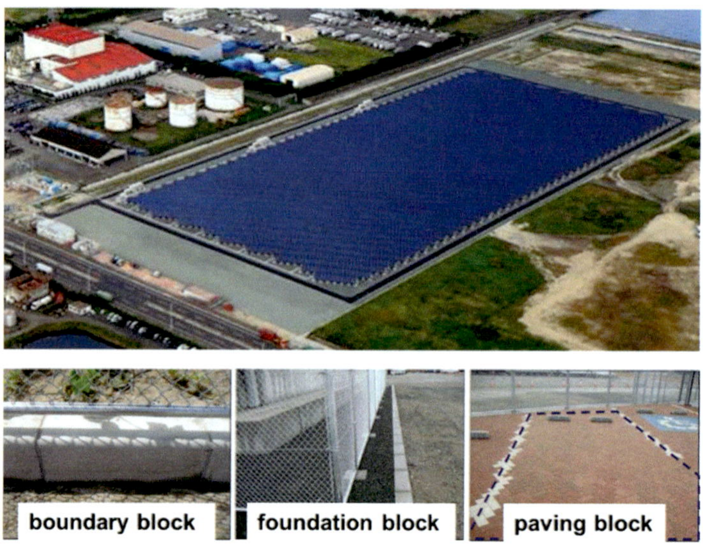

FIGURE 13.7 Application of CO_2-SUICOM in Fukuyama Photovoltaic Power Station (Yoshioka et al., 2013).

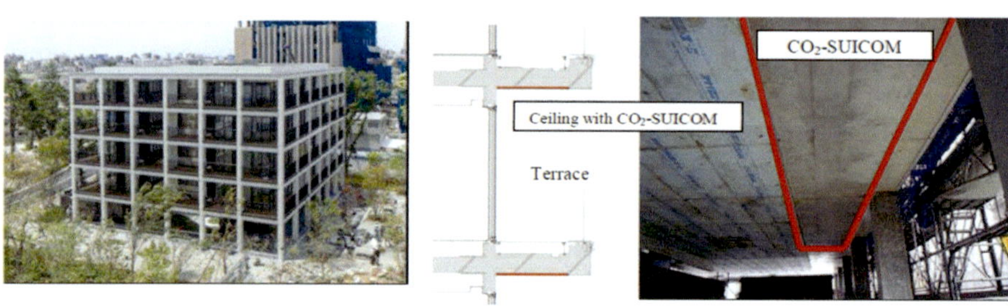

FIGURE 13.8 Application of CO_2-SUICOM in Nakano Central Park Residence (Yoshioka et al., 2013).

13.2.1.5 Solidia technology

In 2021, Solidia Technologies (USA) proposed a low-carbon concrete. Solidia cement is a non-hydraulic, carbonatable, calcium silicate cement composed mainly of $CaSiO_3$ (wollastonite) and $Ca_3Si_2O_7$ (rankinite) along with complex CaAlSiO-based phases (melilite) (Solidia, 2025).. The total lime content of Solidia cement clinker is in the range of 45–50%, which is 30% less than that required for PC. As a result, the total theoretical enthalpy of formation of Solidia cement clinker is about 1.05 GJ/t, which is almost 40% lower than that of PC clinker. Thus, the production temperature of Solidia cement is much lower than that of PC (1200°C vs. 1450°C) (Ashraf et al., 2019). The CO_2 emissions of Solidia cement could be reduced from ~810 kg/ton in PC clinker to ~565 kg/ton (Meyer et al., 2017). During the mineralization process, each ton of Solidia cement can further absorb 220–236 kg of CO_2 (Decristofaro et al., 2017). Solidia technology can reduce the carbon footprint by up to 70%. The Solidia concrete products reach full strength within 24 hours, compared to up to 28 days for conventional concrete products (Jain et al., 2019). Thin Solidia concrete products (~10 mm thick) can reach full strength in less than 6 hours, while larger concrete parts (~250 mm thick) can be within 24 hours (Ashraf et al., 2019; Meyer et al., 2017).

Products made with Solidia technologies are able to meet or exceed ASTM, CSA, and EU standards for all required mechanical properties, including compressive strength, flexural strength, splitting strength, freeze-thaw resistance, water absorption, and abrasion durability. The key applications of Solidia technology include ready-mix and precast products, like blocks, bricks, pavers, and retaining walls, as shown in Figure 13.9.

13.2.1.6 CarbonBuilt technology

The CarbonBuilt team from the University of California, Los Angeles (UCLA), has proposed the Revers™ process, including CO_2 emission-reducing innovations to both the concrete mixture design and its manufacturing process (CarbonBuilt, 2025). The introduction of portlandite leads to a reduction in the amount of traditional cement used and increases the use of waste products such as fly ash. While concrete is prepared via the same process and equipment used today, waste CO_2 is applied to this process without the need for expensive capture, compression, or purification. In 2020, with support from the U.S. Department of Energy and NRG COSIA Carbon XPRIZE, the technology was applied in the production of over 10,000 concrete blocks using CO_2 from a flue gas taken directly from a coal power plant at the Wyoming Integrated Test Center. In 2021, a second demonstration was successfully completed at the National Carbon Capture Center in Alabama, during which CO_2 in flue gases from both coal and natural gas power plants was used to produce more than 5,000 blocks. In 2023, Blair Block of Childersburg, Ala., plant leverages CarbonBuilt technology to establish the world's first ultra-low carbon concrete masonry units (CMU) production lines. The charter production line is estimated to avoid at least 2,000 tons of CO_2 emissions and consume 500-plus tons of atmospheric CO_2 per year (Figure 13.10).

FIGURE 13.9 Precast products using Solidia concrete.

13.2.2 Applications of CO_2 mineralized steel slag

13.2.2.1 CarbiCrete

CarbiCrete has come up with an innovative cement-free, carbon-negative concrete. The material utilizes steel slag to completely replace conventional Portland cement and adopts an accelerated mineralization process for concrete curing where CO_2 is injected into wet steel slag mixtures (CarbiCrete, 2025). The blocks are then CO_2 mineralized for 24 hours at 1.5 bar CO_2 partial pressure and stored in plastic bags for 35 days to allow for further hydration. Within 24 hours, the concrete can already reach full-strength (Mahoutian and Shao, 2016).

For each standard 18 kg concrete block produced, it is estimated that CarbiCrete can save 2 kg of CO_2 emissions by avoiding the usage of cement and can sequester a further 1 kg of CO_2 during curing. A plant adopting CarbiCrete technology producing 25,000 CMUs per day would reduce annual emissions by 20,000 tons.

Compared to the conventional concrete, CarbiCrete products exhibit the same water absorption properties, higher compressive strength, and better freeze-thaw resistance. It has been estimated that CarbiCrete CMUs are up to 30% stronger than

FIGURE 13.10 Blair Block production lines.

cement-based CMUs. CarbiCrete CMUs meet and even exceed ASTM C90 performance requirements. CarbiCrete concrete is more economical than cement-based concrete, as using steel slag to fully substitute cement represents a cost saving of between 10% and 20%.

However, the application potential of CarbiCrete is limited. Currently, its CO_2 mineralization process can only be carried out in a controlled environment at concrete plants and is mainly applied to produce prefabricated components such as CMUs and pavers (Figures 13.11, 13.12).

13.2.2.2 CarbStone

In collaboration between VITO and the Walloon CTP of Belgium, Carbstone Innovation technology was patented, and the products named Carbinox® were introduced. It uses 100% steel slag ground into an ultra-fine powder to react with CO_2 as the binder instead of cement (Carbstone, 2025). It can be developed in different forms, including floor tiles, roofing tiles, clinkers, and building blocks, as shown in Figure 13.13. Mineralization is performed in an autoclave under high pressure and temperature, and products are fully cured and ready for transport and use on site

FIGURE 13.11 Curing chamber and equipment of CarbiCrete.

FIGURE 13.12 Concrete blocks produced using CarbiCrete technology.

FIGURE 13.13 Concrete blocks produced using CarbStone technology.

after 24 hours of curing. The Carbstone blocks are able to sequester up to 150 g of CO_2 per kg of steel slag (Quaghebeur et al., 2010, 2014). The performance of CO_2 mineralized concrete products has the same characteristics as classic brick products (Quaghebeur et al., 2015).

Since 2013, a pilot plant along the Sambre River in Farciennes-Walloon Municipality has used the Carbstone Innovation technology to treat steel slag from a variety of sources to produce high-value construction products.

13.2.3 Applications of CO_2 Mineralized Magnesium-Based Materials

Calera Corporation (USA) developed a unique technology for carbon sequestration through the carbon mineralization by aqueous precipitation method (CMAP). The basic principle of the CMAP method is to capture flue CO_2 gas from gas or coal-fired power plants with an aqueous solution with a high pH value and react with Ca^{2+} and Mg^{2+} from alkaline solid and liquid sources, such as fly ash, wastewaters, and brines to precipitate $CaCO_3$ and $MgCO_3$. The stable $CaCO_3$ and $MgCO_3$ can be used as supplementary cementitious materials or aggregates in concrete.

Previous studies (Monteiro et al., 2023) reported the production of Calera products by adding NaOH solution to a brine of seawater and $CaCl_2$, then circulating CO_2 throughout the brine. Calera has constructed a demonstration plant at Moss Landing (California), capable of capturing 30,000 tons of CO_2 per year, as shown in Figure 13.14. According to the company, this plant could be expanded to capture 300,000 tons of CO_2 and convert it into more than 550,000 tons of valuable building materials per year. Another related initiative, the Calera Yallourn project in Victoria's Latrobe Valley (Australia), is expected to collect more than 300,000 tons of CO_2 and create more than one million tons of building materials every year and two million gallons of fresh water every day (Calera, 2025).

FIGURE 13.14 Pilot plant of Calera technology.

13.2.4 Applications of CO_2 mineralized materials as aggregates

13.2.4.1 Blue Planet technology

Blue Planet (USA) developed low-carbon aggregates by using calcium sourced from waste to react with CO_2. The calcium source, referred to as Geomass, can be a range of materials such as demolished concrete, cement kiln dust, and steel slag. CO_2 can come from any source, even directly from flue gas without being purified or liquefied (Blue Planet, 2025). Blue Planet produces two kinds of core aggregate products: (1) upcycled RCA aggregates, (2) synthetic limestone aggregates. It has been shown that the strength, performance, and cost of Blue Planet synthetic limestone aggregate are comparable to or even better than quarried lightweight aggregate such as pumice and carbon-intensive manufactured aggregate like expanded clay (Constantz et al., 2022). Blue Planet synthetic limestone aggregate exhibits a lower carbon footprint, as each ton of synthetic limestone aggregate can trap 440 kilograms of CO_2. It is estimated that the utilization of Blue Planet aggregate in concrete is capable of transforming concrete into carbon-negative. Concrete made with Blue Planet aggregate can meet or even exceed all ASTM strength tests (Constantz and Bewernitz, 2021).

Pittsburg, CA, plant of San Francisco Bay Aggregates utilizes CO_2 and waste concrete as feedstocks to produce Blue Planet aggregates and were put into use in the boarding area of San Francisco International Airport in 2016.

However, realizing this potential is faced with an obstacle based on the current stage that the lithified $CaCO_3$ aggregates have a lower density than natural limestone, restricting them to lightweight concrete applications (Figure 13.15).

13.2.4.2 O.C.O Technology

In 2010, Carbon8 Systems (UK), also called O.C.O Technology at present, developed a kind of carbon-negative aggregate named Manufactured Limestone (M-LS) aggregate. The M-LS aggregate is produced by treating and stabilizing a wide range of thermal residues such as cement dust, steel slags, oil-shale ash, incinerator ash, paper ash, and contaminated soils with captured CO_2 through Accelerated CO_2 mineralization Technology (ACT) (O.C.O Technology, 2025). The M-LS production process is a multiple-stage process, and its flowchart is shown in Figure 13.14. First, the waste is mixed with liquid CO_2 and water in a pre-treatment mixer for CO_2 mineralization. Then, CO_2 mineralized wastes are sent to a batch mixer to mix with fillers and binders. The mixture is conveyed to the pelletizer, and waste flue gases are injected to accelerate the cementation process and form round aggregates (Hills and Carey, 2019a). A variety of flue gases are captured and directly utilized in the O.C.O accelerated CO_2 mineralization technology, and rainwater is also collected to be utilized. During the M-LS production process, no solid, liquid, or gaseous waste is emitted. The main electricity consumption is only associated with the transportation and mixing of the materials, as the process is exothermic and requires no additional heat input. Thus, the process is potentially carbon-negative (Hills and Carey, 2019b).

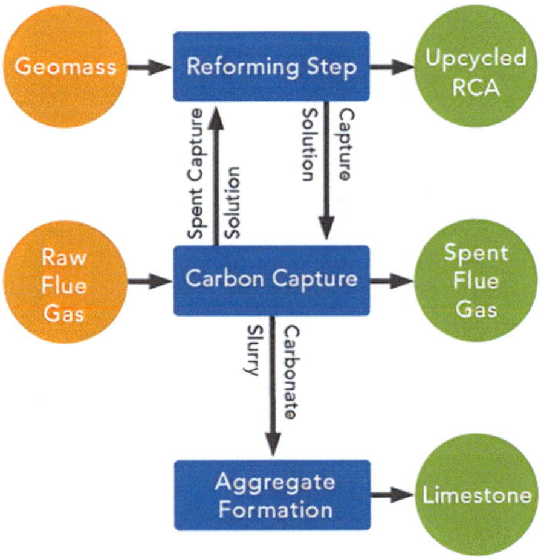

(a)

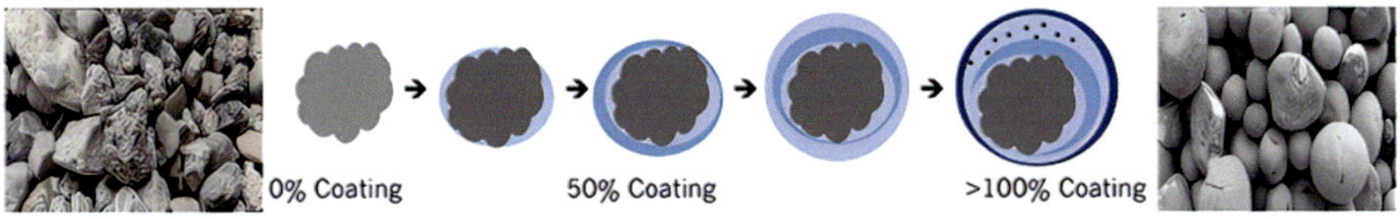

(b)

FIGURE 13.15 Process of Blue Planet aggregates, showing: (a) general flowchart and (b) process of converting flue gas into solid carbonate coating on small pebbles made of demolition debris (Blue Planet, 2025).

At present, O.C.O Technology has established three commercial facilities located in Brandon (Suffolk), Avonmouth (Bristol), and Leeds. These sites process over 100,000 tons of waste that would have otherwise been sent to landfill and produce over 200,000 tons of sustainable aggregate each year.

O.C.O Technology has held End of Waste approval from the Environment Agency in the UK. Their BlockMix aggregate, one of the M-LS products, is certified to BS EN 13055-1 (Lightweight Aggregate for Concrete) and suited for use in masonry. The 6F Series aggregate is certified to BS EN 13242 (Unbound Aggregates) and can be used in earthworks, pavements, and bound materials. M-LS aggregates were used for the first time in an asphalt mix, applied in a 2.5 km cycle path in 2021. This project is claimed to save up to 90% CO_2 when compared to the use of conventional hot AC20 asphalt mixes and reduce the CO_2 emissions by up to 70 tons in its entire construction process. In 2024, the very first concrete blocks made with M-LS aggregates were made in Australia for retaining walls and motorway soundproofing (O.C.O Technology, 2025) (Figure 13.16).

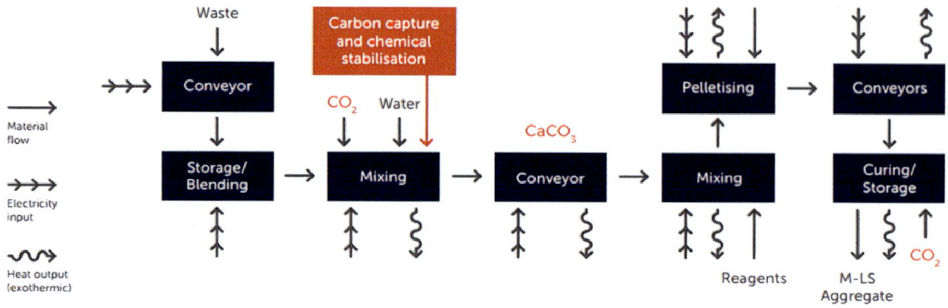

FIGURE 13.16 Flowchart of the M-LS production process (O.C.O Technology, 2025).

13.3 STANDARDIZATION OF CO_2 MINERALIZED MATERIALS AND PRODUCTS

13.3.1 American Society of Testing Materials (ASTM) international standards

In 2023, ASTM C01 Committee on Cement has approved the first two standards addressing non-hydraulic cements that require carbonation curing, ASTM C1905/C1905M - 23, *Specification for Cements that Require Carbonation Curing* and C1910/C1910M, *Standard Test Methods for Cements that Require Carbonation Curing* (ASTM, 2023a, 2023b).

ASTM C1905/C1905M - 23 clearly states that this new group of non-hydraulic cements is qualified for use in non-steel reinforced concrete products and sets no restrictions on the constituents of the cement. These unique cements should conform to the respective standard chemical requirements shown in Table 13.2 and all of the applicable standard physical requirements in Table 13.3.

The next step for implementing standards for CO_2 mineralized materials is to reference them in the list of allowable cementitious materials of existing standards for construction products. For instance, ASTM C15.03 Subcommittee on Concrete Masonry and Related Products has proposed introducing them in ASTM C936 for solid concrete interlocking paving units and in ASTM C1782/C1782M - 23a for segmental concrete paving slabs. It has been reported that pavers manufactured with the C1905 cements are qualified for meeting all of the requirements of C936 such as

TABLE 13.2
Standard chemical requirements

Chemical requirement	Applicable test method	
Sulfur trioxide (SO_3), max, %	C114	3.0
Bound CO_2, min, %	C1910/C1910M	8.0

TABLE 13.3
Standard physical requirements

Physical requirement	Applicable test method	Test limit
Fineness	C2044	Report only
Compressive strength, min, MPa	C1910/C1910M	Pa [4000 psi]

compressive strength, absorption, abrasion, and freeze/thaw testing, and the flexural strength of C1905 concrete paving slabs meets the requirements of C1782.

13.3.2 Technical specification for carbon mineralization of concrete products

In 2022, entrusted by the China Civil Engineering Society (CCES), Prof. Caijun Shi and his team started to compile *Technical specification for carbon mineralization of concrete products* to standardize the application of carbon mineralization preparation technology for concrete products, and to promote CO_2 utilization, solid waste recycling, energy conservation and emission reduction, as well as cost reduction and efficiency improvement.

This specification applies to the production process of concrete products, such as concrete solid bricks, concrete porous bricks, concrete blocks, concrete permeable pavement bricks, and concrete wall panels which are produced by cement, steel slag, calcium carbide slag, magnesium slag and/or other carbonatable solid wastes as the cementitious materials and burnt clayey materials, fly ash, limestone powder as the supplementary cementitious materials, and sand and gravel as the aggregates.

This specification stipulates the requirements of raw material characteristics, requirements of facilities and exhaust gas emission, technical requirements of carbon mineralization process, performance requirements of CO_2 mineralization products, and requirements of operation and maintenance during the production process of CO_2 mineralized cement products.

This specification requires pre-conditioning such as vacuum drying, blast drying, and microwave drying to remove the capillary water therein, which is beneficial for comprehensive, rapid, and uniform CO_2 mineralization of concrete products. This specification clearly states that the internal relative humidity of the concrete product after preconditioning should be maintained within the range of 50–70%. In addition, the carbon mineralization temperature should be maintained within the range of 20–80°C. Carbon mineralized concrete products that still contain hydraulic active minerals can undergo further wet curing for hydration to enhance the strength, compactness, and other properties of carbon mineralized concrete products. In addition to the normal concrete additives, carbon mineralization-enhancing additives are allowed for carbon mineralization concrete products, including alkanolamine-based CO_2 absorbents, as well as metal chelating agents of amino carboxylic acids, hydroxyl carboxylic acids, and hydroxyamino carboxylic acids.

TABLE 13.4
CO_2 fixation grades

	CO_2 fixation degree	
CO_2 fixation grade	CO_2 fixation per unit of product (kgCO_2/t product)	CO_2 fixation rate of cementitious materials (%)
1	10	≥2
2	20	≥4
3	30	≥6
4	40	≥8
5	50	≥10

13.3.3 Technical specification for carbonation curing of cement products

In 2022, China Building Materials Federation (CBMF) organized and led 21 units, including Tongji University and Tsinghua University to co-author the *Technical specification for carbonation curing of cement products.*

Cementitious materials involved in the raw materials of this specification include not only ordinary Portland cement but also some non-conventional cementitious materials such as C_3S_2, CS, and industrial solid waste. The specification specifies that the temperature of the gas used for CO_2 mineralization curing should be controlled between 20°C and 105°C, and the concentration of CO_2 should exceed at least 20%.

This specification not only requires that the performance of CO_2 mineralized products should meet the requirements of the technical indicators of the corresponding products, but also further specifies two CO_2 fixation indexes: CO_2 fixation per unit of product and CO_2 fixation rate of cementitious materials per unit of product, of which CO_2 fixation per unit of product refers to the mass of CO_2 fixed per ton of products, and CO_2 fixation rate of cementitious materials per unit of product refers to the ratio of the mass of CO_2 fixed per unit of products to the mass of the cementitious materials contained therein. In order to facilitate the promotion of CO_2 mineralized products, the specification adopts a hierarchical approach; CO_2 mineralized products will be divided into five carbon sequestration grades according to the two CO_2 fixation indexes, as shown in Table 13.4.

13.4 SUMMARY

This section summarizes the different methods of utilizing CO_2 mineralization technology in producing building materials that are currently being developed or utilized in the industry worldwide. This covered various CO_2 mineralized materials, including Portland cement, industrial waste such as steel slag, demolished concrete, and cement kiln dust, and magnesium-based materials like brines. These CO_2

mineralized materials can act as cementitious materials or produce aggregates and have been applied in different construction products such as bricks and pavers. All applications are reported to achieve higher strength, better durability, and satisfying carbon sequestration capability. These applications are considered as demonstration or small-scale production levels, as they still face various technical and market challenges.

Some standardizations have been set for CO_2 mineralized materials and products. These standardizations are hoped to assist in guiding processing and optimization and commercializing these products on a large scale.

REFERENCES

Ashraf, W., Olek, J., Sahu, S., 2019. Phase evolution and strength development during carbonation of low-lime calcium silicate cement (CSC). *Construction and Building Materials* 210, 473–482.

ASTM, 2023a. C1905/C1905M - 23 Standard specification for cements that require carbonation curing.

ASTM, 2023b. C1910/C1910M - 23 Standard test methods for cements that require carbonation curing.

Blue Planet, 2025. *Blue planet systems corporation webpage*. https://www.blueplanetsystems.com/. Accessed July 4, 2025.

CarbiCrete, 2025. *CarbiCrete webpage*. https://carbicrete.com/technology/. Accessed July 4, 2025.

Carboclave, 2025. *Carboclave webpage*. https://www.carboclave.com/. Accessed July 4, 2025.

CarbonBuilt, 2025. *CarbonBuilt webpage*. https://carbonbuilt.com/. Accessed July 4, 2025.

CarbonCure Technologies, 2025. *CarbonCure technologies webpage*. https://www.carboncure.com/. Accessed July 4, 2025.

Carbstone Innovation, 2025. *Carbstone innovation webpage*. https://www.orbix.be/en/technologies/carbonation. Accessed July 4, 2025.

Constantz, B., Bewernitz, M., 2021. *Carbon sequestration methods and systems*, US Patent: US 11181700 B2. Blue Planet Systems Corporation.

Constantz, B., Bewernitz, M., Camire, C., Kang, S.-H., Schneider, J., 2022. *Continuous carbon sequestration material production methods and systems for practicing the same*, US Patent: US 11344861 B2. Blue Planet Systems Corporation.

Decristofaro, N., Meyer, V., Sahu, S., Bryant, J., 2017. Environmental impact of carbonated calcium silicate cement-based concrete. *Environmental Science, Engineering* 174, 83–89.

GTM, Calera Progress Update: Cement and Fresh Water from CO_2 and Brines, 2025. *Calera webpage*. https://www.greentechmedia.com/articles/read/calera. Accessed July 4, 2025.

Hargest, P.W., Al-Ghouleh, Z., 2020. *CO2-laden concrete precast products and the method of making the same*, US Patent: US 10821629 B2. Carboclave Corp.

Higuchi, T., Morioka, M., Yoshioka, I., Yokozeki, K., 2014. Development of a new ecological concrete with CO2 emissions below zero. *Construction and Building Materials, Concrete Sustainability* 67, 338–343.

Hills, C., Carey, P., 2019a. *Improved production of aggregates*, US Patent Application: US 2019/0119158 A1. Applicant: Carbon8 Systems Limited.

Hills, C., Carey, P., 2019b. *Production of secondary aggregates*, US Patent: US 10343199 B2. University of Greenwich.

Jain, J.A., Seth, A., DeCristofaro, N., 2019. Environmental impact and durability of carbonated calcium silicate concrete. *Proceedings of the Institution of Civil Engineers: Construction Materials* 172, 179–191.

Liu, J., Wang, Y., Li, Y., Tian, J., You, X., Mao, Y., Hu, X., Shi, C., 2022. Carbonated concrete brick capturing carbon dioxide from cement kiln exhaust gas. *Case Studies in Construction Materials* 17, e01474.

Mahoutian, M., Shao, Y., 2016. Production of cement-free construction blocks from industry wastes. *Journal of Cleaner Production* 137, 1339–1346.

Meyer, V., DeCristofaro, N., Bryant, J., Sahu, S., 2017. *Solidia cement an example of carbon capture and utilization.*

Monkman, S., MacDonald, M., 2016. Carbon dioxide upcycling into industrially produced concrete blocks. *Construction and Building Materials* 124, 127–132.

Monkman, S., MacDonald, M., 2017. On carbon dioxide utilization as a means to improve the sustainability of ready-mixed concrete. *Journal of Cleaner Production* 167, 365–375.

Monkman, S., MacDonald, M., Hooton, R.D., Sandberg, P., 2016. Properties and durability of concrete produced using CO_2 as an accelerating admixture. *Cement and Concrete Composites* 74, 218–224.

Monteiro, P.J.M., Clodic, L., Battocchio, F., Kanitpanyacharoen, W., Chae, S.R., Ha, J., Wenk, H.R., 2013, Incorporating carbon sequestration materials in civil infrastructure: A micro and nano-structural analysis. *Cement and Concrete Composites* 40, 14–20.

O.C.O. Technology, 2025. https://oco.co.uk/wp-content/themes/crush-theme/assets/pdf/OCO_Broch. Accessed July 4, 2025.

Quaghebeur, M., Nielsen, P., Horckmans, L., Van Mechelen, D., 2015. Accelerated carbonation of steel slag compacts: development of high-strength construction materials. *Frontiers in Energy Research* 3, 1–12.

Quaghebeur, M., Nielsen, P., Nielsen, B., 2014. *Production of a mainly carbonate bonded article by carbonation of alkaline materials.* Google Patents.

Quaghebeur, M., Nielsen, P., Nielsen, B., Nguyen, E., 2010. Carbstone: Sustainable valorisation technology for fine grained steel slags and CO2. *Refractories Worldforum* 2(2), 75–79.

Shi, C., He, F., Wu, Y., 2012. Effect of pre-conditioning on CO_2 curing of lightweight concrete blocks mixtures. *Construction and Building Materials* 26, 257–267.

Shi, C., Wu, Y., 2008. Studies on some factors affecting CO_2 curing of lightweight concrete products. *Resources, Conservation and Recycling* 52, 1087–1092.

Solidia Technologies, 2025. *Solidia technologies webpage.* https://www.solidiatech.com/. Accessed July 4, 2025.

Yoshioka, K., Obata, D., Nanjo, H., Yokozeki, K., Torichigai, T., Morioka, M., Higuchi, T., 2013. *New ecological concrete that reduces CO_2 emissions below zero level ~ new method for CO2 capture and storage ~.* Energy Procedia, GHGT-11 Proceedings of the 11th International Conference on Greenhouse Gas Control Technologies, November 18–22, 2012, Kyoto, Japan 37, 6018–6025.

Index

For Product Safety Concerns and Information please contact our EU representative GPSR@taylorandfrancis.com Taylor & Francis Verlag GmbH, Kaufingerstraße 24, 80331 München, Germany

Batch number: 10615536

Printed by Printforce, the Netherlands